# Achieving Competence in Mathematics

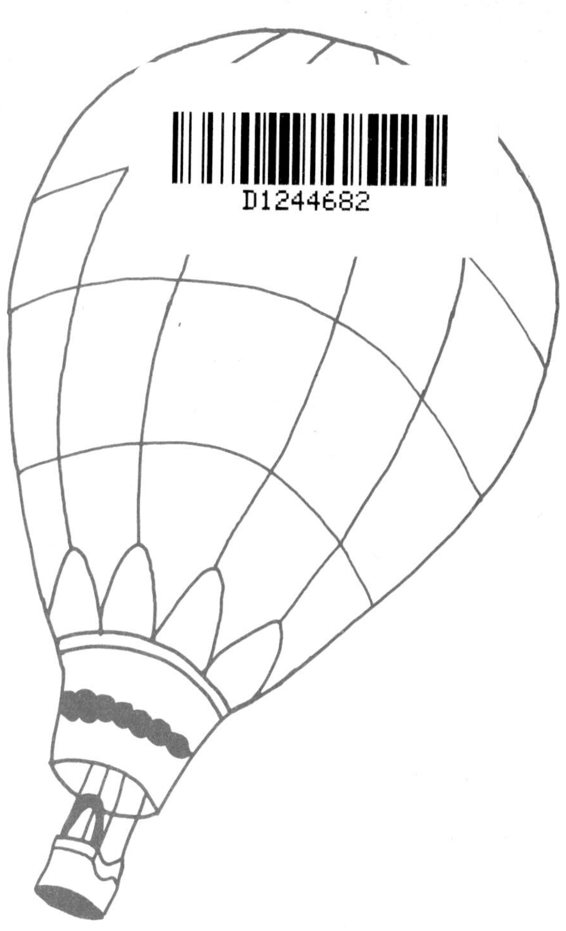


**Mathew M. Mandery**
Principal
Jericho Senior H.S.
Jericho, New York

**Marvin Schneider**
Assistant Principal, Mathematics
Edward R. Murrow H.S.
New York City

When ordering this book please specify: *either* **R 446 P** *or*
ACHIEVING COMPETENCE IN MATHEMATICS, PAPERBACK

**AMSCO SCHOOL PUBLICATIONS, INC.**
**315 Hudson Street New York, N.Y. 10013**

ISBN 0-87720-204-4



PRINTED IN THE UNITED STATES OF AMERICA

# About this textbook

ACHIEVING COMPETENCE IN MATHEMATICS reflects the following basic beliefs:

- Success with computational skills is founded on good quality practice.
- Explanations of procedures and their underlying concepts must be mathematically sound, but presented in language that is understandable to the students.
- Sufficient time must be given to work with concepts, both in the classroom and at home.
- Applications should be presented in as realistic a framework as possible.
- Problem-solving approaches need to be made explicit.
- To achieve retention, students must be given the opportunity to use previously learned skills through review exercises.

## THE DESIGN OF THE TEXTBOOK

### Theme, Module, Unit

The text is divided into 11 themes. Ten of these themes are subdivided into a number of modules, usually 2 or 3. A module represents a broad topic within the theme and provides a convenient break for review and testing.

A unit represents one (or more) day's instruction. Flexibility is deliberately built into this unit approach, since there are classes in which a unit may take longer than a day to cover, and those in which progress may be more rapid. For the 10 themes, the total number of units is 100.

### Problem-Solving Strand

The introductory theme, divided into 3 units, treats problem solving as a skill in itself. Selected strategies and key vocabulary are presented, as well as exercises designed to sharpen decision-making skills.

After this initial presentation, a problem-solving approach is woven throughout the text. Attention is directed to the application of the strategies and the reinforcement of critical thinking by specifically treating appropriate illustrative examples under the head "Thinking About the Problem."

### The Main Idea

At the outset of each unit, the essential concepts are explained in simple, but mathematically correct, language. Students should be encouraged to read "The Main Idea" aloud in class and again at home. The Main Idea presents all the basic information necessary for the successful completion of each unit.

In some cases, to achieve maximum clarity and flexibility, the topic of a unit is presented in 2 or more subunits, each with its own Main Idea.

### Illustrative Examples

Examples of every type covered in the unit are presented with detailed explanations and step-by-step solutions. These examples can be used as the basis for classroom illustration of the concepts. Students can refer to them as models while they work in class and at home.

When there is more than one Main Idea presentation for a unit, the accompanying illustrative examples are numbered continuously.

### Exercises

To provide a spiraled support system for the student, an abundance of exercise material is presented in 3 groups.

*Class Exercises:* These exercises immediately follow the illustrative examples and offer the first opportunity for students to work independently, still under the guidance of the teacher.

*Homework Exercises:* Armed with the detailed illustrative examples and the completed class exercises, students are further supported when they find the required homework exercises presented as follow-ups to those already worked out in class.

Units that have 2 or more subunits, each with its own examples and class exercises, close with a single comprehensive set of homework exercises.

*Spiral Review Exercises:* Here is a daily opportunity to stay in touch with previously presented material.

Allowing for maximum flexibility, this wealth of exercise material is available for assignment at the teacher's discretion.

### Challenge Questions

Each unit ends with a "Challenge Question." This feature takes students beyond the practice exercises and encourages their use of critical thinking and problem-solving skills. In some cases, the Challenge Question is a more challenging application of recent concepts. In other cases, the Challenge Question points to concepts that are taken up in the next unit. These Challenge Questions may be used as enrichment activities for students who are ready to go beyond the average level of the class, or they may be used as thought questions for the whole class, generally at the end of a lesson.

## THE COMPANION WORKBOOK

An accompanying workbook contains:

***Module Reviews*** A comprehensive review is provided for each of the 20 modules.

***Theme Reviews*** A theme review is furnished for each of the 10 themes. These theme reviews are cumulative in nature, each one covering the work from the beginning of the course.

***Model Competency Tests*** The questions and format of the 5 competency tests included in the workbook are patterned after standardized tests widely administered by states and cities.

Students can monitor their achievement on the model competency tests by completing the Personal Progress Chart, which is provided at the end of the workbook.

# CONTENTS

## INTRODUCTORY THEME Problem Solving



## THEME 1 Whole Numbers

### MODULE 1



### MODULE 2



## THEME 2 Integers

### MODULE 3

## THEME 3 Fractions and Rational Numbers

### MODULE 4



### MODULE 5



### MODULE 6



## THEME 4 Decimals

### MODULE 7



### MODULE 8



## THEME 5 Percents

### MODULE 9

### MODULE 16



## THEME 9 Introduction to Algebra

### MODULE 17



### MODULE 18



## THEME 10 Introduction to Geometry

### MODULE 19



### MODULE 20

# Problem Solving

# UNIT P–1 Reading Problems for Information

## THE MAIN IDEA

1. Every problem *gives* you some information and asks you to *find* other information. The information that you have to find is called a ***solution*** or an ***answer.***
2. Often, the given information:
   - follows the word "if."
   - is described as something that someone *has* or *had.*
   - is mentioned as a fact at the beginning of a problem.
   - is given as numbers in a chart (table) or diagram.
   - can be found by counting.
3. Sometimes, some information is not stated, but you are expected to know it. For example, you should know that there are:
   - 12 months in a year.
   - 3 sides to a triangle.
   - 3 feet in a yard.
   - 12 items in a dozen.
   - 16 ounces in a pound.
   - 25 cents in a quarter.
   - 4 aces in a standard deck of playing cards.
4. Often, the information that you have to find:
   - follows the word "find."
   - follows words like "how many . . . ?" "how much . . . ?" "what is . . . ?"
5. Some information that is given may not be needed to find the solution. This is extra information that you ignore.

**EXAMPLE 1** For each problem, state the given information and the information that you are asked to find.

| | Problem | Given Information | Information To Be Found |
|---|---|---|---|
| **a.** | Mr. Harris earns $5.75 per hour. If he worked for 8 hours, how much did he earn? | The hourly rate is $5.75 (given as a fact at the beginning). The number of hours worked is 8 (following the word "if"). | *How much* money was earned? |
| **b.** | Susan had a piece of ribbon 83 inches long. She wanted to cut it into 5-inch pieces of ribbon. How many 5-inch pieces could she make? | There are 83 inches of ribbon. Each piece is to be 5 inches long. (Both are given as facts at the beginning.) | *How many* pieces can be made? |

| Problem | Given Information | Information To Be Found |
|---|---|---|
| c. Find the total weight of a shipment of boxes if each box weighed 32 pounds and there were 144 boxes. | Each box weighed 32 pounds. There were 144 boxes. (Both facts follow the word "if.") | *Find* the total weight. |
| d. What is the average height shown in the table? | The respective heights are 60 in., 58 in., 62 in., 60 in., and 55 in. (given as numbers in the chart). There are 5 children. (Count names in the chart.) | *What is* the average height? |
| e. Find the area of the rectangular room shown in the diagram. | The dimensions are 9 ft. and 20 ft. (given as numbers in the diagram). | *Find* the area. |

| Name | Height |
|---|---|
| Joyce | 60 in. |
| Barbie | 58 in. |
| Mel | 62 in. |
| Scott | 60 in. |
| Jack | 55 in. |

9 ft.

20 ft.

**EXAMPLE 2** For each problem, write the information that you are expected to know but that is not stated in the problem.

| Problem | Unstated Information |
|---|---|
| a. Oranges cost $1.32 a dozen. What is the cost of one orange? | There are 12 oranges in a dozen. |
| b. Mr. Ross earns $55 a day. He does not work on weekends. How much does he earn in a week? | There are 5 working days in a week, not counting the weekend. |
| c. What is the total amount that Jenny can save in her Holiday Club at the bank in one year if she saves $5 a week? | There are 52 weeks in a year. |
| d. Each side of a square measures 12.5 cm. What is the total distance around the square? | The square has 4 sides of equal length. |
| e. Fred has 5 standard decks of playing cards. How many kings does he have in all? | There are 4 kings in a standard deck of playing cards. |

**EXAMPLE 3** For each problem, state which information is extra, if any.

| *Problem* | *Extra Information* |
|---|---|
| **a.** For the past 5 years, Mrs. Harris has been earning $17,000 a year. How much does she earn in 3 years? | "5 years" is extra. You only need to know that she earned $17,000 a year for 3 years. |
| **b.** Find the sales tax that Mrs. Santos paid on her 1980 used car if the ticket price of the car was $750 and the sales tax rate was 8%. | "1980" is extra. You only need to know the price and the sales tax rate. |
| **c.** Find the distance around the rectangle. | "13 cm" is extra. You only need to know the lengths of the sides of the rectangle. |

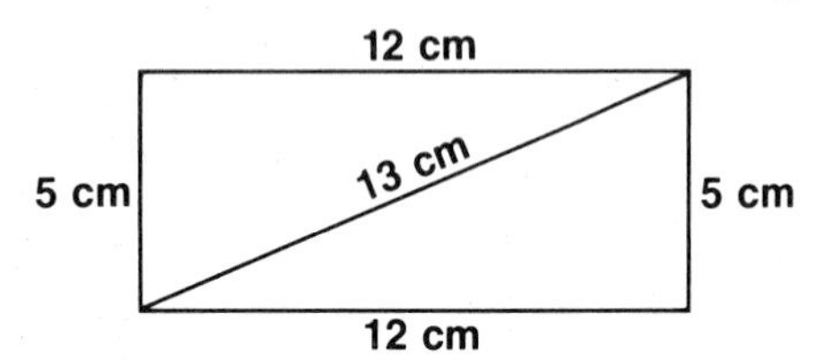


**d.** What is the total number of newspapers that Susan sold this week?

All the numbers except those in Susan's row are extra.

**Number of Newspapers Sold Each Day**

| Salesperson | M | T | W | Th | F | S |
|---|---|---|---|---|---|---|
| Sally | 35 | 37 | 34 | 30 | 29 | 39 |
| James | 16 | 27 | 20 | 18 | 22 | 25 |
| Susan | 23 | 17 | 29 | 35 | 18 | 19 |
| Hank | 14 | 18 | 20 | 13 | 17 | 22 |
| Bill | 20 | 15 | 17 | 19 | 21 | 18 |

**e.** Mr. Wilson made purchases of $12.50, $7.85, and $27.95. If the rate of sales tax was 5%, how much sales tax did he have to pay?

There is no extra information.

## CLASS EXERCISES

For problems 1–8, state the given information and the information that you have to find.

**1.** If the area of a rectangle is 24 sq. yd. and its width measures 4 yd., what is the measure of its length?

**2.** Find the average of 11, 17, 18, 9, and 10.

**3.** The gas tank of a car had 15 gallons of gas in it. The car used one gallon of gas every 12 miles. If the car was driven 60 miles, how much gas was left?

**4.** What is 17% of 300? **5.** Find $\frac{3}{4}$ of 128. **6.** Find the largest value: 51%, $\frac{1}{2}$, .49

7. Find the total rainfall in Centerville for the five months shown in the table.

**Rainfall in Centerville**

| Month | Rainfall |
|---|---|
| January | $1\frac{1}{4}$ in. |
| February | $\frac{1}{2}$ in. |
| March | $\frac{3}{4}$ in. |
| April | $2\frac{1}{2}$ in. |
| May | 3 in. |

8. Find the area of the triangular plot shown.

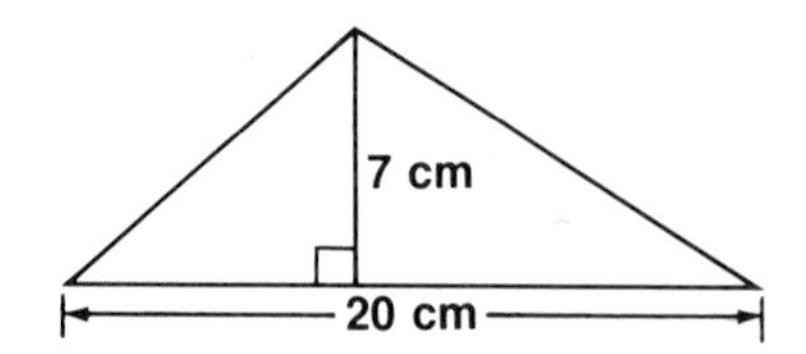


For problems 9–14, write the information that is not stated in the problem but that you are expected to know.

9. The Maintown supermarket added 5 new sales items every day for a week. What was the total number of items on sale that week?

10. A dozen pens costs $2.40. What is the cost of one pen?

11. Mr. Dale earns $940 a month. How much does he earn in a year?

12. Each side of a triangle is 9 inches long. What is the total distance around the triangle?

13. What fractional part of the number of months of the year has names that begin with the letter "J"?

14. If an oil truck can pump 35 gallons of oil in a minute, how many gallons can it pump in a half-hour?

For problems 15–20, state which information is extra, if any.

15. A class of 28 students raised charity money by selling a total of 159 boxes of candy at $1.50 per box. How much money did they raise?

16. Jill bought 12 cans of soup that cost 39¢ per can and 8 pounds of tomatoes that cost 69¢ per pound. What was the total cost of her purchases?

17. Three children bought a 24-ounce bag of candy for $1.79. If they shared the candy equally, how many ounces of candy did each child get?

18. Mrs. Farr drove 350 miles and used 30 gallons of gasoline. If gasoline costs $1.19 a gallon, how much did she spend on gasoline?

19. What is the average weight of these children?

**Medical Information for Class 4–1**

| Name | Height | Weight | Age |
|---|---|---|---|
| Stan | 62 in. | 125 lb. | 10 |
| Phillip | 56 in. | 87 lb. | 11 |
| Marcia | 59 in. | 91 lb. | 10 |
| Nancy | 60 in. | 100 lb. | 10 |
| Barbara | 58 in. | 94 lb. | 11 |

20. What is the total height of this house?

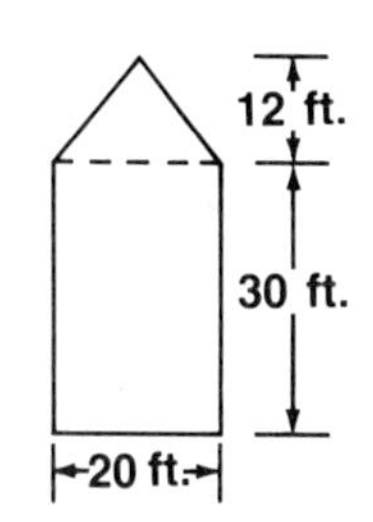

## HOMEWORK EXERCISES

For problems 1–16, state the given information and the information that you have to find.

1. If the perimeter of a square is 48 cm, what is the measure of each side?
2. Find the sum of 28 and 57.
3. What is the product of 9 and 6?
4. Find the total amount that John earned if he was paid $4.80 an hour and he worked for 5 hours?
5. Robin bought 3 blouses that cost $12.50 each. How much did she spend?
6. Find the smallest value: $\frac{7}{8}$, 80%, .79
7. What is 25% of 400?
8. In Northern State High School, there are 29 classes with 30 students in each class. What is the total number of students in these classes?
9. William bought 3 pounds of potatoes at 42 cents per pound and 2 quarts of milk at 79 cents a quart. How much did he spend?
10. In a class, there are 10 boys and 20 girls. What fractional part of the total number of students in the class is the number of girls?
11. Last year, Dr. Jones paid $144,000 as rent for his office. What was the monthly rent?
12. Ms. Jacobs saves $\frac{7}{50}$ of her salary. What percent of her salary does she save?
13. Find the total number of newspapers that Jane delivered for the 5 days shown in the table.

**Jane's Newspaper Record**

| Day | Newspapers Delivered |
|---|---|
| Monday | 105 |
| Tuesday | 111 |
| Wednesday | 108 |
| Thursday | 115 |
| Friday | 102 |

14. Find the average number of points scored by the opponents of Westwood High School basketball team.

**Westwood High School Basketball Team Record**

| Opponent | Points Scored |
|---|---|
| Clark H.S. | 58 |
| Franklin H.S. | 72 |
| Jackson H.S. | 48 |
| Randolph H.S. | 52 |

15. Find the area of the right triangle shown.

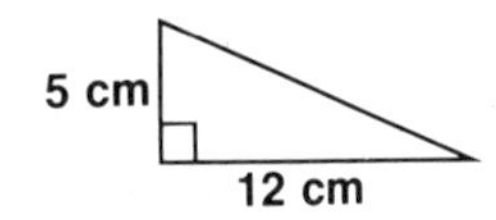


16. Find the amount of fencing needed to enclose the rectangular yard shown.

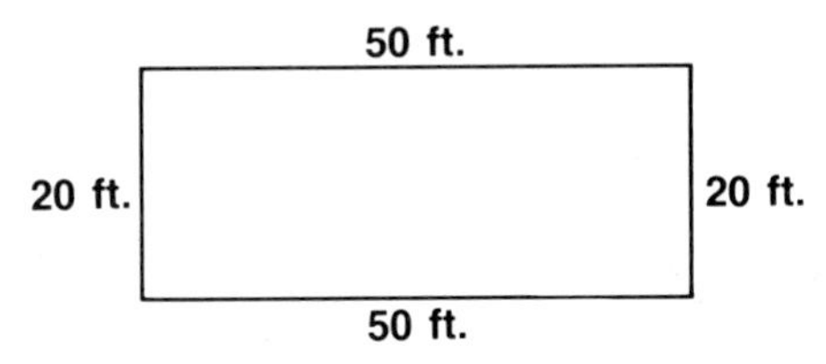

For problems 17–24, write the information that is not stated in the problem but that you are expected to know.

**17.** Sandra spends an average of $7.50 a day. How much does she spend in a week?

**18.** What fractional part of a pound is 10 ounces?

**19.** How many quart bottles can be filled from a 10-gallon can of milk?

**20.** Janet earns $4.50 per hour. How much did she earn if she worked 450 minutes?

**21.** If 3 oranges cost 59 cents, what is the cost of one dozen oranges?

**22.** What is the total distance around a square each of whose sides measures 10 cm?

**23.** A machine can produce a toy every 3 minutes. How many toys can it produce in 5 hours?

**24.** A man walked 3 miles. How many yards did he walk?

For problems 25–30, state which information is extra, if any.

**25.** John bought 5 shirts that cost $10 each and 3 pairs of pants that cost $18 each. How much did he spend on the shirts?

**26.** Mr. Jackson paid 90 cents for a 3-mile bus ride to the train station. He then paid $4.50 for a 24-mile train ride to his job. What was the total cost of his trip?

**27.** James spent $4.99 for a 2-pound bag of peanuts and $2.45 for a gallon bottle of juice. How much change did he receive from a $10 bill?

**28.** Mr. Smith's car averages 25 miles per gallon of gasoline. How many gallons did he use to make a 250-mile trip?

**29.** How far is it from New Place to Saint Albane?

**30.** How much did Ms. Williams spend on food?

**Ms. Williams' Expense Account**

| Date | Hotel | Food | Telephone |
|---|---|---|---|
| Monday 5/1 | $70 | $35 | $8.50 |
| Tuesday 5/2 | $68 | $42 | $12.75 |
| Wednesday 5/3 | $68 | $50 | $23.45 |
| Thursday 5/4 | $80 | $60 | $9.80 |

# UNIT P–2 Deciding Which Operation to Use

## THE MAIN IDEA

1. The four arithmetic operations are ***addition*** (+), ***subtraction*** (−), ***multiplication*** (×), and ***division*** (÷).
2. Some problems contain key words that show which operation to use. For example:

| ***Addition*** | ***Subtraction*** | ***Multiplication*** | ***Division*** |
|---|---|---|---|
| plus | minus | times | divide |
| added to | subtracted from | multiplied by | divided by |
| sum | difference | product | divided into |
| and | deduct | of (a fraction *of* a number; a percent *of* a number) | quotient |
| | remains | | each |
| | | | per |

3. Use certain operations in certain situations. For example:

*Use addition to find:*
- a total amount when each part is given.
- a result when a number is increased.
- an amount after tax is included.

*Use subtraction to find:*
- how much greater or less one number is than another.
- a result when a number is decreased.
- by how much a number has been increased.
- an amount after a discount is made.
- how much is left, or remains, after an amount is spent or used up.

*Use multiplication to find:*
- a total amount when given several equal amounts.
- percents.

*Use division to find:*
- an amount for each item when given the total amount and the number of equal items.
- the percent that is the equivalent of a given fraction.
- a probability.
- an average or mean.

**EXAMPLE 1** For each problem, identify the key word(s) that help you to decide which operation to use.

| | *Problem* | *Key Words* |
|---|---|---|
| **a.** | Mrs. Oates made purchases of \$12.50, \$17.98, and \$8.75. Find the total amount that she spent. | total amount |
| **b.** | If a dozen eggs costs \$1.08, what is the cost of each egg? | cost of each |
| **c.** | After a 20% discount, what is the final cost of a coat that is priced at \$85? | discount, final cost |
| **d.** | In a class of 32 children, each child donated \$1.50 to a toy drive. Find the total amount of money that was collected. | total amount |

**EXAMPLE 2** For each problem, tell which operation you would use to find the answer and write a number phrase to show this operation.

| | *Problem* | *Operation* | *Number Phrase* |
|---|---|---|---|
| **a.** | Find the product of 85 and 153. | Multiplication, because of the word "product." | $85 \times 153$ |
| **b.** | If Mr. Harris earns \$30,000 a year, how much does he earn each month? | Division, because a year is made up of 12 months with equal pay in each month. | $\$30{,}000 \div 12$ |
| **c.** | Find the total length of these line segments:<br>1 cm ——<br>4 cm ————————<br>3.8 cm ——————— | Addition, because you are finding the total of the 3 different parts. | 1 cm + 4 cm + 3.8 cm |
| **d.** | In one day, the temperature outside Jim's room went from 75° to 98°. By how many degrees did the temperature increase? | Subtraction, because you are finding by how much an amount increased. | $98° - 75°$ |
| **e.** | Find the total cost of 7 newspapers if each newspaper costs 35¢. | Multiplication, because you are finding a total amount when you are given one of 7 equal amounts. | 7 × 35¢ |

**EXAMPLE 3** Which number phrase fits the following problem? Mrs. McGill's car gets 14.5 miles per gallon of gasoline. If her car used 7 gallons of gasoline, how far did she travel?

(a) $14.5 + 7$ (b) $14.5 - 7$ (c) $14.5 \times 7$ (d) $14.5 \div 7$

Since "per" means "for each," you have to find the total number of miles traveled when you are given the number of miles traveled for each gallon. Since the number of miles for each gallon is always the same, use multiplication.

*Answer:* (c)

## CLASS EXERCISES

For problems 1–5, identify the key word(s) that help you to decide which operation to use.

1. A 50-foot length of wire is to be divided into 5-foot pieces. How many pieces can be cut?
2. David is 5 ft. 5 in. tall and Murray is 5 ft. 3 in. tall. How much taller is David than Murray?
3. Susan read 120 pages in 5 hours. How many pages did she read per hour?
4. The oil in 15 barrels is to be emptied into a storage tank. If each barrel contains 55 gallons of oil, how many gallons must the storage tank be able to hold?
5. How much change from a $20 bill will Joseph get if he is buying an item that costs $17.98?

For problems 6–13, tell which operation you would use to find the answer and write a number phrase to show this operation.

6. What is the selling price of a $120 bicycle if a $25 discount is given?
7. Find the sum of the even numbers from 30 through 40.
8. If each egg weighs 2.3 oz., find the weight of a dozen eggs.
9. If 24 students want to donate equally and give a total of $200 to the school athletic field fund, how much should each student donate?
10. Mrs. Jameson used $3\frac{1}{2}$ oz. of butter from a full 1-pound container. How much butter did she have left?
11. The Paragon Perfume Company wanted to fill up $2\frac{1}{2}$-oz. bottles of the Right Stuff perfume from a container that held 200 ounces. How many bottles could be filled?
12. The book collection in a library that owned 2,175 books was increased by 450 books. How many books did the library then have?
13. If an apartment requires 40 square yards of carpeting, how much carpeting do 25 identical apartments require?

For problems 14 and 15, tell which number phrase fits the problem.

14. Robert lost an average of $2\frac{1}{2}$ pounds per week when he was on a diet. He lost a total of 30 pounds. How many weeks did it take?
    (a) $30 + 2\frac{1}{2}$ (b) $30 - 2\frac{1}{2}$ (c) $30 \times 2\frac{1}{2}$ (d) $30 \div 2\frac{1}{2}$
15. 240 students voted in the school election. If there were 410 students in the school, how many students did not vote?
    (a) $410 + 240$ (b) $410 - 240$ (c) $410 \times 240$ (d) $410 \div 240$

## HOMEWORK EXERCISES

For problems 1–8, identify the key word(s) that help you to decide which operation to use.

1. If 5 pens cost $1.60, what is the cost of each pen?
2. A box of candy weighs 1.5 kilograms. What is the total weight of 20 boxes?

3. Ms. Diamond bought a car for $8,600. If she paid for the car in 24 equal monthly payments, how much was each payment?

4. Mr. Warren borrowed $2,700 and has repaid $1,900 of his loan. How much does he still owe?

5. The Fast Car Company manufactured 5,348 two-door cars and 7,589 four-door cars last month. What is the total number of cars that it produced?

6. What is the total cost of 12 gallons of gasoline at $1.25 per gallon?

7. What is the total amount of money that Ms. James spent on her business trip?

**Ms. James' Expense Account**

| Item | Cost |
|---|---|
| Meals | $48.50 |
| Hotel | $75.80 |
| Travel | $96.00 |

8. Mr. Arnold bought his house for $45,000 and sold it 10 years later for $83,000. What is the difference in the prices?

For problems 9–17, tell which operation you would use to find the answer and write a number phrase to show this operation.

9. Find the sum of 58 and 29.

10. Find the difference between 100 and 75.

11. Find the sum of the lengths of the sides of the triangle.

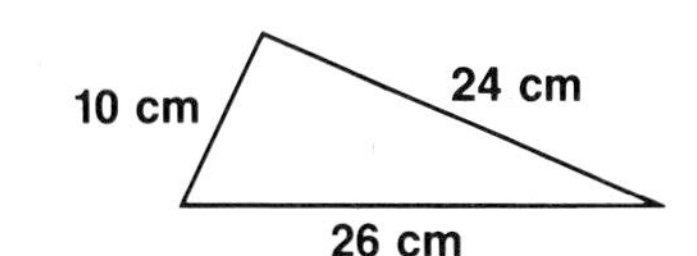


12. If the measure of the length of a rectangle is 9 inches and the measure of the width of the rectangle is 6 inches, how much longer is the length than the width?

13. A recipe calls for 2 cups of flour to make a cake. How much flour should be used to make 8 such cakes?

14. A man paid $4 for a 20-pound bag of potatoes. How much did one pound of potatoes cost him?

15. Before he dieted, John weighed 189 pounds. If his present weight is 165 pounds, how many pounds did he lose?

16. Wilma ran 3.8 miles on Monday and 4.6 miles on Tuesday. Find the total number of miles that she ran.

17. Mr. Adams' son bought 3 records that cost $6.99 each. Find the total cost of the records.

For problems 18–25, tell which number phrase fits the problem.

18. A crate of oranges contains 50 oranges. How many oranges are there in 25 crates?
(a) $50 + 25$ (b) $50 \div 25$ (c) $50 \times 25$ (d) $50 - 25$

19. A $1,000 prize was shared equally by 5 people. How much was each person's share?
(a) $1,000 × 5 (b) $1,000 ÷ 5 (c) $1,000 − 5 (d) $1,000 + 5

20. The population of Smithville was 3,485 in 1979. If the present population is 4,240, find the amount of increase.
(a) 3,485 + 1979 (b) 3,485 − 1979 (c) 4,240 + 3,485 (d) 4,240 − 3,485

**21.** At the beginning of a trip, Mr. Wilson's odometer (the mileage indicator) read 28,940 miles. If he traveled 289 miles, what did the odometer read at the end of the trip?
(a) 28,940 − 289 (b) 28,940 + 289 (c) 28,940 × 289 (d) 28,940 ÷ 289

**22.** 384 cars are parked in rows containing 16 cars each. How many rows are there?
(a) 384 + 16 (b) 384 − 16 (c) 384 × 16 (d) 384 ÷ 16

**23.** In Holtstown, 1,200 people voted in the election for mayor. If $\frac{2}{3}$ of the voters were women, how many women voted?
(a) $1{,}200 \div \frac{2}{3}$ (b) $1{,}200 \times \frac{2}{3}$ (c) $1{,}200 + \frac{2}{3}$ (d) $1{,}200 - \frac{2}{3}$

**24.** The Fantastic Boutique advertised a $90 jacket at a $15 discount. What is the sale price of the jacket?
(a) $90 + $15 (b) $90 − $15 (c) $90 × $15 (d) $90 ÷ $15

**25.** Jenny's height is 162.5 cm and her younger sister, Beth, is 152 cm tall. What is the difference between their heights?
(a) 162.5 cm − 152 cm (c) 162.5 cm ÷ 152 cm
(b) 162.5 cm + 152 cm (d) 162.5 cm × 152 cm

# UNIT P–3 Using Problem-Solving Strategies

## THE MAIN IDEA

There are procedures called ***strategies*** that can help you get the solution to a problem. For example:

1. ***Estimation*** is a strategy to use to get a "rough" answer quickly. You replace the given numbers by rounded values, thus simplifying the calculations. Since the rough answer tells you in advance an *approximate* answer, you are able to tell if your final answer is *reasonable.*
2. ***Breaking a problem into smaller problems*** is a strategy to use to simplify a problem that must have several steps. Instead of beginning by thinking about the final answer, you first work on each smaller problem and then use the results to get the answer.
3. ***Drawing a diagram*** is a strategy to use to help you *see* how numbers in a problem are related to each other. By studying the diagram, you can get ideas for arriving at the solution.
4. ***Guessing and checking*** is a strategy to use when you are given a few possible solutions and want to find out which one is correct. Also, you can use your judgment to guess an answer and then check to see if the result obtained makes sense according to the given information. In this way, you narrow the given choices.
5. ***Working backward*** is a strategy to use when you know a result and are trying to find an unknown piece of information that led to that result. In order to get back to the beginning of the problem, you perform the operation of arithmetic that is the *opposite* of the operation you would have performed if you had been looking for the result.

**EXAMPLE 1** Tell which strategy *is being used* to solve the problem.

*Problem:* You are given the populations of five towns and have to find the total population. Each number has several digits. There is a given set of choices.

You replace the given populations by rounded numbers and you get an approximate total. Then you choose the answer that is closest to your approximation.

### THINKING ABOUT THE PROBLEM

In order to find quickly the most reasonable choice, you used numbers that were easier than those given and you got an approximate answer.

*Answer:* The strategy used is estimation.

**EXAMPLE 2** Tell which strategy *is being used* to solve the problem.

*Problem:* Find the total price of several items if you are given the price of each of three different kinds of items and the number of items of each kind.

You first find the total price for each kind of item. Then you use these three results to find the total price for all the items.

### THINKING ABOUT THE PROBLEM

You have broken this problem into four separate, smaller problems.

*Answer:* The strategy used is breaking the problem into smaller problems.

**EXAMPLE 3** Tell which strategy *is being used* to solve the problem.

*Problem:* You are told the dimensions of a square and a rectangle, and you have to find how many times the square can fit into the rectangle.

You draw a diagram in which you place squares inside the rectangle. You count the number of squares that fit.

### THINKING ABOUT THE PROBLEM

While the problem can be solved by working only with the given numbers, drawing a diagram helped you to see the solution.

*Answer:* The strategy used is drawing a diagram.

**EXAMPLE 4** Tell which strategy *is being used* to solve the problem.

*Problem:* You are told the weight of one can of peas and the weight of one box of rice. You are also told the total weight of several cans of peas and boxes of rice. You have to find how many cans of peas and boxes of rice there are in the total.

You try different combinations of numbers—for example, 2 cans of peas and 3 boxes of rice. For each combination, you compute the total weight. You do this until you find a combination that gives the correct total.

### THINKING ABOUT THE PROBLEM

Each combination that you tried was a guess. You then used this guess to see if it made sense by checking to see if your guess gave the correct total. If your guess gave the wrong total, you were able to tell if your next guess should have been higher or lower.

*Answer:* The strategy used is guessing and checking.

**EXAMPLE 5** Tell which strategy *is being used* to solve the problem.

*Problem:* You are given the present temperature and the amounts of increase and decrease in the temperature for each of the past five hours. You have to find what the temperature was five hours ago.

You begin with the present temperature and make it lower by the amounts of the increases and higher by the amounts of the decreases.

### THINKING ABOUT THE PROBLEM

You knew the final temperature and were trying to find the original temperature. You used the opposites of the operations that you would have used if you had been trying to find the final temperature.

*Answer:* The strategy used is working backward.

**EXAMPLE 6** Tell which strategy *you could use* to solve the problem.

*Problem:* You are given three of Robert's test scores and the average of four of his test scores. You have to find his fourth test score.

### THINKING ABOUT THE PROBLEM

You could make a guess about Robert's fourth test score and use the guess to calculate the average score. If your guess leads to the same average that was given, you have found the solution. If not, guess again and check again. By comparing the wrong average with the known average, you would know if your next guess should be higher or lower.

*Answer:* The strategy to use is guessing and checking.

**EXAMPLE 7** Tell which strategy *you could use* to solve the problem.

*Problem:* You are told a total price and the price of one of the two items that make up this total price. You have to find the price of the second item.

### THINKING ABOUT THE PROBLEM

You could work backward. Start with the total price and perform the arithmetic operation that is the opposite of the operation you would have used if you were trying to find the total price.

*Answer:* The strategy to use is working backward.

**EXAMPLE 8** Tell which strategy *you could use* to solve the problem.

*Problem:* You have to find the total cost of gasoline used for Mr. Meyers' automobile trip. You are given three different amounts of gasoline that he bought at different times during the trip, and different prices per gallon for the three purchases.

### THINKING ABOUT THE PROBLEM

This problem has to be solved in four different parts. First, find the cost of each of the three purchases. Finally, use the three separate results to find the total cost of the gasoline.

*Answer:* The strategy to use is breaking the problem into smaller problems.

## CLASS EXERCISES

In 1–5, tell which strategy *is being used* to solve the problem.

1. You are given the total weight of six students and the individual weights of five of them. You have to find the weight of the sixth student.

   You try a number for the weight of the sixth student and see if this gives you the correct total.

2. You are given the exact measures of the sides of a plot of land; each measure has several digits. You want to find how much fencing is needed to go around the plot.

   You replace the given measures by rounded numbers to find an approximate answer.

3. Two different rectangles are described verbally, and the measures of their sides are given. You have to find the total distance around both rectangles.

   You sketch the rectangles and write all the measures on the sketch.

4. You are told your aunt's age now. You are asked to find her age a certain number of years ago.

   You begin with the present age.

5. You are asked to find how much carpeting is needed to cover the floor of a hotel lobby that has a very complicated shape.

   You sketch the given shape so that you can see the individual geometric figures that make up the shape.

In 6–10, tell which strategy *you could use* to solve the problem.

6. The front of a house is described verbally as a square topped by a triangle, and the dimensions of the square and triangle are given. You have to find the height of the house.
7. You have to find the final price of a purchase of two differently priced items after a discount is made.
8. You are given the final balance in a savings account. You are also given the amounts of deposits and withdrawals that have been made before this final balance was calculated. You have to find the original balance.
9. A high school student's age has been doubled and that result has been subtracted from 100. You are given the final answer. You have to find the student's age.
10. The number of students in a school district is to be multiplied by a dollar-and-cents amount budgeted for each student. You are to choose the correct answer for the total amount spent on the students from four very different choices.

## HOMEWORK EXERCISES

In 1–8, tell which strategy *is being used* to solve the problem.

1. You have to find the total of five dollar-and-cents amounts and choose the correct answer from four given choices.

   You replace the given amounts by rounded values to get an approximate answer that helps you select the correct choice.

2. You are given the present price of a stock and the amounts by which the price has increased and decreased over the past several days. You have to find the original price of the stock.

   You begin with the present price and reduce it by the amounts of increase and raise it by the amounts of decrease.

3. You are told how much water two different containers hold. After water has been poured back and forth from one container to another, you have to find how much water is left in one of them.

   You make a picture of the two containers and draw lines to show the increase and decrease in water levels.

4. When a number is multiplied by 7, and 5 is added to the product, the result is 54. You have to find the original number.

   You keep trying numbers, multiplying by 7 and adding 5, until you find one that gives the correct result.

5. Two students each took five exams. You are asked to find the difference between their exam score averages.

   You first find each student's average, and then you find the difference of the averages.

6. You are given Mr. Manning's hourly rate of pay, his overtime rate of pay, and the number of hours he worked during the week (regular time and overtime). You have to find his total pay for the week.

   You separately calculate his regular pay and his overtime pay, then find the sum.

7. To find out quickly about how many eggs there are in 98 dozen, you multiply 100 by 12.

8. You are told how much change Mark received from a ten-dollar bill after buying a certain number of pens, all at the same price. In order to find the price of one pen, you subtract the amount of change from $10 and divide that result by the number of pens.

In 9–16, tell which strategy *you could use* to solve the problem.

9. You are told the total price of a certain number of pounds of ham and the total price of a different number of pounds of turkey. You are asked to compare the price of one pound of ham to the price of one pound of turkey.

10. After being told that a number of bicycles of equal weight together weigh a certain amount, you want to find the approximate weight of one bicycle. The given numbers contain many digits.

11. After being given Joseph's present weight and his changes in weight over the past eight weeks, you want to find what his weight was eight weeks ago.

12. You are asked to find how the area of a large rectangle compares to the area of a small rectangle, given a relationship between the dimensions of each rectangle.

13. You quickly want to find the approximate number of students in Midtown High School. You are given the exact number of freshmen, sophomores, juniors, and seniors.

14. On a certain exam, each correct answer on part I is worth a given number of points, and each correct answer on part II is worth a different given number of points. You want to find the number of correct answers a student had on each part if her total score is known.

15. Mrs. Smith buys a candy mixture. You are given the different weights and prices of three different kinds of candy that make up the mixture. You want to find the price of the mixture.

16. You want to find the total cost of a given number of books, each of which costs between two given dollar amounts. An exact answer is not necessary.

# Whole Numbers

## MODULE 1



## MODULE 2

# UNIT 1–1 Using a Number Line to Represent and Compare Whole Numbers

## USING A NUMBER LINE TO REPRESENT WHOLE NUMBERS

### THE MAIN IDEA

1. A ***whole number*** is a member of the set {0, 1, 2, 3, 4, . . .}.
2. The smallest whole number is 0.
3. There is no largest whole number.
4. A ***number line*** is used to draw the ***graph*** of the whole numbers. Each whole number is represented by a point on the number line.

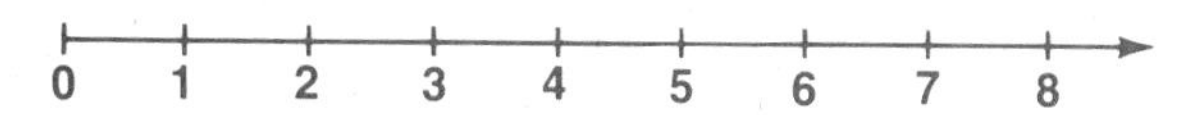


**EXAMPLE 1** For each whole number, give the next whole number on a number line.

The next whole number on a number line can be found by adding 1 to the original whole number.

| | *Original Whole Number* | *Answer* |
|---|---|---|
| **a.** | 16 | 16 + 1 = 17 |
| **b.** | 57 | 57 + 1 = 58 |
| **c.** | 189 | 189 + 1 = 190 |
| **d.** | 2,006 | 2,006 + 1 = 2,007 |
| **e.** | 52,199 | 52,199 + 1 = 52,200 |

**EXAMPLE 2** Write all the whole numbers that are between each pair of whole numbers.

| | *Pair of Whole Numbers* | *Answer* |
|---|---|---|
| **a.** | 29 and 33 | 30, 31, 32 |
| **b.** | 231 and 236 | 232, 233, 234, 235 |
| **c.** | 4,195 and 4,197 | 4,196 |
| **d.** | 5,781 and 5,782 | none |

### *CLASS EXERCISES*

**1.** For each whole number, give the next whole number on a number line.

**a.** 0 **b.** 27 **c.** 99 **d.** 164 **e.** 9,999

**2.** Write all the whole numbers that are between each pair of whole numbers.

**a.** 0; 2 **b.** 9; 12 **c.** 98; 103 **d.** 122; 129 **e.** 912; 913

## USING A NUMBER LINE TO COMPARE WHOLE NUMBERS

### THE MAIN IDEA

1. The symbol $>$ means ***is greater than***. If one number is *to the right of* a second number on a number line, then the first number is greater than the second number.
2. The symbol $<$ means ***is less than***. If one number is *to the left of* a second number on a number line, then the first number is less than the second number.
3. The symbols $>$ and $<$ always point to the smaller number.
4. If a number is ***between*** two other numbers, then we can write the three numbers in *increasing order* of size separated by $<$ symbols, or we can write the three numbers in *decreasing order* of size separated by $>$ symbols.

**EXAMPLE 3** Use the symbol $>$ or $<$ to rewrite each statement based on the number line shown.

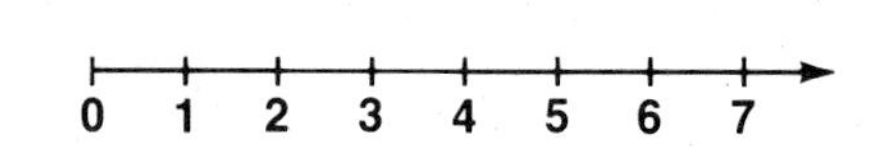


| *Statement* | *Answer* |
|---|---|
| **a.** 5 is greater than 2 | $5 > 2$ |
| **b.** 2 is less than 5 | $2 < 5$ |
| **c.** 2 is between 0 and 5 | $0 < 2 < 5$ or $5 > 2 > 0$ |

**EXAMPLE 4** Tell which of the numbers in each pair is farther to the *right* on a number line.

| *Number Pair* | *Answer* |
|---|---|
| **a.** 6 and 9 | 9 |
| **b.** 22 and 19 | 22 |
| **c.** 111 and 101 | 111 |

**EXAMPLE 5** Tell which of the numbers in each set is *between* the others. Then write the three numbers in order using $>$ or $<$.

| *Number Set* | *Answer* |
|---|---|
| **a.** 2, 3, 5 | 3 is between 2 and 5<br>$2 < 3 < 5$ or $5 > 3 > 2$ |
| **b.** 4, 0, 7 | 4 is between 0 and 7<br>$0 < 4 < 7$ or $7 > 4 > 0$ |

When you use more than one inequality symbol in a number statement, the symbols must point in the same direction.

**EXAMPLE 6** Tell whether each statement is *true* or *false*.

| *Statement* | *Answer* |
|---|---|
| **a.** $26 > 32$ | false |
| **b.** $32 > 26$ | true |
| **c.** $19 < 20 < 21$ | true |
| **d.** $20 < 19 < 18$ | false |
| **e.** $32 > 16 > 3$ | true |

## CLASS EXERCISES

**1.** Use the symbol $>$ or $<$ to rewrite each statement based on the number line shown.

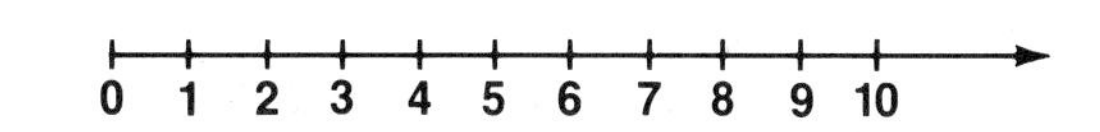


**a.** 1 is greater than 0  **b.** 0 is less than 1
**c.** 1 is less than 6  **d.** 6 is greater than 1
**e.** 1 is between 0 and 6  **f.** 5 is between 2 and 7

**2.** Tell which of the numbers in each pair is farther to the *right* on a number line.
**a.** 2; 3  **b.** 27; 19  **c.** 706; 607  **d.** 454; 445  **e.** 1,000; 999

**3.** Tell which of the numbers in each pair is farther to the *left* on a number line.
**a.** 5; 4  **b.** 121; 123  **c.** 423; 342  **d.** 7,003; 3,007  **e.** 9,999; 10,103

**4.** Tell which of the numbers in each set is farthest to the *right* on a number line.
**a.** 6; 8; 1  **b.** 13; 17; 19  **c.** 441; 404; 440

**5.** Tell whether each comparison is *true* or *false*.
**a.** $1 > 0$  **b.** $100 < 97$  **c.** $263 > 189$  **d.** $2,007 < 2,008$  **e.** $2,742 > 2,740$
**f.** $332 < 471 < 512$

**6.** Each letter on the number line shown represents a number. Replace each ? by $<$ or $>$ to write a true statement.

**a.** E ? O  **b.** I ? A  **c.** A ? U  **d.** U ? O ? E  **e.** A ? E ? U  **f.** I ? O ? U

## HOMEWORK EXERCISES

**1.** For each whole number, give the next whole number on a number line.
**a.** 9  **b.** 57  **c.** 139  **d.** 999  **e.** 1,063

**2.** Write all the whole numbers that are between each pair of whole numbers.
**a.** 1; 5  **b.** 8; 11  **c.** 98; 101  **d.** 482; 483  **e.** 999; 1,011

**3.** Use the symbol $>$ or $<$ to rewrite each statement based on the number line shown.

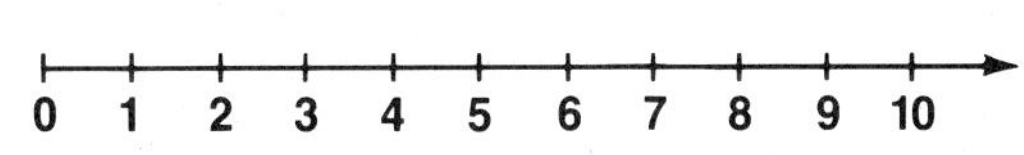


**a.** 8 is greater than 4  **b.** 4 is less than 8
**c.** 0 is less than 3  **d.** 4 is between 0 and 8

4. Tell which of the numbers in each pair is farther to the *right* on a number line.

 **a.** 10; 0  **b.** 17; 27  **c.** 92; 89  **d.** 304; 403  **e.** 1,001; 1,000

5. Tell which of the numbers in each pair is farther to the *left* on a number line.

 **a.** 0; 1  **b.** 20; 19  **c.** 222; 333  **d.** 996; 699  **e.** 2,999; 3,000

6. Tell which of the numbers in each set is farthest to the *left* on a number line.

 **a.** 5; 1; 9  **b.** 21; 19; 7  **c.** 101; 111; 100

7. Tell whether each comparison is *true* or *false*.

 **a.** $19 > 21$  **b.** $189 < 263$  **c.** $4{,}639 > 4{,}641$  **d.** $2{,}741 < 2{,}742$

 **e.** $8{,}999 > 9{,}000$  **f.** $701 > 711 > 747$

8. Referring to the number line shown, replace each ? by $<$ or $>$ to write a true statement about each pair of numbers.

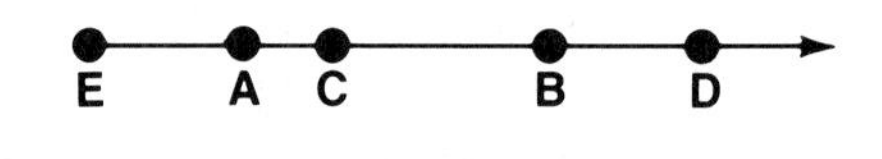


 **a.** A ? C  **b.** C ? A  **c.** D ? E  **d.** B ? D  **e.** D ? C  **f.** A ? C ? B

 **g.** D ? B ? E  **h.** E ? B ? D

9. Complete each statement with $>$ or $<$ to make it true.
 **a.** If A is to the left of B on a number line, then A __ B.
 **b.** If A is to the right of B on a number line, then A __ B.
 **c.** If B is between A and C on a number line, then $A < B$ __ C or C __ $B > A$.

## CHALLENGE QUESTION

**a.** What is the smallest whole number that can replace ? to make this statement true?

$$157 < ? < 379$$

**b.** What is the largest whole number that can replace ? to make the preceding statement true?

# UNIT 1–2 Place Value; Writing and Reading Whole Numbers

## MEANING OF PLACE VALUE

### THE MAIN IDEA

1. A ***numeral*** is the mathematical name for a number. Each of the whole numbers can be written as a numeral by using the ***digits*** 0, 1, 2, 3, 4, 5, 6, 7, 8, and 9.
2. The ***position*** or ***place*** of each digit in a numeral is named as follows:

| *Billions* | | | | *Millions* | | | | *Thousands* | | | | *Ones* | | |
|---|---|---|---|---|---|---|---|---|---|---|---|---|---|---|
| hundred billions | ten billions | billions | | hundred millions | ten millions | millions | | hundred thousands | ten thousands | thousands | | hundreds | tens | ones |
| 3 | 8 | 7 | , | 2 | 0 | 4 | , | 9 | 3 | 6 | , | 8 | 1 | 5 |

   The positions are grouped into sets of three, separated by commas as shown above. Each group is called a ***period,*** and each period has a name such as billions, millions, thousands, and ones.
3. The ***place value of a position*** is the numerical value taken from its name. *As you move to the left, each position has a place value of ten times the previous position.*
4. To find the ***place value of a digit*** in a numeral, multiply the digit by the value of the place in which the digit appears.

**EXAMPLE 1** Name the position of 9 in each numeral.

| *Numeral* | *Answer* |
|---|---|
| **a.** 28<u>9</u> | 9 is in the ones position. |
| **b.** <u>9</u>28 | 9 is in the hundreds position. |
| **c.** 8<u>9</u>2 | 9 is in the tens position. |
| **d.** <u>9</u>0,075 | 9 is in the ten thousands position. |

**EXAMPLE 2** Name the position and tell the value of each of the underlined digits.

| *Numeral* | *Answer* |
|---|---|
| **a.** 784 | 8 is in the tens position and has the value of $8 \times 10$ or 80. |
| **b.** 23,491 | 3 is in the thousands position and has the value of $3 \times 1{,}000$ or 3,000. |
| **c.** 175,892 | 1 is in the hundred thousands position and has the value of $1 \times 100{,}000$ or 100,000. |
| **d.** 2,576,493 | 4 is in the hundreds position and has the value of $4 \times 100$ or 400. |

## CLASS EXERCISES

1. Name the position of 7 in each numeral.
   **a.** 972 **b.** 67,431 **c.** 8,517 **d.** 375,913
2. Name the position and tell the value of each of the underlined digits.
   **a.** 5,473 **b.** 3,528,916 **c.** 83,419 **d.** 5,961,482
3. Complete each statement to make it true.
   **a.** In 375, the digit 3 is in the ________ position.
   **b.** In 28,972, the digit 8 is in the ________ position and has the value of ________.
   **c.** The value of 2 in 7,241 is ________ times greater than the value of 2 in 623.

## WRITING WHOLE NUMBERS AS NUMERALS

### THE MAIN IDEA

To write as a numeral a whole number that is expressed in words:

1. Determine the number of places needed. (Use the name of the position farthest to the left.)
2. Moving from right to left, separate the number of places into periods of three by commas. The period farthest to the left may have fewer than three places.
3. Fill each period with numerals from left to right.
4. If there is no digit for a place in the whole number, use 0 as a placeholder.

**EXAMPLE 3** Write seven hundred nine as a numeral.

__ __ __ "Hundred" tells us that we need three places.

7 0 9 Since there is no digit for the tens place, use 0 as a placeholder.

Notice that if 0 were not used, the number represented would be seventy-nine (79).

**EXAMPLE 4** Write three million, fifty-two thousand as a numeral.

| | |
|---|---|
| _,_ _ _,_ _ _ | "Million" tells us that we need seven places. |
| 3,_ _ _,_ _ _ | Fill periods from left to right. |
| 3,0 5 2,_ _ _ | Since there is no digit for the hundred thousands place, use 0 as a placeholder there. |
| 3,0 5 2,0 0 0 | Since there are no digits for the hundreds, tens, or ones places, use three 0's as placeholders there. |

## CLASS EXERCISES

Write as a numeral:

1. four hundred twenty-three
2. seven thousand, eight hundred sixty-two
3. fifteen thousand, three hundred twelve
4. four million, three hundred sixty-two thousand, one hundred twenty
5. nine hundred twenty-one thousand, two hundred seventeen
6. six thousand, fifty-two
7. two hundred eight thousand
8. one million, four hundred sixty-two
9. two hundred thousand, five hundred
10. seven million, seventeen thousand

## READING WHOLE NUMBERS THAT ARE WRITTEN AS NUMERALS

### THE MAIN IDEA

To read a whole number that is written as a numeral:

1. Begin with the largest period, the period farthest to the left. Read the one-, two-, or three-digit number followed by the name of the period, dropping the final "s."
2. Repeat this process for each of the remaining periods. When reading the ones period, do not say the name "ones."

**EXAMPLE 5** Read: 2,573

| 2, | 573 |
|---|---|
| thousands | ones |

Begin with the largest period, the period farthest to the left. Read the number in this period (2) followed by the name of this period, dropping the final "s": "thousand." — two thousand,

Read the number in the next period (573). Since this is the ones period, do not say the period name. — five hundred seventy-three

*Answer:* 2,573 is read: "two thousand, five hundred seventy-three."

**EXAMPLE 6** Read: 3,000,809

| 3, | 000, | 809 |
|---|---|---|
| millions | thousands | ones |

Begin with the largest period, the period farthest to the left. Read the number in this period (3) followed by the word "million." — three million,

Since the next period, thousands, has only zeros that are used as placeholders, skip it and read the following period.

Read the number in the ones period (809). — eight hundred nine

*Answer:* 3,000,809 is read: "three million, eight hundred nine."

## CLASS EXERCISES

Read each whole number.

**1.** 397 **2.** 5,146 **3.** 23,412 **4.** 7,006 **5.** 352,000

**6.** 507,904 **7.** 4,370,600 **8.** 67,000,092

## WRITING A WHOLE NUMBER AS A WORD NAME

### THE MAIN IDEA

To write a whole number as a word name:

1. Write each period as you read it, separating period names with commas.
2. Use a hyphen to separate tens from ones in each period.
3. Use 0 as a placeholder when needed.

**EXAMPLE 7** Write the word name for 541,096.

$$\underbrace{541,}_{\text{thousands}} \quad \underbrace{096}_{\text{ones}}$$

Begin with the largest period, thousands. Read the number in this period (541) and write its name followed by the word "thousand." — five hundred forty-one thousand,

Read the number in the next period (96). Since this is the ones period, we omit the period name. — ninety-six

Notice that the 0 is used as a placeholder, but it is not written in the word name.

Notice also that the tens and ones in each period are separated by a hyphen.

*Answer:* The word name for 541,096 is five hundred forty-one thousand, ninety-six.

## CLASS EXERCISES

Write a word name for each whole number.

**1.** 48 **2.** 783 **3.** 2,475 **4.** 5,081 **5.** 257,402

**6.** 706,000 **7.** 3,000,400 **8.** 9,470,082

## HOMEWORK EXERCISES

**1.** Name the position of 9 in each number.

**a.** 93 **b.** 198 **c.** 4,009 **d.** 907 **e.** 97,832 **f.** 9,637,728

**g.** 1,903,275 **h.** 9,134,000,000

**2.** Determine the value of each of the underlined digits.

**a.** $4\underline{9}2$ **b.** $4{,}\underline{5}82$ **c.** $27\underline{6}$ **d.** $8\underline{1}{,}079$ **e.** $\underline{3}{,}457{,}892$ **f.** $2\underline{7}5{,}342$

**g.** $\underline{4}{,}132{,}789{,}076$

**3.** Write as a numeral:

**a.** two hundred six **b.** one thousand, ninety-two

**c.** fifteen thousand, seventy **d.** one hundred thirty thousand, nine

**e.** two million, seventy-three thousand, one hundred eleven

**f.** seventy thousand, six hundred eighty-six

**g.** five billion, three hundred thousand

**4.** Write a word name for each whole number.

**a.** 251 **b.** 45,075 **c.** 3,010 **d.** 4,150,000 **e.** 505,028

**f.** 7,002 **g.** 205,918 **h.** 1,005,870 **i.** 209 **j.** 970,091

**5.** In which place is the digit 6 in the numeral 986,452?

6. What is the value of 7 in 1,732,584?

7. In which numeral does the digit 4 have the value of forty?
   (a) 4,580 (b) 17,410 (c) 40,000 (d) 92,040

8. In the numeral 782,639, the digit 2 has the value
   (a) 20 (b) 200 (c) 2,000 (d) 20,000

## SPIRAL REVIEW EXERCISES

1. Tell whether each comparison is *true* or *false*.
   **a.** $109 > 190$ **b.** $1{,}111 < 1{,}211$
   **c.** $2{,}050 < 2{,}500$ **d.** $70{,}900 > 79{,}000$
   **e.** $100{,}000 > 99{,}999$
   **f.** $1{,}000{,}000 < 100{,}895$
   **g.** $50 < 40 < 100$ **h.** $79 > 75 > 70$

2. Referring to the number line shown, replace each ? by $>$ or $<$ to write a true statement about each pair of numbers.

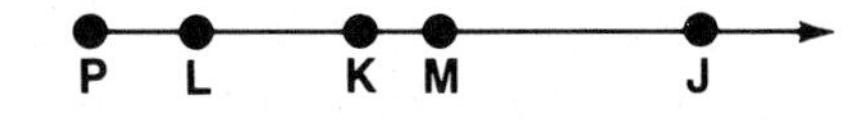


   **a.** M ? L **b.** K ? J **c.** L ? J
   **d.** J ? L **e.** L ? K ? J **f.** M ? K ? P

### CHALLENGE QUESTION

By how much will the number 26,493 be increased if you change the digit 6 to an 8?

# UNIT 1–3 Rounding Whole Numbers

## THE MAIN IDEA

1. A *rounded number* is an *approximation* for an exact whole number.
2. To *round* a whole number to a desired place:
   a. Circle the digit in the place to be rounded.
   b. Look at the digit to the immediate right of the circled digit.
      (1) If the digit to the immediate right is 5 or more, add 1 to the circled digit *(round up)*.
      (2) If the digit to the immediate right is less than 5, keep the circled digit as is *(round down)*.
   c. Replace all the digits to the right of the rounded digit by zeros.

**EXAMPLE 1** Round 8,273 to the nearest ten.

8,2⑦3 "The nearest ten" tells you that the digit to be rounded is in the tens place. Circle the 7. The digit to the immediate right of the circled digit is 3.

8,2⑦_ Since the digit to the immediate right is less than 5, *round down* (leave the circled digit as is).

8,2⑦0 Replace all the digits to the right of the rounded digit by zeros.

*Answer:* 8,273 rounded to the nearest ten is 8,270.

**EXAMPLE 2** Round 21,895 to the nearest thousand.

2①,895 "The nearest thousand" tells you that the digit to be rounded is in the thousands place. Circle the 1. The digit to the immediate right of the circled digit is 8.

2②,_ _ _ Since the digit to the immediate right is greater than 5, *round up* (add 1 to the circled digit).

2②,000 Replace all the digits to the right of the rounded digit by zeros.

*Answer:* 21,895 rounded to the nearest thousand is 22,000.

**EXAMPLE 3** In one year, 3,951,386 people were born in the United States. Round the number of births to the nearest hundred thousand.

3,⑨51,386 "The nearest hundred thousand" tells you that the digit to be rounded is in the hundred thousands place. Circle the 9. The digit to the immediate right of the circled digit is 5.

4,⓪_ _,_ _ _ Since the digit to the immediate right is exactly 5, *round up* (in this case, 39 + 1 = 40).

4,⓪00,000 Replace all the digits to the right of the rounded digit by zeros.

*Answer:* 3,951,386 rounded to the nearest hundred thousand is 4,000,000.

**EXAMPLE 4** 1,111,111 rounded to the nearest million is:
(a) 1,000,000 (b) 1,100,000 (c) 1,110,000 (d) 1,111,000

*Answer:* (a)

## CLASS EXERCISES

1. Round 1,465 to the nearest hundred.
2. Round 293,124 to the nearest ten thousand.
3. Round 82,563 to the nearest ten.
4. Round 4,986,416 to the nearest hundred thousand.
5. In 1982, Star Company had total sales of $31,418,000. Round the total sales to the nearest million.
6. The population of Belgium is 9,835,000. Round the population to the nearest ten thousand.
7. 55,555,555 rounded to the nearest ten million is
   (a) 50,000,000 (b) 56,000,000 (c) 55,600,000 (d) 60,000,000
8. The area of Japan is approximately 370,325 square kilometers. Round the area to the nearest thousand square kilometers.

## HOMEWORK EXERCISES

1. Round to the nearest ten:
   **a.** 187 **b.** 47,095 **c.** 3,063 **d.** 583,498
2. Round to the nearest hundred:
   **a.** 592 **b.** 15,813 **c.** 819,457 **d.** 1,596,962
3. Round to the nearest thousand:
   **a.** 4,527 **b.** 153,697 **c.** 73,918 **d.** 1,249,897
4. Write in your own words how you would round 592,483 to the nearest thousand.
5. 372,516 rounded to the nearest hundred thousand is
   (a) 300,000 (b) 370,000 (c) 400,000 (d) 470,000
6. The Empire State Building is 1,250 feet tall. Round its height to the nearest hundred feet.
7. In 1980, Ronald Reagan received 43,899,248 votes in the presidential election. Round the number of votes to the nearest ten thousand.
8. The Nile River is 4,145 miles long. Round the length of the Nile River to the nearest ten miles.
9. In one season, 34,540,000 households watched the Super Bowl. Round this number to the nearest hundred thousand.

## SPIRAL REVIEW EXERCISES

**1.** Replace each ? by $<$ or $>$ to make a true comparison.

**a.** 241 ? 214

**b.** 103 ? 301

**c.** 1,000 ? 1,100 ? 11,000

**d.** twenty-seven ? twenty-six

**e.** nine hundred ninety thousand ? one million

**2.** Tell whether each statement is *true* or *false*.

**a.** 101 is to the left of 111 on a number line.

**b.** There are three whole numbers between 7 and 10.

**c.** If $a < b$, then $a$ is to the right of $b$ on a number line.

**d.** 5,782,979 is the largest whole number.

**e.** There is no smallest whole number.

### CHALLENGE QUESTION

Each of the letters A, B, and C represents a different digit in this addition example.

$$\begin{array}{r} 4\,9 \\ +\ 5\,\mathrm{A} \\ \hline \mathrm{C}\,\mathrm{B}\,7 \end{array}$$

Find the value of each of the three letters.

# UNIT 1–4 Adding Whole Numbers; Properties of Addition

## ADDING WHOLE NUMBERS

### THE MAIN IDEA

1. The result of an addition is called the *sum*.
2. The numbers in an addition are named in the following way:

$$\begin{array}{rl} 2 & \leftarrow \text{addend} \\ +5 & \leftarrow \text{addend} \\ \hline 7 & \leftarrow \text{sum} \end{array} \qquad \underset{\text{addend}}{2} + \underset{\text{addend}}{5} = 7 \leftarrow \text{sum}$$

3. To find the sum of whole numbers:
   a. Write the numerals in vertical columns, lining up the ones digits.
   b. Add the digits in the columns, beginning with the ones column.
   c. Remember to carry when necessary.

**EXAMPLE 1** Find the sum of 253 and 42.

Write the numerals in vertical columns, lining up the ones digits.

$$\begin{array}{r} 253 \\ +42 \\ \hline \end{array}$$

Add in the ones column (3 + 2 = 5).

$$\begin{array}{r} 253 \\ +42 \\ \hline 5 \end{array}$$

Add in the tens column (5 + 4 = 9).

$$\begin{array}{r} 253 \\ +42 \\ \hline 95 \end{array}$$

Add in the hundreds column (2 + 0 = 2).

$$\begin{array}{r} 253 \\ +42 \\ \hline 295 \end{array}$$

*Answer:* The sum is 295.

**EXAMPLE 2** Add 2,765, 842, and 98.

Add in the ones column (5 + 2 + 8 = 15). Write 5 and carry 1.

$$\begin{array}{r} 1 \\ 2{,}765 \\ 842 \\ +98 \\ \hline 5 \end{array}$$

Add in the tens column (1 + 6 + 4 + 9 = 20). Write 0 and carry 2.

$$\begin{array}{r} 21 \\ 2{,}765 \\ 842 \\ +98 \\ \hline 05 \end{array}$$

Add in the hundreds column (2 + 7 + 8 = 17). Write 7 and carry 1.

$$\begin{array}{r} 1\ 21 \\ 2{,}765 \\ 842 \\ +98 \\ \hline 705 \end{array}$$

Add in the thousands column (1 + 2 = 3). Write 3.

$$\begin{array}{r} 1\ 21 \\ 2{,}765 \\ 842 \\ +98 \\ \hline 3{,}705 \end{array}$$

*Answer:* The sum is 3,705.

**EXAMPLE 3** Steve wants to buy a stereo system. The receiver costs $479, the speakers cost $396, and the turntable costs $82. Find the total cost of the system.

**THINKING ABOUT THE PROBLEM**

The given pieces of information are the three amounts, $479, $396, and $82. The key words "total cost" tell you to use addition.

Find the total cost by adding the price of each of the separate parts.

$$\begin{array}{r} 479 \\ 396 \\ +82 \\ \hline 957 \end{array}$$

*Answer:* The system costs $957.

**EXAMPLE 4** At a football game, 15,283 fans bought general admission tickets, 2,815 fans bought box seats, and 11,017 fans bought bleacher seats. How many fans attended the game?

(a) 28,015 (b) 39,115 (c) 19,015 (d) 29,115

**THINKING ABOUT THE PROBLEM**

You can use estimation to help you choose the correct answer. Replace 15,283 by 15,000, replace 2,815 by 3,000, and replace 11,017 by 11,000.

Add the rounded numbers.

$$\begin{array}{r} 15{,}000 \\ 3{,}000 \\ +11{,}000 \\ \hline 29{,}000 \end{array}$$

Choose the answer that is closest to 29,000.

*Answer:* (d)

**EXAMPLE 5** The Smiths are driving from their home in Titus to attend a high school class reunion in Laren.

**a.** They have friends in Stotsburg. If they travel by way of Stotsburg, how many miles must they drive from Titus to Laren?

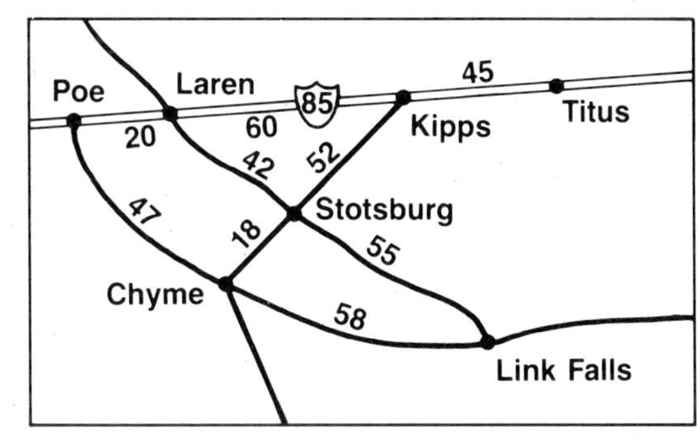


To travel from Titus to Laren by way of Stotsburg, the Smiths must drive from Titus to Kipps to Stotsburg to Laren. Add the distances.

$$45 + 52 + 42 = 139$$

*Answer:* The Smiths must drive 139 miles.

**b.** If they go back to Titus from Laren directly along Route 85, how many miles will that drive be?

Add the two distances along Route 85.

$$60 + 45 = 105$$

*Answer:* The drive back will be 105 miles.

## CLASS EXERCISES

**1.** Find the sum of each pair of whole numbers.

**a.** 215 and 381 **b.** 472 and 749 **c.** 1,896 and 4,528 **d.** 805 and 96

**e.** 5,947 and 859 **f.** 40,816 and 92,458

**2.** Add:

| a. | b. | c. | d. |
|---:|---:|---:|---:|
| 57 | 1,092 | 579 | 26,468 |
| 962 | 4,842 | 6,842 | 5,904 |
| 439 | 658 | 39,275 | 17,629 |

**3.** Add: 2,581, 698, and 57. **4.** Add: 5,476, 2,398, and 11,465.

**5.** At a football game between Beefburg High School and Millard Fillmore High School, 378 students from Beefburg High School attended and 589 students from Fillmore High School attended. What was the total number of students at the game?

**6.** The base price of a car is $8,752. The cost of a stereo radio is $286, and the cost of air conditioning is $637. Find the total cost of the car.

**7.** In March, Mr. Harris drove a total of 3,729 miles. In April, he drove 4,162 miles, and in May, 5,236 miles.

**a.** Round each number to the nearest thousand.

**b.** By adding the rounded numbers, estimate the total number of miles he drove in this 3-month period.

**c.** Use the estimate to choose the exact answer:
(a) 30,127 (b) 23,127 (c) 13,127 (d) 10,127

In 8 and 9, use the road distances shown on the map on page 33.

**8.** How far is it from Chyme to Kipps?

**9.** What is the shortest distance between Poe and Stotsburg?

## PROPERTIES OF ADDITION

### THE MAIN IDEA

1. The sum of two whole numbers is always a whole number.
2. The order in which you add two whole numbers does not change the sum.
   $$5 + 9 = 9 + 5$$
3. The way you group numbers in order to perform a series of additions does not change the sum.
   $$(6 + 3) + 7 = 6 + (3 + 7)$$
4. The sum of 0 and a whole number is that same whole number.
   $$9 + 0 = 9$$

**EXAMPLE 6** Tell whether each statement is *true* or *false*.

| ***Statement*** | ***Answer*** |
|---|---|
| **a.** $11 + 5 = 5 + 11$ | *True*. The order in which you add numbers does not change the sum. |
| **b.** $(28 + 10) + 2 > 28 + (10 + 2)$ | *False*. The way you group numbers does not change the sum. |
| **c.** $0 + 37 < 37$ | *False*. The sum of 0 and a number is that number. |
| **d.** If $\triangle$ and $\square$ are whole numbers, then $\triangle + \square$ is a whole number. | *True*. The sum of two whole numbers is a whole number. |

**EXAMPLE 7** Show that $(7 + 2) + 8 = 7 + (2 + 8)$ is a true statement.

$$(7 + 2) + 8 \stackrel{?}{=} 7 + (2 + 8)$$

Add first. ↗ ↖ Add first.

$$9 + 8 \stackrel{?}{=} 7 + 10$$

$$17 = 17 \quad \textit{True}$$

**EXAMPLE 8** Replace each ? by a number to make a true statement.

| ***Statement*** | ***Answer*** |
|---|---|
| **a.** $(17 + 6) + 8 = 17 + (? + 8)$ | $(17 + 6) + 8 = 17 + (6 + 8)$ |
| **b.** $352 + 178 = ? + 352$ | $352 + 178 = 178 + 352$ |
| **c.** $2{,}511 + ? = 2{,}511$ | $2{,}511 + 0 = 2{,}511$ |
| **d.** $? + 47 = 47 + ?$ | any number (the same number in both places) |

## CLASS EXERCISES

1. Show that (8 + 11) + 17 = 8 + (11 + 17) is a true statement.
2. Tell whether each statement is *true* or *false*.
   **a.** (12 + 11) + 9 = 12 + (11 + 9) **b.** 26 + 5 < 5 + 26 **c.** 79 + 1 = 79
   **d.** 73 + 0 = 73
3. Replace □ by a number to form a true statement.
   **a.** □ + 296 = 296 + 85 **b.** (32 + 40) + □ = 32 + (40 + 19) **c.** □ + 512 = 512
   **d.** (11 + 17) + 19 = □ + (17 + 19) **e.** 2,575 + 137 = 137 + □ **f.** 1,720,000 + 0 = □

## HOMEWORK EXERCISES

1. Add:

| a. | b. | c. | d. | e. | f. | g. |
|---|---|---|---|---|---|---|
| 638 | 467 | 1,240 | 3,087 | 739 | 401 | 2,003 |
| +275 | +82 | +760 | +13 | 31 | 97 | 4,884 |
| | | | | +251 | +3,967 | +3,127 |

2. Find each sum.
   **a.** 22 + 49 **b.** 113 + 99 **c.** 279 + 88 **d.** 109 + 202 + 0 **e.** 196 + 28 + 42
   **f.** 97 + 246 + 524 **g.** 28 + 563 + 72 **h.** 398 + 47 + 53 **i.** 2,756 + 39 + 842
3. Tell whether each statement is *true* or *false*.
   **a.** (18 + 9) + 6 = 18 + (9 + 6) **b.** 17 + 3 > 3 + 17 **c.** 15 + 19 < 19 + 15
   **d.** 262 + 0 = 262 **e.** 48 + 0 = 0 + 48 **f.** 345 + 0 > 345
4. Show that (29 + 42) + 56 = 29 + (42 + 56) is a true statement.
5. Replace □ by a number that will make a true statement.
   **a.** 9 + 3 = □ + 9 **b.** 27 + □ = 17 + 27 **c.** (8 + □) + 11 = 8 + (7 + 11)
   **d.** (3 + 15) + 6 = □ + (15 + 6) **e.** (9 + 22) + 83 = 9 + (□ + 83)
6. John's weekly salary is $138. Last week he earned $53 more by working overtime. How much did he earn last week?
7. A car costs $7,980 and the sales tax is $634. What is the cost of the car, including the sales tax?
8. Mary delivered 132 newspapers on Monday, 143 newspapers on Tuesday, and 118 newspapers on Wednesday. Find the total number of papers Mary delivered on these three days.

   In 9 and 10, use the road distances shown on the map on page 33.
9. How far is it from Poe to Titus?
10. What is the shortest distance between Laren and Chyme?

## SPIRAL REVIEW EXERCISES

**1.** Replace ? by $<$, $>$, or $=$ to make a true comparison.

**a.** 8 + 9 ? 17 + 5

**b.** 15 + 32 ? 42 + 18

**c.** 96 + 73 ? 84 + 68

**d.** 112 + 28 ? 46 + 94

**e.** 230 + 0 ? 0 + 197

**f.** 2,248 + 3,741 ? 3,741 + 2,248

**2.** Write as a numeral:

**a.** one million, sixty thousand

**b.** five hundred fifty thousand, two hundred eleven

**c.** forty-seven thousand, thirty-six

**d.** five thousand, seven

**3.** In the election for governor, Mr. Adams received 123,798 votes, Ms. Berkeley received 204,078 votes, and Mr. Jackson received 112,218 votes.

**a.** Round each number to the nearest thousand.

**b.** By adding the rounded numbers, estimate the total number of votes cast for the three candidates.

**c.** Use the estimate to choose the exact answer:

(a) 440,094 (c) 340,094
(b) 400,094 (d) 300,094

**4.** Round each number to the nearest ten thousand.

**a.** 68,418 **b.** 175,906 **c.** 7,287,199

**5.** In 1981, the United States produced 2,437,828 gallons of gasoline. Round the number of gallons to the nearest hundred thousand.

**6.** Replace ? by $<$ or $>$ to make a true statement.

**a.** 123 ? 132 **b.** 86,340 ? 9,999

**c.** 24 + 23 ? 48 **d.** 60 + 3 ? 63 + 1

**e.** the sum of ninety-nine and one ? 101

**f.** the value of 2 in 29 ? the value of 9 in 339

### CHALLENGE QUESTION

Peggy sells souvenirs at a baseball stadium. The chart shows her sales of buttons, posters, and pennants for three weeks.

| Item | Week 1 | Week 2 | Week 3 |
|---|---|---|---|
| Buttons | 133 | 137 | 152 |
| Posters | 148 | 112 | 128 |
| Pennants | 123 | 144 | 139 |

**a.** For which type of souvenir was the total number of sales the greatest?

**b.** For which type of souvenir was the total number of sales the least?

**c.** How many more buttons did she sell than pennants?

# UNIT 1–5 Subtracting Whole Numbers; Addition and Subtraction as Inverse Operations

## SUBTRACTING WHOLE NUMBERS

### THE MAIN IDEA

1. The result of a subtraction is called the ***difference.***
2. The numbers in a subtraction are named in the following way:

$$\begin{array}{rl} 7 & \longleftarrow \textit{minuend} \\ \underline{-5} & \longleftarrow \textit{subtrahend} \\ 2 & \longleftarrow \textit{difference} \end{array}$$

3. To subtract two whole numbers:
   a. Write the numerals in vertical columns, lining up the ones digits. (The "from" number goes on top.)
   b. Subtract the digits in the columns, beginning with the ones column.
   c. If a digit in the subtrahend is greater than the digit above it in the minuend, rename the minuend.

**EXAMPLE 1** Subtract 72 from 495.

The "from" number is the minuend.

Write the numerals in vertical columns, lining up the ones digits.

$$\begin{array}{r} 495 \\ \underline{-72} \end{array}$$

Subtract in the ones column $(5 - 2 = 3)$.

$$\begin{array}{r} 495 \\ \underline{-72} \\ 3 \end{array}$$

Subtract in the tens column $(9 - 7 = 2)$.

$$\begin{array}{r} 495 \\ \underline{-72} \\ 23 \end{array}$$

Subtract in the hundreds column $(4 - 0 = 4)$.

$$\begin{array}{r} 495 \\ \underline{-72} \\ 423 \end{array}$$

*Answer:* The difference is 423.

**EXAMPLE 2** Subtract 97 from 238.

The "from" number is the minuend. Write the numerals in vertical columns.

$$\begin{array}{r} 238 \\ \underline{-97} \end{array}$$

Subtract in the ones column $(8 - 7 = 1)$.

$$\begin{array}{r} 238 \\ \underline{-97} \\ 1 \end{array}$$

In the tens column, the digit in the subtrahend is greater than the digit in the minuend (9 is greater than 3).

You must rename the minuend: Write one of the hundreds as 10 tens. Thus, 238 is renamed as 1 hundred + 13 tens + 8 ones.

$$\begin{array}{rrr} {\scriptstyle 1} & {\scriptstyle 13} & \\ \not{2} & \not{3} & 8 \\ - & 9 & 7 \\ \hline & 4 & 1 \end{array}$$

Subtract in the tens column $(13 - 9 = 4)$.

Subtract in the hundreds column $(1 - 0 = 1)$.

$$\begin{array}{rrr} {\scriptstyle 1} & {\scriptstyle 13} & \\ \not{2} & \not{3} & 8 \\ - & 9 & 7 \\ \hline 1 & 4 & 1 \end{array}$$

*Answer:* The difference is 141.

**EXAMPLE 3** From 807 subtract 498.

The "from" number is the minuend.
Write the numerals in vertical columns.

$$\begin{array}{r} 807 \\ -498 \\ \hline \end{array}$$

In the ones column, since 8 is greater than 7, rename 807 as 7 hundreds + 9 tens + 17 ones.

Subtract in the ones column (17 − 8 = 9).

$$\begin{array}{rrr} 7 & 9 & 17 \\ \cancel{8} & \cancel{0} & \cancel{7} \\ -4 & 9 & 8 \\ \hline & & 9 \end{array}$$

Subtract in the tens column (9 − 9 = 0).

$$\begin{array}{rrr} 7 & 9 & 17 \\ \cancel{8} & \cancel{0} & \cancel{7} \\ -4 & 9 & 8 \\ \hline & 0 & 9 \end{array}$$

Subtract in the hundreds column (7 − 4 = 3).

$$\begin{array}{rrr} 7 & 9 & 17 \\ \cancel{8} & \cancel{0} & \cancel{7} \\ -4 & 9 & 8 \\ \hline 3 & 0 & 9 \end{array}$$

*Answer:* The difference is 309.

**EXAMPLE 4** Mr. Smith borrowed $7,542 to buy a car. He has repaid $4,789. How much does he still owe?

**THINKING ABOUT THE PROBLEM**

Since Mr. Smith has repaid some of the money he borrowed, the original amount has been decreased. This tells you to subtract.

You must find the difference between 7,542 and 4,789.

$$\begin{array}{rrrr} 6 & 14 & 13 & 12 \\ \cancel{7}, & \cancel{5} & \cancel{4} & \cancel{2} \\ -4, & 7 & 8 & 9 \\ \hline 2, & 7 & 5 & 3 \end{array}$$

*Answer:* Mr. Smith still owes $2,753.

**EXAMPLE 5** Of 150 students in a school, 7 were absent from school because of illness, and 25 were away on a school trip. How many students remained in school?

**THINKING ABOUT THE PROBLEM**

Break the problem into simpler problems.

Using addition, you should first find the *total* number of students who were not in school. Then, subtract to find the number of students left.

Add 7 and 25 to find the total number of students not in school.

$$\begin{array}{r} 7 \\ +25 \\ \hline 32 \end{array}$$

Subtract 32 from 150 to find the number of students left.

$$\begin{array}{r} 150 \\ -32 \\ \hline 118 \end{array}$$

*Answer:* 118 students remained in school.

## CLASS EXERCISES

**1.** Subtract:

**a.** $\begin{array}{r} 27 \\ -13 \\ \hline \end{array}$ **b.** $\begin{array}{r} 89 \\ -54 \\ \hline \end{array}$ **c.** $\begin{array}{r} 586 \\ -271 \\ \hline \end{array}$ **d.** $\begin{array}{r} 9{,}873 \\ -7{,}452 \\ \hline \end{array}$ **e.** $\begin{array}{r} 803 \\ -486 \\ \hline \end{array}$ **f.** $\begin{array}{r} 961 \\ -875 \\ \hline \end{array}$ **g.** $\begin{array}{r} 1{,}009 \\ -546 \\ \hline \end{array}$

**2.** Subtract:

**a.** 96 − 43 **b.** 294 − 147 **c.** 811 − 472 **d.** 709 − 546 **e.** 5,002 − 3,895

**3.** From 4,971 subtract 875. **4.** Subtract 279 from 352.

**5.** From the sum of 461 and 29, subtract 113.

**6.** A hospital needs $425,600 to buy new equipment. It has received $283,700 in donations. How much money is still needed?

**7.** Teresa had $35. She spent $12 for a blouse and $9 for a belt. How much money did she have left?

## ADDITION AND SUBTRACTION AS INVERSE OPERATIONS

### THE MAIN IDEA

1. Addition and subtraction are called ***inverse operations*** because each can be used to undo the other.
2. Use the inverse of an operation to check the operation.

**EXAMPLE 6** Use 7 + 9 = 16 to show that subtraction is the inverse of addition.

original number → 7 + 9 = 16 ← sum

(9: add 9)

To undo this addition, subtract 9 from the sum.

sum → 16 − 9 = 7 ← original number

(9: subtract 9)

**EXAMPLE 7** Use 20 − 12 = 8 to show that addition is the inverse of subtraction.

original number → 20 − 12 = 8 ← difference

(12: subtract 12)

To undo this subtraction, we can add 12 to the difference.

difference → 8 + 12 = 20 ← original number

(12: add 12)

**EXAMPLE 8** Find the sum of 235 and 472. Check by using the inverse operation.

$$\begin{array}{r} 235 \\ +472 \\ \hline 707 \end{array}$$

235 ⟵ original number
+472 ⟵ add
707 ⟵ The sum is 707.

Check by using subtraction, the inverse of addition.

707 ⟵ sum
−472 ⟵ subtract
235 ⟵ Since 235 is the original number, the addition is correct.

**EXAMPLE 9** Find the difference between 327 and 712. Check by using the inverse operation.

712 ⟵ original number
−327 ⟵ subtract
385 ⟵ The difference is 385.

Check by using addition, the inverse of subtraction.

385 ⟵ difference
+327 ⟵ add
712 ⟵ Since 712 is the original number, the subtraction is correct.

## CLASS EXERCISES

1. Use 11 + 7 = 18 to show that subtraction is the inverse of addition.
2. Use 57 − 23 = 34 to show that addition is the inverse of subtraction.
3. Perform each operation. Check by using the inverse operation.
   **a.** 197 + 56 **b.** 2,006 − 492 **c.** 83 + 126 **d.** 549 − 387
4. Show that 4,197 + 384 = 4,581 is correct by using the inverse operation as a check.
5. Show that 851 − 675 = 176 is correct by using the inverse operation as a check.
6. Write in your own words why 53 + 15 − 15 = 53.

## HOMEWORK EXERCISES

1. Subtract:

| **a.** 96 <br> −43 | **b.** 487 <br> −236 | **c.** 2,948 <br> −528 | **d.** 5,689 <br> −2,627 | **e.** 706 <br> −253 | **f.** 3,002 <br> −993 |
|---|---|---|---|---|---|
| **g.** 6,219 <br> −1,474 | **h.** 3,223 <br> −1,911 | **i.** 2,016 <br> −49 | **j.** 9,000 <br> −111 | **k.** 5,000 <br> −2,784 | **l.** 2,975 <br> −1,897 |

2. Subtract:
   **a.** 32 − 9 **b.** 164 − 38 **c.** 400 − 76 **d.** 222 − 123 **e.** 746 − 239
   **f.** 2,841 − 726 **g.** 7,049 − 3,237 **h.** 5,004 − 2,475
3. Subtract 89 from 275. **4.** From the sum of 306 and 64, subtract 101.
5. Find the difference between 593 and 479.
6. How much larger is 976 than 842? **7.** How much smaller is 246 than 783?
8. Perform each operation. Check by using the inverse operation.
   **a.** 549 − 474 **b.** 74 + 89 **c.** 187 − 96 **d.** 231 − 0 **e.** 2,007 + 3,049
   **f.** 502 − 496 **g.** 1,807 − 1,698

9. Determine whether each statement is correct or incorrect by using the inverse operation as a check.

   a. 893 − 486 = 407   b. 125 + 236 = 351   c. 1,209 − 675 = 434

10. There were 13 states in the United States in 1790. Now there are 50 states. How many more states are there now than there were in 1790?

11. Christopher Columbus sailed to America in 1492. The Declaration of Independence was signed in 1776. How many years apart were the two events?

12. Alaska is our largest state. It has 586,400 square miles of land. Our smallest state, Rhode Island, has 1,200 square miles of land.
    a. How much bigger is our largest state than our smallest state?
    b. What is the total number of square miles in Alaska and Rhode Island?

13. John, Betty, and Sam ran for president of their class. John received 40 votes, and Betty received 53 votes. If 200 votes were cast, how many votes did Sam receive?

14. Joseph's car weighs 3,175 pounds and Marcia's car weighs 2,950 pounds.
    a. Round each number to the nearest hundred.
    b. By subtracting the rounded numbers, estimate how many pounds heavier Joseph's car is.
    c. Use the estimate to choose the exact answer:
       (a) 3,205   (b) 325   (c) 2,255   (d) 225

## SPIRAL REVIEW EXERCISES

1. Replace ? by <, >, or = to make a true comparison.

   a. 15 + 7 ? 11 + 11
   b. 18 + 9 ? 22 + 8
   c. 38 + 15 ? 14 + 37
   d. (12 + 9) + 14 ? 12 + (9 + 14)
   e. 2,999 + 0 ? 0 + 2,999

2. State University has 5,927,773 books in its library. Round the number of books to the nearest ten thousand.

3. Round to the nearest thousand:

   a. 95,641   b. 8,947   c. 9,578

4. To the sum of 289 and 97, add 158.

5. Each letter on the number line shown represents a number. Replace each ? by < or > to make a true statement.

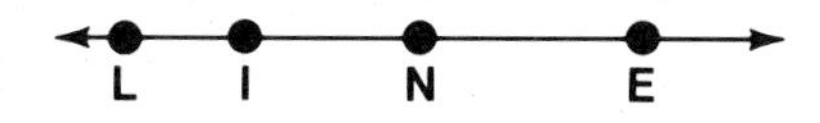


   a. E ? I   b. L ? N
   c. E ? N ? L   d. L ? I ? E

### CHALLENGE QUESTION

Carlos and Jamie each bowled three games. Each boy forgot to enter the score of one of his games. Use the chart shown to find the missing scores.

| | Game 1 | Game 2 | Game 3 | Total |
|---|---|---|---|---|
| Carlos | | 152 | 135 | 434 |
| Jamie | 175 | 132 | | 494 |

# UNIT 2–1 Multiplying Whole Numbers; Properties of Multiplication

## MULTIPLYING WHOLE NUMBERS

### THE MAIN IDEA

1. Multiplication is a way to accomplish repeated additions of the same number.

$$5 \times 9 = \underbrace{9 + 9 + 9 + 9 + 9}_{\text{9 is written 5 times}} = 45$$

2. The result of a multiplication is called the ***product.***
3. Each number in a multiplication is named in the following way:

$$\underset{\text{multiplier}}{3} \times \overset{\text{multiplicand}}{2} = 6 \leftarrow \text{product}$$

$$\begin{array}{r} 2 \leftarrow \text{multiplicand} \\ \times 3 \leftarrow \text{multiplier} \\ \hline 6 \leftarrow \text{product} \end{array}$$

4. To multiply a whole number by a one-digit multiplier, multiply *each digit* of the whole number by the multiplier, starting with the ones digit, and carry when necessary.

**EXAMPLE 1** Jessie earns $8 for each lawn that she mows. How much will Jessie earn if she mows 4 lawns?

You can solve this problem by adding $8 + 8 + 8 + 8 = 32$.
A quicker solution is to multiply $4 \times 8$.

*Answer:* Jessie will earn $32.

In order to perform multiplications, you must know the multiplication facts in the following table:

| × | 0 | 1 | 2 | 3 | 4 | 5 | 6 | 7 | 8 | 9 | 10 |
|---|---|---|---|---|---|---|---|---|---|---|---|
| 0 | 0 | 0 | 0 | 0 | 0 | 0 | 0 | 0 | 0 | 0 | 0 |
| 1 | 0 | 1 | 2 | 3 | 4 | 5 | 6 | 7 | 8 | 9 | 10 |
| 2 | 0 | 2 | 4 | 6 | 8 | 10 | 12 | 14 | 16 | 18 | 20 |
| 3 | 0 | 3 | 6 | 9 | 12 | 15 | 18 | 21 | 24 | 27 | 30 |
| 4 | 0 | 4 | 8 | 12 | 16 | 20 | 24 | 28 | 32 | 36 | 40 |
| 5 | 0 | 5 | 10 | 15 | 20 | 25 | 30 | 35 | 40 | 45 | 50 |
| 6 | 0 | 6 | 12 | 18 | 24 | 30 | 36 | 42 | 48 | 54 | 60 |
| 7 | 0 | 7 | 14 | 21 | 28 | 35 | 42 | 49 | 56 | 63 | 70 |
| 8 | 0 | 8 | 16 | 24 | 32 | 40 | 48 | 56 | 64 | 72 | 80 |
| 9 | 0 | 9 | 18 | 27 | 36 | 45 | 54 | 63 | 72 | 81 | 90 |
| 10 | 0 | 10 | 20 | 30 | 40 | 50 | 60 | 70 | 80 | 90 | 100 |

**EXAMPLE 2** Find the product of 42 and 3.

Multiply each digit of the multiplicand by the multiplier.

Multiply the ones digit of the multiplicand by the multiplier ($3 \times 2 = 6$).

$$\begin{array}{r} 42 \\ \times 3 \\ \hline 6 \end{array}$$

Multiply the tens digit of the multiplicand by the multiplier ($3 \times 4 = 12$).

$$\begin{array}{r} 42 \\ \times 3 \\ \hline 126 \end{array}$$

*Answer:* The product is 126.

**EXAMPLE 3** Multiply 59 by 8.

Multiply the ones digit of the multiplicand by the multiplier. Since $8 \times 9 = 72$, write 2 in the ones place and carry 7.

$$\begin{array}{r} {\scriptstyle 7} \\ 59 \\ \times 8 \\ \hline 2 \end{array}$$

Multiply the tens digit of the multiplicand by the multiplier ($8 \times 5 = 40$). Add the number carried ($40 + 7 = 47$).

$$\begin{array}{r} {\scriptstyle 7} \\ 59 \\ \times 8 \\ \hline 472 \end{array}$$

*Answer:* The product is 472.

## CLASS EXERCISES

1. Write each addition as a multiplication.
   **a.** 7 + 7 + 7 **b.** 3 + 3 + 3 + 3 + 3 + 3 **c.** 8 + 8 + 8 + 8 + 8 + 8 + 8 + 8 + 8
2. Find each product mentally.
   **a.** 5 × 3 **b.** 8 × 2 **c.** 7 × 9 **d.** 9 × 4 **e.** 6 × 6
3. Find each product.
   **a.** 28 × 7 **b.** 96 × 9 **c.** 135 × 4 **d.** 476 × 7

## MULTIPLYING A WHOLE NUMBER BY A MULTIPLIER THAT HAS MORE THAN ONE DIGIT

### THE MAIN IDEA

1. To multiply a whole number by a two-digit multiplier:
   a. Start with the ones digit of the multiplier and multiply each digit of the multiplicand by that digit. The result is called a ***partial product***.
   b. Next use the tens digit of the multiplier to multiply each digit of the multiplicand. Write a zero in the ones place of this partial product to maintain place value.
2. When the multiplier has more than two digits, continue the procedure.

**EXAMPLE 4** Find the product of 358 and 46.

To obtain the first partial product, multiply each digit of the multiplicand by the ones digit of the multiplier, carrying as necessary.

```
  358
  ×46
2,148 ← first partial product
```

Begin the second partial product by writing a zero in the ones place. Then multiply each digit of the multiplicand by the tens digit of the multiplier, carrying as necessary.

```
   358
   ×46
 2,148
14,320 ← second partial product
```

Add the two partial products, carrying as necessary. (In this case, there is no carrying.)

```
   358
   ×46
 2,148
14,320
16,468 ← product
```

*Answer:* The product is 16,468.

**EXAMPLE 5** Find the product of 1,628 and 435.

```
   1,628
   ×435
   8,140 ← partial product of 1,628 × 5 ← ones digit
  48,840 ← partial product of 1,628 × 3 ← tens digit
           (0 is used as a placeholder)
 651,200 ← partial product of 1,628 × 4 ← hundreds digit
           (two zeros are needed as placeholders)
 708,180 ← product (sum of the partial products)
```

*Answer:* The product is 708,180.

**EXAMPLE 6** Multiply 519 by 806.

```
     519
    ×806
   3,114 ← partial product of 519 × 6 ← ones digit
         ← partial product of 519 × 0 ← tens digit
           (does not have to be written because it equals 0)
 415,200 ← partial product of 519 × 8 ← hundreds digit
 418,314   (two zeros are needed as placeholders)
```

*Answer:* The product is 418,314.

**EXAMPLE 7** Seaside Aquarium has 37 fish tanks. If each tank holds 25 gallons of water, what is the total amount of water the tanks hold?

**THINKING ABOUT THE PROBLEM**

The key words "each" and "total" tell you to use multiplication. You are given 37 equal amounts and you are asked for the *total amount.*

Multiply 37 by 25.

```
  37
 ×25
 185
 740
 925
```

*Answer:* The tanks hold a total of 925 gallons.

**EXAMPLE 8** A theater had 100 rows of seats, with 30 seats in each row. If 2,500 seats were occupied, how many seats were not occupied?

**THINKING ABOUT THE PROBLEM**

Use the strategy of breaking the problem into two smaller problems. First, find the total number of seats in the theater. Then, from this total, subtract the number of occupied seats.

Find the total number of seats.

```
   100
   ×30
 3,000
```

Subtract 2,500 occupied seats from the total.

```
  3,000
 −2,500
    500
```

*Answer:* 500 seats were not occupied.

## CLASS EXERCISES

1. Multiply:

a. $\begin{array}{r} 29 \\ \times 58 \\ \hline \end{array}$ b. $\begin{array}{r} 76 \\ \times 32 \\ \hline \end{array}$ c. $\begin{array}{r} 932 \\ \times 65 \\ \hline \end{array}$ d. $\begin{array}{r} 90 \\ \times 38 \\ \hline \end{array}$ e. $\begin{array}{r} 504 \\ \times 87 \\ \hline \end{array}$ f. $\begin{array}{r} 375 \\ \times 402 \\ \hline \end{array}$

g. $\begin{array}{r} 562 \\ \times 212 \\ \hline \end{array}$ h. $\begin{array}{r} 967 \\ \times 38 \\ \hline \end{array}$ i. $\begin{array}{r} 897 \\ \times 564 \\ \hline \end{array}$ j. $\begin{array}{r} 989 \\ \times 879 \\ \hline \end{array}$

2. Find the product of 567 and 54.
3. Find the product of 487 and 608.
4. New Mode clothes shop ordered 36 boxes of T-shirts. Each box contained 24 T-shirts. How many T-shirts were in the order?
5. Channel 79 received 189 calls from listeners, each promising to donate $75 to the station. How much money was promised?
6. A sports stadium has 275 sections. Each section contains 157 seats. How many seats are there in the stadium?
7. A carton contains 2,715 pieces of candy. Find the number of pieces of candy in 58 cartons.
8. For a school trip, 5 buses each took 48 students to the museum. Thirty-eight students took the train to get there. How many students from the school were at the museum?

## PROPERTIES OF MULTIPLICATION

### THE MAIN IDEA

1. The product of two whole numbers is a whole number.
2. The order in which you multiply two whole numbers does not change the product.

$$2 \times 9 = 9 \times 2$$

3. The way you group numbers in a series of multiplications does not change the final product.

$$(2 \times 5) \times 8 = 2 \times (5 \times 8)$$

4. The product of 1 and a whole number is that same whole number.

$$9 \times 1 = 9$$

5. The product of 0 and a whole number is 0.

$$7 \times 0 = 0$$

**EXAMPLE 9** Tell whether each statement is *true* or *false*.

| *Statement* | *Answer* |
|---|---|
| **a.** $11 \times 89 < 89 \times 11$ | *False*. The order in which you multiply does not change the product. |
| **b.** $(5 \times 7) \times 6 = 5 \times (7 \times 6)$ | *True*. The way you group numbers does not change the product. |
| **c.** $193 \times 1 = 193$ | *True*. The product of a whole number and 1 is that same whole number. |
| **d.** $17 \times 0 = 0$ | *True*. The product of zero and any whole number is zero. |

**EXAMPLE 10** Show that $(5 \times 3) \times 7 = 5 \times (3 \times 7)$ is a true statement.

$$(5 \times 3) \times 7 \stackrel{?}{=} 5 \times (3 \times 7)$$

Multiply first. ↗ ↖ Multiply first.

$$15 \times 7 \stackrel{?}{=} 5 \times 21$$

$$105 = 105 \quad \textit{True}$$

**EXAMPLE 11** Replace each ? by a number to make a true statement.

| *Statement* | *Answer* |
|---|---|
| **a.** $15 \times 8 = ? \times 15$ | $15 \times 8 = 8 \times 15$ |
| **b.** $2{,}351 \times ? = 2{,}351$ | $2{,}351 \times 1 = 2{,}351$ |
| **c.** $? \times 0 = 0$ | any number $\times\ 0 = 0$ |
| **d.** $(17 \times 9) \times 6 = ? \times (9 \times 6)$ | $(17 \times 9) \times 6 = 17 \times (9 \times 6)$ |

**EXAMPLE 12** A dealer sold 37 boats at $4,921 each. What is the total amount of the sale?

### THINKING ABOUT THE PROBLEM

Since this problem gives one of many equal amounts, the word "total" tells you to multiply.

You can use estimation to get an approximate answer: 37 is close to 40; $4,921 is close to $5,000. A reasonable solution would be about 40 × $5,000 = $200,000.

You must multiply 37 by $4,921 to find the total amount of the sale. Since the order in which you multiply does not change the product, you can write:

$$\begin{array}{r} 37 \\ \times 4{,}921 \\ \hline \end{array} \quad \text{or} \quad \begin{array}{r} 4{,}921 \\ \times 37 \\ \hline \end{array}$$

It is easier to use the smaller number as the multiplier.

$$\begin{array}{r} 4{,}921 \\ \times 37 \\ \hline 34{,}447 \\ 147{,}630 \\ \hline 182{,}077 \end{array}$$

This answer rounded to the nearest hundred thousand dollars is $200,000. It is a reasonable answer because it agrees with the estimate.

*Answer:* The total amount of the sale is $182,077.

## CLASS EXERCISES

1. Show that $(7 \times 9) \times 4 = 7 \times (9 \times 4)$ is a true statement.
2. Tell whether each statement is *true* or *false*.
   **a.** $837 \times 592 = 592 \times 837$ **b.** $1 \times 474 = 474$ **c.** $(9 \times 8) \times 7 > 9 \times (8 \times 7)$
   **d.** $0 \times 13 = 13$ **e.** $57 \times 0 < 57 \times 1$ **f.** $385 \times 1 > 385$
3. Replace each ? by a number to make a true statement.
   **a.** $42 \times 847 = ? \times 42$ **b.** $129 \times 1 = ?$ **c.** $(15 \times ?) \times 9 = 15 \times (12 \times 9)$
   **d.** $? \times 1 = 1{,}300{,}000$ **e.** $5{,}200 \times ? = 4{,}500 \times 5{,}200$ **f.** $1 \times 0 = 2 \times ?$
4. A theater has 128 rows. Each row has 40 seats. How many seats are there in the theater?
5. What is the total number of pieces of candy in 105 boxes if each box contains 36 pieces?
6. A microcomputer is on sale for $995. What is the cost of 18 such microcomputers?

## HOMEWORK EXERCISES

1. Find each product.
   **a.** $9 \times 7$ **b.** $13 \times 5$ **c.** $82 \times 9$ **d.** $47 \times 3$ **e.** $125 \times 6$
   **f.** $99 \times 7$ **g.** $211 \times 5$ **h.** $49 \times 8$
2. Multiply:
   **a.** $52 \times 37$ **b.** $46 \times 54$ **c.** $83 \times 71$ **d.** $57 \times 48$ **e.** $106 \times 42$ **f.** $412 \times 55$
   **g.** $652 \times 92$ **h.** $829 \times 88$ **i.** $315 \times 95$ **j.** $227 \times 109$ **k.** $415 \times 223$ **l.** $946 \times 385$
3. Find the product of 712 and 75.
4. What is the product of 437 and 52?
5. Find the product of 39 and 516.
6. Tell whether each statement is *true* or *false*.
   **a.** $(17 \times 2) \times 3 = 17 \times (2 \times 3)$ **b.** $87 \times 3 > 3 \times 87$ **c.** $0 = 18 \times 0$
   **d.** $1 \times 253 > 253$ **e.** $1 \times 54 < 0 \times 54$ **f.** $95 \times 1 > 95$
7. Replace each ? by a number that will make a true statement.
   **a.** $4 \times 3 = 3 \times ?$ **b.** $(8 \times ?) \times 7 = 8 \times (3 \times 7)$ **c.** $7 \times ? = 1 \times 7$ **d.** $312 \times ? = 0$
8. Kevin earns $58 a week. How much will he earn in 32 weeks?
9. A dealer has 58 cases of soda. If each case contains 24 bottles, how many bottles of soda does the dealer have?
10. The manager of a store bought 290 shirts. Each shirt cost $12. What was the total cost?
11. The library in Midtown High School has 418 shelves. Each shelf can hold 36 books. What is the greatest number of books that the library can have on all its shelves?
12. Jennifer bought 7 records at $4 each. If she started with $50, how much money did she have left?

## SPIRAL REVIEW EXERCISES

1. Add: 21,974 + 845 + 1,437

2. Find the difference of 729 and 346.

3. Round each number to the nearest hundred.

   a. 2,746   b. 5,929   c. 897

4. Subtract 192 from the sum of 518 and 359.

5. Replace each ? by <, >, or = to make a true statement.

   a. 357 − 82 ? 199 + 76

   b. 273 + 1 ? 273 × 1

   c. 5,873 − 2,401 ? 225 × 17

6. Which is a larger number, one hundred seven thousand, or 7,100?

7. What is the shortest distance from Hastings to Evern?

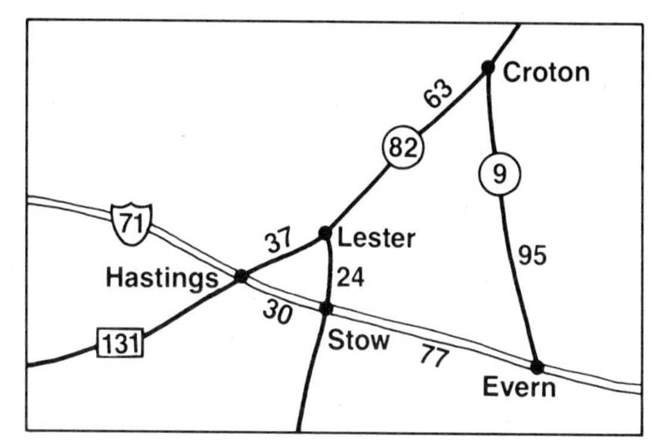


**CHALLENGE QUESTION**

The total attendance at a baseball game was 3,245. In the stadium, 48 sections of seats each had 65 fans who bought tickets. The rest of the fans received free passes. How many fans received free passes?

# UNIT 2–2 Dividing Whole Numbers; Division Involving Zero

## DIVIDING WHOLE NUMBERS

### THE MAIN IDEA

1. Division is the inverse of multiplication.

   $$\begin{array}{r} 9 \\ 7\overline{)63} \end{array}$$ means $7 \times 9 = 63$

   (In $7\overline{)63}$: 9 ← quotient, 7 ← divisor, 63 ← dividend. In $7 \times 9 = 63$: 7 ← divisor, 9 ← quotient, 63 ← dividend.)

2. The result of a division is called the ***quotient.***
3. Division can be shown in other ways.

   $35 \div 7 = 5$ (35 ← dividend, 7 ← divisor, 5 ← quotient)

   $\frac{35}{7} = 5$ (35 ← dividend, 7 ← divisor, 5 ← quotient)

4. If the divisor does not divide the dividend exactly, the "extra" is called the ***remainder.***
5. To check a division, first find the product of the divisor and the quotient. Then add the remainder (if there is one). The resulting sum should equal the dividend.

   ***Quotient × Divisor + Remainder = Dividend***

**EXAMPLE 1** Find each quotient and check.

| | *Divide* | *Answer* | *Check* |
|---|---|---|---|
| **a.** | $4\overline{)28}$ | $\begin{array}{r} 7 \\ 4\overline{)28} \\ \underline{28} \\ 0 \end{array}$ | $7 \times 4 = 28$ |
| **b.** | $72 \div 9$ | $72 \div 9 = 8$ | $8 \times 9 = 72$ |
| **c.** | $\frac{45}{9}$ | $\frac{45}{9} = 5$ | $5 \times 9 = 45$ |

**EXAMPLE 2** Divide 96 by 8 and check.

*Solution:*

(1) *Divide* $8\overline{)9}$. 1 is the largest whole-number quotient.

$$\begin{array}{r} 1 \\ 8\overline{)96} \\ \underline{8} \\ 1 \end{array}$$

(2) *Multiply* $1 \times 8 = 8$.

(3) *Subtract* $\begin{array}{r} 9 \\ \underline{-8} \\ 1 \end{array}$

(4) *Bring down* the next digit from the dividend.

$$\begin{array}{r} 1 \\ 8\overline{)96} \\ \underline{8\downarrow} \\ 16 \end{array}$$

This four-step process is repeated.

(1) *Divide* $8\overline{)16} = 2$.

(2) *Multiply* $2 \times 8 = 16$.

(3) *Subtract* $\begin{array}{r} 16 \\ \underline{-16} \\ 0 \end{array}$

$$\begin{array}{r} 12 \\ 8\overline{)96} \\ \underline{8\downarrow} \\ 16 \\ \underline{16} \\ 0 \end{array}$$

(4) *Bring down.* There is no digit to bring down.

*Check:* Since division and multiplication are inverse operations, we check a division by using multiplication.

Quotient × Divisor + Remainder = Dividend

$12 \times 8 + 0 = 96$

*Answer:* The quotient is 12.

**EXAMPLE 3** Divide 2,819 by 7 and check.

*Solution:*

(1) *Divide* $7\overline{)2}$. This division is not possible. Thus, we include the next digit.

$7\overline{)28} = 4$.

(2) *Multiply* $4 \times 7 = 28$.

$$\begin{array}{r} 4\phantom{19} \\ 7\overline{)2,819} \\ \underline{2\,8}\phantom{19} \\ 0\phantom{19} \end{array}$$

(3) *Subtract* $28 - 28 = 0$.

(4) *Bring down* the next digit from the dividend.

$$\begin{array}{r} 4\phantom{19} \\ 7\overline{)2,819} \\ \underline{2\,8\downarrow}\phantom{9} \\ 01\phantom{9} \end{array}$$

Repeat this four-step process.

(1) *Divide* $7\overline{)1}$. This division is not possible. Use 0 as a placeholder in the quotient, and bring down the next digit from the dividend.

$$\begin{array}{r} 40\phantom{9} \\ 7\overline{)2,819} \\ \underline{2\,8\downarrow\downarrow} \\ 019 \end{array}$$

$7\overline{)19} = 2$.

(2) *Multiply* $2 \times 7 = 14$.

(3) *Subtract* $19 - 14 = 5$.

$$\begin{array}{r} 402 \\ 7\overline{)2,819} \\ \underline{2\,8\downarrow\downarrow} \\ 019 \\ \underline{14} \\ 5 \end{array}$$

← remainder

(4) *Bring down.* There is no digit to bring down.

*Check:*

$$\begin{array}{rl} 402 & \leftarrow \text{quotient} \\ \underline{\times 7} & \leftarrow \text{divisor} \\ 2,814 & \\ \underline{+5} & \leftarrow \text{Add the remainder.} \\ 2,819 & \leftarrow \text{dividend} \end{array}$$

*Answer:* The quotient is 402, and the remainder is 5.

**EXAMPLE 4** A group of 210 students is to be divided into 15 teams with an equal number of students on each team. Find how many students should be on each team, and check.

### THINKING ABOUT THE PROBLEM

Since each team will be smaller than the whole group, the answer must be a number smaller than 210. The problem uses the key words "divided into" and "each."

Divide 210 by 15.

$$\begin{array}{r} 14 \\ 15\overline{)210} \\ \underline{15\downarrow} \\ 60 \\ \underline{60} \\ 0 \end{array}$$

*Check:* Multiply 14 by 15.

$$\begin{array}{rl} 14 & \leftarrow \text{divisor} \\ \underline{\times 15} & \leftarrow \text{quotient} \\ 70 & \\ \underline{140} & \\ 210 & \\ \underline{+0} & \leftarrow \text{Add the remainder.} \\ 210 & \leftarrow \text{dividend} \end{array}$$

*Answer:* There should be 14 students on each team.

## CLASS EXERCISES

1. Find each quotient.

   **a.** $9\overline{)54}$ **b.** $3\overline{)18}$ **c.** $64 \div 8$ **d.** $32 \div 8$ **e.** $\frac{42}{7}$ **f.** $\frac{18}{9}$

2. Divide and check.

   **a.** $8\overline{)400}$ **b.** $9\overline{)207}$ **c.** $6\overline{)1{,}830}$ **d.** $4\overline{)2{,}024}$ **e.** $7\overline{)42{,}007}$

3. Divide and check.

   **a.** $16\overline{)192}$ **b.** $48\overline{)672}$ **c.** $76\overline{)856}$ **d.** $83\overline{)6{,}308}$ **e.** $76\overline{)18{,}336}$

4. Mr. Coles made 300 cookies. He wanted to make packages that would each contain 25 cookies. Find how many packages he could make, and check.

5. Show that $(36 \div 6) \div 2 = 36 \div (6 \div 2)$ is a false statement.

## DIVISION INVOLVING ZERO

### THE MAIN IDEA

1. When zero is divided by any nonzero number, the quotient is zero.
2. Dividing a number by zero is not possible.

**EXAMPLE 5** Perform each division and check.

| | *Divide* | *Answer* | *Check* |
|---|---|---|---|
| a. | $0 \div 5$ | $0 \div 5 = 0$ | $\underset{\text{quotient}}{0} \times \underset{\text{divisor}}{5} = \underset{\text{dividend}}{0}$ |
| b. | $0 \div 1{,}000$ | $0 \div 1{,}000 = 0$ | $\underset{\text{quotient}}{0} \times \underset{\text{divisor}}{1{,}000} = \underset{\text{dividend}}{0}$ |

**EXAMPLE 6** Show that $5 \div 0$ is not possible.

*Solution:*

Since division is the inverse of multiplication, $5 \div 0 = ?$ means that $? \times 0 = 5$. There is no number to replace ? since any number $\times\ 0 = 0$, not 5.

## CLASS EXERCISES

1. Find the quotient if it can be found, or write "the division is not possible."

   **a.** $0 \div 15$ **b.** $18 \div 0$ **c.** $37\overline{)0}$ **d.** $0\overline{)37}$ **e.** $0 \div 0$

2. Replace ? with a number to make a true statement.

   **a.** $\frac{?}{5} = 0$ **b.** $\frac{0}{9} = ?$ **c.** $\frac{17}{17} = ?$ **d.** $\frac{?}{2{,}752} = 1$

## HOMEWORK EXERCISES

1. Find each quotient.

   **a.** $7\overline{)49}$ **b.** $8\overline{)16}$ **c.** $19\overline{)38}$ **d.** $15\overline{)60}$ **e.** $25 \div 5$ **f.** $50 \div 10$

   **g.** $0 \div 8$ **h.** $27 \div 3$ **i.** $\frac{36}{9}$ **j.** $\frac{39}{13}$ **k.** $\frac{0}{15}$ **l.** $\frac{15}{3}$

2. Using mathematical symbols, write "60 divided by 15" in three different ways.

3. Divide and check.

   **a.** $5\overline{)85}$ **b.** $7\overline{)742}$ **c.** $9\overline{)252}$ **d.** $6\overline{)1{,}530}$ **e.** $8\overline{)24{,}040}$ **f.** $7\overline{)325}$ **g.** $8\overline{)1{,}557}$

4. Divide and check.

   **a.** $12\overline{)144}$ **b.** $17\overline{)119}$ **c.** $28\overline{)336}$ **d.** $12\overline{)1{,}080}$ **e.** $53\overline{)10{,}865}$

   **f.** $27\overline{)875}$ **g.** $43\overline{)2{,}559}$ **h.** $81\overline{)5{,}450}$

5. Replace ? by a whole number to make a true statement.

   **a.** $15 \div 5 = ?$ means $5 \times ? = 15$.

   **b.** $? \div 9 = 5$ means $9 \times 5 = ?$.

   **c.** $48 \div ? = 4$ means $? \times 4 = 48$.

6. Tell whether each statement is *true* or *false*.

   **a.** $\frac{0}{357} = 0$ **b.** $\frac{211}{211} = 1$ **c.** $\frac{137}{0} = 0$ **d.** $\frac{5{,}283}{5{,}283} = 1$ **e.** $\frac{864}{27} = 32$

7. Divide 1,564 by 17.
8. What is the quotient of 9,546 and 43?
9. If 52 cards are dealt to 4 players, how many cards does each player receive?
10. If there are 8 maps on each page of an atlas and 256 maps are shown, how many pages are there in the atlas?
11. The Manville Public Library has 6,045 books on its shelves. Each shelf holds 65 books. How many shelves are there in the library?
12. Forty-eight pounds of food were distributed equally among 16 campers. Each camper carried an additional 17 pounds of equipment. What was the total weight carried by each camper?

## SPIRAL REVIEW EXERCISES

1. Replace ? by <, >, or = to make a true comparison.

   **a.** $35 + 5 \ ? \ 35 \div 5$

   **b.** $147 + 29 \ ? \ 147 \times 29$

   **c.** $375 - 125 \ ? \ 375 + 125$

   **d.** $450 - 30 \ ? \ 400 + 20$

   **e.** $7 \times 0 \ ? \ \frac{0}{7}$ **f.** $8 + 0 \ ? \ 8 \times 0$

   **g.** $9 \times 1 \ ? \ \frac{9}{1}$ **h.** $1 + 0 \ ? \ \frac{0}{1}$

2. Find the product of 25 and 107.
3. Find the sum of 1,496 and 258.
4. Find the difference of 4,096 and 3,584.
5. Perform each operation.

   **a.** $1,204 + 87 + 192$ **b.** $75 \times 12$

   **c.** $412 \div 4$ **d.** $4,009 - 2,027$

6. At a zoo, 38 pounds of birdseed are used each week. The number of pounds of birdseed used in a year is
   (a) 2,236 (c) 2,976
   (b) 1,876 (d) 1,976

## CHALLENGE QUESTION

Ms. Edmunds, the supply secretary of Northeast High School, forgot to enter all the information in her book records. Use the book record card below to find:

**a.** the number of mathematics books.

**b.** the total cost of the English books.

**c.** the cost per book of the science books.

| Type of Book | Number of Books | Cost per Book in Dollars | Total Cost in Dollars |
|---|---|---|---|
| Mathematics | | 15 | 630 |
| English | 65 | 8 | |
| Science | 91 | | 910 |
| | | | $2,060 |

Total Cost of Order

# UNIT 2–3 Factors; Prime and Composite Numbers; Greatest Common Factor

## MEANING OF FACTOR

### THE MAIN IDEA

1. When two nonzero whole numbers are multiplied, each is called a *factor* of the product.

$$\underbrace{2 \times 5}_{\text{factors}} = \underbrace{10}_{\text{product}}$$

2. If a number is divided by one of its factors, the remainder is 0 and the quotient is the other factor.

$$\underset{}{10} \div \underset{\text{factor}}{2} = \underset{\text{factor}}{5}$$

This means that a factor of a number is an ***exact divisor*** of the number.

**EXAMPLE 1** Show that 18 is a factor of 144.

```
     8 ← quotient
18)144
   144
     0 ← remainder
```

Since the remainder is 0, 18 is a factor of 144. The quotient, 8, is the other factor. That is, $8 \times 18 = 144$.

**EXAMPLE 2** Determine if 19 is a factor of 647.

```
    34
19)647
   57↓
    77
    76
     1 ← remainder not 0
```

Since the remainder is not 0, 19 is not a factor of 647.

**EXAMPLE 3** Determine if 42 is an exact divisor of 2,394.

```
      57
42)2,394
   2 10↓
     294
     294
       0 ← remainder
```

Since the remainder is 0, 42 is an exact divisor of 2,394.

**EXAMPLE 4** Find all the factors of 40.

The exact divisors of 40 are 1, 2, 4, 5, 8, 10, 20, and 40. To be sure that you are getting *all* the factors, think of the factors in *pairs:*

1, 2, 4, 5, 8, 10, 20, 40

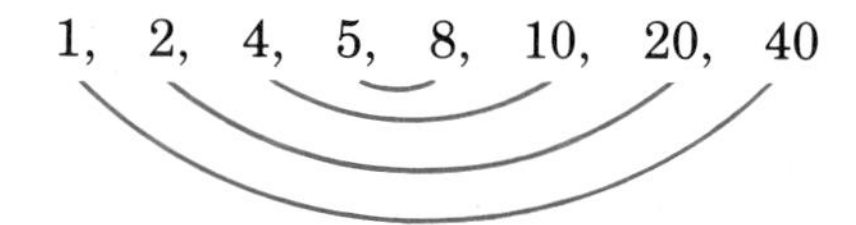

## CLASS EXERCISES

1. Tell whether the first number is a factor of the second.
   **a.** 5; 75 **b.** 7; 56 **c.** 9; 73 **d.** 30; 90 **e.** 11; 121 **f.** 15; 45
   **g.** 21; 84 **h.** 17; 33
2. Tell whether each statement is *true* or *false.*
   **a.** 19 is a factor of 608. **b.** 27 is an exact divisor of 1,161.
   **c.** 7 is not a factor of 1,428. **d.** 42 and 37 are factors of 1,554.
3. Write all the pairs of factors of each number.
   **a.** 15 **b.** 36 **c.** 81 **d.** 100

## MEANING OF PRIME AND COMPOSITE NUMBERS

### THE MAIN IDEA

1. If a whole number has 1 and itself as its only factors, then that number is called a ***prime number.*** 2, 3, 5, 7, 11, 13, 17, 19, 23, 29, 31, 37 are examples of primes.
2. If a whole number also has factors other than 1 and itself, then that whole number is called a ***composite number.*** 8, 9, 14, 15, 25, 26 are examples of composite numbers.
3. The number 1 is not called prime or composite.

**EXAMPLE 5** Tell whether each number is prime or composite.

| | *Number* | *Answer* |
|---|---|---|
| **a.** | 2 | prime |
| **b.** | 12 | composite |
| **c.** | 17 | prime |
| **d.** | 28 | composite |
| **e.** | 35 | composite |

**EXAMPLE 6** Show that 123 is not prime.

```
   41
3)123
  12↓
    3
    3
    0
```

*Answer:* Since 3 and 41 are factors of 123, 123 is not prime.

## CLASS EXERCISES

1. Tell whether each number is prime or composite.
   **a.** 4 **b.** 7 **c.** 22 **d.** 37 **e.** 55 **f.** 81 **g.** 99 **h.** 29 **i.** 87 **j.** 91
2. Show that each number is not prime.
   **a.** 40 **b.** 328 **c.** 125 **d.** 500 **e.** 729

## GREATEST COMMON FACTOR

### THE MAIN IDEA

1. A *common factor* of two numbers is a number that is a factor of both of them.
2. The *greatest common factor* (GCF) of two numbers is the largest common factor of both numbers.
3. To find the greatest common factor (GCF):
   a. Find all the factors of each of the numbers.
   b. Determine the factors common to both numbers.
   c. Select the largest common factor.

**EXAMPLE 7** Find the greatest common factor of 8 and 12.

The factors of 8 are 1, 2, 4, and 8.
The factors of 12 are 1, 2, 3, 4, 6, and 12.
The common factors of 8 and 12 are 1, 2, and 4.
The greatest common factor of 8 and 12 is 4. *Ans.*

**EXAMPLE 8** Find the greatest common factor of 36 and 42.

The factors of 36 are 1, 2, 3, 4, 6, 9, 12, 18, and 36.
The factors of 42 are 1, 2, 3, 6, 7, 14, 21, and 42.
The common factors of 36 and 42 are 1, 2, 3, and 6.
The greatest common factor of 36 and 42 is 6. *Ans.*

**EXAMPLE 9** Find the greatest common factor of 28 and 55.

Factors of 28: 1, 2, 4, 7, 14, 28
Factors of 55: 1, 5, 11, 55
GCF is 1. *Ans.*

**EXAMPLE 10** Factor 100 until all the factors are primes.

You may begin with several different pairs of factors for 100.

$2 \times 50$ → $2 \times 25 \times 2$ → $2 \times 5 \times 5 \times 2$

$4 \times 25$ → $2 \times 2 \times 5 \times 5$

$10 \times 10$ → $5 \times 2 \times 5 \times 2$

Notice that no matter how you begin to factor 100, if you continue factoring until you reach prime numbers, the result is the same. That is, $100 = 2 \times 2 \times 5 \times 5$.

## CLASS EXERCISES

1. Find the common factors of each pair of numbers.

   **a.** 10 and 20 **b.** 24 and 36. **c.** 12 and 30

2. Find the greatest common factor of each pair of numbers.

   **a.** 22 and 33 **b.** 16 and 18 **c.** 12 and 21 **d.** 25 and 100

3. Factor each number until all the factors are primes.

   **a.** 30 **b.** 27 **c.** 48 **d.** 500

## HOMEWORK EXERCISES

1. Tell whether the first number is a factor of the second number.

   **a.** 12; 36 **b.** 5; 30 **c.** 18; 72 **d.** 14; 29 **e.** 24; 144 **f.** 50; 250

   **g.** 19; 82 **h.** 46; 178

2. Write in your own words how you can tell if 6 is a factor of 24.

3. Find all the factors of each number.

   **a.** 12 **b.** 20 **c.** 48 **d.** 72 **e.** 144

4. Tell whether each statement is *true* or *false*.

   **a.** 23 is a factor of 1,403. **b.** 37 is a factor of 221. **c.** 13 is not a factor of 3,913.

   **d.** 19 and 42 are factors of 798. **e.** 56 is an exact divisor of 225.

5. Tell whether each number is prime or composite.

   **a.** 2 **b.** 6 **c.** 19 **d.** 30 **e.** 31 **f.** 100 **g.** 333 **h.** 51 **i.** 81 **j.** 47

6. Find the common factors of each pair of numbers.

   **a.** 20 and 40 **b.** 18 and 32 **c.** 12 and 36 **d.** 15 and 18

   **e.** 10 and 25 **f.** 24 and 30

7. Find the greatest common factor of each pair of numbers.

   **a.** 24 and 42 **b.** 45 and 50 **c.** 9 and 15 **d.** 16 and 32

   **e.** 56 and 72 **f.** 14 and 42

8. Factor each number until all the factors are primes.

   **a.** 18 **b.** 36 **c.** 60 **d.** 121 **e.** 50

## SPIRAL REVIEW EXERCISES

**1.** Perform each operation.

**a.** Add: 2,752
1,496
847

**b.** Subtract: 7,809
− 2,489

**c.** Multiply: 53
× 29

**d.** Divide: 72)14,472

**2.** Subtract 358 from 4,027 and check.

**3.** Find the product of 435 and 36 and check.

**4.** Show that "1,305 ÷ 29 = 45" is a true statement by using the inverse operation.

**5.** Write as a numeral:

**a.** thirty-five thousand, sixty-two

**b.** thirty thousand, six

**c.** three hundred thousand, six hundred twenty

**d.** three million, three hundred sixty-two

**6.** Round 47,496 to the nearest thousand.

**7.** Round 42,552 to the nearest hundred.

**8.** Replace ? by $<$, $>$, or $=$ to make a true comparison.

**a.** 1 ? the product of 475 and 0

**b.** the quotient of 0 and 295 ? 295

**c.** the quotient of 357 and 357 ? 0

**d.** the difference of 139 and 129 ? the product of 20 and 1

**9.** Kathy earns $5 per hour. She works 35 hours every week. How much does she earn in a week?

**10.** Sixteen identical pieces of lumber weigh 240 pounds. How much does one piece of lumber weigh?

### CHALLENGE QUESTION

Mary earns $3 an hour baby-sitting. On Friday night, she worked 4 hours and on Saturday night, she worked 5 hours. She calculated her total earnings as follows: 3 × 4 + 5, and got the answer $17. Explain what Mary did wrong and find the correct answer.

# UNIT 2–4 Using the Order of Operations to Evaluate Numerical Expressions

## THE MAIN IDEA

1. A *numerical expression* contains numerals and operational symbols.
2. We *evaluate* a numerical expression by performing the operations shown to obtain a single numerical value.
3. To evaluate a numerical expression that contains two or more operations, we use the following *order of operations*:
   a. First, do all the multiplications and divisions, working from left to right.
   b. Next, do all the additions and subtractions, working from left to right.

**EXAMPLE 1** Evaluate $2 + 8 \times 5$.

The given numerical expression contains two operations: addition and multiplication. $\quad 2 + 8 \times 5$

According to the order of operations, multiply first. $\quad 2 + 40$

Next, add. $\quad 42$ *Ans.*

**EXAMPLE 2** Evaluate $22 - 56 \div 7$.

The given numerical expression contains two operations: subtraction and division. $\quad 22 - 56 \div 7$

According to the order of operations, divide first. $\quad 22 - 8$

Next, subtract. $\quad 14$ *Ans.*

**EXAMPLE 3** Evaluate $9 \times 5 - 3 \times 7$.

The given numerical expression contains two multiplications and a subtraction. $\quad 9 \times 5 - 3 \times 7$

Do all the multiplications first. $\quad 45 - 21$

Next, subtract. $\quad 24$ *Ans.*

**EXAMPLE 4** Evaluate $100 \div 50 \times 2$.

The given numerical expression contains both a division and a multiplication. $\quad 100 \div 50 \times 2$

Remember to work from left to right. Otherwise, you could get different answers. $\quad 2 \times 2$

$4$ *Ans.*

**EXAMPLE 5** Evaluate $50 - 20 \div 5 + 3 \times 2$.

The given numerical expression contains a subtraction, a division, an addition, and a multiplication. $\quad 50 - 20 \div 5 + 3 \times 2$

Do all the multiplications and divisions first. $\quad 50 - 4 + 6$

Next, do the additions and subtractions. Remember to work from left to right. $\quad 46 + 6$

$52$ *Ans.*

## CLASS EXERCISES

1. Tell which operation you would perform first.

   a. $9 + 8 \times 15$ b. $12 \div 2 + 10 \div 5$ c. $30 - 15 + 8$ d. $81 - 12 \times 3$

2. Evaluate each numerical expression.

   a. $12 + 5 \times 2$ b. $30 \div 5 + 8$ c. $100 - 80 \div 4$ d. $8 \times 9 + 7$

   e. $75 - \frac{20}{4}$ f. $52 - 12 \times 4$

3. Evaluate each numerical expression.

   a. $72 \div 9 - 15 \div 3$ b. $20 + 5 \times 8 - 16$ c. $12 \times 3 + 3 \times 9$ d. $42 - 28 \div 7 + 30$

   e. $30 \div 10 + 5 \times 7$ f. $15 + 3 \times 7 - 8 \times 4$

## HOMEWORK EXERCISES

1. Evaluate each numerical expression.

   a. $36 - 14 \times 2$ b. $18 + 12 \div 6$ c. $20 \times 3 - 40$ d. $18 \div 6 + 12$

   e. $28 - 14 \div 7$ f. $3 \times 0 + 7$

2. Explain in your own words why we must all follow the same order of operations.

3. Evaluate each numerical expression.

   a. $18 + 9 \times 2$ b. $\frac{15}{3} - 3$ c. $9 \times 7 - 40$ d. $36 + 8 \times 5$

   e. $50 - \frac{100}{5}$ f. $6 \times 1 - 6$

4. Evaluate each numerical expression.

   a. $5 \times 8 + 3 \times 5$ b. $47 - 4 \times 7 + 6$ c. $8 + 5 \times 7 + 12$ d. $50 \div 5 + 10 \times 4$

   e. $5 \times 9 + 7 \times 6$ f. $\frac{20}{4} + \frac{18}{6}$ g. $90 \div 6 - 15 \div 3$ h. $100 - 90 \div 10 \times 9$

   i. $35 - 18 \div 9 + 2$ j. $90 - 14 \times 2 \times 3$ k. $7 - 12 \times 0 + 8 \div 1$

   l. $72 \div 3 - 8 \times 3$

5. Evaluate each numerical expression.

   a. $18 \times 2 + 17 \times 3$ b. $49 - 7 \times 8 \div 4$ c. $37 - 18 \times 2 + 12$ d. $29 - 3 \times 8$

   e. $15 \times 4 + 8 \times 3$ f. $32 + 25 \div 5 \times 2$ g. $18 \div 6 - \frac{90}{45}$ h. $42 \div 7 + 9$

   i. $96 \div 16 - 3 \times 2$ j. $18 + 40 \div 8 - 9$

## SPIRAL REVIEW EXERCISES

1. Find the sum: 12,848 + 3,497 + 23,575
2. Find the product of 157 and 29.
3. Divide 4,320 by 48 and check.
4. Subtract 7,909 from 11,823.
5. 18,965 rounded to the nearest ten is
   (a) 19,000 (c) 18,960
   (b) 18,900 (d) 18,970
6. Fifty thousand three hundred, written as a numeral, is
   (a) 5,300 (c) 500,300
   (b) 50,300 (d) 503,000
7. Adding 1 to which digit of the number 9,476,523 will increase the number by one hundred thousand?
   (a) 9 (b) 4 (c) 7 (d) 6
8. Which statement is not true?
   (a) $5 + (9 + 6) = (5 + 9) + 6$
   (b) $18 \times 72 = 72 \times 18$
   (c) $(40 - 18) - 7 = 40 - (18 - 7)$
   (d) $320 + 89 = 89 + 320$
9. A section of a stadium has 125 rows with 20 seats in each row. If 2,140 seats were occupied, how many seats were not occupied?

## CHALLENGE QUESTION

Replace $\square$ by $<$, $>$, or $=$ to make each statement a true comparison.

**a.** $30 - 7 \times 4 - 1 \square 1 \times 0$

**b.** $176 - 95 \square 6 \times 12 + 8$

**c.** $28 - 4 \times 7 \square 17 \div 17$

**d.** $36 - 6 \times 6 + 11 \square 121 \div 11$

**e.** $21 \div 3 \times 11 \square 616 \div 8 + 0$

**f.** $1{,}001 - 1 \times 763 \square 27 - 13 \times 2$

# UNIT 2-5 Evaluating Numerical Expressions Containing Parentheses; Distributing Multiplication Over Addition or Subtraction

## EVALUATING NUMERICAL EXPRESSIONS CONTAINING PARENTHESES

### THE MAIN IDEA

1. *Parentheses* ( ) are used as a grouping symbol. The numerical expression inside them names a single number.
2. To evaluate numerical expressions that contain parentheses:
   a. First, evaluate within the parentheses. Be sure to follow the order of operations.
   b. Perform whatever operations remain, following the order of operations.

**EXAMPLE 1** Evaluate $(3 + 5) \times 7$.

The given numerical expression contains parentheses. Do the work inside the parentheses first.

Perform the remaining operation.

$(3 + 5) \times 7$
$(8) \times 7$
$56$ *Ans.*

**EXAMPLE 2** Evaluate $12 - (40 \div 4 - 6)$

The given numerical expression contains parentheses. Do the work inside the parentheses first. Follow the order of operations: divide, then subtract.

Perform the remaining operation.

$12 - (40 \div 4 - 6)$
$12 - (\ 10\ - 6)$
$12 - (4)$
$8$ *Ans.*

**EXAMPLE 3** Evaluate:

**a.** $4 + (5 \times 3)$
$4 + (15)$
$19$ *Ans.*

**b.** $(4 + 5) \times 3$
$(9) \times 3$
$27$ *Ans.*

Notice how the use of parentheses gave different instructions in each expression.

**EXAMPLE 4** Evaluate $11 \times (15 - 12) + 5 \times (12 - 2 \times 5)$.

The given numerical expression has two pairs of parentheses. Do the work within each pair of parentheses, carefully following the order of operations as needed. Then perform the remaining operations in their proper order.

$11 \times (15 - 12) + 5 \times (12 - 2 \times 5)$
$11 \times (3) + 5 \times (12 - 10)$
$11 \times (3) + 5 \times (2)$
$33 + 10$
$43$ *Ans.*

**EXAMPLE 5** Joseph had 20 pieces of candy. After eating 4 of them, he divided the remaining candy equally between 2 friends. Which number phrase can be used to find the number of pieces of candy that each friend got?
(a) $(20 \div 4) - 2$ (b) $(20 + 4) \div 2$ (c) $(20 - 2) \div 4$ (d) $(20 - 4) \div 2$

### THINKING ABOUT THE PROBLEM

This is a two-step problem in which the order of operations is important. First, use subtraction to find how much candy was left. Then, use division to find how much candy each friend got. Choices (c) and (d) both show subtraction done first, and then division. But the numbers in choice (c) are wrong.

*Answer:* (d)

**EXAMPLE 6** Martin sells T-shirts for \$7 apiece. Before lunch he sold 13 T-shirts, and after lunch he sold 28. How much money did he earn?

### THINKING ABOUT THE PROBLEM

Break the problem into simple problems. First, find the total number of T-shirts sold, using addition. Then, multiply this number by \$7 to find the amount of money earned.

Think of the number phrase $(13 + 28) \times 7$.

Add 13 and 28 to find the total number of T-shirts sold.

$$\begin{array}{r} 13 \\ +28 \\ \hline 41 \end{array}$$

Multiply 41 by \$7 to find the amount of money earned.

$$\begin{array}{r} 41 \\ \times 7 \\ \hline 287 \end{array}$$

*Answer:* Martin earned \$287.

## CLASS EXERCISES

Evaluate each numerical expression.

**1.** **a.** $5 + 3 \times 8 + 7$ **b.** $(5 + 3) \times 8 + 7$

**2.** **a.** $(47 - 11) \times (3 + 5)$ **b.** $47 - 11 \times 3 + 5$

**3.** **a.** $(5 + 9) \times 7$ **b.** $9 \times (12 - 8)$

**4.** **a.** $5 + (3 \times 8 \div 2)$ **b.** $17 + 2 \times (4 + 3 \times 5)$

**5.** **a.** $(17 - 11) \times (3 + 5)$ **b.** $50 \div (8 + 2) \times 6$

**6.** Find the value of $(10 + 5 \times 8) - (20 + 42 \div 7)$.

**7.** Mary earns \$45 a day from her employer. She earns an additional \$5 a day in tips. Which number phrase can be used to find the amount of money that Mary earns in 6 days?
(a) $45 + 5 + 6$ (b) $(45 + 5) \times 6$ (c) $45 + (5 \times 6)$ (d) $(45 + 5) \div 6$

**8.** Mr. Murray travels 15 miles by bus and 12 miles by train to get to work. How many miles does he travel in 5 trips to work?

## DISTRIBUTING MULTIPLICATION OVER ADDITION OR SUBTRACTION

### THE MAIN IDEA

The operation of multiplication can be *distributed* over the operations of addition or subtraction.

$$6 \times (5 + 7) = 6 \times 5 + 6 \times 7$$

$$9 \times (8 - 2) = 9 \times 8 - 9 \times 2$$

**EXAMPLE 7** Replace ? by a number so that each statement illustrates distributing multiplication.

| | *Statement* | *Answer* |
|---|---|---|
| **a.** | $6 \times (11 + 8) = 6 \times 11 + 6 \times ?$ | $6 \times (11 + 8) = 6 \times 11 + 6 \times \underline{8}$ |
| **b.** | $7 \times (14 + 12) = ? \times 14 + ? \times 12$ | $7 \times (14 + 12) = \underline{7} \times 14 + \underline{7} \times 12$ |
| **c.** | $9 \times (6 - ?) = 9 \times 6 - 9 \times 2$ | $9 \times (6 - \underline{2}) = 9 \times 6 - 9 \times 2$ |
| **d.** | $14 \times (? - 7) = 14 \times 20 - 14 \times 7$ | $14 \times (\underline{20} - 7) = 14 \times 20 - 14 \times 7$ |

**EXAMPLE 8** Find the product $5 \times 23$ by distributing multiplication.

$$\begin{aligned} 5 \times (20 + 3) &= 5 \times 20 + 5 \times 3 \\ &= 100 + 15 \\ &= 115 \quad \textit{Ans.} \end{aligned}$$

**EXAMPLE 9** The seats in a theater are divided into two sections. Section A has 10 rows of seats, and section B has 8 rows of seats. If each row has 9 seats, find the number of seats in the theater.

*Method I*

Multiply to find the number of seats in section A. $9 \times 10$

Multiply to find the number of seats in section B. $9 \times 8$

Add to find the total number of seats.

$$\begin{aligned} & 9 \times 10 + 9 \times 8 \\ = & \ 90 + 72 \\ = & \ 162 \quad \textit{Ans.} \end{aligned}$$

*Method II*

Add to find the number of rows. $10 + 8$

Multiply to find the total number of seats.

$$\begin{aligned} & 9 \times (10 + 8) \\ = & \ 9 \times 18 \\ = & \ 162 \quad \textit{Ans.} \end{aligned}$$

Notice that both methods give the same result and, therefore, they show that multiplication distributes over addition: $9 \times (10 + 8) = 9 \times 10 + 9 \times 8$.

## CLASS EXERCISES

1. Show that $11 \times (8 + 5) = 11 \times 8 + 11 \times 5$ is a true statement.
2. Replace ? by a number so that each statement illustrates distributing multiplication.
   a. $5 \times (9 + 2) = 5 \times ? + 5 \times 2$
   b. $5 \times (? + 8) = 5 \times 7 + 5 \times 8$
   c. $12 \times (4 - ?) = 12 \times 4 - 12 \times 2$
   d. $? \times (7 + 5) = ? \times 7 + 3 \times 5$
   e. $14 \times (12 - 3) = ? \times 12 - ? \times 3$

## HOMEWORK EXERCISES

1. Evaluate each numerical expression.
   a. $(8 + 5) \times 7$
   b. $8 + 5 \times 7$
   c. $36 \div 9 + 3$
   d. $36 \div (9 + 3)$
   e. $(24 - 8) \div 2$
   f. $24 - 8 \div 2$
   g. $15 \times 2 + 3 \times 5$
   h. $15 \times (2 + 3) \times 5$
   i. $30 \div (5 - 2) \times 8$
   j. $30 \div 5 + 2 \times 8$
   k. $12 + 8 \times 3 + 5$
   l. $(12 + 8) \times (3 + 5)$
   m. $(90 - 60) \div (20 + 10)$
   n. $90 - 60 \div 20 + 10$
   o. $(32 + 8) \div 4 - 2$
   p. $32 + 8 \div (4 - 2)$
2. Replace ? so that each expression is an illustration of distributing multiplication.
   a. $8 \times (5 + 9) = 8 \times 5 + 8 \times ?$
   b. $6 \times (3 + 4) = ? \times 3 + ? \times 4$
   c. $2 \times (? + 9) = 2 \times 7 + 2 \times 9$
   d. $5 \times (2 + ?) = 5 \times 2 + 5 \times 11$
3. A company produced 3,000 toys. It sold 22 boxes, each containing 100 toys. Which number phrase can be used to find the number of toys that were not sold?
   (a) $(3{,}000 - 100) \times 22$
   (b) $(3{,}000 \times 22) - 100$
   (c) $(3{,}000 - 22) \times 100$
   (d) $3{,}000 - (22 \times 100)$
4. Mr. Harris said he would contribute three times the total amount that his two children donated to the charity drive. If one child donated \$12 and the other donated \$17, which number phrase can be used to find Mr. Harris' donation?
   (a) $3 \times (12 + 17)$ (b) $(3 \times 12) + 17$ (c) $3 + (12 \times 17)$ (d) $(3 + 12) \times 17$
5. Isaac needed \$188 to buy a new bicycle. He saved \$15 a week for 8 weeks. How much more money did he need?
6. Each of six children donated \$3 toward a gift. Their parents gave them \$25 toward the gift. How much money was collected in all?

## SPIRAL REVIEW EXERCISES

1. Find the difference: $4{,}846 - 2{,}797$
2. Find the quotient: $41{,}654 \div 59$
3. Find the product: $742 \times 26$
4. Written as a numeral, seventy thousand, one hundred two is
   (a) 7,102
   (b) 70,102
   (c) 700,102
   (d) 70,002

5. The value of "9" in the number 795,418 is
   (a) 900 (c) 90,000
   (b) 9,000 (d) 900,000

6. 875,483 rounded to the nearest ten thousand is
   (a) 875,000 (c) 870,000
   (b) 876,000 (d) 880,000

7. At a Mets baseball game, 22,481 people paid for admission. Another 9,576 persons attended on free passes. What was the total attendance for that game?

8. Each ticket to a baseball game costs $6. A total of $89,688 was collected. How many tickets were sold?

**CHALLENGE QUESTION**

If parentheses are not used, the numerical expression $15 + 9 \times 2 + 3$ has a value 36. Rewrite the expression, inserting ( ), so that the expression will have the given value:

**a.** 51 **b.** 120 **c.** 60

# Integers

## MODULE 3

# UNIT 3–1 The Meaning and Use of Signed Numbers; the Set of Integers

## THE MAIN IDEA

1. Numbers that have the signs + (positive) and – (negative) are called *signed numbers*. Zero (0) is neither positive nor negative. If a nonzero number is written without a sign, it is understood to be positive.
2. When we attach the signs + and – to the whole numbers, we obtain the set of *integers*.

   $$\{\text{Integers}\} = \{. . . , -3, -2, -1, 0, +1, +2, +3, . . .\}$$

3. To show the set of integers on a graph, we use a number line that contains 0 and that extends without end to the right of 0 for the positive integers and to the left of 0 for the negative integers.

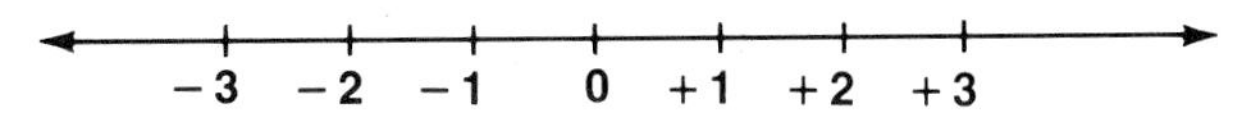


4. Every signed number has an opposite. *Opposites* are two signed numbers that are the *same distance from 0 on a number line but are in different directions,* right or left. Zero is its own opposite.

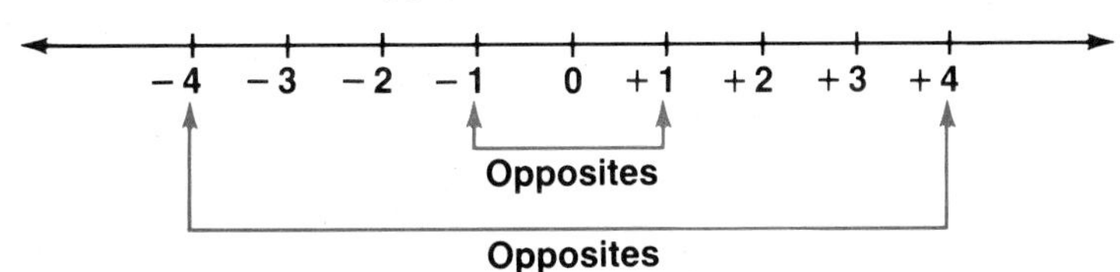


5. Signed numbers are also called *directed numbers*. These numbers allow us to work with situations in which there are two opposite directions, such as:

| + | – | + | – |
|---|---|---|---|
| increasing, | decreasing | deposit, | withdrawal |
| rising, | falling | profit, | loss |
| gaining, | losing | above, | below |
| forward, | backward | future, | past |

**EXAMPLE 1** Name the integer that is represented by each of the letters shown on the number line.

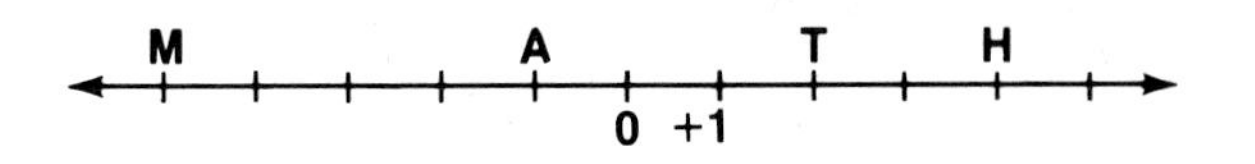


| | Letter | | Answer |
|---|---|---|---|
| a. | M | M is 5 units to the left of 0. | −5 |
| b. | A | A is 1 unit to the left of 0. | −1 |
| c. | T | T is 2 units to the right of 0. | 2 |
| d. | H | H is 4 units to the right of 0. | 4 |

**EXAMPLE 2** Tell how many units on a number line each of the given integers is from zero. Also tell whether the integer is to the left or to the right of zero.

| | Integer | Answer |
|---|---|---|
| a. | +5 | 5 units from 0, to the right of 0 |
| b. | −11 | 11 units from 0, to the left of 0 |
| c. | 9 | 9 units from 0, to the right of 0 |

**EXAMPLE 3** Graph each integer on a number line.

| | Integer | Answer |
|---|---|---|
| a. | +3 | −2 −1 0 +1 +2 +3 +4 |
| b. | −4 | −5 −4 −3 −2 −1 0 +1 +2 +3 |
| c. | 0 | −4 −3 −2 −1 0 +1 +2 +3 +4 |
| d. | 2 | −3 −2 −1 0 +1 +2 +3 |

**EXAMPLE 4** Tell the opposite of each integer.

| | Integer | Answer |
|---|---|---|
| a. | +7 | −7 |
| b. | −50 | +50 |
| c. | −1,700 | +1,700 |
| d. | 0 | 0 |
| e. | 12 | −12 |

**EXAMPLE 5** Use a signed number to show each situation.

| | Situation | Answer |
|---|---|---|
| a. | a loss of 9 pounds | −9 |
| b. | a profit of $15 | +15 |
| c. | 5 years ago | −5 |
| d. | a $20 increase in price | +20 |
| e. | 17° above zero | +17 |

## CLASS EXERCISES

1. Name the integer that is represented by each of the letters shown on the number line.

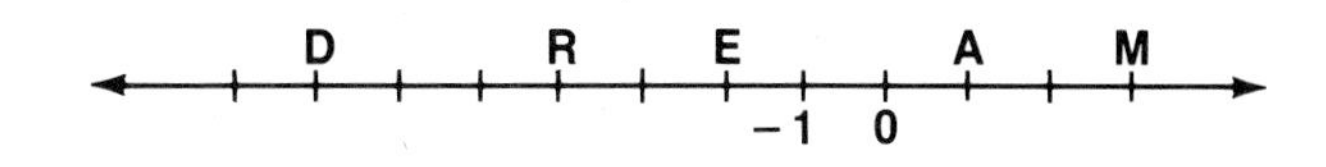


2. Tell how many units on a number line each of the given integers is from zero. Also tell whether the integer is to the left or to the right of zero.

   a. −7  b. +3  c. 5  d. −10  e. +10

3. Graph each integer on a number line.

   a. +6  b. −2  c. −8  d. 4  e. −3

4. Tell the opposite of each integer.

   a. 2  b. −8  c. −12  d. +15  e. −30

5. Use a signed number to show each situation.

   a. 7 years from now  b. 10° below zero  c. a gain of 14 pounds

   d. a $50 deposit  e. a fall of 25 feet  f. a $10 decrease in price

   g. 5 seconds before launch time  h. 1,000 meters above sea level

## HOMEWORK EXERCISES

1. Name the integer that is represented by each of the letters shown on the number line.

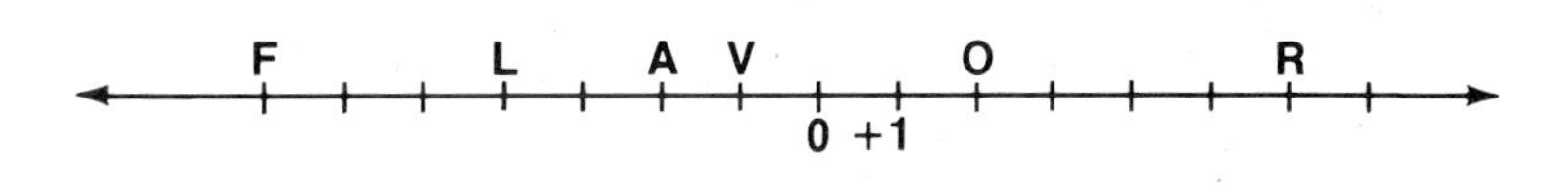


2. Tell how many units on a number line each of the given integers is from zero. Also tell whether the integer is to the left or to the right of zero.

   a. 15  b. −11  c. +11  d. −17  e. 29  f. 0

3. Graph each integer on a number line.

   a. −1  b. −5  c. +7  d. 2  e. −6  f. +6  g. −2  h. −11

4. Tell the opposite of each integer.

   a. −17  b. +8  c. −9  d. 9  e. 3  f. 0

5. Use a signed number to show each situation.

   a. a loss of 5 pounds  b. 50 feet below sea level

   c. a 10-yard gain in football  d. a $30 decrease in price

   e. 10 seconds before takeoff  f. 20° above zero

   g. 12 years ago  h. a $10 withdrawal

## SPIRAL REVIEW EXERCISES

1. Name the place and tell the value of each of the underlined digits.

   **a.** $3\underline{7}1$ **b.** $5,\underline{9}42$ **c.** $\underline{2}96,417$

2. Write twenty-three thousand, four hundred seven as a numeral.

3. Round each of the following to the nearest hundred.

   **a.** 2,791 **b.** 42,547 **c.** 568

4. Add: $3,760 + 946 + 47,869$

5. Subtract 4,782 from 6,891.

6. From 7,813 subtract 908.

7. Find the product of 702 and 83.

8. Divide 391 by 17.

9. What is the total number of ice-cream bars in 15 cartons if each carton contains 36 bars?

10. How many newspapers will each newsstand receive if 1,200 newspapers are to be equally divided among 8 newsstands?

11. Frank sold 73 magazines on Monday and 35 magazines on each of the next four days. What was the total number of magazines that he sold during this five-day period?

12. Find the value:

    **a.** $8 + 5 \times 4$ **b.** $(8 + 5) \times 4$

### CHALLENGE QUESTION

Which value is larger, $-17$ or $-23$?

# UNIT 3–2 Using a Number Line to Compare Integers

## THE MAIN IDEA

To compare two integers:

1. Think of their locations on a number line.
2. The number to the right is the larger number.

**EXAMPLE 1** Tell which of the integers in each pair is farther to the right on a number line.

| | *Number Pair* | *Answer* |
|---|---|---|
| **a.** | $-7$ and 5 | 5 |
| **b.** | $-4$ and $-5$ | $-4$ |
| **c.** | $-50$ and $-51$ | $-50$ |
| **d.** | 0 and 3 | 3 |
| **e.** | $-4$ and 0 | 0 |

**EXAMPLE 2** Replace each ? with $<$ or $>$ to make a true comparison.

| | *Number Pair* | *Placement on a Number Line* | *Answer* |
|---|---|---|---|
| **a.** | 3 ? $-12$ | 3 is to the right of $-12$ | $3 > -12$ |
| **b.** | $-4$ ? $-2$ | $-4$ is to the left of $-2$ | $-4 < -2$ |
| **c.** | $-5$ ? $-8$ | $-5$ is to the right of $-8$ | $-5 > -8$ |

**EXAMPLE 3** Tell which of the numbers in each set is between the others. Next, write the three numbers in order, using $<$. Then, write the three numbers in order, using $>$.

| | *Number Set* | *Answer* |
|---|---|---|
| **a.** | $-4, 3, 0$ | 0 is between $-4$ and 3.<br>$-4 < 0 < 3$;  $3 > 0 > -4$ |
| **b.** | $5, -6, -2$ | $-2$ is between $-6$ and 5.<br>$-6 < -2 < 5$;  $5 > -2 > -6$ |

**EXAMPLE 4** Tell whether the given statement is *true* or *false*.

| | *Statement* | *Answer* |
|---|---|---|
| **a.** | $-6 > -11$ | true |
| **b.** | $2 < -9$ | false |
| **c.** | $-40 > 1$ | false |
| **d.** | $0 > -2$ | true |
| **e.** | $-5 < 2 < 5$ | true |
| **f.** | $5 > -2 > -1$ | false |

**EXAMPLE 5** The chart shows the temperature in Detroit over an eight-hour period on January 15.

| Time | Temperature |
|---|---|
| 6 A.M. | $-11°$ |
| 7 A.M. | $-5°$ |
| 8 A.M. | 0° |
| 9 A.M. | 5° |
| 10 A.M. | 3° |
| 11 A.M. | 6° |
| 12 Noon | 10° |
| 1 P.M. | 2° |
| 2 P.M. | $-8°$ |

**a.** What was the lowest temperature during this period?

**b.** What was the highest temperature during this period?

**c.** Was the temperature higher at 7 A.M. or at 9 A.M.?

**d.** Was the temperature higher at 6 A.M. or at 2 P.M.?

**e.** Did the temperature increase or decrease from 7 A.M. to 8 A.M.?

## THINKING ABOUT THE PROBLEM

Draw a diagram in the form of a number line. This strategy will help you to compare the sizes of the given numbers.

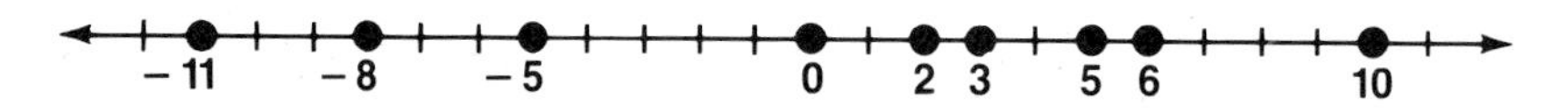


| *Reasoning From the Number Line* | *Conclusion* |
|---|---|
| **a.** Since $-11$ is farthest to the left, | then the lowest temperature was $-11°$. |
| **b.** Since 10 is farthest to the right, | then the highest temperature was 10°. |
| **c.** Since 5 is to the right of $-5$, 5 is greater than $-5$. | Thus, the 9 A.M. temperature, 5°, was higher than the 7 A.M. temperature, $-5°$. |
| **d.** Since $-8$ is to the right of $-11$, $-8$ is greater than $-11$. | Thus, the 2 P.M. temperature, $-8°$, was higher than the 6 A.M. temperature, $-11°$. |
| **e.** Since 0 is to the right of $-5$, 0 is greater than $-5$, and the 8 A.M. temperature, 0°, was higher than the 7 A.M. temperature, $-5°$. | Thus, from 7 A.M. to 8 A.M., the temperature increased. |

## CLASS EXERCISES

**1.** Tell which of the integers in each pair is farther to the right on a number line.

**a.** 2 and $-5$ **b.** $-10$ and 4 **c.** $-7$ and $-8$ **d.** $-24$ and $-30$ **e.** $-8$ and 0

**2.** Replace ? with < or > to make a true comparison.

**a.** $-3$ ? 1 **b.** 0 ? $-2$ **c.** 5 ? $-10$ **d.** $-7$ ? $-11$ **e.** $-10$ ? $-9$ **f.** $-5$ ? 0

**3.** Tell which of the numbers in each set is between the others. Next, write the three numbers in order, using <. Then, write the three numbers in order, using >.

**a.** $-2, 3, 1$ **b.** $-6, -9, -2$ **c.** $-5, 6, -8$ **d.** $3, -6, 0$ **e.** $-10, -30, -20$

4. Tell whether the given statement is *true* or *false*.

   **a.** $-10 < -12$ **b.** $-6 > 0$ **c.** $5 > -15$ **d.** $-30 > -20$ **e.** $28 < -50$

   **f.** $-2 < 0 < 1$ **g.** $-6 < -4 < -10$ **h.** $8 > -2 > -3$ **i.** $-10 > -8 > 0$

5. For ten weeks, Robert made a chart by recording his weight change at the end of each week.

   **a.** At the end of which week did Robert's weight increase the most?

   **b.** At the end of which week did Robert's weight decrease the most?

   **c.** Did Robert's weight increase more at the end of week 8 or at the end of week 1?

   **d.** Did Robert's weight decrease more at the end of week 3 or at the end of week 1?

   **e.** At the end of which week(s) did Robert's weight change the least?

| Week | Weight Change (in lb.) |
|---|---|
| 1 | +2 |
| 2 | +3 |
| 3 | −1 |
| 4 | −4 |
| 5 | 0 |
| 6 | −1 |
| 7 | +2 |
| 8 | +1 |
| 9 | 0 |
| 10 | −3 |

## HOMEWORK EXERCISES

1. Tell which of the integers in each pair is farther to the right on a number line.

   **a.** $-7$ and 6 **b.** $-12$ and $-8$ **c.** $-10$ and 12 **d.** $-14$ and $-18$

   **e.** $-100$ and $-101$ **f.** $-9$ and 0

2. Replace ? with $<$ or $>$ to make a true comparison.

   **a.** $-5$ ? $-4$ **b.** $-10$ ? 0 **c.** 9 ? $-10$ **d.** $-12$ ? $-11$ **e.** $-17$ ? $-20$

   **f.** $-26$ ? 18 **g.** $-19$ ? $-18$ **h.** $-25$ ? $-35$ **i.** $-98$ ? $-99$

3. Tell which of the numbers in each set is between the others. Next, write the three numbers in order, using $<$. Then, write the three numbers in order, using $>$.

   **a.** 3, $-1$, 0 **b.** $-8$, 4, $-2$ **c.** $-11$, $-7$, $-15$ **d.** 4, 7, $-5$ **e.** $-6$, $-4$, $-1$

   **f.** $-18$, $-100$, $-25$ **g.** 85, $-72$, $-71$ **h.** $-33$, $-30$, $-35$ **i.** 0, $-11$, $-7$

4. Tell whether the given statement is true or false.

   **a.** $5 > -5$ **b.** $-10 > -6$ **c.** $-50 < -60$ **d.** $-100 > 10$ **e.** $0 > -12$

   **f.** $-18 > -38$ **g.** $-21 < -22$ **h.** $-10 < -12 < -14$ **i.** $-5 < -4 < 1$

**5.** During each quarter of the last 2 games, Coach Sherman kept a record of his football team's total yardage.

| Game | Quarter | Yardage |
|---|---|---|
| October 5 | 1 | 45 |
| | 2 | 0 |
| | 3 | 30 |
| | 4 | −15 |
| October 12 | 1 | −25 |
| | 2 | 50 |
| | 3 | 25 |
| | 4 | −30 |

**a.** During which quarter did the team gain the greatest yardage?

**b.** During which quarter did the team lose the greatest yardage?

**c.** Was the team more successful in the 4th quarter of the October 5 game or in the 1st quarter of the October 12 game?

**d.** During which quarter of the October 5 game did the team do the worst?

**e.** Did the team do better in the 1st quarter or in the 4th quarter of the October 12 game?

**6.** Mrs. Price kept a record of deposits and withdrawals in her savings account.

| Date | Deposit + / Withdrawal − (in dollars) |
|---|---|
| 1/12 | +150 |
| 1/19 | −75 |
| 1/25 | −50 |
| 1/30 | +200 |
| 2/2 | −100 |
| 2/12 | +100 |
| 2/20 | +75 |
| 2/25 | −200 |

**a.** When did Mrs. Price withdraw the greatest amount of money?

**b.** When did she deposit the greatest amount?

**c.** Did she take a greater amount of money from her account on 1/19 or on 1/25?

**d.** Did she decrease her balance more on 2/2 or on 2/25?

## SPIRAL REVIEW EXERCISES

**1.** Use a signed number to show each of the following situations.

**a.** a $50 deposit **b.** 10° below zero

**c.** a 20° decrease in temperature

**d.** a loss of $10

**2.** Name the opposite of each signed number.

**a.** −15 **b.** 8 **c.** 23 **d.** 0

**3.** 5,864,293 rounded to the nearest ten thousand is
(a) 5,860,000 (c) 5,900,000
(b) 5,870,000 (d) 5,800,000

**4.** Which comparison is true?
(a) $0 > 7$ (c) $8 < 5 < 3$
(b) $9 < 8$ (d) $15 > 9 > 3$

**5.** Perform each operation. Check your answer by using the inverse operation.

**a.** $384 - 295$ **b.** $906 \times 53$

**c.** $5{,}384 + 12{,}957$ **d.** $646 \div 19$

**6.** Evaluate each expression.

**a.** $5 \times 7 - 10$ **b.** $30 + 6 \times 9$

**c.** $5 \times 8 + 3 \times 11$

**d.** $144 \div 36 + 24 \div 8$

**7.** The value of the expression $(8 + 5) \times (11 - 7)$ is
(a) 234 (b) 56 (c) 52 (d) 42

**8.** The greatest common factor of 36 and 54 is
(a) 9 (b) 12 (c) 15 (d) 18

**9.** Which number is composite?
(a) 3 (b) 5 (c) 7 (d) 9

**10.** A prize of $609 will be shared equally among Fran, Bill, and Karen. How much money will each person receive?

**11.** The chart shows the number of meters Jane jogged last week.

**a.** What is the total number of meters that she jogged?

**b.** On which day did she jog the longest distance?

**c.** On which day did she jog the shortest distance?

**d.** What is the difference between the longest and shortest distances that she jogged?

| Day | Meters Jogged |
|---|---|
| Monday | 3,550 |
| Tuesday | 2,980 |
| Wednesday | 1,840 |
| Thursday | 3,050 |
| Friday | 4,700 |

**CHALLENGE QUESTION**

In Anchorage, Alaska, the afternoon temperature rose 15° from the morning temperature, which was $-12°$. What was the afternoon temperature?

# UNIT 3–3 Adding Integers

## THE MAIN IDEA

1. To add two integers on a number line:
   a. Locate the first number.
   b. From the location of the first number, move the number of units indicated by the second number.
      i. If the second number is $+$, move to the right.
      ii. If the second number is $-$, move to the left.
   c. The sum is the number at the last location.
2. To add two integers without using a number line:
   a. If both numbers have the same sign (both $+$ or both $-$), add the numbers as if they had no signs. Give to the sum the original sign.
   b. If the numbers have different signs (one $+$ and one $-$), subtract the numbers as if they had no signs. Give to the sum the sign of the number that is farther from zero.

**EXAMPLE 1** Use a number line to add $-5$ and 8.

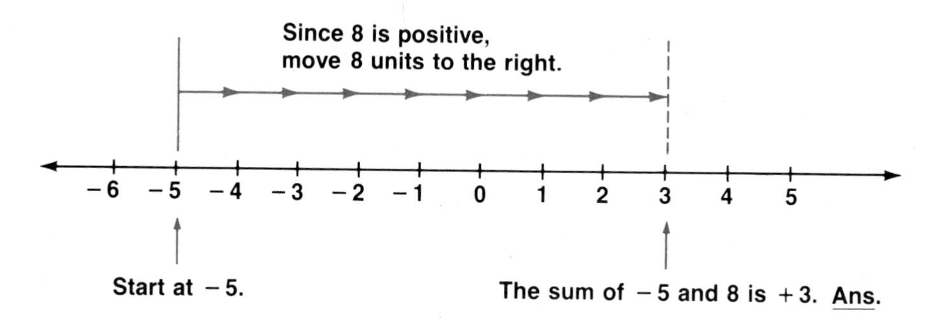


**EXAMPLE 2** Using a number line, add: $-2 + (-3)$

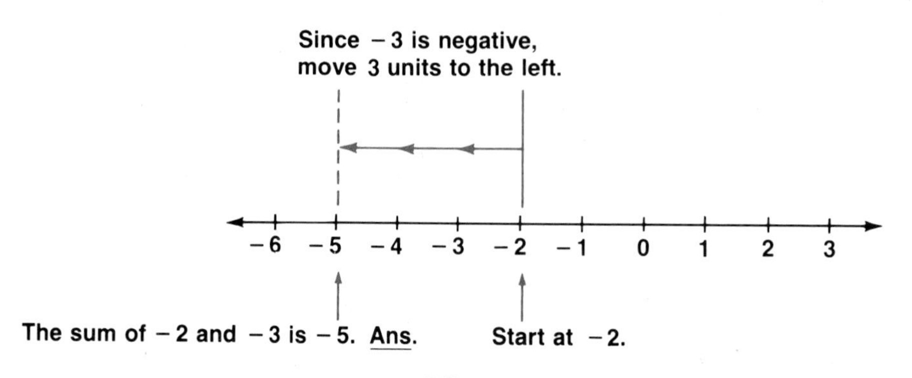

**EXAMPLE 3** Add: $-6 + (-3)$

The signs are the same, both $-$. $-6 + (-3)$

Add as if the numbers had no signs. $6 + 3 = 9$

For the sum, use the original sign, $-$. $= -9$ *Ans.*

**EXAMPLE 4** Add: $+12 + (-8)$

The signs are different, one $+$ and one $-$. $+12 + (-8)$

Subtract the numbers as if they had no signs. $12 - 8 = 4$

For the sum, use the sign of the number that is farther from zero. In this case, use $+$. $= +4$ *Ans.*

**EXAMPLE 5** Add: $-5 + (+1) + (-3)$

**THINKING ABOUT THE PROBLEM**

Use the strategy of breaking the problem into two smaller problems.

Add the first two signed numbers. $-5 + (+1) = -4$

To this sum, add the third signed number. $-4 + (-3) = -7$

*Answer:* $-5 + (+1) + (-3) = -7$

**EXAMPLE 6** The Mudville Tigers football team gained 35 yards during the first quarter and lost 40 yards during the second quarter of their game. Write a number phrase that can be used to find the total number of yards gained or lost during the two quarters.

**THINKING ABOUT THE PROBLEM**

The key words in this problem tell us to:

(1) Use signed numbers: "gain" $(+)$, "loss" $(-)$.

(2) Use addition: "total."

Show a gain of 35 yards by a positive number. $+35$

Show a loss of 40 yards by a negative number. $-40$

Show the total by an addition sign, $+$. $+35 + (-40)$ *Ans.*

**EXAMPLE 7** In January, a Cub Scout troop gained 11 members, and in February, it lost 13 members. For the two months, find the total gain or loss of members.

**THINKING ABOUT THE PROBLEM**

Write a number phrase that can be used to find the sum of two signed numbers, and find the sum.

A gain of 11 members. $+11$

A loss of 13 members. $-13$

Write a number phrase that shows the total gain or loss. $+11 + (-13)$

Find the sum. $+11 + (-13) = -2$

*Answer:* In the two months, the troop lost 2 members.

## CLASS EXERCISES

1. Use a number line to add:

   a. $3 + (-1)$ b. $5 + (-7)$ c. $-2 + (-5)$ d. $-4 + (-6)$ e. $-5 + (8)$

   f. $8 + (-5)$ g. $10 + (-10)$ h. $-10 + (10)$

2. Add:

   a. $-7 + (6)$ b. $-7 + (-2)$ c. $-11 + (-15)$ d. $-7 + (-6)$ e. $-5 + (9)$

   f. $-18 + (7)$ g. $-3 + (9)$ h. $11 + (-6)$ i. $-22 + (0)$ j. $-12 + (5)$

   k. $-8 + (-4)$ l. $28 + (-30)$

3. Add:

   a. $\begin{array}{r} -10 \\ \underline{+8} \end{array}$ b. $\begin{array}{r} -12 \\ \underline{-17} \end{array}$ c. $\begin{array}{r} -19 \\ \underline{+24} \end{array}$ d. $\begin{array}{r} -50 \\ \underline{-27} \end{array}$

4. Add:

   a. $50 + (-17)$ b. $-150 + (70)$ c. $-120 + (55)$ d. $-225 + (-145)$

   e. $-29 + (101)$ f. $-175 + (+175)$ g. $-47 + (-133)$ h. $250 + (-250)$

   i. $-8 + (+11) + (+9)$ j. $+4 + (-9) + (-6)$ k. $-12 + (-8) + (-7)$

In 5–8, write a number phrase that can be used to find the total gain or loss.

5. The temperature dropped 12 degrees and then dropped another 7 degrees.
6. Mr. Borman made a withdrawal of $87 and then made a deposit of $150 in his savings account.
7. Marie gained 5 pounds and then lost 6 pounds.
8. Rex Clothing Store reduced the price of a suit by $8 and then reduced it further by $3.

## HOMEWORK EXERCISES

1. Use a number line to add:

   a. $8 + (-6)$ b. $2 + (-7)$ c. $3 + (+4)$ d. $-5 + (-2)$ e. $-6 + (9)$

   f. $-7 + (0)$ g. $-1 + (-7)$ h. $4 + (-4)$ i. $-7 + (-5)$ j. $3 + (-4)$

   k. $5 + (-1) + (3)$ l. $-2 + (0) + (4)$

2. Add:

   a. $-4 + (9)$ b. $-6 + (-3)$ c. $-8 + (3)$ d. $7 + (-10)$ e. $0 + (-7)$

   f. $-15 + (15)$ g. $12 + (-7)$ h. $-16 + (8)$ i. $-20 + (-5)$ j. $-7 + (7)$

   k. $-22 + (10)$ l. $24 + (-30)$

**3.** Add:

| **a.** | **b.** | **c.** | **d.** | **e.** | **f.** |
|---|---|---|---|---|---|
| $-10$ | $-15$ | $+36$ | $-40$ | $+80$ | $-15$ |
| $+20$ | $-18$ | $-21$ | $-35$ | $-90$ | $+19$ |

| **g.** | **h.** | **i.** | **j.** |
|---|---|---|---|
| $+42$ | $-21$ | $+48$ | $-100$ |
| $-42$ | $-21$ | $-50$ | $+100$ |

**4.** Add:

**a.** $-35 + (-50)$ **b.** $-20 + (42)$ **c.** $-81 + (81)$ **d.** $46 + (-25)$ **e.** $32 + (-50)$

**f.** $-75 + (-25)$ **g.** $-19 + (43)$ **h.** $-17 + (-82)$ **i.** $-29 + (-80)$

**j.** $-100 + (0)$ **k.** $+200 + (-200)$ **l.** $-96 + (-24)$ **m.** $51 + (0) + (-27)$

**n.** $-32 + (25) + (-61)$ **o.** $-70 + (41) + (-28)$

**5.** Find the sum of $-48$ and $-24$.

**6.** The sum of $-50$ and 35 is (a) $-15$ (b) 15 (c) $-85$ (d) 85

In 7–9, write a number phrase that can be used to find the total gain or loss.

**7.** A kite was flying at a height of 250 m. It was raised 50 m and then lowered 125 m.

**8.** Dennis had $18 and spent $10.

**9.** On a winter morning, the temperature was 5°F. In the afternoon, the temperature rose 17°, and in the evening fell 9°.

In 10–12, use signed numbers to find the final result.

**10.** A football team:

**a.** gained 20 yards and lost 6 yards.

**b.** lost 5 yards and lost 9 yards.

**c.** lost 11 yards, gained 8 yards, and lost 3 yards.

**d.** gained 15 yards, lost 12 yards, and gained 8 yards.

**11.** In a game, Ben started with 35 points. Then he won 9 points and lost 12 points.

**12.** The temperature one day was $-4$°F and it dropped 9°.

## SPIRAL REVIEW EXERCISES

**1.** Write the opposite of each signed number.

**a.** $-4$ **b.** $-9$ **c.** 7 **d.** 23

**e.** $-68$

**2.** Tell whether each given statement is *true* or *false*.

**a.** $-10 > -8$ **b.** $0 > -4$

**c.** $-13 < 13$ **d.** $-9 > -5 > -2$

**3.** Evaluate each expression.

**a.** $100 - 22 \times 3$

**b.** $(100 - 22) \times 3$

**c.** $4 \times 9 + 8 \times 3$

**d.** $4 \times (9 + 8) \times 3$

**4.** Find the product of 240 and 52.

**5.** Find the sum of 472 and 5,948.

6. Replace each ? by a whole number to make a true statement.

   a. $79 + 62 = ? + 79$

   b. $71 \times ? = 71$

   c. $93 - ? = 93$

   d. $(27 + 46) + 85 = ? + (46 + 85)$

7. Which is a true statement?
   (a) $-17 + (+17) = 0$
   (b) $-17 + (-17) = 0$
   (c) $+17 + (+17) = 0$
   (d) $-17 + (0) = 0$

8. Replace each ? by an integer to make a true statement.

   a. $9 + ? = 0$

   b. $-12 + (-17) = ? + (-12)$

   c. $+3 + ? = -5 + (+3)$

   d. $(-7 + 2) + ? = -7 + (2 + 9)$

9. Mr. Conte had $12,500 in his bank account. He took $8,560 from his account to pay for a car. How much money was left in his account?

**CHALLENGE QUESTION**

One morning, the temperature in Detroit was 8°F and the temperature in Cleveland was −5°F. How much warmer was it in Detroit?

## UNIT 3–4 Subtracting Integers

### THE MAIN IDEA

Subtracting a number means adding its opposite. To subtract:

1. Replace the number being subtracted by its opposite and change the subtraction symbol to addition.
2. Then add, remembering the rules for addition of integers.

**EXAMPLE 1** Using a number line, subtract: $2 - (5)$

Subtracting a number means adding its opposite.

$$2 - (5) = 2 + (-5)$$

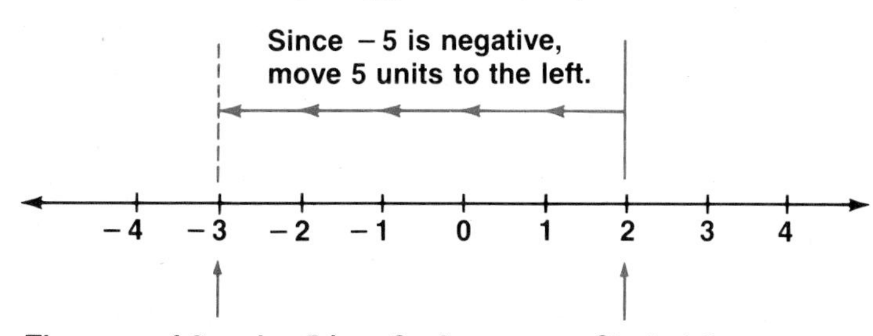


**EXAMPLE 2** Subtract: $-9 - (8)$

Replace 8, the number being subtracted, by its opposite, and change to addition. $\quad -9 - (8)$ $= -9 + (-8)$

Add, remembering the rules for addition of integers. $\quad = -17$ *Ans.*

**EXAMPLE 3** Subtract $-6$ from $-7$.

The number being subtracted is $-6$. $\quad -7 - (-6)$

Replace $-6$ by its opposite, and change to addition. $\quad = -7 + (+6)$ $= -1$ *Ans.*

**EXAMPLE 4** From the sum of $+5$ and $-7$, subtract $-1$.

**THINKING ABOUT THE PROBLEM**

Break this problem into two smaller problems.

Add $+5$ and $-7$. $\quad +5 + (-7) = -2$

To subtract $-1$, add its opposite, $+1$. $\quad -2 + (+1) = -1$

*Answer:* $+5 + (-7) - (-1) = -1$

**EXAMPLE 5** Find the change in temperature between a morning temperature of $-12°F$ and an afternoon temperature of $+15°F$.

**THINKING ABOUT THE PROBLEM**

The word "change" tells you to subtract to find the difference. Begin with the most recent temperature, the afternoon temperature, and subtract the earlier temperature, the morning temperature.

Write a number phrase. $+15 - (-12)$

Find the difference by adding the opposite. $+15 + (+12)$
$= +27$

*Answer:* The temperature increased 27 degrees.

## CLASS EXERCISES

1. Use a number line to do each subtraction.

**a.** $2 - (5)$ **b.** $6 - (-3)$ **c.** $-7 - (-2)$ **d.** $-9 - (+3)$ **e.** $-10 - (-5)$

**f.** $-3 - (-3)$ **g.** $8 - (-4)$ **h.** $-4 - (8)$

2. Subtract:

**a.** $-9 - (5)$ **b.** $4 - (+7)$ **c.** $-2 - (-18)$ **d.** $7 - (-3)$ **e.** $-6 - (2)$

**f.** $-15 - (9)$ **g.** $-9 - (-7)$ **h.** $-10 - (5)$ **i.** $25 - (-25)$ **j.** $5 - (-8)$

**k.** $-10 - (-5)$ **l.** $-25 - (25)$

3. Subtract:

**a.** $\begin{array}{r} -14 \\ \underline{9} \end{array}$ **b.** $\begin{array}{r} +76 \\ \underline{-12} \end{array}$ **c.** $\begin{array}{r} 16 \\ \underline{-12} \end{array}$ **d.** $\begin{array}{r} -38 \\ \underline{-38} \end{array}$ **e.** $\begin{array}{r} +10 \\ \underline{+25} \end{array}$ **f.** $\begin{array}{r} -51 \\ \underline{+39} \end{array}$ **g.** $\begin{array}{r} -42 \\ \underline{-19} \end{array}$ **h.** $\begin{array}{r} -82 \\ \underline{-47} \end{array}$

4. From the sum of $-7$ and 3, subtract 2.

5. Subtract $-3$ from the sum of 1 and $-4$.

6. Subtract the sum of 5 and $-1$ from the sum of $-2$ and 6.

7. **a.** Find the change in temperature from $+22°C$ to $-12°C$.

   **b.** Find the change in altitude from 1,200 feet above sea level to 200 feet below sea level.

## HOMEWORK EXERCISES

1. Using a number line, subtract:

**a.** $3 - (5)$ **b.** $2 - (-4)$ **c.** $-6 - (-3)$ **d.** $-7 - (4)$ **e.** $-8 - (-9)$

**f.** $-4 - (-2)$ **g.** $6 - (-3)$ **h.** $-3 - (-3)$

2. Subtract:

**a.** $-10 - (6)$ **b.** $-15 - (-9)$ **c.** $-22 - (-30)$ **d.** $17 - (23)$ **e.** $26 - (-18)$

**f.** $-9 - (-5)$ **g.** $-4 - (4)$ **h.** $-4 - (-4)$ **i.** $20 - (-10)$ **j.** $-10 - (20)$

**k.** $-25 - (-15)$ **l.** $-15 - (-25)$

**3.** Subtract:

| **a.** | **b.** | **c.** | **d.** | **e.** | **f.** | **g.** | **h.** |
|---|---|---|---|---|---|---|---|
| $-19$ | $42$ | $-40$ | $58$ | $-36$ | $-58$ | $50$ | $-18$ |
| $-20$ | $50$ | $+18$ | $-32$ | $-15$ | $+32$ | $-30$ | $-18$ |

**4.** Subtract $-17$ from 26. **5.** From $-26$ subtract $-9$. **6.** Subtract 51 from $-80$.

**7.** From $-75$ subtract 25. **8.** Subtract 9 from the sum of $-2$ and $-8$.

**9.** From the sum of 6 and $-8$, subtract $-15$. **10.** Subtract $-12$ from the sum of $-9$ and 14.

**11.** Subtract the sum of $-2$ and $-5$ from the sum of $-4$ and 8.

**12.** Find the difference between an altitude of 250 feet below sea level and an altitude of 3,000 feet above sea level.

**13.** Find the change in time between 4 minutes before a rocket blast-off and 1 minute before the blast-off.

## *SPIRAL REVIEW EXERCISES*

**1.** The temperature at 8 A.M. was $-6$°F and it rose 19 degrees. Find the new temperature.

**2.** The temperature at noon was 8°C and it dropped 12 degrees by midnight. Find the temperature at midnight.

**3.** Barry's science class weighed 10 packages of food to see how close the actual weight of the food came to the advertised weight on the packages. The chart shows how the weights compared.

| **Package** | A | B | C | D | E | F | G | H | I | J |
|---|---|---|---|---|---|---|---|---|---|---|
| **Comparison (ounces)** | +1 | +1 | +2 | 0 | −1 | +2 | 0 | 0 | −3 | −4 |

**a.** Which package(s) had the most extra weight?

**b.** Which package was closer to the advertised weight, package C or package E?

**c.** Which package was further below its advertised weight, package I or package J?

**4.** Add:

**a.** $-9 + (-16)$ **b.** $-19 + (12)$

**c.** $32 + (-18)$ **d.** $-27 + (30)$

**e.** $-14 + 0 + (-12)$

**f.** $-32 + (-1) + (16)$

**5.** Find the sum of $-22$ and $-45$.

**6.** Which number is prime?
(a) 81 (b) 21 (c) 31 (d) 121

**7.** The value of 9 in the numeral 79,523 is
(a) 90 (c) 9,000
(b) 900 (d) 90,000

**8.** Which number rounded to the nearest hundred has the value 2,700?
(a) 2,740 (c) 2,640
(b) 2,790 (d) 2,649

**9.** When 23 is divided by 5, the remainder is
(a) 1 (b) 2 (c) 3 (d) 4

**10.** Find the product of 75 and 11.

**11.** How many packages of 16 cookies each can be made from a batch of 400 cookies?

**12.** On Tuesday, Philip ate food containing 2,070 calories and Marcos ate food containing 1,450 calories. How many more calories did Philip consume than Marcos?

**CHALLENGE QUESTION**

Add: $-50 + (+17) + (+42) + (-21) + (+31) + (-29) + (+10)$

# UNIT 3–5 Multiplying Integers

## THE MAIN IDEA

To multiply two integers:

1. Multiply as if the numbers had no signs.
2. a. If both numbers have the same sign (both + or both −), then the product is positive.

   b. If the two numbers have different signs (one + and one −), then the product is negative.

**EXAMPLE 1** Multiply: $-3 \times (-5)$

Multiply as if the numbers had no signs. $3 \times 5 = 15$

Since the signs are the same (both −), the product is positive. $-3 \times (-5) = +15$ or $15$ *Ans.*

**EXAMPLE 2** Find the product of 9 and −4.

"Find the product" means multiply. $9 \times (-4)$

Multiply as if the numbers had no signs. $9 \times 4 = 36$

Since the signs are different (one + and one −), the product is negative. $+9 \times (-4) = -36$ *Ans.*

**EXAMPLE 3** Multiply: $-4 \times (-6) \times (-3)$

**THINKING ABOUT THE PROBLEM**

Use the strategy of breaking the problem into smaller problems.

Find the product of the first two integers. $-4 \times (-6) = +24$

Multiply this product by the third integer. $+24 \times (-3) = -72$

*Answer:* $-4 \times (-6) \times (-3) = -72$

## CLASS EXERCISES

**1.** Multiply:

**a.** $9 \times (-3)$ **b.** $-9 \times (-3)$ **c.** $-5 \times (7)$ **d.** $8 \times (-4)$ **e.** $-7 \times (-8)$

**f.** $-6 \times (9)$ **g.** $-4 \times (3)$ **h.** $3 \times (-4)$ **i.** $-2 \times (-5)$ **j.** $6 \times (-4)$

**k.** $-6 \times (-5)$ **l.** $-9 \times (7)$

2. Find the product of each of the given pairs of integers.

   a. 3 and −5  b. −2 and 7  c. −8 and −4  d. −6 and +3  e. −9 and −7
   f. 5 and −2  g. −7 and −4  h. −4 and 7  i. −2 and +2  j. −5 and −5
   k. 6 and −4  l. −7 and 0

3. Multiply:

   a. $-10 \times (+2) \times (-4)$  b. $+5 \times (-6) \times (7)$  c. $+1{,}467 \times (-7{,}631) \times (0)$

## HOMEWORK EXERCISES

1. Multiply:

   a. $-9 \times (5)$  b. $-10 \times (-2)$  c. $4 \times (-7)$  d. $-8 \times (-3)$  e. $-12 \times (4)$
   f. $0 \times (-2)$  g. $-6 \times (-6)$  h. $-7 \times (9)$  i. $3 \times (-9)$  j. $-10 \times (5)$
   k. $5 \times (-10)$  l. $-9 \times (0)$

2. Find the product of each of the given pairs of integers.

   a. 7 and −2  b. −10 and 6  c. −9 and −8  d. −3 and −3  e. −14 and 0
   f. −8 and 7  g. −4 and −5  h. −9 and 2  i. −11 and −5  j. −2 and −7
   k. −8 and 4  l. 4 and −8

3. Multiply:

   a. $-8 \times (+3) \times (+4)$  b. $-8 \times (-2) \times (-3)$  c. $5 \times (-6) \times (-4)$

## SPIRAL REVIEW EXERCISES

1. The chart shows how membership increased and decreased at the Wonderbody Health Club each month.

| Month | Jan. | Feb. | Mar. | Apr. | May | June | July | Aug. | Sept. | Oct. | Nov. | Dec. |
|---|---|---|---|---|---|---|---|---|---|---|---|---|
| Change in Membership | +4 | +18 | +21 | +18 | +11 | +6 | 0 | 0 | −4 | −10 | −18 | −6 |

   a. In which month did membership decrease the most?
   b. Did the membership decrease more in July or in December?

2. Add:

   a. $-9 + (-12)$  b. $-17 + (3)$
   c. $24 + (-11)$  d. $-15 + (-10)$
   e. $-63 + (-42) + (8)$
   f. $101 + (-11) + (-100)$

3. Subtract:

   a. $-7 - (-8)$  b. $-11 - (4)$
   c. $19 - (-4)$  d. $-20 - (-10)$

4. Find the sum of −15 and 10.

5. Subtract −9 from −16.

**6.** From the sum of $+7$ and $-2$, subtract $-23$.

**7.** Replace each ? by $<$ or $>$ to make a true statement.

**a.** $-14 \; ? \; -16$ **b.** $-2 \; ? \; 0$

**c.** $-14 + (+14) \; ? \; -1$

**d.** $20 + (-5) \; ? \; 20 - (-5)$

**8.** Replace each ? by an integer to make a true statement.

**a.** $+2 \times (-8) = ? \times (+2)$

**b.** $-10 \times ? = 0$

**c.** $-7 \times ? = -7$

**d.** $(? \times -2) \times 4 = 6 \times (-2 \times 4)$

**e.** $-15 \times ? = +15$

**9.** Nine thousand, twenty-three written as a numeral is
(a) 92,300 (c) 9,230
(b) 90,230 (d) 9,023

**10.** 6,542,183 rounded to the nearest million is
(a) 6,000,000 (c) 6,500,000
(b) 7,000,000 (d) 6,600,000

**11.** The value of 7 in 57,382 is
(a) 70,000 (c) 700
(b) 7,000 (d) 70

**12.** The greatest common factor of 24 and 48 is
(a) 2 (b) 6 (c) 12 (d) 24

**13.** An oil tank that holds 5,000 gallons of oil loses 1,750 gallons. How many gallons of oil are left?

**14.** The price of a ticket for a school concert is $4. How much money was collected if 250 people bought tickets?

**15.** The value of $6 + 4 \times 3$ is
(a) 30 (b) 13 (c) 18 (d) 72

## CHALLENGE QUESTION

Replace each □ by an integer to make a true statement.

**a.** $+5 \times \square = +20$ **b.** $+7 \times \square = -35$ **c.** $-9 \times \square = 36$ **d.** $-8 \times \square = -64$

## UNIT 3–6 Dividing Integers

### THE MAIN IDEA

To divide two integers:

1. Divide as if the numbers had no signs.
2. a. If both numbers have the same sign (both + or both −), then the quotient is positive.

   b. If the two numbers have different signs (one + and one −), then the quotient is negative.

**EXAMPLE 1** Divide: $-18 \div (-9)$

| | |
|---|---|
| Divide as if the numbers had no signs. | $18 \div 9 = 2$ |
| Since the signs are the same (both −), the quotient is positive. | $-18 \div (-9) = +2$ or 2 *Ans.* |

**EXAMPLE 2** Divide: $20 \div (-4)$

| | |
|---|---|
| Divide as if the numbers had no signs. | $20 \div 4 = 5$ |
| Since the signs are different (one + and one −), the quotient is negative. | $+20 \div (-4) = -5$ *Ans.* |

**EXAMPLE 3** Divide the sum of −6 and 30 by −3.

**THINKING ABOUT THE PROBLEM**

Use the strategy of breaking the problem into smaller problems.

| | |
|---|---|
| Find the sum of −6 and 30. | $-6 + (+30) = +24$ |
| Divide this sum by −3. | $+24 \div (-3) = -8$ *Ans.* |

### CLASS EXERCISES

1. Divide:

**a.** $-10 \div (2)$ **b.** $-15 \div (-3)$ **c.** $20 \div (-5)$ **d.** $-12 \div (-4)$ **e.** $8 \div (-2)$

**f.** $-16 \div (8)$ **g.** $-50 \div (10)$ **h.** $-48 \div (6)$ **i.** $-28 \div (-7)$ **j.** $-40 \div (10)$

**k.** $-100 \div (-20)$ **l.** $-81 \div (-9)$

2. Divide:

a. $120 \div (-6)$ b. $56 \div (-7)$ c. $-32 \div (-8)$ d. $-96 \div (-12)$ e. $100 \div (-25)$

f. $-150 \div (75)$ g. $-108 \div (9)$ h. $225 \div (-15)$ i. $-144 \div (-8)$

3. Divide the sum of $-12$ and $+20$ by $-2$.
4. Divide the difference of $+15$ and $-15$ by $+3$.
5. Divide the product of $-12$ and $-2$ by $-3$.

## HOMEWORK EXERCISES

1. Divide:

a. $20 \div (-5)$ b. $-14 \div (7)$ c. $-24 \div (-4)$ d. $-12 \div (6)$ e. $-36 \div (-9)$

f. $-90 \div (10)$ g. $0 \div (-8)$ h. $-45 \div (9)$ i. $-63 \div (-7)$ j. $-80 \div (8)$

k. $56 \div (-7)$ l. $-49 \div (7)$

2. Divide:

a. $-88 \div (11)$ b. $-100 \div (-10)$ c. $-64 \div (16)$ d. $48 \div (-12)$

e. $-96 \div (-24)$ f. $200 \div (-50)$ g. $-99 \div (33)$ h. $-120 \div (-40)$

i. $-54 \div (-18)$ j. $-78 \div (26)$ k. $99 \div (-11)$ l. $-100 \div (-25)$

3. Divide the sum of $-14$ and $+18$ by $+4$.
4. Divide the difference of $-28$ and $-21$ by 7.
5. Divide the product of 0 and $-14$ by $-7$.
6. Divide the sum of $-9$ and $-5$ by the product of $-2$ and $+1$.

## SPIRAL REVIEW EXERCISES

1. Multiply:

a. $3 \times (-7)$ b. $-6 \times (-9)$

c. $-7 \times (8)$ d. $-9 \times (-6)$

e. $-6 \times (-3) \times (-4)$

f. $12 \times (+10) \times (-2)$

2. Find the sum of 29 and $-17$.
3. Subtract $-18$ from 18.
4. Perform the indicated operation.

a. $-6 + (-10)$ b. $-6 - (-10)$

c. $-6 \times (-10)$ d. $-16 \div (-4)$

e. $28 \div (-7)$ f. $28 - (-7)$

5. Find the product of $-18$ and $-5$.
6. From $-27$ subtract $-43$.
7. Which is a true statement?
(a) $8 \times (-1) = 8$
(b) $5 - 7 = 7 - 5$
(c) $-9 + 1 < -8$
(d) $-7 \div (-1) = 7$
8. Replace each ? by $<$ or $>$ to make a true statement.

a. $17 - 12 \; ? \; 12 - 17$

b. $-17 + (+1) \; ? \; -17 \times (+1)$

c. $-2 \times (-3) \times (-1) \; ? \; 0$

d. $-5 + 0 \; ? \; -5 \times 0$

9. When 494 is divided by 16, the remainder is
(a) 14 (b) 12 (c) 10 (d) 0

10. At the beginning of a trip, the odometer of a car read 23,760 miles. At the end of the trip, it read 24,095 miles. How long was the trip?

11. Mr. Harris travels 37 miles round-trip each day to work. How many miles will he travel in 20 workdays?

12. At 8 P.M., the temperature was 9°C. The temperature dropped 5° between 8 P.M. and 9 P.M., and 6° during the next hour. Find the temperature at 10 P.M.

13. Mel had \$50 and spent \$32. Find the amount of money that he still has.

14. A football team gained 15 yards, lost 12 yards, and then gained 8 yards. Find the total number of yards gained or lost.

## CHALLENGE QUESTION

Replace each □ by a number to make a true statement.

**a.** $\square \times 24 = -24$ **b.** $\square + 24 = -24$ **c.** $\square \div 24 = -24$ **d.** $\square - 24 = -24$

# THEME 3

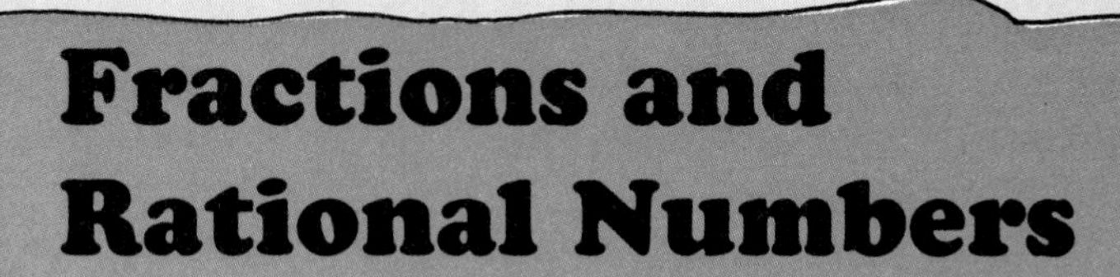


## MODULE 4



## MODULE 5



## MODULE 6

MODULE 4

# UNIT 4–1 Kinds of Fractions; Representing Rational Numbers on a Number Line

## KINDS OF FRACTIONS

### THE MAIN IDEA

1. A ***fraction*** is written with two numerals separated by a division line:

$$\frac{\textit{numerator}}{\textit{denominator}\ \text{(not zero)}}$$

   In arithmetic, a fraction represents a positive number.

2. One use for a fraction is to represent *part of a whole quantity*. The meaning of $\frac{3}{4}$ can be seen in the following diagram:

| The whole quantity. | The denominator 4 divides the whole quantity into 4 equal parts. | The numerator 3 shows the number of equal parts represented by the fraction. |
| --- | --- | --- |

3. The number 1 represents a whole quantity.

4. A ***proper fraction*** represents less than a whole quantity.
   In a proper fraction, the numerator is less than the denominator, and the value of the fraction is less than 1.

5. An ***improper fraction*** represents either more than a whole quantity or exactly a whole quantity.
   In an improper fraction, the numerator is greater than or equal to the denominator, and the value of the fraction is greater than or equal to 1.

6. Any number that can be written as a fraction in which the numerator and the denominator are integers is called a ***rational number***. The set of rational numbers also includes the whole numbers and the integers.

**EXAMPLE 1** Name the numerator and the denominator of each fraction.

| | *Fraction* | *Numerator* | *Denominator* |
|---|---|---|---|
| a. | $\frac{7}{9}$ | 7 | 9 |
| b. | $\frac{81}{113}$ | 81 | 113 |
| c. | $\frac{0}{27}$ | 0 | 27 |
| d. | $\frac{12}{9}$ | 12 | 9 |
| e. | $\frac{16}{16}$ | 16 | 16 |

**EXAMPLE 2** Name each of the *shaded* regions as a fraction.

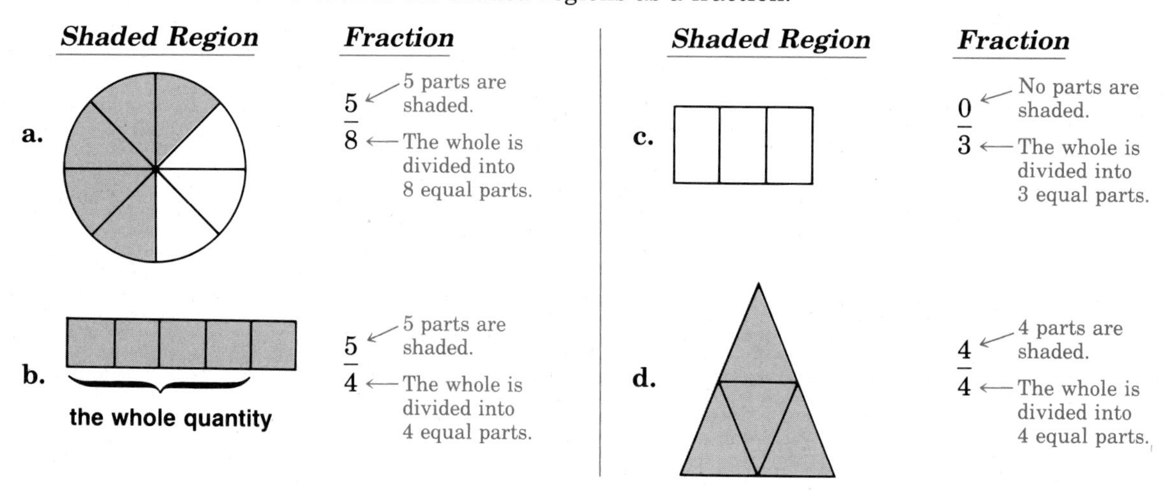


**EXAMPLE 3** Tell whether each fraction is proper or improper.

| | *Fraction* | *Type of Fraction* | |
|---|---|---|---|
| a. | $\frac{5}{7}$ | proper | (numerator $<$ denominator) |
| b. | $\frac{9}{9}$ | improper | (numerator $=$ denominator) |
| c. | $\frac{5}{3}$ | improper | (numerator $>$ denominator) |
| d. | $\frac{3}{5}$ | proper | (numerator $<$ denominator) |
| e. | $\frac{234}{234}$ | improper | (numerator $=$ denominator) |
| f. | $\frac{0}{4}$ | proper | (numerator $<$ denominator) |

**EXAMPLE 4** Use $<$, $>$, or $=$ to compare each fraction with 1.

| | *Fraction* | *Comparison With 1* |
|---|---|---|
| **a.** | $\frac{3}{11}$ | $\frac{3}{11} < 1$ |
| **b.** | $\frac{12}{17}$ | $\frac{12}{17} < 1$ |
| **c.** | $\frac{15}{7}$ | $\frac{15}{7} > 1$ |
| **d.** | $\frac{22}{22}$ | $\frac{22}{22} = 1$ |
| **e.** | $\frac{0}{4}$ | $\frac{0}{4} < 1$ |

## CLASS EXERCISES

**1.** Name the numerator and the denominator of each fraction.

**a.** $\frac{11}{12}$ **b.** $\frac{7}{512}$ **c.** $\frac{157}{121}$ **d.** $\frac{231}{231}$ **e.** $\frac{12}{10}$ **f.** $\frac{0}{426}$

**2.** Name each of the *shaded* regions as a fraction.

**a.**

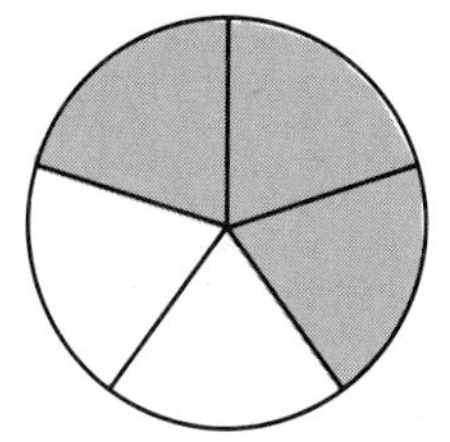

**b.**

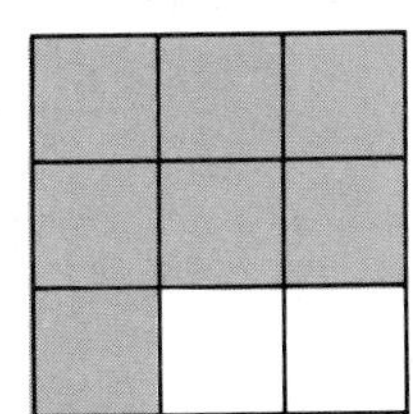

**c.**

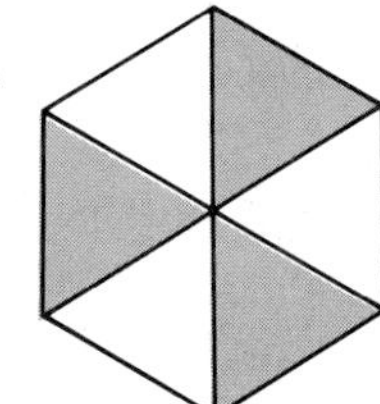

**d.**

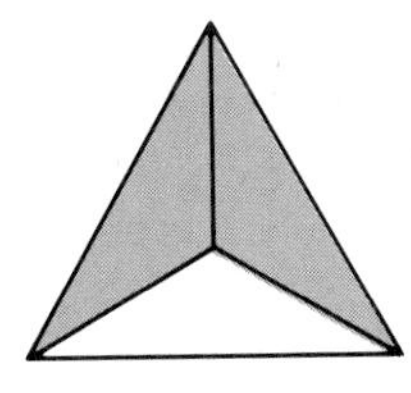

**e.**

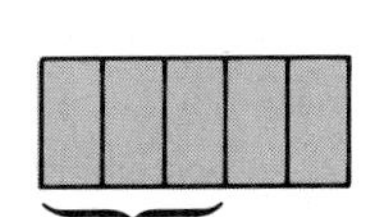

the whole quantity

**f.**

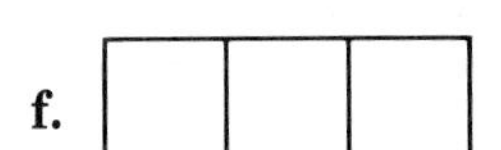

**3.** Tell whether each fraction is proper or improper.

**a.** $\frac{9}{11}$ **b.** $\frac{15}{17}$ **c.** $\frac{13}{11}$ **d.** $\frac{81}{81}$ **e.** $\frac{50}{75}$ **f.** $\frac{0}{17}$ **g.** $\frac{112}{113}$ **h.** $\frac{113}{112}$

**4.** Replace ? by $<$, $>$, or $=$ to make a true comparison with 1.

**a.** $\frac{5}{7}$ ? 1 **b.** $\frac{7}{5}$ ? 1 **c.** $\frac{7}{7}$ ? 1 **d.** $\frac{0}{9}$ ? 1 **e.** $\frac{121}{120}$ ? 1 **f.** $\frac{397}{397}$ ? 1 **g.** $\frac{1000}{1001}$ ? 1

## REPRESENTING RATIONAL NUMBERS ON A NUMBER LINE

### THE MAIN IDEA

Just as we represent whole numbers as points on a number line, we also represent rational numbers as points on a number line.

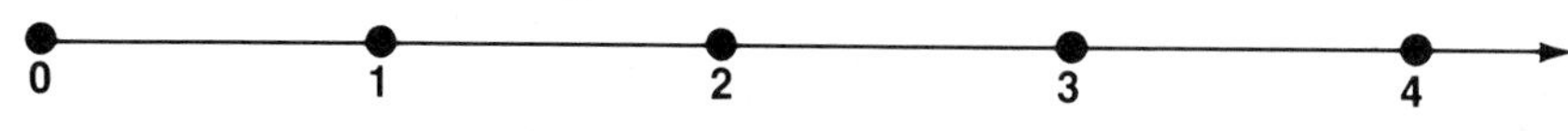

Whole Numbers on a Number Line

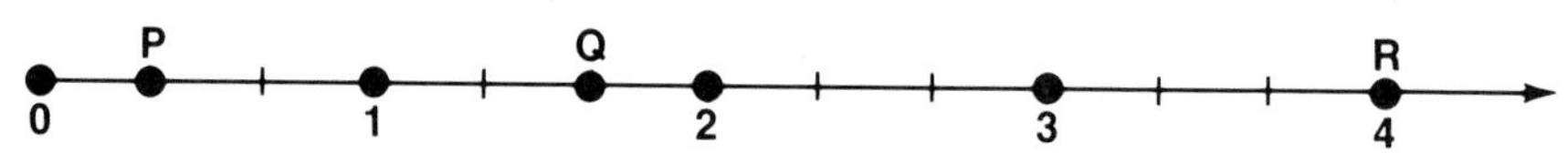

Rational Numbers on a Number Line That Is Divided Into Thirds

The number line above has been divided so that there are 3 equal parts between any two whole numbers. Point $P$ is on the first such division. Thus, point $P$ represents $\frac{1}{3}$. Point $Q$ represents $\frac{5}{3}$. Point $R$ represents $\frac{12}{3}$.

**EXAMPLE 5** Use a rational number to name each point on the number lines shown.

| | *Point* | *Rational Number* |
|---|---|---|
| a. | A (0, 1, 2) | $\frac{1}{4}$ |

*Think:* A whole quantity has been divided into 4 equal parts. Thus, 4 is the *denominator* of the rational number. Since point $A$ is on the first division, 1 is the *numerator* of the rational number.

| | *Point* | *Rational Number* |
|---|---|---|
| b. | B (0, 1, 2) | $\frac{2}{3}$ |
| c. | C (0, 1, 2) | $\frac{7}{8}$ |
| d. | D (0, 1, 2) | $\frac{5}{4}$ |
| e. | E (0, 1, 2) | $\frac{12}{6}$ or 2 |
| f. | F (0, 1, 2) | $\frac{0}{3}$ or 0 |

## CLASS EXERCISES

Use a rational number to name each point on the number lines shown.

1., 2., 3.

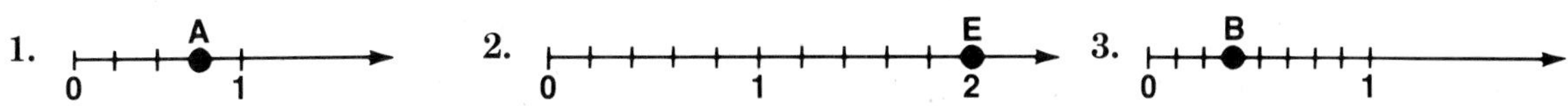


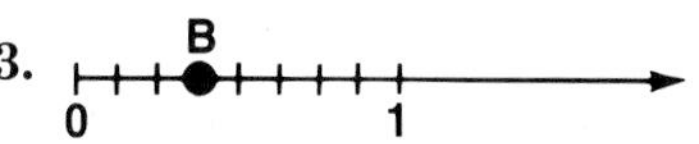

4.

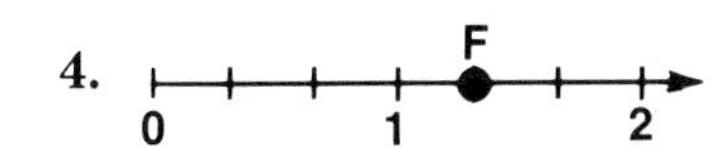


5.

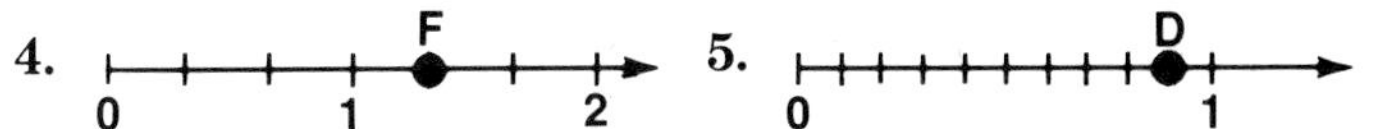


6.

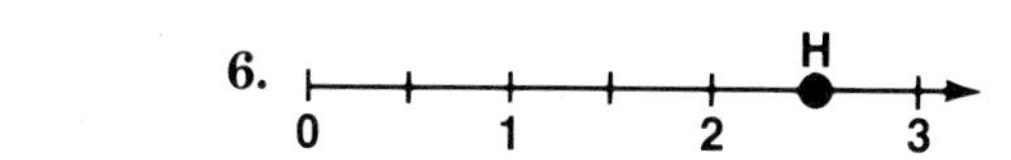


## HOMEWORK EXERCISES

1. Name the numerator and the denominator of each fraction.

   a. $\frac{17}{19}$ b. $\frac{22}{23}$ c. $\frac{8}{27}$ d. $\frac{14}{37}$ e. $\frac{193}{111}$ f. $\frac{85}{85}$ g. $\frac{192}{250}$ h. $\frac{72}{70}$ i. $\frac{0}{100}$

2. Name each of the *shaded* regions as a fraction.

   a. 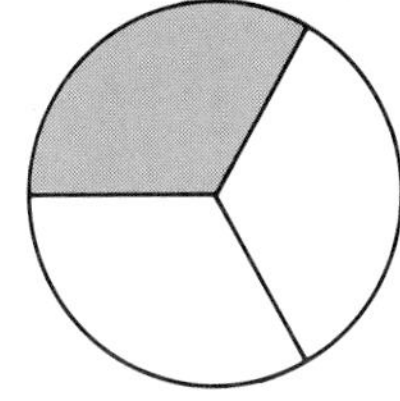

   b. 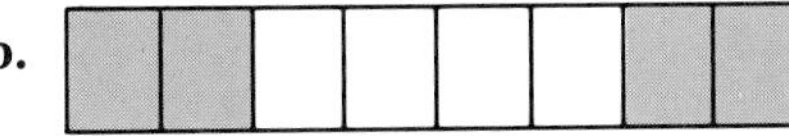

   c. 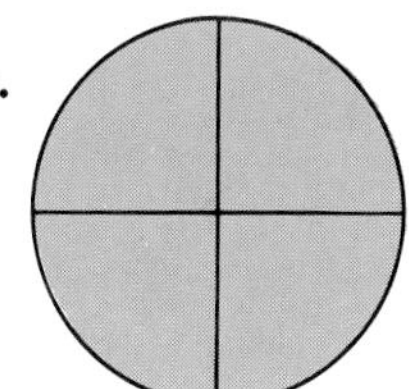

   d. 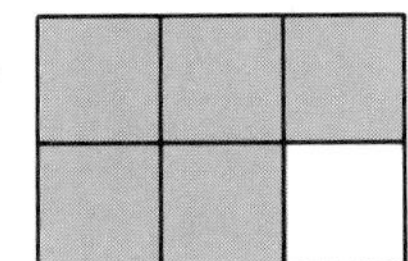

   e. 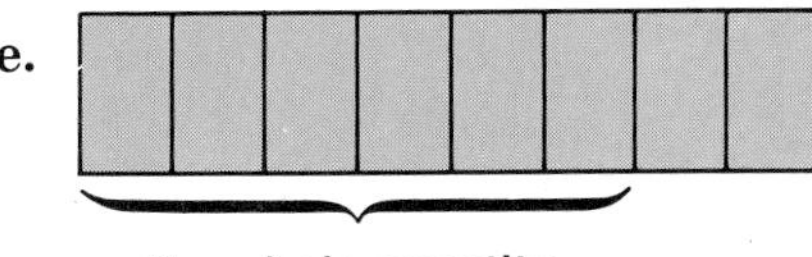


   f.  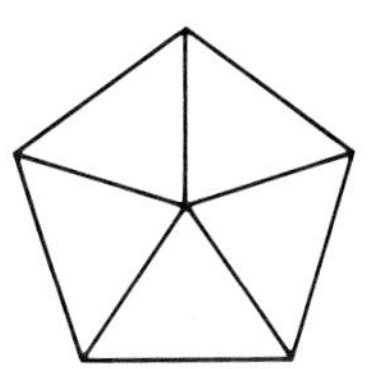

3. Tell whether each fraction is proper or improper.

   a. $\frac{5}{7}$ b. $\frac{3}{4}$ c. $\frac{0}{16}$ d. $\frac{17}{15}$ e. $\frac{12}{21}$ f. $\frac{21}{12}$ g. $\frac{75}{75}$ h. $\frac{75}{74}$

4. Replace ? by $<$, $>$, or $=$ to make a true comparison with 1.

   a. $\frac{9}{11}$ ? 1 b. $\frac{0}{6}$ ? 1 c. $\frac{7}{8}$ ? 1 d. $\frac{8}{7}$ ? 1 e. $\frac{20}{17}$ ? 1 f. $\frac{17}{20}$ ? 1 g. $\frac{52}{52}$ ? 1

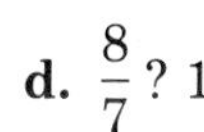

5. Explain why an improper fraction must always be greater than a proper fraction.

**6.** Use a rational number to name each point on the number lines shown.

a. B (0, 1) b. C (0, 1) c. E (0, 1, 2)

d. F (0, 1, 2) e. G (0, 1, 2, 3) f. H (0, 1, 2)

**7.** The fraction that represents the point on the number line shown is

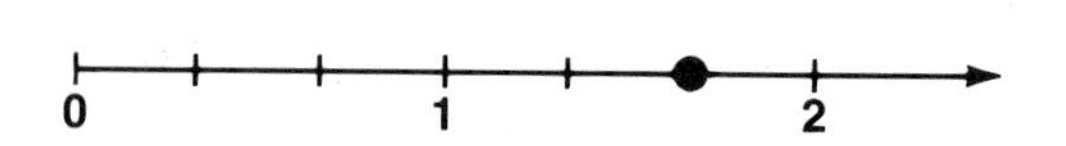


(a) $\frac{5}{6}$ (b) $\frac{5}{3}$ (c) $\frac{7}{4}$ (d) $\frac{5}{5}$

## SPIRAL REVIEW EXERCISES

**1.** Round 27,592 to the nearest thousand.

**2.** The product of $-8$ and $-7$ is
(a) $-15$ (c) $+56$
(b) $+15$ (d) $-56$

**3.** Which comparison is true?
(a) $-11 < -3$ (c) $-5 > 0 > 2$
(b) $-25 > -19$ (d) $0 < 5 < -10$

**4.** The value of $9 \times 4 - 5 \times 3$ is
(a) 93 (b) 51 (c) 41 (d) 21

**5.** Find the greatest common factor of 24 and 60.

**6.** In the month of April, Ms. Johnson earned $1,746 and Mr. Frank earned $1,498. How much more did Ms. Johnson earn?

**7.** Which whole number is prime?
(a) 1 (b) 15 (c) 17 (d) 21

**8.** Fifty thousand, three hundred seven written as a numeral is
(a) 5,307 (c) 53,007
(b) 50,307 (d) 53,070

**9.** Mr. Gorham bought 6 shirts at $12 a shirt and 4 ties at $9 a tie. What is the total amount of money that he spent?

**10.** Francine uses 18 strips of balsa wood to make a model. She bought 2 boxes of balsa wood that contain 36 strips each. How many models can she make?

## CHALLENGE QUESTION

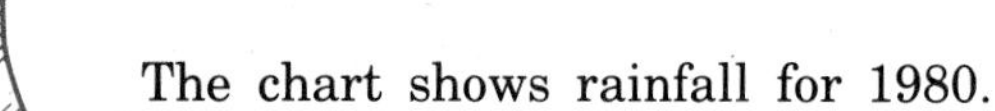

The chart shows rainfall for 1980.

**a.** 1980 had 366 days. Express as a fraction the part of the year shown on the chart.

**b.** In 1980, there was a total of 13 inches of rain. Express as a fraction the part of the yearly rain that fell during March and May together.

| Month | Days in Month | Inches of Rain |
|---|---|---|
| February | 29 | 1 |
| March | 31 | 2 |
| April | 30 | 3 |
| May | 31 | 1 |

# UNIT 4–2 Mixed Numbers

## THE MEANING OF A MIXED NUMBER

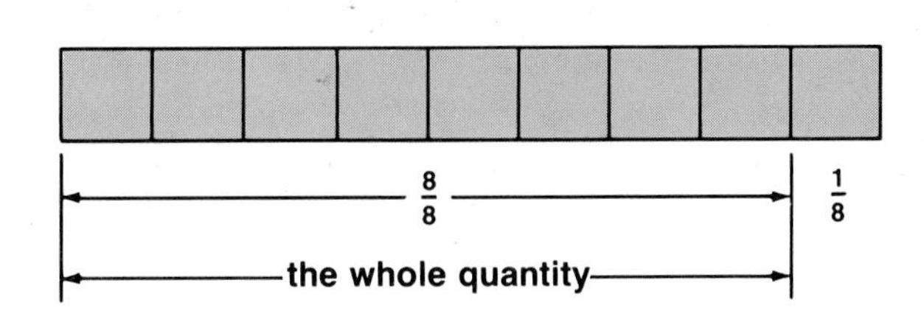


1. From the diagram, we can see that the improper fraction $\frac{9}{8}$ is equal to $\frac{8}{8} + \frac{1}{8}$ or $1 + \frac{1}{8}$. A simpler way to write $1 + \frac{1}{8}$ is to use the mixed number $1\frac{1}{8}$ (read as "one and one-eighth").
2. A *mixed number* represents the sum of a whole number and a proper fraction.

$$9\frac{1}{8} = 9 + \frac{1}{8}$$

whole-number part (9) — fraction part ($\frac{1}{8}$)

**EXAMPLE 1** Write each sum as a mixed number.

| | *Sum* | *Mixed Number* |
|---|---|---|
| **a.** | $3 + \frac{7}{8}$ | $3\frac{7}{8}$ |
| **b.** | $7 + \frac{5}{16}$ | $7\frac{5}{16}$ |
| **c.** | $12 + \frac{9}{11}$ | $12\frac{9}{11}$ |

**EXAMPLE 2** Name the whole-number part and the fraction part of each mixed number.

| | *Mixed Number* | *Whole-Number Part* | *Fraction Part* |
|---|---|---|---|
| **a.** | $7\frac{2}{9}$ | 7 | $\frac{2}{9}$ |
| **b.** | $12\frac{3}{5}$ | 12 | $\frac{3}{5}$ |
| **c.** | $19\frac{11}{17}$ | 19 | $\frac{11}{17}$ |
| **d.** | $65\frac{110}{111}$ | 65 | $\frac{110}{111}$ |

## CLASS EXERCISES

1. Write each sum as a mixed number.

a. $6 + \frac{5}{9}$ b. $3 + \frac{6}{13}$ c. $8 + \frac{3}{7}$ d. $10 + \frac{1}{6}$ e. $15 + \frac{8}{11}$ f. $19 + \frac{5}{12}$

2. Name the whole-number part and the fraction part of each mixed number.

a. $5\frac{3}{4}$ b. $6\frac{7}{12}$ c. $7\frac{9}{11}$ d. $10\frac{5}{8}$ e. $4\frac{2}{5}$ f. $17\frac{8}{19}$ g. $11\frac{12}{19}$ h. $23\frac{7}{8}$

## CHANGING AN IMPROPER FRACTION INTO A MIXED NUMBER

### THE MAIN IDEA

To change an improper fraction into an equivalent mixed number, divide the denominator of the improper fraction into the numerator.

The quotient will be the whole-number part of the mixed number.

The $\frac{\text{remainder}}{\text{divisor}}$ will be the fraction part of the mixed number.

**EXAMPLE 3** Change $\frac{19}{8}$ into a mixed number.

Divide the denominator 8 into the numerator 19.

$8\overline{)19}$

Obtain a quotient and a remainder.

$$\begin{array}{r} 2 \leftarrow \text{quotient} \\ 8\overline{)19} \\ \underline{16} \\ 3 \leftarrow \text{remainder} \end{array}$$

The quotient is the whole-number part of the mixed number, and the $\frac{\text{remainder}}{\text{divisor}}$ is the fraction part of the mixed number.

$\frac{19}{8} = 2\frac{3}{8}$ *Ans.*

## CLASS EXERCISES

Change each improper fraction into an equivalent mixed number.

1. $\frac{15}{2}$ 2. $\frac{17}{5}$ 3. $\frac{23}{4}$ 4. $\frac{5}{3}$ 5. $\frac{81}{7}$ 6. $\frac{109}{9}$ 7. $\frac{58}{5}$ 8. $\frac{143}{10}$ 9. $\frac{99}{8}$ 10. $\frac{120}{11}$

## CHANGING A MIXED NUMBER INTO AN IMPROPER FRACTION

### THE MAIN IDEA

To change a mixed number into an equivalent improper fraction:

1. Multiply the whole-number part by the denominator of the fraction part.
2. Add this product to the numerator of the fraction part.
3. The improper fraction is:

$$\frac{\text{the sum obtained in Step 2}}{\text{the denominator of the fraction part of the mixed number}}$$

A good way to remember this procedure is to start at the denominator and work in a clockwise direction as follows:

$3\frac{2}{5}$ ← start (+, ×)

$$3\frac{2}{5} = \frac{5 \times 3 + 2}{5} = \frac{17}{5}$$

**EXAMPLE 4** Change $5\frac{3}{4}$ into an equivalent improper fraction.

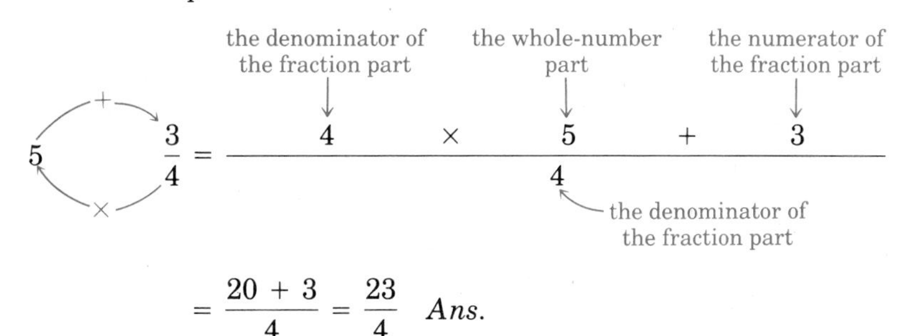


$$5\frac{3}{4} = \frac{4 \times 5 + 3}{4}$$

$$= \frac{20 + 3}{4} = \frac{23}{4} \quad \textit{Ans.}$$

### CLASS EXERCISES

Change each mixed number into an equivalent improper fraction.

**1.** $4\frac{1}{9}$ **2.** $3\frac{4}{7}$ **3.** $1\frac{7}{13}$ **4.** $8\frac{4}{5}$ **5.** $6\frac{7}{10}$ **6.** $9\frac{5}{7}$ **7.** $6\frac{9}{11}$ **8.** $12\frac{3}{4}$

**9.** $20\frac{2}{3}$ **10.** $18\frac{2}{5}$

## HOMEWORK EXERCISES

1. Write each sum as a mixed number.

   a. $3 + \frac{2}{7}$ b. $5 + \frac{2}{3}$ c. $7 + \frac{3}{5}$ d. $2 + \frac{15}{16}$ e. $11 + \frac{4}{9}$ f. $9 + \frac{3}{11}$ g. $10 + \frac{11}{12}$

2. Name the whole-number part and the fraction part of each mixed number.

   a. $1\frac{5}{8}$ b. $3\frac{4}{5}$ c. $9\frac{2}{7}$ d. $7\frac{8}{11}$ e. $10\frac{1}{2}$ f. $12\frac{7}{8}$ g. $20\frac{15}{17}$ h. $42\frac{17}{25}$

3. Change each improper fraction into an equivalent mixed number.

   a. $\frac{5}{4}$ b. $\frac{9}{5}$ c. $\frac{11}{2}$ d. $\frac{20}{3}$ e. $\frac{15}{7}$ f. $\frac{16}{3}$ g. $\frac{25}{8}$ h. $\frac{17}{2}$

   i. $\frac{26}{5}$ j. $\frac{35}{6}$ k. $\frac{42}{9}$ l. $\frac{56}{11}$ m. $\frac{107}{20}$ n. $\frac{91}{9}$ o. $\frac{135}{11}$

4. Change each mixed number into an equivalent improper fraction.

   a. $1\frac{3}{4}$ b. $1\frac{7}{16}$ c. $2\frac{5}{8}$ d. $3\frac{4}{7}$ e. $5\frac{9}{10}$ f. $8\frac{2}{3}$ g. $10\frac{1}{2}$ h. $14\frac{3}{5}$

   i. $20\frac{5}{6}$ j. $9\frac{7}{16}$ k. $6\frac{10}{11}$ l. $11\frac{5}{9}$ m. $24\frac{1}{3}$ n. $30\frac{5}{7}$ o. $10\frac{10}{13}$

5. $\frac{19}{3}$ is equivalent to (a) $4\frac{2}{3}$ (b) $5\frac{1}{3}$ (c) $6\frac{1}{3}$ (d) $6\frac{2}{3}$

6. $6\frac{3}{4}$ is equivalent to (a) $\frac{27}{4}$ (b) $\frac{25}{4}$ (c) $\frac{25}{3}$ (d) $\frac{22}{3}$

## SPIRAL REVIEW EXERCISES

1. Find the product of 209 and 96.

2. Divide: $28\overline{)11{,}396}$

3. What is the fraction that represents the shaded region in the diagram?

   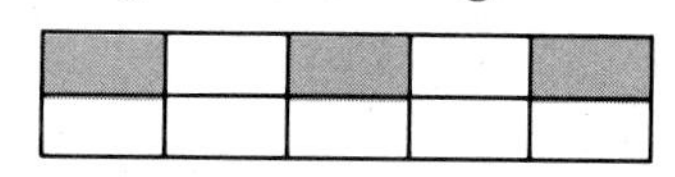

   (a) $\frac{1}{3}$ (b) $\frac{3}{10}$ (c) $\frac{1}{4}$ (d) $\frac{1}{5}$

4. Subtract 738 from 926.

5. Felix earned \$29 on Monday, \$42 on Tuesday, \$52 on Wednesday, \$18 on Thursday, and \$37 on Friday. What is the total amount of money he earned that week?

6. What is the rational number that is represented on the number line shown?

   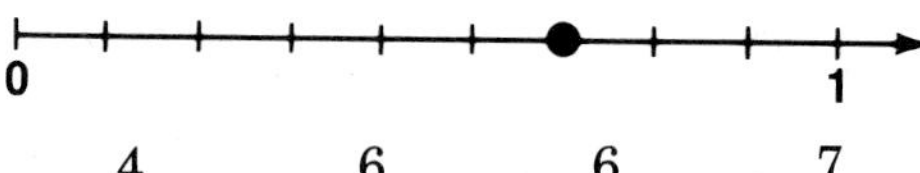


   (a) $\frac{4}{9}$ (b) $\frac{6}{10}$ (c) $\frac{6}{9}$ (d) $\frac{7}{9}$

7. Perform each of the indicated operations.

   a. $-18 + (+9)$ b. $-18 \times (+9)$

   c. $-18 - (+9)$ d. $-18 \div (+9)$

8. In a nine-inning baseball game, Jim pitched four innings. What fractional part of the game did he pitch?

9. Last year, Ms. Pepper earned \$11,856. If she worked 52 weeks at the same wage, how much did she earn each week?

10. A baker had 126 pounds of flour. He made 50 cakes and used 2 pounds of flour for each cake. How much flour did he have left?

11. Evaluate: $58 - 4 \times 9$

12. At the beginning of a trip, the odometer on Mrs. Wilson's car read 32,897 miles. At the end of the trip, the odometer showed 33,124 miles. How many miles were traveled?

## CHALLENGE QUESTION

Using each of the three numbers 2, 7, and 9 exactly once in each answer, write:

**a.** three proper fractions **b.** three improper fractions **c.** three mixed numbers

# UNIT 4-3 Equivalent Fractions

## THE MEANING OF EQUIVALENT FRACTIONS

### THE MAIN IDEA

1. Fractions that represent the same quantity are called *equivalent fractions.*
$\frac{3}{6}, \frac{4}{8}, \frac{5}{10}$, etc. are equivalent fractions since each represents the same amount, $\frac{1}{2}$ of a whole quantity:

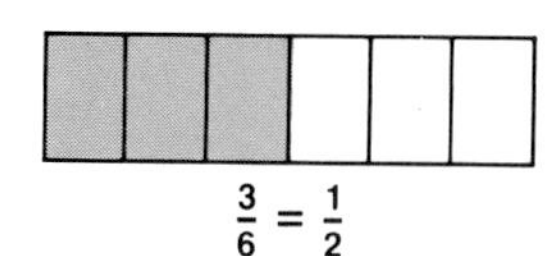

$\frac{3}{6} = \frac{1}{2}$

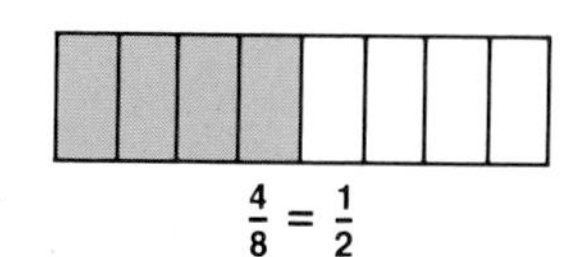

$\frac{4}{8} = \frac{1}{2}$

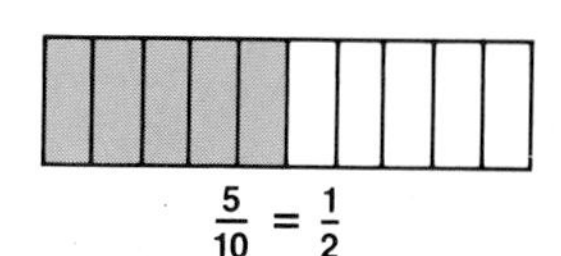

$\frac{5}{10} = \frac{1}{2}$

2. If we multiply or divide both the numerator and the denominator of a fraction by the same nonzero number, an equivalent fraction will result:

$$\frac{1 \times 3}{2 \times 3} = \frac{3}{6} \qquad \frac{3}{6} \text{ is equivalent to } \frac{1}{2}$$

$$\frac{9 \div 3}{12 \div 3} = \frac{3}{4} \qquad \frac{3}{4} \text{ is equivalent to } \frac{9}{12}$$

**EXAMPLE 1** Find the fraction with a denominator of 21 that is equivalent to $\frac{2}{3}$.

$$\frac{2}{3} = \frac{?}{21}$$

Notice that we must multiply 3 by 7 in order to obtain 21. Therefore, we must multiply 2 by 7 to obtain the numerator of the equivalent fraction.

$$\frac{2 \times 7}{3 \times 7} = \frac{14}{21} \quad \textit{Ans.}$$

**EXAMPLE 2** Which fraction is *not* equivalent to $\frac{3}{5}$?

| | *Fraction* | *Has the Same Multiplier Been Used?* | *Is the Fraction Equivalent to $\frac{3}{5}$?* |
|---|---|---|---|
| (a) | $\frac{6}{10}$ | $\frac{3 \times 2}{5 \times 2}$ | yes |
| (b) | $\frac{12}{20}$ | $\frac{3 \times 4}{5 \times 4}$ | yes |
| (c) | $\frac{27}{45}$ | $\frac{3 \times 9}{5 \times 9}$ | yes |
| (d) | $\frac{18}{60}$ | $\frac{3 \times 6}{5 \times 12}$ | no |

*Answer:* (d)

## CLASS EXERCISES

1. Change each fraction into an equivalent fraction that has the denominator shown in parentheses.

   a. $\frac{1}{2}$, (40) b. $\frac{3}{8}$, (32) c. $\frac{2}{5}$, (35) d. $\frac{5}{16}$, (96) e. $\frac{7}{12}$, (36)

   f. $\frac{7}{9}$, (72) g. $\frac{4}{11}$, (66) h. $\frac{9}{10}$, (90)

2. Replace ? to make a true statement.

   a. $\frac{3}{5} = \frac{?}{20}$ b. $\frac{7}{11} = \frac{?}{55}$ c. $\frac{5}{8} = \frac{?}{72}$ d. $\frac{5}{12} = \frac{?}{144}$ e. $\frac{11}{20} = \frac{?}{80}$ f. $\frac{13}{50} = \frac{?}{150}$

3. Which fraction is not equivalent to $\frac{4}{9}$? (a) $\frac{12}{27}$ (b) $\frac{28}{63}$ (c) $\frac{32}{81}$ (d) $\frac{400}{900}$

4. Which fraction is not equivalent to $\frac{8}{15}$? (a) $\frac{16}{30}$ (b) $\frac{32}{75}$ (c) $\frac{24}{45}$ (d) $\frac{80}{150}$

5. Explain why $\frac{2}{3}$ and $\frac{12}{18}$ are names for the same number.

## DETERMINING IF TWO FRACTIONS ARE EQUIVALENT

### THE MAIN IDEA

A short way to tell if two fractions are equivalent is to find the ***cross products***.

1. If the cross products are equal, then the fractions are equivalent.
2. If the cross products are not equal, then the fractions are not equivalent.

**EXAMPLE 3** Are $\frac{3}{10}$ and $\frac{15}{50}$ equivalent?

$\frac{3}{10}$ ⟷ $\frac{15}{50}$

first cross product: $3 \times 50 = 150$
second cross product: $10 \times 15 = 150$

*Answer:* Since the cross products are equal, the fractions $\frac{3}{10}$ and $\frac{15}{50}$ are equivalent.

**EXAMPLE 4** Are $\frac{7}{14}$ and $\frac{5}{8}$ equivalent?

$\frac{7}{14}$ ⟷ $\frac{5}{8}$

first cross product: $7 \times 8 = 56$
second cross product: $14 \times 5 = 70$

*Answer:* Since the cross products are *not* equal, $\frac{7}{14}$ and $\frac{5}{8}$ are *not* equivalent.

## CLASS EXERCISES

In each pair, tell if the fractions are equivalent.

1. $\frac{2}{3}$ and $\frac{10}{15}$ 2. $\frac{7}{11}$ and $\frac{5}{8}$ 3. $\frac{30}{40}$ and $\frac{9}{12}$ 4. $\frac{8}{9}$ and $\frac{24}{25}$ 5. $\frac{3}{7}$ and $\frac{15}{28}$

6. $\frac{5}{10}$ and $\frac{15}{30}$ 7. $\frac{9}{10}$ and $\frac{27}{30}$ 8. $\frac{7}{11}$ and $\frac{20}{33}$ 9. $\frac{8}{30}$ and $\frac{4}{15}$

## HOMEWORK EXERCISES

1. Change each fraction into an equivalent fraction that has the denominator shown in parentheses.

a. $\frac{2}{3}$, (12) b. $\frac{7}{8}$, (16) c. $\frac{4}{5}$, (25) d. $\frac{5}{9}$, (72) e. $\frac{9}{16}$, (48) f. $\frac{11}{12}$, (72)

g. $\frac{3}{10}$, (80) h. $\frac{6}{11}$, (99) i. $\frac{4}{13}$, (52) j. $\frac{12}{25}$, (75) k. $\frac{17}{20}$, (100) l. $\frac{11}{30}$, (120)

2. Replace ? to make a true statement.

a. $\frac{4}{7} = \frac{?}{28}$ b. $\frac{3}{11} = \frac{?}{44}$ c. $\frac{7}{9} = \frac{?}{54}$ d. $\frac{3}{4} = \frac{?}{40}$ e. $\frac{11}{16} = \frac{?}{80}$

f. $\frac{5}{12} = \frac{?}{96}$ g. $\frac{9}{25} = \frac{?}{125}$ h. $\frac{13}{20} = \frac{?}{160}$ i. $\frac{17}{50} = \frac{?}{200}$

3. Which fraction is not equivalent to $\frac{1}{2}$? (a) $\frac{7}{14}$ (b) $\frac{15}{30}$ (c) $\frac{20}{40}$ (d) $\frac{30}{50}$

4. Which fraction is not equivalent to $\frac{7}{12}$? (a) $\frac{21}{36}$ (b) $\frac{30}{60}$ (c) $\frac{56}{96}$ (d) $\frac{63}{108}$

5. Explain why $\frac{3}{7}$ and $\frac{24}{56}$ are names for the same number.

6. Explain why $\frac{7}{10}$ and $\frac{42}{50}$ are *not* names for the same number.

7. In each pair, tell if the fractions are equivalent.

a. $\frac{1}{2}$ and $\frac{15}{30}$ b. $\frac{3}{4}$ and $\frac{27}{36}$ c. $\frac{5}{12}$ and $\frac{20}{46}$ d. $\frac{7}{11}$ and $\frac{70}{110}$ e. $\frac{7}{16}$ and $\frac{42}{92}$

f. $\frac{8}{15}$ and $\frac{40}{75}$ g. $\frac{7}{10}$ and $\frac{35}{60}$ h. $\frac{9}{13}$ and $\frac{18}{25}$ i. $\frac{7}{24}$ and $\frac{13}{48}$ j. $\frac{9}{40}$ and $\frac{6}{30}$

k. $\frac{20}{25}$ and $\frac{12}{15}$ l. $\frac{14}{20}$ and $\frac{21}{30}$ m. $\frac{12}{18}$ and $\frac{6}{8}$ n. $\frac{27}{36}$ and $\frac{12}{16}$ o. $\frac{25}{40}$ and $\frac{20}{36}$

## SPIRAL REVIEW EXERCISES

1. Change $\frac{47}{3}$ into an equivalent mixed number.

2. Change $5\frac{7}{15}$ into an equivalent improper fraction.

3. Perform the indicated operation.

   a. $-8 + (-6)$  b. $-12 - (-9)$

   c. $-8 \times (-4)$  d. $-8 \div (-4)$

4. Round 298,472 to the nearest ten thousand.

5. John bought $\frac{99}{10}$ gallons of gasoline. Express the number of gallons as a mixed number.

6. At a fund-raising dinner, each person paid $25. If the total amount collected was $4,050, how many people attended the dinner?

7. A dealer bought 14 cars that cost $6,500 each. If he sold the cars at $6,800 each, what was the total amount of his profit?

8. Replace ? by $<$, $>$, or $=$ to make a true statement.

   a. $5 + 10 \times 2$ ? $(5 + 10) \times 2$

   b. $8 + (3 \times 2)$ ? $8 + 3 \times 2$

   c. $(15 + 6) \times 5 + 8$ ? $15 + 6 \times 5 + 8$

### CHALLENGE QUESTION

Mike walked $\frac{7}{8}$ of a mile, while Marilyn walked $\frac{23}{24}$ of a mile. Did they walk the same distance? If not, who walked farther?

# UNIT 4–4 Comparing Fractions; Least Common Multiple and Lowest Common Denominator

## COMPARING LIKE FRACTIONS

### THE MAIN IDEA

1. *Like fractions* are fractions that have a *common* denominator:

$\frac{2}{7}$ and $\frac{5}{7}$ are like fractions

2. To compare two like fractions, look at the numerators. The fraction with the larger numerator is the larger fraction:

$\frac{5}{7}$ is larger than $\frac{2}{7}$

**EXAMPLE 1** Select the largest fraction from $\frac{9}{31}$, $\frac{21}{31}$, $\frac{7}{31}$, and $\frac{4}{31}$.

Since the fractions have a common denominator, the fractions are like fractions.

Since 21 is the largest numerator, $\frac{21}{31}$ is the largest fraction.

*Answer:* $\frac{21}{31}$

**EXAMPLE 2** Select the smallest fraction from $\frac{8}{41}$, $\frac{21}{41}$, and $\frac{9}{41}$.

Since 8 is the smallest numerator of the like fractions, $\frac{8}{41}$ is the smallest fraction.

*Answer:* $\frac{8}{41}$

**EXAMPLE 3** Replace ? by $<$ or $>$ to make a true comparison.

| | *Like Fractions* | *True Comparison* |
|---|---|---|
| **a.** | $\frac{5}{19} ? \frac{11}{19}$ | $\frac{5}{19} < \frac{11}{19}$ |
| **b.** | $\frac{18}{47} ? \frac{21}{47}$ | $\frac{18}{47} < \frac{21}{47}$ |
| **c.** | $\frac{17}{35} ? \frac{16}{35}$ | $\frac{17}{35} > \frac{16}{35}$ |
| **d.** | $\frac{91}{99} ? \frac{93}{99}$ | $\frac{91}{99} < \frac{93}{99}$ |

### CLASS EXERCISES

**1.** Select the larger fraction from each pair of fractions.

**a.** $\frac{7}{11}$ and $\frac{5}{11}$ **b.** $\frac{5}{8}$ and $\frac{7}{8}$ **c.** $\frac{9}{57}$ and $\frac{12}{57}$ **d.** $\frac{11}{15}$ and $\frac{9}{15}$ **e.** $\frac{34}{91}$ and $\frac{42}{91}$

**2.** Select the smallest fraction from each group of fractions.

**a.** $\frac{5}{7}$ and $\frac{6}{7}$ **b.** $\frac{9}{11}$ and $\frac{7}{11}$ **c.** $\frac{14}{17}$ and $\frac{16}{17}$ **d.** $\frac{5}{27}$, $\frac{8}{27}$, and $\frac{7}{27}$ **e.** $\frac{17}{35}$, $\frac{21}{35}$, and $\frac{32}{35}$

**3.** Replace ? by < or > to make a true comparison.

**a.** $\frac{3}{11} ? \frac{7}{11}$ **b.** $\frac{9}{31} ? \frac{6}{31}$ **c.** $\frac{14}{81} ? \frac{26}{81}$ **d.** $\frac{43}{44} ? \frac{37}{44}$ **e.** $\frac{59}{100} ? \frac{61}{100}$ **f.** $\frac{77}{112} ? \frac{59}{112}$

## THE MEANING OF LEAST COMMON MULTIPLE AND LOWEST COMMON DENOMINATOR

### THE MAIN IDEA

1. The ***least common multiple*** (LCM) of two or more whole numbers is the *smallest* whole number that is a multiple of each.

   Some multiples of 4 are: 4, 8, 12, 16, 20, 24, 28, 32, 36.

   Some multiples of 6 are: 6, 12, 18, 24, 30, 36.

   Of those shown, the common multiples of 4 and 6 are 12, 24, and 36. The *least* common multiple, LCM, of 4 and 6 is 12.

2. To find the LCM of two or more numbers, work with the largest number.

   a. If the other numbers are exact divisors of the largest number, then the largest number is the LCM.

   b. If the largest number is not the LCM, then try, in order, multiples of the largest number, until you find a multiple that all the other numbers divide exactly.

3. The ***lowest common denominator*** (LCD) of two or more fractions is the LCM of the denominators.

   The LCD of $\frac{1}{4}$ and $\frac{5}{6}$ is 12, the LCM of 4 and 6.

**EXAMPLE 4** Find the LCM of 14 and 21.

Work with the larger number, 21, and see if 14 is an exact divisor of 21.

$$\begin{array}{r} 1 \\ 14\overline{)21} \\ \underline{14} \\ 7 \end{array} \leftarrow \text{remainder not 0}$$

Since 14 is not an exact divisor of 21, try 21 × 2 or 42, the next multiple of 21, and see if 14 is an exact divisor of 42.

$$\begin{array}{r} 3 \\ 14\overline{)42} \\ \underline{42} \\ 0 \end{array} \checkmark$$

*Answer:* 42 is the LCM of 14 and 21.

**EXAMPLE 5** Find the LCD of $\frac{4}{9}$ and $\frac{5}{12}$.

To find the LCD, find the LCM of 9 and 12.
Work with the larger number, 12.
Try multiples of 12: 9 is not an exact divisor of 12.
9 is not an exact divisor of 24.
9 is an exact divisor of 36.

*Answer:* 36 is the LCD of $\frac{4}{9}$ and $\frac{5}{12}$.

**EXAMPLE 6** Find the LCD of $\frac{1}{4}$, $\frac{2}{5}$, and $\frac{7}{20}$.

Find the LCM of 4, 5, and 20.

Work with the largest number, 20. Since 4 is an exact divisor of 20 and 5 is an exact divisor of 20, 20 is the LCD.

*Answer:* 20 is the LCD of $\frac{1}{4}$, $\frac{2}{5}$, and $\frac{7}{20}$.

## CLASS EXERCISES

1. Find the least common multiple (LCM) for each group of whole numbers.

**a.** 4 and 8 **b.** 6 and 15 **c.** 7 and 5 **d.** 27 and 54

**e.** 6, 9, and 12 **f.** 8, 10, and 20

2. Find the lowest common denominator (LCD) for each group of fractions.

**a.** $\frac{3}{5}$ and $\frac{9}{35}$ **b.** $\frac{1}{3}$ and $\frac{5}{12}$ **c.** $\frac{1}{6}$ and $\frac{7}{8}$ **d.** $\frac{7}{10}$ and $\frac{3}{4}$

**e.** $\frac{9}{24}$, $\frac{11}{36}$, and $\frac{7}{12}$ **f.** $\frac{8}{15}$, $\frac{11}{30}$, and $\frac{3}{5}$

## COMPARING UNLIKE FRACTIONS

### THE MAIN IDEA

1. *Unlike fractions* are fractions that do not have a common denominator.
2. To compare unlike fractions:
   a. Find the LCD for the fractions.
   b. Change each fraction into an equivalent fraction that has the LCD as the denominator.
   c. Compare the resulting like fractions.

**EXAMPLE 7** Which is larger, $\frac{3}{7}$ or $\frac{11}{28}$?

Find the LCD.

Since 7 is an exact divisor of 28, the LCD is 28.

Rewrite $\frac{3}{7}$ as an equivalent fraction with denominator 28.

$$\frac{3 \times 4}{7 \times 4} = \frac{12}{28}$$

Since $\frac{12}{28}$ and $\frac{11}{28}$ are like fractions, the fraction with the larger numerator is the larger fraction.

*Answer:* $\frac{12}{28}$ or $\frac{3}{7}$ is the larger fraction.

**EXAMPLE 8** Roger walked $\frac{7}{10}$ of a mile and Grace walked $\frac{5}{8}$ of a mile. Who walked the shorter distance?

To find the LCD, try multiples of 10. 8 is not an exact divisor of 10, 20, or 30. Since 8 is an exact divisor of 40, the LCD is 40.

Rewrite $\frac{7}{10}$ and $\frac{5}{8}$ with denominator 40.

$$\frac{7 \times 4}{10 \times 4} = \frac{28}{40}$$

$$\frac{5 \times 5}{8 \times 5} = \frac{25}{40}$$

Compare $\frac{28}{40}$ and $\frac{25}{40}$.

The smaller fraction is $\frac{25}{40}$ or $\frac{5}{8}$.

*Answer:* Grace walked the shorter distance.

## CLASS EXERCISES

**1.** Select the largest fraction from each group of fractions.

**a.** $\frac{2}{5}$ and $\frac{12}{35}$ **b.** $\frac{5}{6}$ and $\frac{29}{36}$ **c.** $\frac{7}{12}$ and $\frac{5}{8}$ **d.** $\frac{15}{36}$ and $\frac{21}{24}$

**e.** $\frac{3}{5}, \frac{7}{15}$, and $\frac{28}{45}$ **f.** $\frac{5}{7}, \frac{13}{14}$, and $\frac{10}{21}$

**2.** Replace ? by < or > to make a true comparison.

**a.** $\frac{2}{3} ? \frac{3}{4}$ **b.** $\frac{3}{8} ? \frac{7}{16}$ **c.** $\frac{23}{25} ? \frac{7}{10}$ **d.** $\frac{13}{20} ? \frac{5}{8}$ **e.** $\frac{17}{36} ? \frac{5}{12}$ **f.** $\frac{3}{5} ? \frac{5}{9}$

**3.** James ran $\frac{8}{10}$ of a mile, and Mario ran $\frac{7}{8}$ of a mile. Who ran farther?

**4.** One bottle contained $\frac{3}{4}$ of a quart of juice, and another bottle contained $\frac{6}{10}$ of a quart of juice. Which bottle contained less juice?

## HOMEWORK EXERCISES

1. Select the largest fraction from each group of fractions.

   a. $\frac{3}{17}$ and $\frac{9}{17}$ b. $\frac{5}{16}$ and $\frac{3}{16}$ c. $\frac{4}{9}$ and $\frac{7}{9}$ d. $\frac{21}{51}, \frac{13}{51}$, and $\frac{24}{51}$ e. $\frac{47}{101}, \frac{39}{101}$, and $\frac{42}{101}$

2. Select the smallest fraction from each group of fractions.

   a. $\frac{9}{11}$ and $\frac{6}{11}$ b. $\frac{7}{19}$ and $\frac{10}{19}$ c. $\frac{14}{27}$ and $\frac{12}{27}$ d. $\frac{18}{67}, \frac{22}{67}$, and $\frac{20}{67}$ e. $\frac{81}{91}, \frac{85}{91}$, and $\frac{87}{91}$

3. Replace ? by < or > to make a true comparison.

   a. $\frac{3}{5} ? \frac{2}{5}$ b. $\frac{5}{7} ? \frac{6}{7}$ c. $\frac{9}{17} ? \frac{12}{17}$ d. $\frac{18}{25} ? \frac{21}{25}$ e. $\frac{75}{91} ? \frac{72}{91}$ f. $\frac{50}{99} ? \frac{53}{99}$

   g. $\frac{81}{111} ? \frac{71}{111}$ h. $\frac{92}{101} ? \frac{97}{101}$

4. Find the least common multiple (LCM) for each group of whole numbers.

   a. 3 and 6 b. 7 and 21 c. 8 and 12 d. 20 and 15 e. 18 and 27

   f. 24 and 16 g. 10, 20, and 25 h. 6, 15, and 18 i. 4, 6, and 24

5. Find the lowest common denominator (LCD) for each group of fractions.

   a. $\frac{5}{12}$ and $\frac{3}{4}$ b. $\frac{7}{20}$ and $\frac{4}{5}$ c. $\frac{11}{18}$ and $\frac{25}{36}$ d. $\frac{2}{3}$ and $\frac{1}{2}$ e. $\frac{5}{8}$ and $\frac{5}{6}$

   f. $\frac{7}{10}$ and $\frac{8}{15}$ g. $\frac{5}{6}, \frac{2}{9}$, and $\frac{7}{18}$ h. $\frac{7}{12}, \frac{1}{4}$, and $\frac{3}{8}$ i. $\frac{5}{6}, \frac{11}{12}$, and $\frac{1}{2}$

6. Select the largest fraction from each group of fractions.

   a. $\frac{2}{3}$ and $\frac{7}{12}$ b. $\frac{3}{8}$ and $\frac{1}{4}$ c. $\frac{5}{18}$ and $\frac{8}{27}$ d. $\frac{11}{12}$ and $\frac{4}{5}$

   e. $\frac{3}{4}, \frac{11}{16}$, and $\frac{5}{8}$ f. $\frac{3}{5}, \frac{2}{3}$, and $\frac{7}{10}$

7. Select the smallest fraction from each group of fractions.

   a. $\frac{5}{6}$ and $\frac{13}{18}$ b. $\frac{7}{9}$ and $\frac{29}{36}$ c. $\frac{2}{3}$ and $\frac{3}{5}$ d. $\frac{5}{8}$ and $\frac{7}{10}$

   e. $\frac{8}{27}, \frac{7}{18}$, and $\frac{1}{3}$ f. $\frac{5}{6}, \frac{2}{3}$, and $\frac{7}{8}$

8. Replace ? by < or > to make a true comparison.

   a. $\frac{3}{5} ? \frac{7}{10}$ b. $\frac{1}{2} ? \frac{3}{8}$ c. $\frac{11}{20} ? \frac{3}{4}$ d. $\frac{9}{16} ? \frac{2}{3}$ e. $\frac{2}{5} ? \frac{3}{8}$ f. $\frac{7}{9} ? \frac{5}{7}$

   g. $\frac{7}{12} ? \frac{3}{5}$ h. $\frac{5}{18} ? \frac{1}{4}$ i. $\frac{7}{15} ? \frac{5}{12}$

**9.** Helen memorized $\frac{4}{5}$ of a vocabulary list, and Bob memorized $\frac{5}{8}$ of the list. Who memorized more of the list?

**10.** Which oil tank is less full, one that is $\frac{5}{16}$ full or one that is $\frac{1}{4}$ full?

## SPIRAL REVIEW EXERCISES

**1.** Change $\frac{29}{7}$ to an equivalent mixed number.

**2.** Perform each of the indicated operations and check.

**a.** $2{,}752 + 947$ **b.** $837 - 592$

**c.** $58 \times 43$ **d.** $507 \div 3$

**3.** Evaluate: $15 + 9 \times 3 - 8$

**4.** Write five thousand, twenty-one as a numeral.

**5.** The value of 9 in 792,465 is
(a) 900,000 (c) 9,000
(b) 90,000 (d) 900

**6.** Replace ? by < or > to make a true comparison.

**a.** $-9 \; ? \; -10$ **b.** $-115 \; ? \; 2$

**c.** $0 \; ? \; -75$ **d.** $-1{,}001 \; ? \; -1{,}010$

**7.** Subtract $-50$ from 92.

**8.** The greatest common factor of 56 and 28 is
(a) 7 (b) 14 (c) 28 (d) 56

**9.** Juan pitched $\frac{22}{3}$ innings without allowing a hit. Express the number of hitless innings as a mixed number.

**10.** Mr. Jones, a traveling salesman, is paid \$4 a mile traveling expenses by his company. If he traveled 178 miles, how much was he paid?

**11.** A baseball bat costs \$11 and a baseball glove costs \$24. What is the total cost of supplying bats and gloves for the 15 members of the Bantam little league team?

## CHALLENGE QUESTION

If 1 is added to the numerator and to the denominator of the fraction $\frac{7}{9}$, is the resulting fraction less than, equal to, or greater than the original fraction?

# UNIT 4–5 Simplifying Fractions

## THE MAIN IDEA

1. To write a fraction in *simplest form* or *lowest terms:*
   a. Find a number that is a common factor (exact divisor) of the numerator and the denominator.
   b. Divide the numerator and the denominator by this common factor.
   c. Repeat this process until 1 is the only common factor of the numerator and the denominator.
2. If the numerator and the denominator of a fraction are each divided by the greatest common factor (GCF), the resulting fraction will be in simplest form.

**EXAMPLE 1** Simplify: $\frac{15}{21}$

Divide the numerator and the denominator by the common factor 3. $\frac{15 \div 3}{21 \div 3} = \frac{5}{7}$

Since 1 is the only common factor of 5 and 7, the fraction is in simplest form.

*Answer:* $\frac{15}{21} = \frac{5}{7}$

**EXAMPLE 2** Simplify: $\frac{24}{36}$

Divide the numerator and the denominator by the common factor 4. $\frac{24 \div 4}{36 \div 4} = \frac{6}{9}$

Divide again by the common factor 3. $\frac{6 \div 3}{9 \div 3} = \frac{2}{3}$

Since 1 is the only common factor of 2 and 3, the fraction is in simplest form.

*Answer:* $\frac{24}{36} = \frac{2}{3}$

We could have simplified $\frac{24}{36}$ in one step by dividing both the numerator and the denominator by 12, the greatest common factor of 24 and 36.

Dividing immediately by 12, the greatest common factor, shortens the problem. $\frac{24 \div 12}{36 \div 12} = \frac{2}{3}$

**EXAMPLE 3** For each fraction, find the GCF of the numerator and the denominator. Use the GCF to write the fraction in simplest form.

| | *Fraction* | *GCF of the Numerator and the Denominator* | *Fraction in Simplest Form* |
|---|---|---|---|
| **a.** | $\frac{8}{20}$ | 4 | $\frac{2}{5}$ |
| **b.** | $\frac{36}{48}$ | 12 | $\frac{3}{4}$ |
| **c.** | $\frac{63}{72}$ | 9 | $\frac{7}{8}$ |
| **d.** | $\frac{70}{210}$ | 70 | $\frac{1}{3}$ |

**EXAMPLE 4** A baseball team won 18 out of 27 games. Express in simplest form the fractional part of the games won.

**THINKING ABOUT THE PROBLEM**

The key words "out of" tell you to compare the numbers 18 and 27.

Translate the words into a fraction. $\frac{18}{27}$ ← number of games won / ← total number of games

Simplify the fraction by dividing the numerator and the denominator by the GCF. $\frac{18 \div 9}{27 \div 9} = \frac{2}{3}$

*Answer:* The team won $\frac{2}{3}$ of the games.

## CLASS EXERCISES

**1.** Name the GCF of the numerator and the denominator of each fraction. Then write each fraction in simplest form.

**a.** $\frac{4}{8}$ **b.** $\frac{42}{63}$ **c.** $\frac{24}{36}$ **d.** $\frac{56}{72}$ **e.** $\frac{77}{99}$ **f.** $\frac{120}{130}$ **g.** $\frac{60}{72}$ **h.** $\frac{16}{64}$ **i.** $\frac{80}{200}$ **j.** $\frac{70}{105}$

**2.** Simplify:

**a.** $\frac{3}{6}$ **b.** $\frac{11}{33}$ **c.** $\frac{8}{12}$ **d.** $\frac{20}{25}$ **e.** $\frac{9}{18}$ **f.** $\frac{6}{10}$ **g.** $\frac{12}{48}$ **h.** $\frac{21}{28}$ **i.** $\frac{75}{100}$ **j.** $\frac{88}{121}$

**3.** Jane delivers newspapers to 72 homes on her paper route. If she has already delivered papers to 60 homes, express as a fraction in simplest form the part of the route that she has completed.

**4.** Roberto has \$100 and plans to spend \$55 on a sports jacket. Express as a fraction in simplest form the part of his money that he plans to spend on the sports jacket.

## HOMEWORK EXERCISES

1. Name the GCF of the numerator and the denominator of each fraction. Then write the fraction in simplest form.

   a. $\frac{20}{25}$ b. $\frac{18}{36}$ c. $\frac{36}{48}$ d. $\frac{40}{60}$ e. $\frac{42}{54}$ f. $\frac{40}{64}$ g. $\frac{75}{100}$ h. $\frac{24}{32}$

   i. $\frac{72}{81}$ j. $\frac{48}{72}$ k. $\frac{75}{120}$ l. $\frac{50}{200}$ m. $\frac{90}{120}$ n. $\frac{54}{90}$ o. $\frac{80}{400}$

2. Simplify:

   a. $\frac{12}{14}$ b. $\frac{15}{20}$ c. $\frac{30}{48}$ d. $\frac{25}{50}$ e. $\frac{20}{24}$ f. $\frac{21}{28}$ g. $\frac{40}{100}$ h. $\frac{30}{32}$

   i. $\frac{48}{96}$ j. $\frac{50}{75}$ k. $\frac{12}{18}$ l. $\frac{77}{88}$ m. $\frac{42}{63}$ n. $\frac{70}{90}$ o. $\frac{40}{45}$

3. Franklin answered 42 of 50 questions correctly. Express as a fraction in simplest form the part of the test that he got correct.

4. Sheila is saving money to buy a stereo that costs $200. If she has saved $150, express as a fraction in simplest form the part of the money that she has already saved.

## SPIRAL REVIEW EXERCISES

1. Explain why $\frac{15}{28}$ is in simplest form.

2. Simplify each improper fraction. Do not change the result to a mixed number.

   a. $\frac{24}{18}$ b. $\frac{14}{10}$ c. $\frac{25}{20}$ d. $\frac{120}{16}$

3. Write each of your answers to exercise 2 as a mixed number.

4. Explain why $\frac{3}{7}$ cannot be changed to an equivalent fraction having a denominator of 24.

5. Find the least common multiple of 14 and 21.

6. $40 \div (-8)$ is
   (a) $-320$ (c) $+5$
   (b) $-5$ (d) $-6$

7. Which fraction is not equivalent to $\frac{5}{8}$?
   (a) $\frac{15}{24}$ (b) $\frac{35}{56}$ (c) $\frac{20}{32}$ (d) $\frac{45}{81}$

8. 27,563,418 rounded to the nearest hundred thousand is
   (a) 28,000,000 (c) 27,600,000
   (b) 27,000,000 (d) 27,500,000

9. Mr. Goodstone makes $8 profit for each tire sold. What is the amount of profit if 86 tires are sold?

10. A dealer plans to pack 2,080 ping-pong balls in 40 cases so that each case will contain the same number of ping-pong balls. How many balls will be in each case?

11. Write eight hundred thousand, two hundred five as a numeral.

12. The value of $5 + 8 \times 3 + 2$ is
    (a) 65 (b) 31 (c) 41 (d) 45

## CHALLENGE QUESTION

Match each fraction in Column I with the letter of the number line in Column II that represents the graph of the rational number in simplest form.

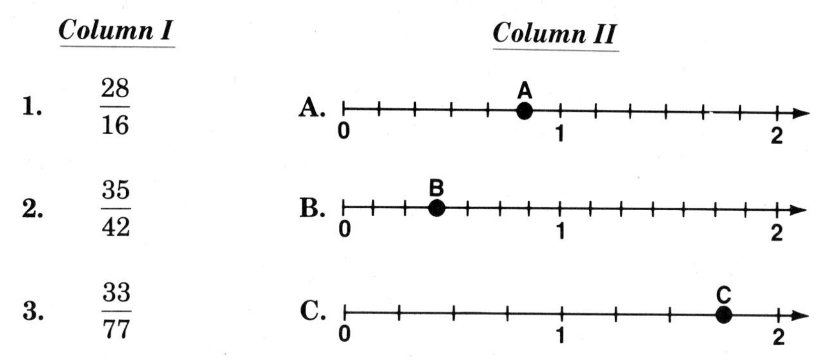

MODULE 5

## UNIT 5–1 Adding and Subtracting Like Fractions

### THE MAIN IDEA

Like fractions are fractions that have the same denominator. To add or subtract like fractions:

1. Add or subtract the numerators.
2. Keep the original denominator.
3. If necessary, simplify the resulting fraction.

**EXAMPLE 1** Add: $\frac{5}{17} + \frac{7}{17}$

Since the fractions are alike (have the same denominator), add the numerators and keep the original denominator:

$$\frac{5}{17} + \frac{7}{17} = \frac{5+7}{17} = \frac{12}{17} \quad \textit{Ans.}$$

**EXAMPLE 2** $\frac{13}{18} + \frac{5}{18} + \frac{7}{18}$ equals (a) $2\frac{5}{18}$ (b) $3\frac{7}{18}$ (c) $2\frac{7}{18}$ (d) $1\frac{7}{18}$

Since the fractions are alike, add the numerators and keep the original denominator:

$$\frac{13}{18} + \frac{5}{18} + \frac{7}{18} = \frac{13+5+7}{18} = \frac{25}{18}$$

Change the improper fraction $\frac{25}{18}$ into a mixed number:

$$\begin{array}{r} 1 \leftarrow \text{quotient} \\ \text{divisor} \rightarrow 18\overline{)25} \\ \underline{18} \\ 7 \leftarrow \text{remainder} \end{array} \qquad \frac{25}{18} = 1\frac{7}{18}$$

*Answer:* (d)

**EXAMPLE 3** John ran $\frac{29}{32}$ of a mile in the morning and $\frac{19}{32}$ of a mile in the evening. How much farther did he run in the morning?

**THINKING ABOUT THE PROBLEM**

The key words "how much farther" tell you to subtract to find how much greater one number is than another.

Write a number phrase. $\frac{29}{32} - \frac{19}{32}$

Since the fractions are alike, subtract the numerators and keep the original denominator.

$$\frac{29}{32} - \frac{19}{32} = \frac{29 - 19}{32}$$

$$= \frac{10}{32}$$

Simplify.

$$\frac{10 \div 2}{32 \div 2} = \frac{5}{16}$$

*Answer:* John ran $\frac{5}{16}$ of a mile farther in the morning.

**EXAMPLE 4** Andy trimmed $\frac{3}{16}$ of an inch from one side of a photograph and $\frac{9}{16}$ of an inch from the opposite side. What was the total amount that he trimmed from the photograph?

**THINKING ABOUT THE PROBLEM**

The key word "total" tells you to add the two fractions.

Write a number phrase.

$$\frac{3}{16} + \frac{9}{16}$$

Add the like fractions.

$$\frac{3}{16} + \frac{9}{16} = \frac{3 + 9}{16}$$

$$= \frac{12}{16}$$

Simplify.

$$\frac{12 \div 4}{16 \div 4} = \frac{3}{4}$$

*Answer:* The total amount trimmed was $\frac{3}{4}$ of an inch.

## CLASS EXERCISES

1. Add. Answers should be in simplest form.

**a.** $\frac{5}{9} + \frac{2}{9}$ **b.** $\frac{6}{17} + \frac{9}{17}$ **c.** $\frac{13}{41} + \frac{27}{41}$ **d.** $\frac{24}{69} + \frac{12}{69} + \frac{15}{69}$

**e.** $\frac{69}{100} + \frac{6}{100}$ **f.** $\frac{5}{23} + \frac{7}{23} + \frac{9}{23}$ **g.** $\frac{8}{25} + \frac{4}{25} + \frac{3}{25}$ **h.** $\frac{7}{24} + \frac{11}{24} + \frac{5}{24}$

2. Subtract. Answers should be in simplest form.

a. $\frac{5}{9} - \frac{2}{9}$ b. $\frac{6}{7} - \frac{4}{7}$ c. $\frac{7}{8} - \frac{1}{8}$ d. $\frac{11}{19} - \frac{7}{19}$ e. $\frac{11}{12} - \frac{5}{12}$ f. $\frac{29}{37} - \frac{22}{37}$

g. $\begin{array}{r} \frac{51}{77} \\ -\frac{31}{77} \\ \hline \end{array}$ h. $\begin{array}{r} \frac{12}{25} \\ -\frac{7}{25} \\ \hline \end{array}$

3. On his bike, James rode $\frac{6}{8}$ of a mile to his friend's house, $\frac{3}{8}$ of a mile to the cleaners, and $\frac{7}{8}$ of a mile to his home. The total mileage that James traveled is

(a) $1\frac{5}{8}$ (b) $\frac{5}{8}$ (c) 2 (d) $\frac{1}{2}$

4. Vera bought $\frac{5}{6}$ of a pizza. Donna bought $\frac{3}{6}$ of a pizza. How much more did Vera buy than Donna?

5. Mrs. Ramos bought $\frac{3}{4}$ of a pound of ham, $\frac{1}{4}$ of a pound of cheese, and $\frac{3}{4}$ of a pound of roast beef. What was the total weight of her purchases?

6. Joseph ran $\frac{9}{10}$ of a mile, and Michael ran $\frac{4}{10}$ of a mile. How much more did Joseph run than Michael?

## HOMEWORK EXERCISES

1. Add. Answers should be in simplest form.

a. $\frac{3}{7} + \frac{2}{7}$ b. $\frac{5}{8} + \frac{2}{8}$ c. $\frac{2}{9} + \frac{6}{9}$ d. $\frac{11}{48} + \frac{14}{48}$ e. $\frac{5}{24} + \frac{11}{24}$ f. $\frac{6}{21} + \frac{10}{21}$

g. $\begin{array}{r} \frac{8}{31} \\ \frac{11}{31} \\ +\frac{7}{31} \\ \hline \end{array}$ h. $\begin{array}{r} \frac{23}{72} \\ \frac{10}{72} \\ +\frac{15}{72} \\ \hline \end{array}$ i. $\frac{21}{100} + \frac{37}{100} + \frac{32}{100}$

2. Betty assembled 3 electronic parts. The first was $\frac{5}{16}$ of an inch long, the second was $\frac{3}{16}$ of an inch long, and the third was $\frac{7}{16}$ of an inch long. What was the total length of the assembly?

3. Michael bought $\frac{1}{4}$ of a yard of cotton, $\frac{3}{4}$ of a yard of rayon, and $\frac{1}{4}$ of a yard of orlon. What was the total number of yards of cloth that he bought?

4. Subtract. Answers should be in simplest form.

a. $\frac{5}{6} - \frac{4}{6}$ b. $\frac{6}{8} - \frac{3}{8}$ c. $\frac{5}{9} - \frac{2}{9}$ d. $\frac{9}{10} - \frac{6}{10}$ e. $\frac{9}{12} - \frac{4}{12}$ f. $\frac{9}{22} - \frac{5}{22}$

g. $\frac{9}{16} - \frac{4}{16}$ h. $\frac{11}{26} - \frac{5}{26}$ i. $\frac{9}{20} - \frac{1}{20}$

5. One coin was $\frac{9}{16}$ of an inch wide, and another was $\frac{15}{16}$ of an inch wide. How much wider was the larger coin than the smaller one?

## SPIRAL REVIEW EXERCISES

1. The value of 7 in 17,428 is
(a) 70,000 (c) 700
(b) 7,000 (d) 7

2. The product of −7 and −3 is
(a) 21 (b) −21 (c) −10 (d) 10

3. The least common multiple (LCM) of 8 and 12 is
(a) 8 (b) 12 (c) 24 (d) 96

4. Of the fractions $\frac{1}{2}$, $\frac{1}{3}$, $\frac{2}{3}$, and $\frac{2}{5}$, the one with the largest value is

(a) $\frac{1}{2}$ (b) $\frac{1}{3}$ (c) $\frac{2}{3}$ (d) $\frac{2}{5}$

5. Simplify the fraction $\frac{12}{75}$.

6. Write $8\frac{2}{9}$ as an improper fraction.

7. On Saturday, the attendance at the ballpark was 9,075 and on Sunday, it was 10,162. How many more people attended the game on Sunday than on Saturday?

8. A case of 288 phonograph records is to be repacked into boxes containing 16 records each. How many boxes will be needed?

9. The price of a shirt is $9, and the price of a tie is $6. Which costs more, 12 shirts or 19 ties?

10. How much greater is 80,109 than 78,865?

### CHALLENGE QUESTION

Replace ? by a fraction that will make the following a true statement.

$$\frac{3}{5} + ? = \frac{24}{25}$$

# UNIT 5–2 Adding and Subtracting Unlike Fractions

## THE MAIN IDEA

To add or subtract fractions that have unlike denominators:

1. Find the LCD.
2. Change the given fractions to equivalent fractions that have the LCD as their denominators.
3. Use the method for adding or subtracting fractions that have like denominators.

**EXAMPLE 1** Add: $\frac{3}{8} + \frac{1}{6}$

Find the LCD of 8 and 6. The LCD is 24.

Change the fractions into equivalent fractions that have 24 as the denominator.

$$\frac{3 \times 3}{8 \times 3} = \frac{9}{24}$$

$$\frac{1 \times 4}{6 \times 4} = \frac{4}{24}$$

Add the equivalent like fractions.

$$\frac{9}{24} + \frac{4}{24} = \frac{13}{24} \quad \textit{Ans.}$$

**EXAMPLE 2** Subtract: $\frac{17}{36} - \frac{1}{18}$

Find the LCD of 36 and 18. The LCD is 36.

Change $\frac{1}{18}$ to an equivalent fraction that has 36 as the denominator.

$$\frac{1 \times 2}{18 \times 2} = \frac{2}{36}$$

Subtract the like fractions.

$$\frac{17}{36} - \frac{2}{36} = \frac{15}{36}$$

Simplify.

$$\frac{15 \div 3}{36 \div 3} = \frac{5}{12} \quad \textit{Ans.}$$

**EXAMPLE 3** $\frac{3}{4} + \frac{5}{6} - \frac{1}{3}$ is equal to (a) $\frac{7}{13}$ (b) $1\frac{1}{4}$ (c) $\frac{5}{12}$ (d) $1\frac{11}{12}$

The LCD of 4, 6, and 3 is 12. $\frac{3}{4} + \frac{5}{6} - \frac{1}{3}$

Rewrite each fraction. $\frac{3 \times 3}{4 \times 3} + \frac{5 \times 2}{6 \times 2} - \frac{1 \times 4}{3 \times 4}$

$\frac{9}{12} + \frac{10}{12} - \frac{4}{12}$

Following the order of operations, work from left to right. Add, then subtract. $\frac{19}{12} - \frac{4}{12} = \frac{15}{12}$

Change $\frac{15}{12}$ to a mixed number and simplify. $\frac{15}{12} = 1\frac{3}{12} = 1\frac{1}{4}$

*Answer:* (b)

**EXAMPLE 4** When Mr. Miser began his car trip, the gas gauge read $\frac{5}{8}$ full. When he returned, the gauge showed $\frac{1}{3}$ full. What part of a full tank of gas did Mr. Miser use on his trip?

**THINKING ABOUT THE PROBLEM**

To find the amount used, subtract the final amount from the original amount.

Write a number phrase. $\frac{5}{8} - \frac{1}{3}$

The LCD of 8 and 3 is 24.

Rewrite each fraction. $\frac{5 \times 3}{8 \times 3} - \frac{1 \times 8}{3 \times 8}$

Subtract the like fractions. $\frac{15}{24} - \frac{8}{24} = \frac{7}{24}$

*Answer:* Mr. Miser used $\frac{7}{24}$ of a tank of gas.

## CLASS EXERCISES

**1.** Add. Answers should be in simplest form.

**a.** $\frac{1}{4} + \frac{2}{3}$ **b.** $\frac{3}{5} + \frac{3}{10}$ **c.** $\frac{15}{27} + \frac{5}{9}$ **d.** $\frac{5}{6} + \frac{3}{11}$ **e.** $\frac{3}{4} + \frac{1}{5} + \frac{5}{8}$ **f.** $\frac{4}{27} + \frac{1}{3} + \frac{5}{9}$

2. Subtract. Answers should be in simplest form.

a. $\frac{3}{4} - \frac{5}{16}$ b. $\frac{5}{6} - \frac{1}{2}$ c. $\frac{4}{5} - \frac{1}{3}$ d. $\frac{12}{7} - \frac{2}{3}$ e. $\frac{9}{10} - \frac{3}{4}$ f. $\frac{8}{12} - \frac{2}{3}$

3. Perform the indicated operations. Answers should be in simplest form.

a. $\frac{1}{3} + \frac{3}{5} - \frac{1}{2}$ b. $\frac{3}{4} - \frac{3}{7} + \frac{1}{4}$ c. $\frac{4}{5} - \frac{1}{2} - \frac{1}{4}$

4. Carol bought $\frac{1}{2}$ of a pound of cashews, $\frac{2}{3}$ of a pound of walnuts, and $\frac{3}{4}$ of a pound of peanuts. Find the total weight of the nuts that Carol bought.

5. Fred had to lose $\frac{7}{8}$ of a pound to qualify for his wrestling match. He dieted and lost $\frac{7}{10}$ of a pound. How much does he still have to lose?

## HOMEWORK EXERCISES

1. Add. Answers should be in simplest form.

a. $\frac{2}{3} + \frac{1}{4}$ b. $\frac{3}{5} + \frac{1}{2}$ c. $\frac{3}{4} + \frac{2}{5}$

d. $\frac{1}{8} + \frac{1}{2}$ e. $\frac{1}{6} + \frac{2}{3}$ f. $\frac{4}{5} + \frac{3}{10}$ g. $\frac{3}{8} + \frac{1}{12} + \frac{5}{24}$ h. $\frac{3}{4} + \frac{7}{20} + \frac{2}{5}$ i. $\frac{4}{15} + \frac{2}{5} + \frac{2}{3}$

2. Subtract. Answers should be in simplest form.

a. $\frac{5}{8} - \frac{1}{4}$ b. $\frac{7}{16} - \frac{1}{8}$ c. $\frac{3}{4} - \frac{1}{8}$

d. $\frac{3}{5} - \frac{3}{10}$ e. $\frac{15}{16} - \frac{3}{4}$ f. $\frac{9}{10} - \frac{4}{5}$ g. $\frac{11}{12} - \frac{2}{5}$ h. $\frac{5}{6} - \frac{2}{5}$ i. $\frac{18}{20} - \frac{9}{10}$

3. Perform the indicated operations. Answers should be in simplest form.

a. $\frac{3}{4} - \frac{1}{2} + \frac{1}{6}$ b. $\frac{2}{5} + \frac{7}{10} - \frac{3}{10}$ c. $\frac{6}{7} - \frac{1}{2} - \frac{1}{4}$

4. A highway crew built a fence $\frac{1}{2}$ of a mile long and then added two more sections, one that was $\frac{1}{5}$ of a mile long and another $\frac{1}{4}$ of a mile long. What was the total length of the fence?

5. A jeweler removed $\frac{3}{16}$ of an inch from the length of a gold wire that was originally $\frac{5}{8}$ of an inch long. What was the final length of the wire?

## SPIRAL REVIEW EXERCISES

1. A prime number between 62 and 70 is
(a) 63 (b) 67 (c) 68 (d) 69

2. The greatest common factor of 75 and 21 is
(a) 1 (b) 3 (c) 5 (d) 21

3. The sum of $\frac{1}{12}$ and $\frac{7}{12}$ is
(a) $\frac{2}{3}$ (b) $\frac{8}{24}$ (c) $\frac{1}{3}$ (d) $\frac{3}{4}$

4. From $\frac{12}{21}$, subtract $\frac{5}{21}$.

5. Simplify the fraction $\frac{42}{70}$.

6. Divide: $-20 \div (-5)$

7. Round 275,892 to the nearest thousand.

8. Which fraction is equivalent to $\frac{2}{3}$?
(a) $\frac{3}{4}$ (b) $\frac{12}{20}$ (c) $\frac{10}{21}$ (d) $\frac{16}{24}$

9. A stadium contains 64 sections with 128 seats in each section. What is the total number of seats in the stadium?

10. How many 12-inch lengths of rope can be cut from a 336-inch length of rope?

11. The value of $24 + 16 \div 8 - 4$ is
(a) 1 (b) 28 (c) 10 (d) 22

### CHALLENGE QUESTION

What part of $\frac{3}{5}$ is $\frac{3}{10}$?

# UNIT 5–3 Multiplying Fractions; Multiplying a Fraction and a Whole Number

## MULTIPLYING FRACTIONS

### THE MAIN IDEA

To multiply two fractions:

1. Try to simplify. If possible, divide a numerator and a denominator by a common factor.
2. Multiply the resulting numerators.
3. Multiply the resulting denominators.
4. If necessary, simplify the resulting fraction.

**EXAMPLE 1** Multiply: $\frac{3}{5} \times \frac{2}{7}$

No simplification is possible.

Multiply the numerators and multiply the denominators.

$$\frac{3}{5} \times \frac{2}{7} = \frac{3 \times 2}{5 \times 7}$$

$$= \frac{6}{35} \quad \textit{Ans.}$$

**EXAMPLE 2** Multiply: $\frac{5}{8} \times \frac{4}{15}$

There is a common factor of 4. Divide 8 and 4 by 4.

$$\frac{5}{\underset{2}{\cancel{8}}} \times \frac{\overset{1}{\cancel{4}}}{15}$$

There is a common factor of 5. Divide 5 and 15 by 5.

$$\frac{\overset{1}{\cancel{5}}}{\underset{2}{\cancel{8}}} \times \frac{\overset{1}{\cancel{4}}}{\underset{3}{\cancel{15}}}$$

Multiply the results.

$$\frac{1 \times 1}{2 \times 3} = \frac{1}{6} \quad \textit{Ans.}$$

**EXAMPLE 3** Find the product of $\frac{42}{75}$ and $\frac{50}{63}$.

"Product" tells you to multiply.

$$\frac{42}{75} \times \frac{50}{63}$$

To simplify, divide 42 and 63 by 7. Divide 75 and 50 by 25.

$$\frac{\overset{6}{\cancel{42}}}{\underset{3}{\cancel{75}}} \times \frac{\overset{2}{\cancel{50}}}{\underset{9}{\cancel{63}}}$$

Before multiplying, you can simplify further. Divide 6 and 3 by 3.

$$\frac{\overset{2}{\cancel{6}}\!\!\!\!\!\!\cancel{42}}{\underset{1}{\cancel{3}}\!\!\!\!\!\!\cancel{75}} \times \frac{\overset{2}{\cancel{50}}}{\underset{9}{\cancel{63}}}$$

Multiply the results.

$$\frac{2 \times 2}{1 \times 9} = \frac{4}{9} \quad Ans.$$

**EXAMPLE 4** Sherry has $\frac{6}{8}$ of a pound of candy. She wants to give half of the candy to her sister. Find the amount of candy that Sherry should give to her sister.

**THINKING ABOUT THE PROBLEM**

A key word in the problem is "of." To find half of $\frac{6}{8}$, multiply $\frac{1}{2}$ by $\frac{6}{8}$.

Write a number phrase.

$$\frac{1}{2} \times \frac{6}{8}$$

To simplify, divide 2 and 6 by 2.

$$\frac{1}{\underset{1}{\cancel{2}}} \times \frac{\overset{3}{\cancel{6}}}{8}$$

Multiply the results.

$$\frac{1 \times 3}{1 \times 8} = \frac{3}{8}$$

*Answer:* Sherry should give $\frac{3}{8}$ of a pound of the candy to her sister.

## CLASS EXERCISES

**1.** Multiply. Answers should be in simplest form.

**a.** $\frac{1}{4} \times \frac{2}{5}$ **b.** $\frac{3}{8} \times \frac{2}{3}$ **c.** $\frac{5}{6} \times \frac{3}{7}$ **d.** $\frac{7}{8} \times \frac{16}{21}$ **e.** $\frac{3}{5} \times \frac{5}{3}$ **f.** $\frac{8}{3} \times \frac{27}{16}$

**2. a.** Find $\frac{3}{7}$ of $\frac{35}{48}$. **b.** Find $\frac{7}{8}$ of $\frac{32}{49}$. **c.** Find $\frac{9}{10}$ of $\frac{45}{72}$.

**3.** Jane's house is $\frac{3}{4}$ of a mile from Marcia's house. They agree to meet at a point that is $\frac{1}{2}$ the distance between the two houses. How far is the meeting point from Jane's house?

**4.** Mr. Norris bought $\frac{8}{10}$ of a pound of meat and wanted to prepare $\frac{1}{4}$ of this amount for his lunch. How much meat should he prepare for lunch?

**5.** A metal part is $\frac{24}{32}$ of an inch long. A mechanic wants to replace it with a part that is $\frac{2}{3}$ the length. How long is the replacement part?

## MULTIPLYING A FRACTION AND A WHOLE NUMBER

### THE MAIN IDEA

To multiply a fraction and a whole number:

1. Write the whole number as an improper fraction with a denominator of 1.
2. Use the method for multiplying two fractions.

**EXAMPLE 5** Multiply: $20 \times \frac{3}{5}$

Write the whole number 20 as a fraction. $\frac{20}{1} \times \frac{3}{5}$

To simplify, divide 20 and 5 by 5. $\frac{\overset{4}{\cancel{20}}}{1} \times \frac{3}{\underset{1}{\cancel{5}}}$

Multiply. $\frac{4 \times 3}{1 \times 1} = \frac{12}{1} = 12$ *Ans.*

**EXAMPLE 6** Two-thirds of the people who took a taste test preferred Zappo Cola. If there were 120 people in the test, how many preferred Zappo?

**THINKING ABOUT THE PROBLEM**

The words "two-thirds of" tell you to multiply.

Write a number phrase. $\frac{2}{3} \times 120$

Write the whole number 120 as a fraction. $\frac{2}{3} \times \frac{120}{1}$

To simplify, divide 3 and 120 by 3. $\frac{2}{\underset{1}{\cancel{3}}} \times \frac{\overset{40}{\cancel{120}}}{1}$

Multiply. $\frac{2 \times 40}{1 \times 1} = \frac{80}{1} = 80$

*Answer:* 80 people preferred Zappo.

## CLASS EXERCISES

1. Multiply. Answers should be in simplest form.

   a. $40 \times \frac{1}{5}$ b. $\frac{1}{6} \times 72$ c. $\frac{2}{3} \times 33$ d. $48 \times \frac{3}{8}$ e. $100 \times \frac{2}{5}$ f. $\frac{5}{6} \times 81$

   g. $98 \times \frac{4}{7}$ h. $144 \times \frac{7}{12}$ i. $\frac{9}{10} \times 100$ j. $55 \times \frac{7}{10}$ k. $64 \times \frac{5}{24}$ l. $108 \times \frac{7}{72}$

2. a. Find $\frac{3}{4}$ of 48. b. Find $\frac{2}{3}$ of 63. c. Find $\frac{7}{8}$ of 40. d. Find $\frac{4}{5}$ of 120.

   e. Find $\frac{3}{10}$ of 25. f. Find $\frac{7}{18}$ of 81.

3. How many pounds of ground beef would be needed to make 60 hamburgers if each hamburger weighs $\frac{1}{4}$ of a pound before cooking?

4. Four hundred fifty students voted for the president of the student government. If $\frac{3}{5}$ of these students voted for Carlton Besk, find the number of students that voted for Carlton.

## HOMEWORK EXERCISES

1. Multiply. Answers should be in simplest form.

   a. $\frac{1}{2} \times \frac{4}{5}$ b. $\frac{2}{3} \times \frac{5}{8}$ c. $\frac{3}{5} \times \frac{5}{7}$ d. $\frac{2}{3} \times \frac{1}{5}$ e. $\frac{7}{24} \times \frac{8}{14}$ f. $\frac{5}{12} \times \frac{8}{15}$

   g. $\frac{12}{24} \times \frac{24}{60}$ h. $\frac{9}{10} \times \frac{20}{45}$

2. Find each product. Answers should be in simplest form.

   a. $5 \times \frac{1}{8}$ b. $\frac{2}{3} \times 4$ c. $12 \times \frac{1}{5}$ d. $\frac{3}{4} \times 11$ e. $\frac{1}{3} \times 24$ f. $50 \times \frac{2}{5}$

   g. $33 \times \frac{3}{22}$ h. $\frac{5}{18} \times 63$

3. a. Find $\frac{5}{8}$ of 40. b. Find $\frac{3}{5}$ of 45. c. Find $\frac{9}{10}$ of 80. d. Find $\frac{3}{7}$ of 42. e. Find $\frac{2}{3}$ of $\frac{4}{5}$.

   f. Find $\frac{1}{8}$ of $\frac{4}{5}$. g. Find $\frac{8}{15}$ of 50. h. Find $\frac{5}{28}$ of 42. i. Find $\frac{2}{75}$ of 100.

4. Two-thirds of a road that is $\frac{9}{10}$ of a mile long is to be repaved. How long is the section to be repaved?

5. Sixty beads, each $\frac{3}{8}$ of an inch long, are strung together on a wire. What is the total length of the 60 beads?

6. Of the 640 students at Oakdale School, $\frac{2}{5}$ are sophomores. How many sophomores are there?

## SPIRAL REVIEW EXERCISES

1. From $-12$, subtract $-3$.

2. When 25 is divided by 7, the remainder is
   (a) 3 (b) 4 (c) 5 (d) 7

3. Add: $\frac{9}{15} + \frac{7}{35}$

4. $\frac{5}{40} + \frac{6}{40} + \frac{1}{40}$ is equivalent to
   (a) $\frac{6}{40}$ (b) $\frac{1}{10}$ (c) $\frac{12}{120}$ (d) $\frac{3}{10}$

5. Mrs. Ruiz baked 24 chocolate chip cookies and 48 oatmeal cookies. What part of the total batch of cookies were chocolate chip cookies?
   (a) $\frac{1}{2}$ (b) $\frac{1}{3}$ (c) $\frac{1}{4}$ (d) $\frac{2}{3}$

6. Mrs. Brady had 144 pieces of candy. She gave 2 pieces to each of her 28 students. How many pieces were left?

7. If each mathematics book contains 273 pages, how many pages are there in 35 such books?

8. Mr. and Mrs. Jonas began their trip with \$1,000. First, they spent \$250 on train fare, and then they spent \$325 on hotel rooms. How much money was left?

9. The value of $-9 - (-9)$ is
   (a) $-18$ (b) $+18$ (c) $+9$ (d) 0

### CHALLENGE QUESTION

How many pieces $\frac{3}{8}$ of an inch long can be cut from a 6-inch length of wire?

# UNIT 5–4 The Meaning of Reciprocal; Using Reciprocals in Dividing Fractions

## THE MEANING OF RECIPROCAL

### THE MAIN IDEA

1. The *reciprocal* of a fraction is formed by exchanging the numerator and the denominator of the original fraction. This is the same as inverting a fraction, or turning it upside down.

   The reciprocal of $\frac{2}{3}$ is $\frac{3}{2}$.

   The reciprocal of $\frac{6}{5}$ is $\frac{5}{6}$.

2. The reciprocal of a whole number is $\frac{1}{\text{whole number}}$.

   The reciprocal of 5 is $\frac{1}{5}$.

3. 0 is the only number that does not have a reciprocal.

4. If we multiply a number and its reciprocal, we always obtain the number 1.

$$\frac{2}{3} \times \frac{3}{2} = 1$$

**EXAMPLE 1** Find the reciprocal of each number.

| | *Number* | *Reciprocal* |
|---|---|---|
| a. | $\frac{5}{9}$ | $\frac{9}{5}$ |
| b. | $\frac{9}{5}$ | $\frac{5}{9}$ |
| c. | $\frac{11}{16}$ | $\frac{16}{11}$ |
| d. | $\frac{13}{2}$ | $\frac{2}{13}$ |
| e. | 21 | $\frac{1}{21}$ |
| f. | 0 | 0 does not have a reciprocal. |

## CLASS EXERCISES

Write the reciprocal of each number.

**1.** $\frac{7}{9}$ **2.** $\frac{1}{6}$ **3.** $\frac{3}{2}$ **4.** 7 **5.** 32 **6.** $\frac{2}{5}$ **7.** 1 **8.** $\frac{22}{15}$ **9.** $\frac{9}{10}$ **10.** $\frac{10}{3}$

## USING RECIPROCALS IN DIVIDING FRACTIONS

### THE MAIN IDEA

1. Division by a number gives the same result as multiplication by the reciprocal of that number:

$$20 \div 5 \text{ gives a result of } 4$$

$$20 \times \frac{1}{5} \text{ gives a result of } 4$$

2. A division can be changed into a multiplication by using the reciprocal of the divisor:

$$\frac{1}{5} \div \frac{2}{3} \text{ is the same as } \frac{1}{5} \times \frac{3}{2}$$

3. To divide two fractions, use the reciprocal of the divisor and change the division to multiplication.

**EXAMPLE 2** Use the reciprocal of the divisor to rewrite each division as an equivalent multiplication.

| | *Division* | *Multiplication* |
|---|---|---|
| **a.** | $\frac{1}{12} \div \frac{2}{5}$ | $\frac{1}{12} \times \frac{5}{2}$ |
| **b.** | $\frac{2}{3} \div \frac{4}{7}$ | $\frac{2}{3} \times \frac{7}{4}$ |
| **c.** | $\frac{4}{9} \div \frac{7}{11}$ | $\frac{4}{9} \times \frac{11}{7}$ |

**EXAMPLE 3** Divide: $\frac{3}{5} \div \frac{2}{3}$

$$\frac{3}{5} \div \frac{2}{3}$$

Use the reciprocal of the divisor and change the division to multiplication.

$$\frac{3}{5} \times \frac{3}{2}$$

Since there are no common factors, multiply.

$$\frac{3 \times 3}{5 \times 2} = \frac{9}{10} \quad \textit{Ans.}$$

**EXAMPLE 4** Divide: $\frac{18}{25} \div \frac{27}{20}$

$$\frac{18}{25} \div \frac{27}{20}$$

Use the reciprocal of the divisor and change the division to multiplication.

$$\frac{18}{25} \times \frac{20}{27}$$

To simplify, divide 18 and 27 by 9. Divide 20 and 25 by 5.

$$\frac{\overset{2}{\cancel{18}}}{\underset{5}{\cancel{25}}} \times \frac{\overset{4}{\cancel{20}}}{\underset{3}{\cancel{27}}}$$

Multiply the results.

$$\frac{2 \times 4}{5 \times 3} = \frac{8}{15} \quad \textit{Ans.}$$

**EXAMPLE 5** Divide: $6 \div \frac{3}{5}$

Write the whole number 6 as a fraction.

$$\frac{6}{1} \div \frac{3}{5}$$

Use the reciprocal of the divisor and change the division to multiplication.

$$\frac{6}{1} \times \frac{5}{3}$$

To simplify, divide 6 and 3 by 3.

$$\frac{\overset{2}{\cancel{6}}}{1} \times \frac{5}{\underset{1}{\cancel{3}}}$$

Multiply the results.

$$\frac{2 \times 5}{1 \times 1} = \frac{10}{1} = 10 \quad \textit{Ans.}$$

**EXAMPLE 6** A piece of wire $\frac{3}{4}$ of a meter long is to be cut into 6 equal pieces. The length of each piece is (a) 2 meters (b) $4\frac{1}{2}$ meters (c) $\frac{1}{6}$ meter (d) $\frac{1}{8}$ meter

**THINKING ABOUT THE PROBLEM**

To separate a given amount into a number of equal parts, you divide.

Write a number phrase.

$$\frac{3}{4} \div 6$$

Write the whole number 6 as a fraction.

$$\frac{3}{4} \div \frac{6}{1}$$

Use the reciprocal of the divisor and change the division to multiplication.

$$\frac{3}{4} \times \frac{1}{6}$$

Simplify and multiply.

$$\frac{\overset{1}{\cancel{3}} \times 1}{4 \times \underset{2}{\cancel{6}}} = \frac{1}{8}$$

*Answer:* (d)

## CLASS EXERCISES

1. Rewrite each division as an equivalent multiplication.

   a. $12 \div 3$ b. $8 \div \frac{1}{12}$ c. $\frac{18}{5} \div \frac{3}{2}$

2. Divide. Answers should be in simplest form.

   a. $\frac{3}{4} \div \frac{1}{8}$ b. $\frac{6}{5} \div \frac{2}{3}$ c. $\frac{5}{3} \div \frac{4}{3}$ d. $\frac{4}{3} \div \frac{5}{3}$ e. $\frac{9}{2} \div \frac{1}{6}$ f. $\frac{2}{5} \div \frac{2}{5}$

3. Divide. Answers should be in simplest form.

   a. $12 \div \frac{2}{3}$ b. $\frac{3}{5} \div 6$ c. $9 \div \frac{3}{4}$ d. $3 \div \frac{9}{5}$ e. $8 \div \frac{3}{7}$ f. $\frac{7}{15} \div 14$

4. The distance around a racetrack is $\frac{3}{8}$ of a mile. How many times around the track must Cynthia run in order to run 27 miles?

5. How many pieces of plastic tubing $\frac{3}{4}$ of an inch long can be cut from a 6-inch piece of tubing?

6. Mr. Adams wants to put a "No Trespassing" sign on every $\frac{1}{4}$-mile section of his 3-mile-long fence. How many signs will he need?

## HOMEWORK EXERCISES

1. Write the reciprocal of each number.

   a. $\frac{2}{5}$ b. 3 c. $\frac{1}{2}$ d. $\frac{1}{25}$ e. $\frac{5}{9}$ f. 1 g. 100 h. $\frac{2}{4}$ i. 48 j. $\frac{4}{3}$

2. Rewrite each division as an equivalent multiplication.

   a. $12 \div \frac{1}{3}$ b. $\frac{2}{5} \div \frac{3}{10}$ c. $\frac{5}{8} \div 2$ d. $\frac{11}{3} \div \frac{3}{4}$

3. Divide. Answers should be in simplest form.

   a. $\frac{3}{4} \div \frac{1}{8}$ b. $\frac{3}{10} \div \frac{4}{5}$ c. $\frac{1}{2} \div \frac{5}{8}$ d. $\frac{5}{12} \div \frac{5}{9}$ e. $\frac{14}{35} \div \frac{2}{7}$ f. $\frac{7}{10} \div \frac{3}{10}$

   g. $\frac{25}{48} \div \frac{16}{30}$ h. $\frac{49}{24} \div \frac{7}{8}$ i. $\frac{11}{17} \div \frac{11}{17}$

4. Divide. Answers should be in simplest form.

   a. $2 \div \frac{1}{2}$ b. $\frac{2}{3} \div 3$ c. $24 \div \frac{1}{6}$ d. $6 \div \frac{4}{5}$ e. $\frac{9}{10} \div 3$ f. $12 \div \frac{9}{4}$

   g. $\frac{44}{50} \div 20$ h. $\frac{45}{60} \div 18$ i. $10 \div \frac{18}{3}$

5. How many postage stamps, each $\frac{11}{16}$ of an inch long, are in a row of stamps 22 inches long?

6. Mrs. Cody needed 15 pounds of cheese to make sandwiches for her company picnic. The cheese comes in packages containing $\frac{5}{8}$ of a pound. How many packages did she need?

## SPIRAL REVIEW EXERCISES

**1.** Round 25,364 to the nearest hundred.

**2.** Find $\frac{3}{4}$ of 48.

**3.** The product of $\frac{4}{12}$ and $\frac{1}{2}$ is

(a) $\frac{8}{12}$ (b) $\frac{5}{14}$ (c) $\frac{1}{6}$ (d) $\frac{2}{3}$

**4.** How much longer is $\frac{5}{8}$ in. than $\frac{1}{4}$ in.?

(a) $\frac{5}{8}$ in. (c) $\frac{3}{8}$ in.

(b) $\frac{1}{2}$ in. (d) $\frac{1}{4}$ in.

**5.** When $-24$ is divided by 3, the quotient is

(a) $\frac{1}{8}$ (b) $-\frac{1}{8}$ (c) 8 (d) $-8$

**6.** Bob bought 3 shirts and 4 pairs of slacks. The price of a shirt was $14 and the price of a pair of slacks was $24. The difference between the cost of the slacks and the cost of the shirts can be represented by the number phrase

(a) $4 \times 24 - 3 \times 14$
(b) $3 \times 24 - 4 \times 14$
(c) $(3 + 4) \times (14 + 24)$
(d) $(14 + 24) - (3 + 4)$

**7.** $\frac{2}{3}$ of 300 is

(a) 100 (b) 200 (c) 300 (d) 400

**8.** Which is a longer run, 8 laps around a 300-yard track or 6 laps around a 360-yard track?

**9.** Mary earned $19 on each of 5 days, and an additional $25 on the sixth day. What was the total of her earnings?

**10.** Evaluate:

**a.** $2 + 3 \times 4$

**b.** $-3 - (-5)$

**c.** $6 \times 8 \div 4 + 12$

**d.** $6 \times 8 \div (4 + 12)$

**11.** Replace ? by < or > to make a true comparison.

**a.** $\frac{7}{10} ? \frac{3}{5}$ **b.** $\frac{4}{7} ? \frac{5}{9}$ **c.** $\frac{7}{18} ? \frac{5}{12}$

**12.** There are three thousand, seventeen students at Westwood High School, and two thousand, eight hundred nine students at McKinley High School. How many more students are at Westwood?

### CHALLENGE QUESTION

Explain why the reciprocal of an improper fraction is not always a proper fraction. Give an example.

## UNIT 6–1 Adding Mixed Numbers

### ADDING MIXED NUMBERS WITH LIKE-FRACTION PARTS

**THE MAIN IDEA**

To add mixed numbers with like-fraction parts:

1. Add the like-fraction parts. If possible, simplify the resulting fraction.
2. Add the whole-number parts.
3. If the fraction is an improper fraction, change it into an equivalent mixed number and add it to the whole-number part.

**EXAMPLE 1** Add: $5\frac{7}{16} + 3\frac{2}{16}$

First add the like-fraction parts. Then add the whole-number parts. No simplification is possible since $\frac{9}{16}$ is in lowest terms.

$$\begin{array}{r} 5\frac{7}{16} \\ +3\frac{2}{16} \\ \hline 8\frac{9}{16} \end{array} \quad \textit{Ans.}$$

like fractions

**EXAMPLE 2** Add: $2\frac{3}{8} + 5\frac{1}{8}$

Add the like-fraction parts and add the whole-number parts.

$$\begin{array}{r} 2\frac{3}{8} \\ +5\frac{1}{8} \\ \hline 7\frac{4}{8} \end{array}$$

Simplify the resulting fraction part.

$$7\frac{4}{8} = 7\frac{4 \div 4}{8 \div 4} = 7\frac{1}{2} \quad \textit{Ans.}$$

**EXAMPLE 3** Add: $15\frac{7}{13} + 12\frac{9}{13}$

Add the like-fraction parts and add the whole-number parts.

$$\begin{array}{r} 15\frac{7}{13} \\ +12\frac{9}{13} \\ \hline 27\frac{16}{13} \end{array}$$

Since the resulting fraction part is an improper fraction, change it into an equivalent mixed number.

$$\frac{16}{13} = 1\frac{3}{13}$$

Add the resulting mixed number to the whole-number part.

$$27 + 1\frac{3}{13} = 28\frac{3}{13} \quad \textit{Ans.}$$

**EXAMPLE 4** Mrs. Nelson needs a piece of ribbon $11\frac{5}{8}$ inches long for a hat she is making, and another piece $7\frac{5}{8}$ inches long for trim on a blouse. What is the total length of ribbon that she needs?

### THINKING ABOUT THE PROBLEM

The key word "total" tells you to add.

Write a number phrase. $11\frac{5}{8} + 7\frac{5}{8}$

Add the like-fraction parts and add the whole-number parts.

$$\begin{array}{r} 11\frac{5}{8} \\ +7\frac{5}{8} \\ \hline 18\frac{10}{8} \end{array}$$

Change the resulting improper fraction into an equivalent mixed number and simplify.

$$\frac{10}{8} = 1\frac{2}{8} = 1\frac{2 \div 2}{8 \div 2} = 1\frac{1}{4}$$

Add the resulting mixed number to the whole-number part.

$$18 + 1\frac{1}{4} = 19\frac{1}{4}$$

*Answer:* The total ribbon length is $19\frac{1}{4}$ inches.

## CLASS EXERCISES

1. Add. Answers should be in simplest form.

   a. $2\frac{3}{5} + 4\frac{1}{5}$ b. $9\frac{3}{8} + 7\frac{1}{8}$ c. $11\frac{13}{32} + 20\frac{5}{32}$

   d. $\begin{array}{r} 12\frac{3}{4} \\ +5\frac{1}{4} \\ \hline \end{array}$ e. $\begin{array}{r} 22\frac{5}{17} \\ +14\frac{13}{17} \\ \hline \end{array}$ f. $9\frac{5}{7} + 2\frac{3}{7} + 1\frac{6}{7}$ g. $5\frac{5}{12} + 8\frac{7}{12}$ h. $\begin{array}{r} 12\frac{9}{10} \\ +25\frac{7}{10} \\ \hline \end{array}$ i. $\begin{array}{r} 13\frac{4}{15} \\ 12\frac{11}{15} \\ +26\frac{3}{15} \\ \hline \end{array}$

2. X-Ray Industries' stock was priced at $19\frac{7}{8}$. The stock gained $1\frac{5}{8}$ points. What was the new price of the stock?

3. June jogged $3\frac{3}{4}$ miles one morning, and she jogged another $4\frac{3}{4}$ miles that afternoon. What was the total distance she jogged that day?

## ADDING MIXED NUMBERS WITH UNLIKE-FRACTION PARTS

### THE MAIN IDEA

To add mixed numbers with unlike-fraction parts:

1. Find the LCD of the fraction parts.
2. Change the fraction parts into equivalent fractions with the LCD as the denominator.
3. Follow the procedure for adding mixed numbers with like-fraction parts.

**EXAMPLE 5** Add: $3\frac{1}{2} + 7\frac{2}{5}$

The LCD of the unlike fractions $\frac{1}{2}$ and $\frac{2}{5}$ is 10.

Rewrite the unlike-fraction parts as equivalent like fractions.

$$3\frac{1 \times 5}{2 \times 5} = 3\frac{5}{10}$$

$$7\frac{2 \times 2}{5 \times 2} = 7\frac{4}{10}$$

Add the resulting like-fraction parts and the whole-number parts. No simplification is possible.

$$\begin{array}{r} 3\frac{5}{10} \\ +7\frac{4}{10} \\ \hline 10\frac{9}{10} \end{array} \quad \textit{Ans.}$$

**EXAMPLE 6** Add: $6\frac{1}{3} + 5\frac{3}{4} + 2\frac{5}{6}$

The LCD of the unlike fractions $\frac{1}{3}$, $\frac{3}{4}$, and $\frac{5}{6}$ is 12.

Rewrite the unlike fractions as equivalent like fractions.

$$6\frac{1 \times 4}{3 \times 4} = 6\frac{4}{12}$$

$$5\frac{3 \times 3}{4 \times 3} = 5\frac{9}{12}$$

$$2\frac{5 \times 2}{6 \times 2} = 2\frac{10}{12}$$

Add the resulting like-fraction parts and the whole-number parts.

$$13\frac{23}{12}$$

Change the resulting improper fraction into an equivalent mixed number.

$$\frac{23}{12} = 1\frac{11}{12}$$

Add the resulting mixed number to the whole-number part.

$$13 + 1\frac{11}{12} = 14\frac{11}{12} \quad \textit{Ans.}$$

**EXAMPLE 7** A pencil was made from a wooden part that was $6\frac{3}{4}$ inches long and an eraser end that was $\frac{7}{8}$ of an inch long. Find the length of the completed pencil.

**THINKING ABOUT THE PROBLEM**

Finding the length of the "completed" pencil is the same as finding the "total length" of the pencil. "Total" tells you to add.

Write a number phrase. $6\frac{3}{4} + \frac{7}{8}$

The LCD of the unlike fractions $\frac{3}{4}$ and $\frac{7}{8}$ is 8.

Rewrite the unlike fractions as equivalent like fractions and add the like-fraction parts.

$$6\frac{3 \times 2}{4 \times 2} = 6\frac{6}{8}$$
$$\begin{array}{r} 6\frac{6}{8} \\ +\frac{7}{8} \\ \hline 6\frac{13}{8} \end{array}$$

Rewrite the resulting improper fraction as a mixed number. $\frac{13}{8} = 1\frac{5}{8}$

Add the resulting mixed number to the whole-number part. $6 + 1\frac{5}{8} = 7\frac{5}{8}$

*Answer:* The total length of the pencil is $7\frac{5}{8}$ inches.

## CLASS EXERCISES

1. Add. Answers should be in simplest form.

a. $5\frac{2}{3} + 3\frac{1}{6}$ b. $6\frac{2}{5} + 3\frac{2}{3}$ c. $5\frac{3}{4} + 1\frac{3}{8}$

d. $\begin{array}{r} 3\frac{4}{9} \\ +1\frac{2}{3} \\ \hline \end{array}$ e. $\begin{array}{r} 7\frac{5}{6} \\ +2\frac{3}{4} \\ \hline \end{array}$ f. $\begin{array}{r} 3\frac{1}{5} \\ 2\frac{2}{3} \\ +5\frac{3}{10} \\ \hline \end{array}$ g. $\begin{array}{r} 5\frac{7}{8} \\ +2\frac{3}{5} \\ \hline \end{array}$ h. $\begin{array}{r} 6\frac{1}{4} \\ 3\frac{5}{10} \\ +8\frac{1}{2} \\ \hline \end{array}$ i. $\begin{array}{r} 4\frac{5}{12} \\ 5\frac{1}{6} \\ +9\frac{5}{8} \\ \hline \end{array}$

2. Mr. McCarthy bought $2\frac{1}{4}$ pounds of ham, $3\frac{3}{8}$ pounds of cheese, and $1\frac{7}{16}$ pounds of roast beef. What was the total weight of his purchases?

3. An electric cable is made from a connector on each end and a piece of wire between. If one connector is $\frac{5}{8}$ of an inch long, the wire is $18\frac{3}{4}$ inches long, and the second connector is $\frac{1}{2}$ inch long, find the total length of the cable.

## HOMEWORK EXERCISES

1. Add. Answers should be in simplest form.

a. $5\frac{1}{6} + 3\frac{4}{6}$ b. $9\frac{11}{32} + 5\frac{13}{32}$ c. $48\frac{9}{15} + 52\frac{8}{15}$ d. $5\frac{1}{8} + 3\frac{3}{8} + 2\frac{5}{8}$

e. $7\frac{3}{8} + 8\frac{2}{8}$ f. $3\frac{11}{64} + 7\frac{54}{64}$ g. $11\frac{7}{12} + 13\frac{1}{12}$ h. $24\frac{3}{10} + 32\frac{7}{10} + 16\frac{9}{10}$ i. $22\frac{11}{40} + 11\frac{11}{40}$

2. Add. Answers should be in simplest form.

a. $7\frac{1}{2} + 4\frac{1}{3}$ b. $23\frac{1}{12} + 18\frac{2}{3}$ c. $6\frac{1}{8} + 4\frac{5}{12} + 1\frac{9}{24}$ d. $6\frac{2}{3} + 11\frac{3}{4}$

e. $9\frac{7}{8} + 17\frac{5}{12}$ f. $36\frac{7}{36} + 48\frac{5}{48}$ g. $18\frac{3}{8} + 5\frac{1}{4}$ h. $6\frac{1}{2} + 5\frac{1}{3} + 4\frac{1}{4}$ i. $35\frac{7}{15} + 48\frac{7}{12} + 52\frac{7}{20}$

3. To sweeten the lemonade, Mom put in $1\frac{1}{3}$ cups of sugar. Joe thought it was still too sour and added another $\frac{1}{2}$ cup of sugar. How many cups of sugar were in the lemonade?

4. It costs about $7\frac{1}{2}$¢ per mile to operate a car in Miami, Florida. How much would it cost to drive it for 2 miles?

5. Bob caught fish weighing $4\frac{1}{4}$ lb. and $3\frac{5}{8}$ lb. How much did the fish he caught weigh together?

6. After a dinner party ended, there were parts of 3 pies left. There was $\frac{1}{8}$ of the banana pie left, $\frac{3}{8}$ of the peach pie, and $\frac{5}{8}$ of the apple pie. How much pie was left in all?

7. One week Rebecca worked 3 days. The first day she worked $5\frac{2}{3}$ hours, the next day she worked $6\frac{3}{4}$ hours, and the third day she worked $4\frac{1}{2}$ hours. How many hours did she work in all?

## SPIRAL REVIEW EXERCISES

1. Which fraction has the largest value?
   (a) $\frac{2}{3}$ (b) $\frac{3}{4}$ (c) $\frac{3}{5}$ (d) $\frac{4}{5}$

2. Find the sum of $\frac{2}{5}$ and $\frac{3}{10}$.

3. Find the product of $\frac{3}{5}$ and $\frac{10}{27}$.

4. $\frac{22}{3}$ has the same value as
   (a) $8\frac{1}{3}$ (b) $7\frac{1}{3}$ (c) $6\frac{2}{3}$ (d) $5\frac{2}{3}$

5. Subtract $-10$ from $-20$.

6. What is the value of 2 in 524,689?

7. Which is a prime number?
   (a) 21 (b) 31 (c) 51 (d) 81

8. Find the least common multiple (LCM) of 15 and 25.

9. How many 4-ounce portions can be made from 304 ounces of ground beef?

10. A number phrase showing that the total price of 12 books, each costing \$8, has been reduced by \$5 is
    (a) $(12 - 8) \times 5$ (c) $12 \times 8 - 5$
    (b) $(12 - 5) \times 8$ (d) $12 \times (8 - 5)$

11. Hector Blair ordered 14 tires, at \$35 each, for his auto service center. He received credit for 6 tires, costing \$31 each, that he returned to the supplier. How much does he still owe?

## CHALLENGE QUESTION

Copy and complete the chart to obtain the final answer.

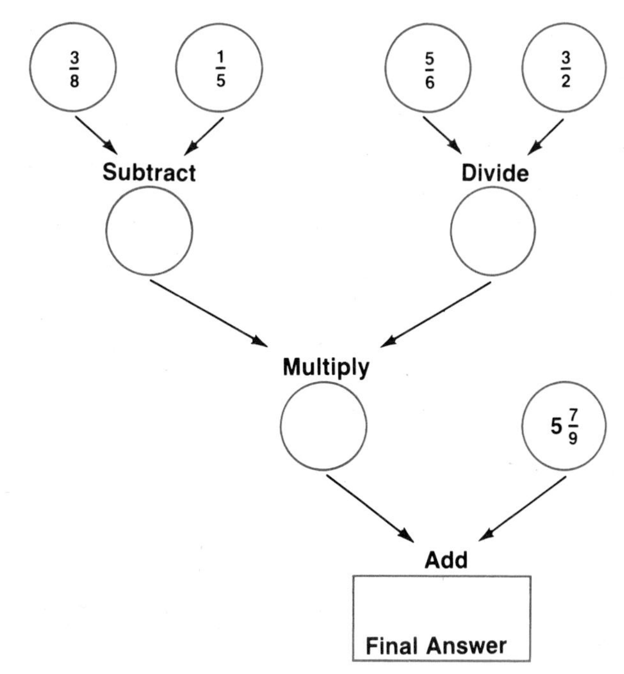

# UNIT 6–2 Subtracting Mixed Numbers

## SUBTRACTING MIXED NUMBERS WITH LIKE-FRACTION PARTS

### THE MAIN IDEA

To subtract mixed numbers with like-fraction parts:

1. Subtract the like-fraction parts.
   a. If necessary, rename the minuend (the first, or top, number).
   b. If possible, simplify the resulting fraction.
2. Subtract the whole-number parts.

**EXAMPLE 1** Subtract: $7\frac{7}{11} - 2\frac{5}{11}$

No renaming is necessary, since $\frac{7}{11}$ is greater than $\frac{5}{11}$.

First subtract the like-fraction parts. No simplification is possible. Then subtract the whole-number parts.

$$\begin{array}{r} 7\frac{7}{11} \\ -\,2\frac{5}{11} \\ \hline 5\frac{2}{11} \end{array} \quad \textit{Ans.}$$

**EXAMPLE 2** Subtract: $7\frac{1}{12} - 5\frac{5}{12}$

Since $\frac{5}{12}$ is greater than $\frac{1}{12}$, rename the minuend.

$$7\frac{1}{12} = 6\frac{13}{12}$$

Subtract the like-fraction parts and subtract the whole-number parts.

$$\begin{array}{r} 6\frac{13}{12} \\ -\,5\frac{5}{12} \\ \hline 1\frac{8}{12} \end{array}$$

Simplify the resulting fraction part.

$$1\frac{8 \div 4}{12 \div 4} = 1\frac{2}{3} \quad \textit{Ans.}$$

**EXAMPLE 3** Jill has 7 meters of fabric. She uses $2\frac{3}{10}$ meters to make a dress. How much fabric does she have left?

**THINKING ABOUT THE PROBLEM**

To find how much is left after an amount is used, you subtract.

Write a number phrase. $7 - 2\frac{3}{10}$

Since 7 is a whole number, rename the minuend. $7 = 6\frac{10}{10}$

Subtract the like-fraction parts and subtract the whole-number parts.

$$\begin{array}{r} 6\frac{10}{10} \\ -\,2\frac{3}{10} \\ \hline 4\frac{7}{10} \end{array}$$

*Answer:* Jill has $4\frac{7}{10}$ meters of fabric left.

## CLASS EXERCISES

1. Subtract. Answers should be in simplest form.

a. $\begin{array}{r} 12\frac{9}{17} \\ -8\frac{3}{17} \\ \hline \end{array}$ b. $\begin{array}{r} 22\frac{9}{10} \\ -18\frac{3}{10} \\ \hline \end{array}$ c. $\begin{array}{r} 7\frac{11}{15} \\ -3\frac{8}{15} \\ \hline \end{array}$ d. $\begin{array}{r} 9\frac{3}{8} \\ -5\frac{1}{8} \\ \hline \end{array}$ e. $2\frac{3}{10} - 1\frac{7}{10}$ f. $8\frac{1}{6} - 2\frac{5}{6}$

g. $9 - 6\frac{3}{4}$ h. $12 - 10\frac{11}{16}$

2. Jenny promised her mother she would practice piano for 7 hours this week. If she has practiced for $4\frac{3}{4}$ hours so far, how many hours does she have left to practice?

3. In 1978, the average American man could expect to live to be $69\frac{1}{2}$ years old. By about 1995, the average American man can expect to live to be 73. How much older is this?

## SUBTRACTING MIXED NUMBERS WITH UNLIKE-FRACTION PARTS

### THE MAIN IDEA

To subtract mixed numbers with unlike-fraction parts:

1. Subtract the unlike-fraction parts.
   a. Use the LCD to change into equivalent like fractions.
   b. If necessary, rename the minuend.
   c. If possible, simplify the resulting fraction.
2. Subtract the whole-number parts.

**EXAMPLE 4** Subtract: $8\frac{2}{3} - 4\frac{1}{5}$

The LCD of the unlike fractions $\frac{2}{3}$ and $\frac{1}{5}$ is 15.

Rewrite the unlike-fraction parts as equivalent like fractions.

$$8\frac{2 \times 5}{3 \times 5} = 8\frac{10}{15}$$

$$4\frac{1 \times 3}{5 \times 3} = 4\frac{3}{15}$$

Subtract the resulting like-fraction parts and subtract the whole-number parts. No simplification is possible.

$$\begin{array}{r} 8\frac{10}{15} \\ -\,4\frac{3}{15} \\ \hline 4\frac{7}{15} \end{array} \quad \textit{Ans.}$$

**EXAMPLE 5** A grocer had $15\frac{1}{4}$ pounds of apples. He sold $8\frac{1}{2}$ pounds of the apples. How many pounds of apples did he still have?

**THINKING ABOUT THE PROBLEM**

To find how much is left after an amount is taken away from a given amount, you subtract.

Write a number phrase. $15\frac{1}{4} - 8\frac{1}{2}$

The LCD of the unlike fractions $\frac{1}{4}$ and $\frac{1}{2}$ is 4.

Rewrite the unlike fractions as equivalent like fractions.

$$15\frac{1}{4} = 15\frac{1}{4}$$

$$8\frac{1 \times 2}{2 \times 2} = 8\frac{2}{4}$$

Since $\frac{2}{4}$ is greater than $\frac{1}{4}$, rename the minuend.

$$15\frac{1}{4} = 14\frac{5}{4}$$

Subtract the like-fraction parts and subtract the whole-number parts.

$$\begin{array}{r} 14\frac{5}{4} \\ -\,8\frac{2}{4} \\ \hline 6\frac{3}{4} \end{array}$$

*Answer:* The grocer still had $6\frac{3}{4}$ pounds of apples.

## CLASS EXERCISES

1. Subtract. Answers should be in simplest form.

a. $7\frac{3}{10} - 2\frac{2}{5}$ b. $6\frac{9}{10} - 4\frac{3}{4}$ c. $4\frac{7}{12} - 1\frac{3}{8}$ d. $10\frac{1}{2} - 4\frac{7}{8}$ e. $9\frac{5}{16} - 6\frac{5}{8}$ f. $12\frac{2}{5} - 3\frac{3}{4}$

g. $14\frac{3}{8} - 5\frac{2}{3}$ h. $20\frac{7}{10} - 9\frac{4}{5}$

2. Mary's home is $2\frac{3}{4}$ miles from school. If Mary walked $1\frac{1}{10}$ miles toward school, how much farther must she walk?

3. Alan had $12\frac{1}{8}$ pounds of peanuts. He sold $6\frac{3}{4}$ pounds. How many pounds of peanuts does Alan still have?

## HOMEWORK EXERCISES

1. Subtract. Answers should be in simplest form.

a. $24\frac{11}{12} - 18\frac{10}{12}$ b. $16\frac{3}{4} - 11\frac{1}{4}$ c. $20\frac{7}{8} - 17\frac{4}{8}$ d. $42\frac{9}{10} - 24\frac{7}{10}$ e. $19\frac{5}{6} - 9\frac{1}{6}$ f. $53\frac{21}{30} - 47\frac{17}{30}$

g. $21\frac{9}{36} - 18\frac{7}{36}$ h. $29\frac{16}{17} - 19\frac{15}{17}$ i. $37 - 26\frac{1}{4}$ j. $12\frac{7}{9} - 9$ k. $21 - 11\frac{3}{8}$

2. Subtract. Answers should be in simplest form.

a. $18\frac{3}{4} - 6\frac{1}{2}$ b. $21\frac{5}{8} - 17\frac{1}{4}$ c. $38\frac{4}{5} - 22\frac{3}{10}$ d. $29\frac{5}{12} - 15\frac{1}{6}$ e. $5\frac{2}{3} - 3\frac{1}{2}$ f. $24\frac{3}{8} - 17\frac{1}{5}$

g. $40\frac{5}{9} - 36\frac{2}{5}$ h. $36\frac{1}{8} - 24\frac{3}{5}$ i. $52\frac{3}{7} - 40\frac{3}{5}$ j. $29\frac{7}{10} - 12\frac{7}{8}$ k. $19\frac{1}{20} - 5\frac{1}{15}$

3. From $13\frac{7}{8}$, subtract $9\frac{1}{5}$. 4. Subtract $32\frac{4}{5}$ from $41\frac{1}{7}$.

5. Steve ran $6\frac{3}{4}$ miles Sunday. After he had run $3\frac{1}{4}$ miles, Joy joined him for the rest of the run. How far did Joy run with Steve?

6. It was $4\frac{4}{5}$ miles from Rebecca's house to school. She walked $\frac{2}{5}$ mile and then took the bus. How far did she ride the bus?

7. The largest brook trout caught so far weighed $14\frac{1}{2}$ pounds. The largest rainbow trout caught weighed $42\frac{1}{8}$ pounds. How much more did the largest rainbow trout weigh than the largest brook trout?

8. A package of hamburger weighing $2\frac{1}{2}$ pounds was cooked. After cooking, the hamburger weighed $1\frac{7}{8}$ pounds. How much less did the cooked hamburger weigh than the raw hamburger?

## SPIRAL REVIEW EXERCISES

1. Find the sum of $9\frac{7}{8}$ and $2\frac{2}{3}$.

2. Jennifer needs $4\frac{1}{3}$ cups of milk for one recipe and $6\frac{5}{6}$ cups of milk for a second recipe. How many cups of milk does she need for both recipes?

3. The sum of $-12$ and $-8$ is
   (a) $-4$ (b) 4 (c) $-20$ (d) 20

4. $47 \div 8$ is closest to
   (a) 5 (b) 6 (c) 7 (d) 8

5. The greatest common factor of 27 and 75 is
   (a) 3 (b) 5 (c) 9 (d) 15

6. Divide 40 by $\frac{1}{2}$.

7. Find $\frac{1}{2}$ of 40.

8. If each page of text contains 250 words, how many words are there on 47 such pages?

9. What is the quotient when 12,060 is divided by 60?
   (a) 21 (b) 210 (c) 200 (d) 201

10. Find all the prime numbers between 70 and 80.

11. Write the word name for 23,802,010.

12. Write each group of numbers in order, from smallest to largest.

    **a.** 0, $-16$, $-5$ **b.** $\frac{11}{4}$, $2\frac{3}{5}$, $\frac{13}{6}$

### CHALLENGE QUESTION

After school, John has 7 hours of time available. The chart shows John's plan to use this time. How much time should he budget to watch television?

| Activity | Time Budgeted (in Hours) |
|---|---|
| Homework | $2\frac{3}{4}$ |
| Track Practice | $2\frac{1}{2}$ |
| Dinner | $\frac{3}{4}$ |
| Television | ? |

## UNIT 6–3 Multiplying Mixed Numbers

### THE MAIN IDEA

To multiply mixed numbers:

1. Change each mixed number to an improper fraction.
2. Follow the procedure for multiplying two fractions.

**EXAMPLE 1** Multiply: $3\frac{1}{2} \times 2\frac{1}{3}$

Change each mixed number to an improper fraction.

$$3\frac{1}{2} = \frac{2 \times 3 + 1}{2} = \frac{7}{2}$$

$$2\frac{1}{3} = \frac{3 \times 2 + 1}{3} = \frac{7}{3}$$

Multiply the fractions.

$$\frac{7}{2} \times \frac{7}{3} = \frac{49}{6}$$

*Answer:* $\frac{49}{6}$ or $8\frac{1}{6}$

**EXAMPLE 2** Multiply: $9 \times 4\frac{2}{3}$

Write the whole number as a fraction.

$$\frac{9}{1} \times 4\frac{2}{3}$$

Change the mixed number to an improper fraction.

$$\frac{9}{1} \times \frac{14}{3}$$

To simplify, divide 9 and 3 by 3.

$$\frac{\overset{3}{\cancel{9}}}{1} \times \frac{14}{\underset{1}{\cancel{3}}}$$

Multiply the results.

$$\frac{3 \times 14}{1 \times 1} = \frac{42}{1} = 42 \quad \textit{Ans.}$$

**EXAMPLE 3** A recipe calls for $2\frac{1}{4}$ cups of flour. How much flour is needed to make $\frac{2}{3}$ of the recipe?

**THINKING ABOUT THE PROBLEM**

The key phrase "$\frac{2}{3}$ of" tells you to multiply.

Write a number phrase.

$$\frac{2}{3} \times 2\frac{1}{4}$$

Change the mixed number to an improper fraction.

$$\frac{2}{3} \times \frac{9}{4}$$

To simplify, divide 2 and 4 by 2. Divide 3 and 9 by 3.

$$\frac{\overset{1}{\cancel{2}}}{\underset{1}{\cancel{3}}} \times \frac{\overset{3}{\cancel{9}}}{\underset{2}{\cancel{4}}}$$

Multiply the results.

$$\frac{1 \times 3}{1 \times 2} = \frac{3}{2}$$

*Answer:* $\frac{3}{2}$ cups or $1\frac{1}{2}$ cups of flour are needed.

## CLASS EXERCISES

1. Multiply. Answers should be in simplest form.

   **a.** $3\frac{1}{2} \times \frac{1}{4}$ **b.** $2\frac{1}{3} \times 2\frac{3}{4}$ **c.** $3\frac{2}{5} \times 5\frac{1}{3}$ **d.** $1\frac{3}{4} \times \frac{1}{5}$ **e.** $5\frac{3}{8} \times 3\frac{1}{2}$ **f.** $8 \times 2\frac{1}{6}$

   **g.** $4\frac{2}{3} \times 9$ **h.** $7\frac{3}{4} \times 1\frac{1}{3}$ **i.** $\frac{5}{8} \times 1\frac{1}{4}$

2. The product of $4\frac{4}{5}$ and $3\frac{5}{8}$ is (a) $15\frac{4}{5}$ (b) $16\frac{3}{5}$ (c) $17\frac{2}{5}$ (d) $18\frac{7}{40}$

3. The distance from Michelle's house to school is $\frac{3}{5}$ of a mile. How far did Michelle walk if she made this trip 20 times?

4. A bug landed on a phonograph record that was making $33\frac{1}{3}$ revolutions per minute. The bug stayed on the record for 6 minutes before flying off. How many complete revolutions did the bug make on the record?

## HOMEWORK EXERCISES

1. Multiply. Answers should be in simplest form.

   **a.** $5\frac{1}{3} \times \frac{1}{2}$ **b.** $3\frac{3}{4} \times \frac{1}{5}$ **c.** $10\frac{1}{2} \times 9$ **d.** $4\frac{1}{2} \times 3\frac{1}{3}$ **e.** $5\frac{5}{8} \times 1\frac{2}{3}$ **f.** $12 \times 3\frac{3}{8}$

   **g.** $\frac{7}{9} \times 3\frac{3}{7}$ **h.** $5\frac{3}{5} \times 2\frac{1}{5}$ **i.** $16\frac{3}{10} \times 4\frac{3}{5}$

2. When $3\frac{2}{3}$ is multiplied by $2\frac{1}{2}$, the product is (a) $\frac{37}{6}$ (b) 11 (c) $\frac{22}{15}$ (d) $9\frac{1}{6}$

3. Find $\frac{7}{16}$ of $2\frac{2}{7}$.

4. A pound of brown sugar fills $2\frac{1}{3}$ cups. How many cups would be filled by $1\frac{1}{2}$ pounds of brown sugar?

5. A bag of grapefruit weighs $8\frac{1}{2}$ pounds. How much would 3 bags of grapefruit weigh?

6. A melon weighed $1\frac{7}{16}$ pounds. If you ate $2\frac{1}{2}$ melons in a week, how many pounds of melon would you have eaten?

## SPIRAL REVIEW EXERCISES

1. John's father baked 3 pies. John and his friends ate $1\frac{1}{6}$ pies. How many pies were left?

2. Add: $12\frac{3}{5} + 8\frac{2}{3}$

3. From 5, subtract $3\frac{3}{8}$.

4. Olga practiced playing the piano $\frac{3}{4}$ hour Monday, $\frac{1}{2}$ hour Tuesday, $\frac{1}{2}$ hour Wednesday, $1\frac{1}{3}$ hours Thursday, and $\frac{2}{3}$ hour Friday. How much time did she spend practicing the piano in all?

5. Find the remainder when 189 is divided by 13.

6. Subtract 13 from $-20$.

7. Simplify: $\frac{48}{80}$

8. Joy's diet allowed her to eat $\frac{1}{8}$ of a pound of cheese in a sandwich. How many sandwiches could she make from $2\frac{3}{4}$ pounds of cheese?

9. On Martin's block, there are 13 garages that can hold 2 cars each and 8 garages that can hold 1 car each. What is the total number of cars that can be garaged on the block?

10. How many blood specimens can be held on 8 trays if each tray has 12 rows of test tubes with 10 test tubes in each row?

## CHALLENGE QUESTION

Kathy had a very profitable year with her investments. Each of her four stocks showed a profit, as indicated on her "stock profit sheet."

**Stock Profit Sheet**

| Name of Stock | Number of Shares | Profit per Share | Total Profit for Stock |
|---|---|---|---|
| Dynamic Dolls | 400 | $2\frac{3}{8}$ | |
| Striking Oil | 1,000 | $10\frac{2}{5}$ | |
| Blue Coal | 800 | $12\frac{3}{16}$ | |
| Fast Bikes | 1,200 | $11\frac{1}{12}$ | |

a. Copy the table and complete each of the entries in the last column.

b. What was Kathy's total profit from the four stocks?

# UNIT 6–4 Dividing Mixed Numbers

## THE MAIN IDEA

To divide mixed numbers:

1. Change each mixed number to an improper fraction.
2. Follow the procedure for dividing two fractions.

**EXAMPLE 1** Divide: $2\frac{3}{5} \div 5\frac{1}{2}$

Change each mixed number to an improper fraction.

$$2\frac{3}{5} = \frac{5 \times 2 + 3}{5} = \frac{13}{5}$$

$$5\frac{1}{2} = \frac{2 \times 5 + 1}{2} = \frac{11}{2}$$

Divide the fractions.

$$\frac{13}{5} \div \frac{11}{2}$$

Use the reciprocal of the divisor and change the division to multiplication.

$$\frac{13}{5} \times \frac{2}{11}$$

There are no common factors. Multiply.

$$\frac{13 \times 2}{5 \times 11} = \frac{26}{55} \quad \textit{Ans.}$$

**EXAMPLE 2** Divide: $2\frac{5}{8} \div 3$

Change the mixed number to an improper fraction.

$$2\frac{5}{8} = \frac{8 \times 2 + 5}{8} = \frac{21}{8}$$

Express the whole number as an improper fraction.

$$3 = \frac{3}{1}$$

Divide the fractions.

$$\frac{21}{8} \div \frac{3}{1}$$

$$\frac{\overset{7}{\cancel{21}}}{8} \times \frac{1}{\underset{1}{\cancel{3}}} = \frac{7 \times 1}{8 \times 1} = \frac{7}{8} \quad \textit{Ans.}$$

**EXAMPLE 3** June bakes cookies and sells them in containers that hold $1\frac{1}{4}$ pounds. How many containers does she need to hold 660 pounds of cookies?

**THINKING ABOUT THE PROBLEM**

To separate a given amount into a number of smaller equal portions, you divide.

Write a number phrase.

$$660 \div 1\frac{1}{4}$$

Change the mixed number to an improper fraction.

$$1\frac{1}{4} = \frac{5}{4}$$

Express the whole number as an improper fraction. $660 = \frac{660}{1}$

Divide the fractions. $\frac{660}{1} \div \frac{5}{4}$

$$\frac{\overset{132}{\cancel{660}}}{1} \times \frac{4}{\underset{1}{\cancel{5}}}$$

$$\frac{132 \times 4}{1 \times 1} = \frac{528}{1} = 528$$

*Answer:* June needs 528 containers.

## CLASS EXERCISES

1. Divide. Answers should be in simplest form.

   a. $1\frac{3}{5} \div 1\frac{1}{3}$ b. $3\frac{1}{3} \div 5\frac{1}{2}$ c. $\frac{5}{6} \div 6\frac{2}{3}$ d. $2\frac{2}{3} \div 1\frac{3}{8}$ e. $2\frac{1}{4} \div 2\frac{2}{5}$ f. $3\frac{3}{8} \div 9$

   g. $1\frac{4}{5} \div \frac{3}{5}$ h. $6\frac{7}{8} \div 1\frac{2}{3}$ i. $42 \div 4\frac{1}{5}$

2. The quotient of $4\frac{2}{3}$ and $1\frac{1}{6}$ is (a) $3\frac{20}{21}$ (b) 4 (c) $4\frac{5}{21}$ (d) $5\frac{1}{21}$

3. Jack has $6\frac{2}{5}$ pounds of raisins that he wants to share equally among himself and three friends. How many pounds should he give to each person?

4. How many pieces of string, each $4\frac{3}{4}$ inches long, can be cut from a length of string that is $85\frac{1}{2}$ inches long?

## HOMEWORK EXERCISES

1. Divide. Answers should be in simplest form.

   a. $2\frac{1}{3} \div \frac{1}{2}$ b. $\frac{3}{5} \div 1\frac{1}{2}$ c. $4\frac{1}{3} \div 2\frac{1}{4}$ d. $5\frac{2}{3} \div 1\frac{3}{5}$ e. $5\frac{1}{4} \div 7$ f. $1\frac{3}{8} \div \frac{22}{32}$

   g. $2\frac{2}{3} \div 3\frac{1}{3}$ h. $7\frac{2}{3} \div 46$ i. $6\frac{1}{2} \div 8\frac{2}{3}$

2. When 48 is divided by $\frac{1}{2}$, the quotient is (a) $48\frac{1}{2}$ (b) 96 (c) 24 (d) $\frac{24}{48}$

3. There were 12 oranges in a bag. The bag weighed $7\frac{1}{2}$ pounds. About how much did each orange weigh?

4. There were $5\frac{1}{2}$ sandwiches for 4 children. How much should each child get?

5. Steve cut a wire that was $20\frac{1}{4}$ feet long into pieces $3\frac{3}{8}$ feet long. How many pieces did he have?

## SPIRAL REVIEW EXERCISES

1. Jane took a trip to Cape Cod. She averaged 50 miles per hour for $4\frac{1}{2}$ hours. How far is Cape Cod from Jane's home?

2. Multiply: $5\frac{1}{2}$ by $3\frac{2}{3}$

3. Find the sum of $8\frac{1}{4}$ and $4\frac{1}{8}$.

4. Subtract $2\frac{3}{7}$ from 5.

5. The opposite of $-11$ is
   (a) 0 (b) 11 (c) $\frac{1}{11}$ (d) $-\frac{1}{11}$

6. What is the reciprocal of 20?

7. Find $\frac{4}{5}$ of $\frac{5}{16}$.

8. Replace ? by a number to make a true statement:
   $$-8 + ? = 11$$

9. Replace ? by <, >, or = to make a true statement:
   **a.** $\frac{7}{16}$ ? $\frac{21}{48}$ **b.** $\frac{1}{5}$ ? $\frac{1}{4}$ **c.** $\frac{2}{3}$ ? $\frac{1}{3}$

10. On a $5\frac{1}{2}$-hour trip, Rich drove $\frac{2}{3}$ of the time. How long did he drive?

11. If one typewriter ribbon can produce 150 clear copies, how many ribbons would be required for 3,300 copies?

12. Write a number phrase that represents the total cost of 8 books costing $5 each and 25 pens costing $2 each.

### CHALLENGE QUESTION

Steve bought 2 packages of hamburger meat. Each package weighed $2\frac{3}{4}$ pounds. When he got home, he divided the meat into $\frac{1}{4}$-pound hamburger patties. How many hamburger patties did he make?

# THEME 4

# Decimals

## MODULE 7



## MODULE 8

# UNIT 7–1 The Meaning of Decimals; Reading and Writing Decimals

## THE MEANING OF DECIMALS

### THE MAIN IDEA

1. A *decimal* is another way of writing a fraction.
2. The place names and place values for decimals are:

| | • | | | | | |
|---|---|---|---|---|---|---|
| **Place Names:** | tenths | hundredths | thousandths | ten-thousandths | hundred-thousandths | millionths |
| **Place Values:** | $\frac{1}{10}$ | $\frac{1}{100}$ | $\frac{1}{1,000}$ | $\frac{1}{10,000}$ | $\frac{1}{100,000}$ | $\frac{1}{1,000,000}$ |

3. As we move to the right from the decimal point, each place stands for a fraction $\frac{1}{10}$ as large as the place before.
4. To find the value of a digit in a decimal, multiply the digit by the value of its place.

**EXAMPLE 1** For each of the underlined digits, write the place name as a word and as a fraction.

| | *Digit* | *Place Name as a Word* | *Place Name as a Fraction* |
|---|---|---|---|
| **a.** | $.82\underline{3}$ | thousandths | $\frac{1}{1,000}$ |
| **b.** | $.\underline{9}41$ | tenths | $\frac{1}{10}$ |
| **c.** | $.6\underline{8}52$ | hundredths | $\frac{1}{100}$ |
| **d.** | $.739\underline{6}$ | ten-thousandths | $\frac{1}{10,000}$ |

**EXAMPLE 2** For each digit of the decimal number .9999, write the value of the digit as a fraction.

| *Digit* | *Value* |
|---|---|
| a. $.\underline{9}999$ | $9 \times \frac{1}{10}$ or $\frac{9}{10}$ |
| b. $.9\underline{9}99$ | $9 \times \frac{1}{100}$ or $\frac{9}{100}$ |
| c. $.99\underline{9}9$ | $9 \times \frac{1}{1{,}000}$ or $\frac{9}{1{,}000}$ |
| d. $.999\underline{9}$ | $9 \times \frac{1}{10{,}000}$ or $\frac{9}{10{,}000}$ |

## *CLASS EXERCISES*

1. For each of the underlined digits, write the name of the place as a word and as a fraction.
   a. $.7\underline{2}6$ b. $.\underline{8}17$ c. $.23\underline{8}9$ d. $.465\underline{7}$ e. $.3\underline{4}68$

2. Write as a fraction the value of 7 in each decimal.
   a. $.2\underline{7}$ b. $.\underline{7}39$ c. $.52\underline{7}8$ d. $.8426\underline{7}$ e. $.493\underline{7}2$

## READING DECIMALS

### THE MAIN IDEA

To read a decimal:

1. Ignore the decimal point and any leading zeros (zeros immediately following the decimal point).
2. Read the digits as if they were naming a whole number.
3. Say the name of the place value of the last digit.
4. If the decimal has a whole-number part:
   a. Read the whole-number part.
   b. Say the word "and."
   c. Read the decimal part.

**EXAMPLE 3** Read each decimal and write its word name.

**a.** .8 Read the digit 8 as "eight" and say the place value of the 8, which is tenths.

*Answer:* eight tenths

**b.** .135 Read 135 as "one hundred thirty-five" and say the place value of the 5, which is thousandths.

*Answer:* one hundred thirty-five thousandths

**c.** .0075 Ignore the two leading zeros. Read 75 as "seventy-five" and say the place value of the 5, which is ten-thousandths.

*Answer:* seventy-five ten-thousandths

**EXAMPLE 4** Read each decimal and write its word name.

| | *Decimal* | *Word Name* |
|---|---|---|
| **a.** | .27 | twenty-seven hundredths |
| **b.** | .0027 | twenty-seven ten-thousandths |
| **c.** | .056 | fifty-six thousandths |
| **d.** | .00056 | fifty-six hundred-thousandths |

**EXAMPLE 5** Read each decimal and write its word name.

| | *Decimal* | *Word Name* |
|---|---|---|
| **a.** | 8.07 | eight and seven hundredths |
| **b.** | 14.3 | fourteen and three tenths |
| **c.** | 72.0005 | seventy-two and five ten-thousandths |
| **d.** | 0.4 or .4 | four tenths |

## CLASS EXERCISES

**1.** Read each decimal and write its word name.

**a.** .14 **b.** 0.5 **c.** 0.07 **d.** 0.27 **e.** .037 **f.** 0.05 **g.** 0.7 **h.** 0.083
**i.** 0.374 **j.** 0.19 **k.** 0.007 **l.** 0.0023

**2.** Read each decimal and write its word name.

**a.** 21.07 **b.** 124.023 **c.** 132.4 **d.** 91.091 **e.** 5.129 **f.** 17.19 **g.** 702.05

## WRITING DECIMALS GIVEN AS WORD NAMES

### THE MAIN IDEA

To write a decimal that is given as a word name:

1. Think of the decimal as a fraction with a denominator whose place name is mentioned.
2. Write the fraction as a decimal.

**EXAMPLE 6** Write twenty-three hundredths as a decimal.

Think of twenty-three *hundredths* as a fraction with a denominator whose place name is mentioned.

twenty-three — The numerator is 23.

hundredths — The denominator is 100.

Write $\frac{23}{100}$ as a decimal.

$$\frac{23}{100} = .23 \quad \textit{Ans.}$$

**EXAMPLE 7** Write each word name as a decimal.

| | *Word Name* | *Decimal* |
|---|---|---|
| **a.** | three and three hundredths | 3.03 |
| **b.** | nine hundred thirty-seven thousandths | 0.937 or .937 |
| **c.** | nine and thirty-seven thousandths | 9.037 |
| **d.** | four hundred seven and nine tenths | 407.9 |
| **e.** | one hundred and three tenths | 100.3 |

## CLASS EXERCISES

Write each word name as a decimal.

**1.** forty-seven hundredths **2.** fifteen thousandths **3.** two hundred and three hundredths

**4.** nine tenths **5.** eleven and seven tenths **6.** two and twelve hundredths

**7.** one hundred forty-five and twenty thousandths

**8.** one hundred forty-five thousandths

**9.** fourteen and eighty-one ten-thousandths

**10.** two hundred twenty-two and two thousandths

## HOMEWORK EXERCISES

**1.** For each of the underlined digits, write the name of the place as a word and as a fraction.

**a.** $.\underline{9}35$ **b.** $0.4\underline{2}9$ **c.** $0.2\underline{6}4$ **d.** $0.82\underline{4}3$ **e.** $0.0\underline{7}$ **f.** $.82\underline{6}$ **g.** $0.375\underline{8}$

**h.** $0.34\underline{9}2$ **i.** $0.\underline{4}759$ **j.** $0.000\underline{7}$

**2.** In each decimal, write the value of 3 as a fraction.

**a.** $0.1\underline{3}5$ **b.** $0.25\underline{3}8$ **c.** $0.\underline{3}17$ **d.** $0.514\underline{3}$ **e.** $0.9\underline{3}65$

**3.** Read each decimal and write its word name.

**a.** .09 **b.** .017 **c.** .213 **d.** .79 **e.** .2 **f.** .9 **g.** .6

**h.** .0213 **i.** .0003 **j.** .002 **k.** 7.21 **l.** 8.09 **m.** 12.015 **n.** 115.003

**o.** 207.702 **p.** 9.3 **q.** 16.57 **r.** 96.08 **s.** 89.03 **t.** 73.5

**4.** Write each word name as a decimal.

**a.** seven tenths **b.** eighty-three hundredths **c.** two hundred thirty-two thousandths

**d.** seventy-five thousandths **e.** five hundredths **f.** six and nine tenths

**g.** twenty-one and fifteen hundredths **h.** two hundred nine thousandths

**i.** two hundred and nine thousandths **j.** two and one hundred nine thousandths

## *SPIRAL REVIEW EXERCISES*

**1.** A prime number between 32 and 40 is
(a) 33 (b) 34 (c) 37 (d) 39

**2.** The least common denominator of the fractions $\frac{3}{4}$ and $\frac{7}{8}$ is
(a) 2 (b) 4 (c) 8 (d) 16

**3.** Divide 414 by 18.

**4.** Divide $-140$ by 7.

**5.** Rearrange the numbers 100, $-100$, 10, and $-10$ in order, from least to greatest.

**6.** If 456 eggs are to be packed into boxes holding 12 eggs per box, how many boxes will be needed?

**7.** From a piece of wood $8\frac{3}{4}$ inches long, $2\frac{3}{8}$ inches are removed. How long is the remaining piece?

**8.** When $\frac{1}{20}$ is divided by $\frac{1}{2}$, the result is
(a) $\frac{1}{10}$ (b) $\frac{1}{40}$ (c) $\frac{2}{10}$ (d) $\frac{2}{40}$

**9.** A rope $13\frac{1}{2}$ feet long is to be cut into 4 pieces of equal length. How long will each piece be?

**10.** What is $\frac{2}{3}$ of 48?

**11.** Mr. James bought a $399 television set on sale for $280. How much money did he save?

**12.** Ryan's father travels a total of 22 miles to and from work. How many miles does he commute in a five-day work week?

**13.** Find the value of $9 \times 4 + 10 \div 2$.

**14.** How much greater is $\frac{3}{4}$ than $\frac{3}{8}$?

### CHALLENGE QUESTION

Write the sum of $\frac{5}{10,000}$, $\frac{3}{1,000}$, $\frac{7}{100}$, and $\frac{1}{10}$ as a decimal.

# UNIT 7–2 Rounding Decimals

## THE MAIN IDEA

To round a decimal:

1. Circle the digit in the place to be rounded.
2. Look at the digit immediately to the right of the circled digit.
   a. If the digit to the right of the circled digit is 5 or more, increase the circled digit by one (round up).
   b. If the digit to the right of the circled digit is less than 5, keep the circled digit (round down).
3. Drop all digits to the right of the circled digit.

**EXAMPLE 1** Round .793 to the nearest tenth.

Circle 7, the digit in the tenths place. 9 is the digit in the next position to the right. — .(7)93

Since 9 > 5, round up. Drop the digits to the right of the tenths place. — .8 *Ans.*

**EXAMPLE 2** Round .952 to the nearest hundredth.

Circle 5, the digit in the hundredths place. 2 is the digit in the next position to the right. — .9(5)2

Since 2 < 5, round down. Drop the digit to the right of the hundredths place. — .95 *Ans.*

**EXAMPLE 3** Round .3965 to the nearest thousandth.

Circle 6, the digit in the thousandths place. 5 is the digit in the next position to the right. — .39(6)5

Since the next digit is 5, round up. Drop the digit to the right of the thousandths place. — .397 *Ans.*

**EXAMPLE 4** Round 3.981 to the nearest tenth.

Since 8 > 5, round up. — 3.(9)81
4.0 or 4 *Ans.*

## CLASS EXERCISES

1. Round each of the decimals to the nearest tenth.
   **a.** 0.93 **b.** 7.25 **c.** 19.375 **d.** 241.9802
2. Round each of the decimals to the nearest hundredth.
   **a.** 0.596 **b.** 6.245 **c.** 42.6842 **d.** 119.5237
3. Round each of the decimals to the nearest thousandth.
   **a.** 0.2468 **b.** 20.1935 **c.** 32.7124 **d.** 116.9465
4. Round 5.395624 to the nearest ten-thousandth.
5. Barbara spent $29.52 for a dress. Round the cost of the dress to the nearest dollar.
6. Mr. Jones needs 125.86 meters of fencing for his yard. Round the amount of fencing needed to the nearest tenth of a meter.

7. Which digit do we circle in 0.592 to round correct to the nearest tenth?

8. When we are rounding, to what number do we compare the digit to the right of the circled digit?

9. When we round 2.943 to the nearest hundredth, which digit do we circle?

## HOMEWORK EXERCISES

1. Round each of the decimals to the nearest tenth.

   **a.** 0.72 **b.** 0.35 **c.** 0.891 **d.** 8.92 **e.** 27.96

2. Round each of the decimals to the nearest hundredth.

   **a.** 0.283 **b.** 0.517 **c.** 0.123 **d.** 2.861 **e.** 6.998

3. Round each of the decimals to the nearest thousandth.

   **a.** 0.8176 **b.** 0.1462 **c.** 0.4739 **d.** 12.8754

4. Bennett spent $42.85 on a pair of shoes. Round the amount of money he paid to the nearest dollar.

5. Rita swam 47.29 meters. Round the distance she swam to the nearest tenth of a meter.

6. 0.549283 rounded to the nearest ten-thousandth is
   (a) 0.549 (b) 0.5492 (c) 0.5493 (d) 0.54928

## SPIRAL REVIEW EXERCISES

1. Write the word name of each decimal.

   **a.** 0.95 **b.** 0.04 **c.** .2 **d.** 0.125

2. For the decimal 10.0625, the value of 2 as a fraction is

   (a) $\frac{2}{10}$ (b) $\frac{2}{100}$ (c) $\frac{2}{1,000}$ (d) $\frac{2}{10,000}$

3. The least common denominator for the fractions $\frac{1}{2}$, $\frac{1}{3}$, and $\frac{1}{5}$ is

   (a) 2 (b) 3 (c) 15 (d) 30

4. Subtract $3\frac{2}{5}$ from $7\frac{3}{4}$.

5. James uses 89 calories for every mile he jogs. If he jogs 12 miles, how many calories does he use?

6. 502 × 98 is closest to
   (a) 5,000,000 (c) 50,000
   (b) 500,000 (d) 5,000

7. The dot on the number line represents the number
   (a) −8 (b) −3 (c) −2 (d) 2

8. Fifteen pounds of meat are to be divided into 4 packages of equal weight. What will be the weight of each package?

9. Find the least common multiple of 16 and 12.

### CHALLENGE QUESTION

John was asked to round .748 to the nearest tenth. He said that the 8 rounds the 4 up to 5, and 5 rounds the 7 up to 8. Thus, John's answer is that .748 rounded to the nearest tenth is .8. Is John correct? Explain.

# UNIT 7–3 Changing Decimals to Fractions

## THE MAIN IDEA

To write a decimal as a fraction, read the decimal using its word name.

1. The denominator of the fraction is the place name that you read.
2. The numerator of the fraction is the whole number that you read.
3. Simplify the fraction by dividing the numerator and the denominator by common factors.

**EXAMPLE 1** Write .37 as a fraction.

Read .37 as "thirty-seven hundredths."

"Hundredths" is the denominator of the fraction: $\frac{\phantom{00}}{100}$

"Thirty-seven" is the numerator of the fraction: $\frac{37}{100}$

*Answer:* $.37 = \frac{37}{100}$

**EXAMPLE 2** Write .059 as a fraction.

Read .059 as "fifty-nine thousandths."

"Thousandths" is the denominator of the fraction: $\frac{\phantom{00}}{1{,}000}$

"Fifty-nine" is the numerator of the fraction: $\frac{59}{1{,}000}$

*Answer:* $.059 = \frac{59}{1{,}000}$

**EXAMPLE 3** Write .8 as a fraction in simplest form.

Read .8 as "eight tenths" and write it as a fraction. $\frac{8}{10}$

Simplify $\frac{8}{10}$ by dividing the numerator and denominator by the common factor 2. $\frac{8 \div 2}{10 \div 2} = \frac{4}{5}$

*Answer:* $.8 = \frac{4}{5}$

**EXAMPLE 4** Write 0.45 as a fraction in simplest form.

Read 0.45 as "forty-five hundredths" and write it as a fraction. $\frac{45}{100}$

Simplify $\frac{45}{100}$ by dividing the numerator and denominator by 5. $\frac{45 \div 5}{100 \div 5} = \frac{9}{20}$

*Answer:* $0.45 = \frac{9}{20}$

**EXAMPLE 5** .0025 names the same point on a number line as

(a) $\frac{1}{40}$ (b) $\frac{1}{400}$ (c) $\frac{1}{4}$ (d) $\frac{25}{1{,}000}$

Change .0025 to a fraction. $\frac{25}{10{,}000}$

Reduce the fraction. $\frac{25 \div 25}{10{,}000 \div 25} = \frac{1}{400}$

*Answer:* (b)

**EXAMPLE 6** 5.4 equals

(a) $5\frac{4}{5}$ (b) $4\frac{5}{10}$ (c) $5\frac{2}{5}$ (d) $6\frac{4}{10}$

Write the decimal number as a whole number plus its decimal part. $5.4 = 5 + .4$

Change the decimal part to a fraction. $.4 = \frac{4}{10}$

Simplify the fraction. $\frac{4 \div 2}{10 \div 2} = \frac{2}{5}$

Thus, $5.4 = 5\frac{2}{5}$.

*Answer:* (c)

## CLASS EXERCISES

1. Write each decimal as a fraction.

   **a.** 0.3 **b.** .051 **c.** 0.417 **d.** 0.019 **e.** 0.27 **f.** .0793 **g.** 0.00027 **h.** 0.3007

2. Write each decimal as a fraction in simplest form.

   **a.** 0.6 **b.** 0.25 **c.** 0.125 **d.** 0.550 **e.** 0.18 **f.** 0.95 **g.** 0.875 **h.** 0.048

3. .045 written as a fraction is (a) $\frac{9}{20}$ (b) $\frac{9}{200}$ (c) $\frac{45}{100}$ (d) $\frac{9}{100}$

4. .0625 names the same point on the number line as (a) $\frac{1}{4}$ (b) $\frac{3}{8}$ (c) $\frac{1}{16}$ (d) $\frac{3}{50}$

5. 8.75 has the same value as (a) $8\frac{3}{5}$ (b) $8\frac{3}{4}$ (c) $8\frac{3}{8}$ (d) $7\frac{3}{8}$

6. Raymond ran 3.85 miles. This distance is the same as

   (a) $3\frac{4}{5}$ miles (b) $3\frac{8}{10}$ miles (c) $2\frac{17}{20}$ miles (d) $3\frac{17}{20}$ miles

## HOMEWORK EXERCISES

1. Write each decimal as a fraction.

   **a.** .29 **b.** .127 **c.** .509 **d.** .31 **e.** .0293 **f.** .7 **g.** .09 **h.** .059 **i.** .003

2. Write each decimal as a fraction in simplest form.

   **a.** .4 **b.** .08 **c.** .375 **d.** .50 **e.** .8750 **f.** .24 **g.** .8 **h.** .58

   **i.** .064 **j.** .05 **k.** .75 **l.** .35 **m.** .625 **n.** .006 **o.** .025

3. .005 written as a fraction is (a) $\frac{1}{2}$ (b) $\frac{1}{20}$ (c) $\frac{1}{200}$ (d) $\frac{1}{2{,}000}$

4. 0.375 names the same point on the number line as (a) $\frac{1}{8}$ (b) $\frac{1}{4}$ (c) $\frac{3}{8}$ (d) $\frac{5}{8}$

5. 9.55 has the same value as (a) $9\frac{1}{2}$ (b) $9\frac{5}{11}$ (c) $9\frac{11}{20}$ (d) $9\frac{3}{4}$

6. Rosa traveled 23.15 miles. This distance is the same as
(a) $23\frac{3}{20}$ (b) $23\frac{3}{10}$ (c) $23\frac{3}{5}$ (d) $23\frac{3}{4}$

## SPIRAL REVIEW EXERCISES

1. Write each word name as a decimal.
   a. seventeen thousandths
   b. three tenths
   c. one hundred and twenty-three hundredths
   d. five and five hundredths

2. .003 is read as
(a) three tenths
(b) three hundredths
(c) three thousandths
(d) three ten-thousandths

3. $\frac{12}{40}$ has the same value as
(a) $\frac{1}{4}$ (b) $\frac{24}{20}$ (c) $\frac{4}{10}$ (d) $\frac{3}{10}$

4. The sum of $-14$ and $+14$ is
(a) 0 (b) $-28$ (c) 28 (d) $-196$

5. Mrs. Kaplan bought $1\frac{1}{4}$ pounds of nuts, $5\frac{1}{2}$ pounds of potatoes, and $2\frac{3}{4}$ pounds of apples. What was the total weight of these items?

6. Three-fourths of the children at Daytime Nursery eat the hot lunch. If there are 68 children at the nursery, how many eat the hot lunch?

7. In 6,417,235, what is the value of the digit 4?
(a) 400,000 (c) 4,000
(b) 40,000 (d) 400

8. Perform the indicated operations:
$(47 + 3) \times (24 - 9)$

9. $3\frac{7}{8}$ written as an improper fraction is
(a) $\frac{10}{8}$ (b) $\frac{24}{8}$ (c) $\frac{21}{8}$ (d) $\frac{31}{8}$

10. What is the total height of a stack of 6 books if each book is $1\frac{3}{8}$ inches thick?

11. The population of Yorkville is 22,430 and the population of Carlsbury is 18,195. How many more people live in Yorkville than live in Carlsbury?

12. Fast Burgers sells 1,250 burgers each day. How many burgers are sold in a seven-day week?

**CHALLENGE QUESTION**

Is .00734 closer to $\frac{73}{10,000}$ or $\frac{74}{10,000}$?

# UNIT 7–4 Changing Fractions to Decimals

## THE MAIN IDEA

To change a fraction to an equivalent decimal, divide the numerator by the denominator.

1. If the division ends, the decimal is called a *terminating decimal.*
2. When the division does not end, we have a *repeating decimal.* We use a bar to show the digits that repeat.

**EXAMPLE 1** Write $\frac{5}{8}$ as a decimal.

Divide the numerator by the denominator. $8\overline{)5}$

Insert a decimal point in the dividend and quotient. $8\overline{)5.}$

Insert as many zeros as necessary in the dividend to complete the division.

$$\begin{array}{r} .625 \\ 8\overline{)5.000} \\ \underline{4\,8\phantom{00}} \\ 20\phantom{0} \\ \underline{16\phantom{0}} \\ 40 \\ \underline{40} \\ 0 \end{array}$$

*Answer:* $\frac{5}{8} = .625$

**EXAMPLE 2** Change $\frac{15}{4}$ to an equivalent decimal.

Notice that $\frac{15}{4}$ is an improper fraction. The equivalent decimal 3.75 has a whole-number part and a fraction part.

$$\begin{array}{r} 3.75 \\ 4\overline{)15.00} \\ \underline{12\phantom{.00}} \\ 3\,0\phantom{0} \\ \underline{2\,8\phantom{0}} \\ 20 \\ \underline{20} \\ 0 \end{array}$$

*Answer:* $\frac{15}{4} = 3.75$

**EXAMPLE 3** Write $\frac{2}{3}$ as a repeating decimal.

The digit 6 repeats.

The remainder is never 0.

$$\begin{array}{r} .666 \\ 3\overline{)2.000} \\ \underline{1\,8\phantom{00}} \\ 20\phantom{0} \\ \underline{18\phantom{0}} \\ 20 \\ \underline{18} \\ 2 \end{array}$$

*Answer:* $\frac{2}{3} = .\overline{6}$

**EXAMPLE 4** Write $\frac{4}{7}$ as a decimal correct to the nearest hundredth.

Since hundredths are in the second decimal place, write 3 zeros so that you can round off to hundredths.

$$\begin{array}{r} .571 \\ 7\overline{)4.000} \\ \underline{3\,5\phantom{00}} \\ 50\phantom{0} \\ \underline{49\phantom{0}} \\ 10 \\ \underline{7} \\ 3 \end{array}$$

Now round .571 to the nearest hundredth. .57

*Answer:* $\frac{4}{7} = .57$, to the nearest hundredth

## CLASS EXERCISES

1. Change to an equivalent decimal.

   a. $\frac{3}{4}$ b. $\frac{7}{16}$ c. $\frac{14}{25}$ d. $\frac{41}{8}$ e. $\frac{2}{5}$ f. $9\frac{1}{2}$ g. $\frac{7}{40}$ h. $\frac{51}{20}$

2. Write each fraction as a repeating decimal.

   a. $\frac{1}{3}$ b. $\frac{5}{6}$ c. $\frac{9}{11}$ d. $\frac{2}{9}$ e. $\frac{11}{3}$

3. Write $\frac{2}{3}$ as a decimal correct to the nearest hundredth.

4. Write $\frac{6}{7}$ as a decimal correct to the nearest tenth.

5. Write $\frac{9}{17}$ as a decimal correct to the nearest thousandth.

## HOMEWORK EXERCISES

1. Change to an equivalent decimal.

   a. $\frac{5}{8}$ b. $\frac{11}{25}$ c. $\frac{23}{50}$ d. $\frac{22}{5}$ e. $\frac{3}{5}$ f. $\frac{1}{4}$ g. $5\frac{1}{2}$ h. $\frac{15}{200}$

   i. $\frac{13}{20}$ j. $\frac{3}{16}$ k. $3\frac{7}{8}$ l. $\frac{40}{16}$

2. Write each fraction as a repeating decimal.

   a. $\frac{2}{3}$ b. $\frac{4}{9}$ c. $4\frac{10}{11}$ d. $\frac{28}{9}$ e. $\frac{5}{11}$ f. $\frac{3}{9}$ g. $14\frac{1}{11}$ h. $\frac{40}{3}$

3. Write $\frac{5}{17}$ as a decimal correct to the nearest hundredth.

4. Write $\frac{6}{11}$ as a decimal correct to the nearest thousandth.

## SPIRAL REVIEW EXERCISES

1. Round .481 to the nearest tenth.

2. Write .724 as a fraction in simplest form.

3. .004 written as a fraction is

   (a) $\frac{1}{25}$ (b) $\frac{1}{250}$ (c) $\frac{1}{2,500}$ (d) $\frac{1}{25,000}$

4. Eli weighs 160.3 pounds. This weight is the same as

   (a) $160\frac{1}{3}$ (c) $160\frac{3}{10}$

   (b) $160\frac{2}{3}$ (d) $160\frac{3}{100}$

**5.** Write as a decimal.

**a.** fifteen hundredths

**b.** three and three tenths

**c.** one thousand and one thousandth

**6.** .007 is read as
(a) seven tenths
(b) seven hundredths
(c) seven thousandths
(d) seven ten-thousandths

**7.** The fraction $\frac{21}{35}$ is equivalent to

(a) $\frac{3}{4}$ (b) $\frac{3}{5}$ (c) $\frac{2}{3}$ (d) $\frac{4}{5}$

**8.** Find the sum of $\frac{2}{3}$ and $\frac{5}{8}$.

**9.** Divide $\frac{20}{63}$ by $\frac{10}{3}$.

**10.** The fraction $\frac{38}{12}$ is equivalent to

(a) $2\frac{1}{12}$ (b) $2\frac{1}{6}$ (c) $3\frac{1}{12}$ (d) $3\frac{1}{6}$

**11.** Multiply $2\frac{1}{3}$ by $3\frac{1}{4}$.

**12.** From $\frac{31}{2}$ subtract $\frac{1}{2}$.

**13.** Find the product of $-12$ and $-4$.

**14.** When 5 is subtracted from $-8$, the result is
(a) $-13$ (b) $-3$ (c) 13 (d) 3

**15.** What is the total length of 5 matchsticks, each of which is $2\frac{1}{4}''$ long?

**16.** Add: 173 + 21,460 + 8

**CHALLENGE QUESTION**

Write a rule for changing proper fractions whose denominators are 9 into decimals.

# UNIT 7–5 Comparing Decimals

## THE MAIN IDEA

To compare decimals:

1. Look at the decimals to see what the largest number of decimal places is.
2. Change each decimal to an equivalent decimal having that same number of decimal places. Do this by writing zeros on the right when necessary.
3. Ignore the decimal point and compare as with whole numbers.

**EXAMPLE 1** Replace ? by < or > to make a true comparison: .7 ? .63

| | |
|---|---|
| The largest number of decimal places is 2. | .7 ? .63 |
| Write .7 as .70 so that both decimals have 2 places. | .70 ? .63 |
| Ignore the decimal point and compare as with whole numbers. | 70 > 63 |

*Answer:* .7 > .63

**EXAMPLE 2** Find the largest decimal: .092, .54, .3

| | | | |
|---|---|---|---|
| The largest number of decimal places is 3. | .092 | .54 | .3 |
| Insert zeros so that each decimal has 3 places. | .092 | .540 | .300 |
| Compare as with whole numbers. 540 is the largest number shown. | 92 | 540 | 300 |

*Answer:* .54 is the largest decimal.

**EXAMPLE 3** Write the numbers 0.25, 0.052, 0.52, and 0.025 in order from smallest to largest.

| | | | | |
|---|---|---|---|---|
| The largest number of decimal places is 3. | 0.25 | 0.052 | 0.52 | 0.025 |
| Insert zeros as needed. | 0.250 | 0.052 | 0.520 | 0.025 |
| Think of the whole numbers. | 250 | 52 | 520 | 25 |
| Arrange the whole numbers in order from smallest to largest. | 3rd | 2nd | 4th | 1st |

*Answer:* 0.025, 0.052, 0.25, 0.52

**EXAMPLE 4** Jeff ran the mile in 5.3 minutes. Jerome's time was 5.27 minutes. Whose time was faster?

| | |
|---|---|
| Compare 5.3 and 5.27. | 5.3 ? 5.27 |
| Write each with 2 decimal places. | 5.30 ? 5.27 |
| | 5.30 > 5.27 |

*Answer:* Jerome's time was faster.

## CLASS EXERCISES

1. Replace ? by $<$ or $>$ to make a true statement.

   **a.** .3 ? .4 **b.** .3 ? .04 **c.** .72 ? .702 **d.** 57.38 ? 58.38 **e.** 23.7 ? 2.37 **f.** .37 ? .5

2. The decimal that has the same value as 0.25 is
   (a) 0.205 (b) 0.025 (c) 0.2500 (d) 0.2005

3. The decimal that has the largest value is
   (a) .7 (b) .707 (c) .077 (d) .77

4. The decimal that has the smallest value is
   (a) .003 (b) .029 (c) .01 (d) .3

5. Write the numbers 7.8, 0.078, 0.78, and 0.807 in order from smallest to largest.

6. At the end of a day of fishing, Carol had caught 4.8 kilograms of fish and Maria had caught 4.68 kilograms of fish. Which girl caught the greater weight of fish?

## HOMEWORK EXERCISES

1. Replace ? by $<$ or $>$ to make a true statement.

   **a.** 0.5 ? 0.7 **b.** 0.302 ? 0.23 **c.** 5.9 ? 6.9 **d.** 0.9 ? 0.6 **e.** 0.517 ? 0.175

   **f.** 8.275 ? 7.975 **g.** 0.7 ? 0.52 **h.** 0.77 ? 0.707 **i.** .52 ? .045 **j.** 0.49 ? 0.94

   **k.** 0.483 ? 0.438 **l.** .6982 ? .7

2. The decimal that has the same value as .37 is
   (a) 0.370 (b) 0.307 (c) 0.037 (d) 0.3007

3. The decimal that has the largest value is
   (a) .404 (b) .0444 (c) .44 (d) .04444

4. The decimal that has the smallest value is
   (a) .05 (b) .049 (c) .009 (d) .1

5. Write the numbers 0.59, 5.9, 9.5, 0.95, and 0.509 in order from largest to smallest.

6. Write the numbers 0.308, 0.803, 0.38, 0.83, 8.3, and 3.8 in order from smallest to largest.

7. At the supermarket, Jim bought 2.5 pounds of apples and Marta bought 2.7 pounds of apples. Who bought the greater weight of apples?

## SPIRAL REVIEW EXERCISES

1. Round 0.837 to the nearest hundredth.

2. Round 2.5894 to the nearest tenth.

3. Round 0.81954 to the nearest thousandth.

4. Round 15.296421 to the nearest ten-thousandth.

5. Write $\frac{5}{9}$ as a repeating decimal.

6. Replace each ? by <, =, or > to make a true statement.

   a. $\frac{5}{8} ? \frac{5}{9}$   b. $\frac{4}{5} ? \frac{3}{5}$

   c. $2\frac{1}{2} ? \frac{2}{5}$   d. $3\frac{3}{4} ? \frac{30}{8}$

7. If each crate of canned hams contains 2,880 cans, how many cans are in $3\frac{1}{4}$ crates?

8. Find the sum of $-23$ and $-43$.

9. By how much is $4\frac{1}{2}$ greater than $3\frac{1}{4}$?

10. Mr. Damsky's truck can safely carry 25,000 pounds. If he loads the truck with 18,750 pounds of bricks, how much more weight can be safely put on the truck?

11. In Mrs. Klein's garden, each row of tomato plants contains 12 plants. How many plants can she plant in 9 rows?

12. Add: $-4 + (-12)$

13. Divide 3,914 by 19.

## CHALLENGE QUESTION

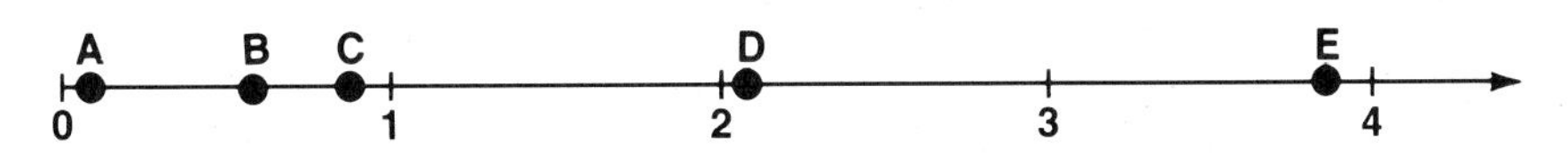


Pair each of the given decimals with the letter that best shows its position on the number line: 2.087, 0.087, 0.87, 3.87, 0.6.

MODULE 8

# UNIT 8–1 Adding and Subtracting Decimals; Using Decimals to Make Change

## ADDING AND SUBTRACTING DECIMALS

### THE MAIN IDEA

To add or subtract decimals:

1. Line up the decimals vertically so that the decimal points are beneath one another. (In a whole number, a decimal point is understood to be at the end.)
2. If necessary, insert zeros so that the decimals have the same number of places.
3. Add or subtract the numbers in the same way that you add or subtract whole numbers.
4. Place the decimal point in the sum or difference directly beneath the points above.

**EXAMPLE 1** Add: 0.7 and 0.2

Line up the decimals vertically. The decimals have the same number of places.

$$\begin{array}{r} 0.7 \\ +0.2 \\ \hline \end{array}$$

Add as with whole numbers.

$$\begin{array}{r} 0.7 \\ +0.2 \\ \hline 9 \end{array}$$

Place the decimal point directly beneath the points above.

$$\begin{array}{r} 0.7 \\ +0.2 \\ \hline 0.9 \end{array} \quad \textit{Ans.}$$

**EXAMPLE 2** Add: 0.3 + 0.27 + 0.009

Line up the decimals vertically.

$$\begin{array}{r} 0.3 \\ 0.27 \\ +0.009 \\ \hline \end{array}$$

Write zeros so that the decimals have the same number of places. Then add, and insert the decimal point.

$$\begin{array}{r} 0.300 \\ 0.270 \\ +0.009 \\ \hline 0.579 \end{array} \quad \textit{Ans.}$$

**EXAMPLE 3** From 12 subtract 9.28 and check.

Line up the decimals vertically. In the whole number 12, the decimal is at the end.

$$\begin{array}{r} 12. \\ -9.28 \\ \hline \end{array}$$

Insert zeros. Then subtract, and insert the decimal point.

$$\begin{array}{r} 12.00 \\ -9.28 \\ \hline 2.72 \end{array} \quad \textit{Ans.}$$

**EXAMPLE 4** Joann spent \$27.80 on a skirt, \$15.99 on a blouse, and \$32.45 on a pair of shoes. Find the total amount of money Joann spent.

The key word "total" tells you to add the three costs.

Since dollars and cents are written as decimals, line up the decimal points when you write them.

$$\begin{array}{r} \$27.80 \\ 15.99 \\ +32.45 \\ \hline \$76.24 \end{array}$$

*Answer:* \$76.24 is the total.

**EXAMPLE 5** Jacqueline ran 14.3 kilometers and Francine ran 8.75 kilometers. How many more kilometers did Jacqueline run than Francine?

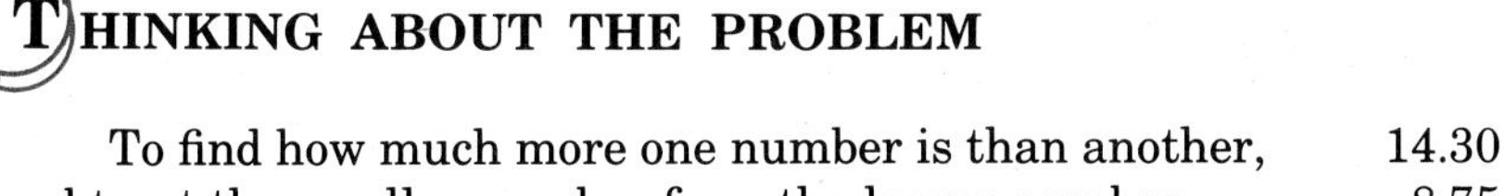


To find how much more one number is than another, subtract the smaller number from the larger number.

$$\begin{array}{r} 14.30 \\ -8.75 \\ \hline 5.55 \end{array}$$

*Answer:* Jacqueline ran 5.55 kilometers more than Francine.

## CLASS EXERCISES

1. Add.

   **a.** 37.6 + 0.015  **b.** 0.9 + 27.05 + 10  **c.** 0.29 + 0.0037 + 0.378

   **d.** $\begin{array}{r} 12.8 \\ 27.38 \\ 0.918 \\ \hline \end{array}$  **e.** $\begin{array}{r} 2.006 \\ 12.61 \\ 29 \\ 0.8 \\ \hline \end{array}$  **f.** $\begin{array}{r} 4.906 \\ 27.53 \\ 0.0962 \\ 352.1 \\ \hline \end{array}$

2. Janette's auto insurance costs \$211.52 for bodily injury liability, \$89.48 for property damage, and \$137.14 for collision coverage. Find the total cost of Janette's auto insurance.
3. Jason mixed 0.46 kilogram of cashews, 0.68 kilogram of walnuts, and 1.2 kilograms of peanuts together. What is the total weight of the mixture?
4. Subtract.

   **a.** 0.75 − 0.32  **b.** 59 − 46.8  **c.** 11.65 − 8.9  **d.** 22.7 − 17.68

   **e.** 126.97 − 89.4  **f.** 96 − 53.94

5. Subtract 0.97 from 2.5.
6. From 37.2 subtract 18.35.
7. Find the difference between 235.8 and 194.25.
8. How much greater is 49.7 than 37.9?
9. Charlene ran 100 meters in 10.9 seconds. Nancy ran the same distance in 12.8 seconds. By how many seconds was Charlene faster than Nancy?
10. Robin saved \$9.70 for a blouse that costs \$12. How much more money does she need to buy the blouse?

## USING DECIMALS TO MAKE CHANGE

### THE MAIN IDEA

1. You receive *change* when the amount of money given is greater than the cost of the purchase.
2. To find the amount of change, subtract the cost of the purchase from the amount of money given.

Amount of Change = Amount Given − Purchase Price

**EXAMPLE 6** If Andrew pays for a $7.29 shopping order with a $20 bill, how much change will he receive?

To find the change, subtract the amount of the purchase from the amount of money given to the cashier.

$$\begin{array}{r} \$20.00 \\ -7.29 \\ \hline \$12.71 \end{array}$$

*Answer:* Andrew will receive $12.71 as change.

**EXAMPLE 7** Joan bought a hamburger for $2.25, a soda for 79¢, and an order of french fries for 65¢. If Joan gave the cashier a $10 bill, how much change did she receive?

**THINKING ABOUT THE PROBLEM**

Use the strategy of breaking the problem into smaller, separate problems.

To find the total amount of the purchase, add the costs of the three items.

$$\begin{array}{r} \$2.25 \\ .79 \\ +.65 \\ \hline \$3.69 \end{array}$$

Subtract the purchase price from the amount given to the cashier.

$$\begin{array}{r} \$10.00 \\ -3.69 \\ \hline \$6.31 \end{array}$$

*Answer:* The amount of change Joan received was $6.31.

**EXAMPLE 8** Karen bought some school supplies. She gave a $5 bill to the cashier and received $1.73 change. How much did Karen spend on the school supplies?

**THINKING ABOUT THE PROBLEM**

If you know the amount of change received, you can find the amount spent by subtracting the change from the amount given to the cashier.

Subtract the change from the amount given to the cashier.

$$\begin{array}{r} \$5.00 \\ -1.73 \\ \hline \$3.27 \end{array}$$

*Answer:* Karen spent $3.27 on the school supplies.

## CLASS EXERCISES

1. Find the amount of change from a $10 bill for each of the following purchases:

   a. $2.75  b. $5.39  c. $.86  d. $9.47

2. Henry bought a shirt for $12.90. How much change did he receive from a $20 bill?
3. The cost of Nancy's breakfast is $1.70. If Nancy gave the cashier a $5 bill, how much change did she receive?
4. Marilyn bought a scarf for $8.99 and a bow for $3.70. How much change did she receive from a $20 bill?
5. Mr. Jones bought $5.25 worth of gasoline and one quart of oil costing $1.80. How much change did he receive from a $10 bill?
6. Frank bought a cake mix costing $1.50, a quart of milk for $.75, and a dozen eggs for $.89. How much change did he receive from a $5 bill?
7. If Julio received $5.90 change from a $10 bill, how much did he spend?
8. How much did Ms. Coles spend on groceries if she received $7.48 change from a $20 bill?
9. Pat paid for a purchase of groceries with a $20 bill. The money shown is what he received as change. What was the amount of Pat's purchase?

## HOMEWORK EXERCISES

1. Add.

a. $\begin{array}{r} 0.6 \\ +0.9 \\ \hline \end{array}$ b. $\begin{array}{r} 0.47 \\ +2.89 \\ \hline \end{array}$ c. $\begin{array}{r} 5.93 \\ 6.02 \\ +0.917 \\ \hline \end{array}$ d. $\begin{array}{r} 8.026 \\ 9.1 \\ +23.25 \\ \hline \end{array}$ e. $\begin{array}{r} 2.6 \\ +.37 \\ \hline \end{array}$

f. $\begin{array}{r} 9.365 \\ +2.09 \\ \hline \end{array}$ g. $\begin{array}{r} 2.94 \\ 3.825 \\ 4.01 \\ \hline \end{array}$ h. $\begin{array}{r} 42.7 \\ 11.43 \\ 240.084 \\ \hline \end{array}$

2. Add.

a. 0.92 + 0.7 b. 2.8 + 0.89 c. 10.75 + 118.073 d. 9.3 + 0.29 + 4.07

e. 842.7 + 24.78 + 208.07 f. 97.45 + 3.719 + 24.9 g. 143.62 + 495.6 + 247.734

h. 47 + 406.215 + 21.59

3. Subtract.

a. $\begin{array}{r} 0.9 \\ -0.6 \\ \hline \end{array}$ b. $\begin{array}{r} 0.7 \\ -0.29 \\ \hline \end{array}$ c. $\begin{array}{r} 14.87 \\ -9.79 \\ \hline \end{array}$ d. $\begin{array}{r} 346.8 \\ -156.264 \\ \hline \end{array}$ e. $\begin{array}{r} 16.0 \\ -8.7 \\ \hline \end{array}$

f. $\begin{array}{r} 31 \\ -19.28 \\ \hline \end{array}$ g. $\begin{array}{r} 24.3 \\ -19.26 \\ \hline \end{array}$ h. $\begin{array}{r} 54.95 \\ -53.99 \\ \hline \end{array}$

4. Subtract.

a. 14.87 − 9.79 b. 9.3 − 7.4 c. 10.6 − 8.47 d. 14.2 − 9.4 e. 33.05 − 18.6

f. 0.3 − 0.28 g. 43.2 − 21.56 h. 19.49 − 12.68

5. Subtract 8.35 from 11 and check.

6. From 18.9 subtract 5.8.

7. How much greater than 6.28 is 9.7?

8. Frank bought a baseball that cost $4.97, a bat that cost $12.75, and a baseball glove that sold for $35.25. Find the total amount of money that Frank spent.

9. Mabel walked 3.1 miles on Monday, 2.8 miles on Tuesday, and 3.7 miles on Wednesday. Find the total distance that she walked

10. Raymond bought a shirt for $15. Joe bought the same shirt on sale for $11.89. How much less did Joe pay?

11. Janice weighs 101.3 pounds. If Yvette weighs 6.8 pounds less than Janice, how much does Yvette weigh?

12. A man gives a $10 bill to the cashier when he buys a hat that costs $6.36. How much change will he receive?

13. Rosario bought two toys. One cost $1.63 and the other cost $.92. How much change did he receive if he paid with a $5 bill?

14. Fran received $1.51 in change when she used a $20 bill to pay for a blouse. How much did the blouse cost?

15. Francis received $13.37 change when he bought a new chess set. How much did the set cost if he paid with a $50 bill?

**16.** Mrs. Corso had ten dollars with her when she went to the company cafeteria. The money shown is what she had left after paying for lunch. How much did her lunch cost?

## SPIRAL REVIEW EXERCISES

1. Write .98 as a fraction in simplest form.
2. Write $\frac{3}{7}$ as a decimal correct to the nearest hundredth.
3. Round 12.397 to the nearest hundredth.
4. 640 × .49 is closest to
   (a) 30  (b) 320  (c) 700  (d) 3,000
5. Which digit is in the hundredths place in 487.132?
6. How many times greater is 431.5 than 4.315?
7. Find the greatest common factor of 75 and 60.
8. From a $12\frac{1}{2}$-pound roast, Mr. Bruno trimmed $\frac{3}{4}$ pound of fat. How much meat was left?

### CHALLENGE QUESTION

Mrs. Ross began the day with $20. She spent $14.30 at the supermarket, found a $5 bill, lost $1.50 in change, and received a $10 refund at the cleaners. How much money did she have at the end of the day?

# UNIT 8–2 Using Decimals in Checking Accounts

## THE MAIN IDEA

1. You can use ***checks*** in place of cash to make payments. In order to write checks, you maintain a ***checking account*** in a bank. In this account, you keep a sum of money called the ***balance***.
2. When you write a check, the amount of the check is *deducted* from the original balance to find the new balance.
3. When you make a ***deposit***, the amount of the deposit is *added* to the original balance to find the new balance.

**EXAMPLE 1** Mr. Darby had a balance of \$492.53 in his checking account. He wrote a check for \$174.49. What is his new balance?

To find the new balance, subtract the amount of the check from the original balance.

$$\begin{array}{r} \$492.53 \\ -\,174.49 \\ \hline \$318.04 \end{array}$$

*Answer:* The new balance is \$318.04.

**EXAMPLE 2** Ms. Carlson had a balance of \$265.87 in her checking account. She deposited \$450.00. What is the new balance?

To find the new balance, add the deposit to the original balance.

$$\begin{array}{r} \$265.87 \\ +\,450.00 \\ \hline \$715.87 \end{array}$$

*Answer:* The new balance is \$715.87.

**EXAMPLE 3** Mary had a balance of \$357.89 in her checking account. She deposited \$182.43 and then wrote a check for \$98.37. How much money does she have left in her account?

**THINKING ABOUT THE PROBLEM**

Use the strategy of breaking the problem into smaller, separate problems.

To find the first new balance, add the deposit to the original balance.

$$\begin{array}{r} \$357.89 \\ +\,182.43 \\ \hline \$540.32 \end{array}$$

To find the final balance, subtract the amount of the check from the previous balance.

$$\begin{array}{r} \$540.32 \\ -\,98.37 \\ \hline \$441.95 \end{array}$$

*Answer:* Mary has \$441.95 left in her checking account.

**EXAMPLE 4** The balance in Mr. Franklin's checking account was $1,452.09. He wrote checks for $378.71 and $532.96. How much money does Mr. Franklin have left in his account?

To find the first new balance, subtract the amount of the first check from the original balance.

$$\begin{array}{r} \$1{,}452.09 \\ -378.71 \\ \hline \$1{,}073.38 \end{array}$$

To find the present balance, subtract the amount of the second check from the previous balance.

$$\begin{array}{r} \$1{,}073.38 \\ -532.96 \\ \hline \$540.42 \end{array}$$

*Answer:* The present balance is $540.42.

**EXAMPLE 5** After writing a check for $159.87, John had a balance of $207.33 in his checking account. What was the balance in John's account before writing this check?

**THINKING ABOUT THE PROBLEM**

Use the strategy of working backwards. Before John wrote the check, he had more money in his account.

To find John's balance before writing the check, add the amount of the check to his present balance.

$$\begin{array}{r} \$207.33 \\ +159.87 \\ \hline \$367.20 \end{array}$$

*Answer:* The original balance was $367.20.

## CLASS EXERCISES

1. The balance in Mr. Apple's checking account was $709.12. Find the new balance after Mr. Apple deposited $246.
2. Ms. Pear had $1,042.82 in her checking account. If she wrote a check for $798.99, how much money does she have left in her account?
3. Dr. Malonski deposited $842.00 into his checking account, which had a previous balance of $59.67. What is the new balance in Dr. Malonski's account?
4. Mrs. Plummer deposited $417.49 into her checking account. Find the amount of money in Mrs. Plummer's account if she had a balance of $389.84 before making the deposit.
5. Before writing a check for $295.07, Mr. Lemmon had $327.08 in his checking account. How much money remains in Mr. Lemmon's account?
6. Ms. Redd deposited $45.29 and then wrote a check for $27.63. If Ms. Redd's checking account originally had a balance of $56.82, what is the present balance in her account?

## HOMEWORK EXERCISES

1. Mr. Happy had a balance of $636.13 in his checking account. Find the new balance after he wrote a check for $129.17.
2. To buy a new refrigerator, Mrs. Bouvier wrote a check for $599.78. If her original balance was $2,363.49, what is her new balance after writing the check?

3. George wrote a check for \$63.17 and another for \$22.11. Find his new balance if his old balance was \$279.38.

4. Susan has a balance of \$93.16 in her checking account. Find her new balance after she made a deposit of \$167.34.

5. How much money does Mr. Adams have in his account if he deposited \$273.09 and he had an old balance of \$564.18?

6. Mrs. Wilson receives a check for \$97.61 each week. If she deposited her check directly into her account for two consecutive weeks and she had an old balance of \$31.11, what was her balance at the end of two weeks?

7. Mrs. Jones had a balance of \$243.97 in her account. If she made a deposit of \$191.73 and then wrote a check for \$362.38, what was her new balance?

8. With a balance of \$636.66 in his account, Franklin wrote a check for \$499.22 and then made a deposit of \$67.22. What was his new balance?

9. Mr. Ray deposited \$480.27 into his checking account, which had a previous balance of \$96.40. If he then wrote a check for \$158.92, what was the new balance?

## SPIRAL REVIEW EXERCISES

1. Ms. Foster buys a handbag that costs \$18.95. How much change will she receive from a \$20 bill?

2. Jonathan paid for a record with a \$10 bill and received \$2.37 change. How much did the record cost?

3. Find the sum of 0.295, 3.7, and 18.19.

4. From 3.47 subtract 2.49.

5. Write 0.125 as a fraction in simplest form.

6. Write $\frac{5}{11}$ as a decimal correct to the nearest hundredth.

7. Round 29.4783 to the nearest thousandth.

8. $\frac{4}{5}$ is closest to
   (a) .75 (b) .81 (c) .85 (d) .9

9. By how much is 2 greater than $-5$?

10. Add: $\frac{5}{12} + \frac{2}{3}$

11. Add: $-12 + (-17)$

12. From $12\frac{3}{4}$ subtract $5\frac{3}{8}$.

## CHALLENGE QUESTION

A check "bounces," or a checking account is "overdrawn," if a check is written for an amount greater than the balance in the account. Each of the following persons made a deposit, then wrote a check. Which person's account is overdrawn?

| Name | Old Balance | Deposit | Check |
|---|---|---|---|
| Mr. Arthur | \$222.17 | \$63.92 | \$281.11 |
| Mrs. Brooks | \$122.64 | \$387.19 | \$593.71 |
| Mr. Czerny | \$273.81 | \$15.68 | \$287.49 |
| Ms. Donaldson | \$1,369.00 | \$2,463.00 | \$3,821.00 |

# UNIT 8–3 Multiplying Decimals

## MULTIPLYING A DECIMAL BY A WHOLE NUMBER

### THE MAIN IDEA

To multiply a decimal by a whole number:

1. Multiply as if both numbers were whole numbers.
2. Count the number of decimal places in the decimal.
3. Put a decimal point in the product so that it has the same number of decimal places as the decimal.

**EXAMPLE 1** Multiply: 25 × .017

Multiply as if both numbers were whole numbers.

```
  .017
  ×25
  085
 034
 0425
```

Since .017 has 3 decimal places, the product must have 3 decimal places. Begin at the right, count off 3 decimal places, and insert a decimal point.

0 . 4 2 5 (← ← ←)

*Answer:* 0.425

**EXAMPLE 2** Multiply: 120 × 3.08

Multiply as with whole numbers.

```
   120
 ×3.08
   960
  000
 360
 36960
```

Since 3.08 has 2 decimal places, the product must have 2 decimal places. Begin at the right, count off 2 decimal places, and insert a decimal point.

3 6 9 . 6 0 (← ←)

*Answer:* 369.60 or 369.6

**EXAMPLE 3** What is the total cost of 3 pounds of roast beef if the price is $4.29 per pound?

**THINKING ABOUT THE PROBLEM**

The key words in this problem are "total" and "per." "Per" means "each."

To find the total cost when the price per pound is given, multiply the price per pound by the number of pounds.

```
 $4.29 ← price per pound
    ×3 ← number of pounds
$12.87 ← total cost
```

*Answer:* $12.87 is the total cost.

**EXAMPLE 4** If an ice-cream cone costs $.60, what is the cost of 5 cones?

Multiply the *item price* by the number of items.

```
 $.60 ← item price
   ×5 ← number of items
$3.00 ← total cost
```

*Answer:* The total cost is $3.00.

## CLASS EXERCISES

1. Multiply.

   a. $\begin{array}{r} 2.75 \\ \times 3 \\ \hline \end{array}$ b. $\begin{array}{r} 38.41 \\ \times 12 \\ \hline \end{array}$ c. $\begin{array}{r} 71.08 \\ \times 25 \\ \hline \end{array}$ d. $\begin{array}{r} 15 \\ \times .03 \\ \hline \end{array}$ e. $28 \times .042$ f. $175 \times 1.4$ g. $9.4 \times 28$

2. Find the total cost of 5 shirts if each shirt costs $15.85.
3. What is the cost of 8 pens if each pen costs $.49?

## MULTIPLYING A DECIMAL BY A DECIMAL

### THE MAIN IDEA

To multiply a decimal by another decimal:

1. Multiply as if both numbers were whole numbers.
2. Count the total number of decimal places in both decimals.
3. Place a decimal point in the product so that it has the same number of decimal places as the total number of decimal places in both decimals.
4. If necessary, write zeros to the left of the digits in the result to allow for the correct number of decimal places.

**EXAMPLE 5** Multiply: $.7 \times .3$

Multiply as with whole numbers.

$$\begin{array}{r} .3 \\ \times .7 \\ \hline 21 \end{array}$$

Count the total number of decimal places in the two decimals.

.3 contains 1 decimal place.
.7 contains 1 decimal place.
The product must contain 2 decimal places.

Begin at the right and count off 2 decimal places.

. 2 1 (← ←)

*Answer:* .21

**EXAMPLE 6** Multiply: $0.47 \times .08$

Multiply as with whole numbers.

$$\begin{array}{r} .047 \\ \times .08 \\ \hline 376 \end{array}$$

Since .047 contains 3 decimal places and .08 contains 2 decimal places, the product must contain 3 + 2, or 5 decimal places.

Begin at the right and count off 5 decimal places. To do this, insert 2 zeros.

. 0 0 3 7 6 (← ← ← ← ←)

*Answer:* .00376

**EXAMPLE 7** The length of a floor tile is 30.6 cm. What is the length of 3.5 tiles placed side by side?

**THINKING ABOUT THE PROBLEM**

Each floor tile has the same length.

Find the total length by multiplication.

$$\begin{array}{r} 30.6 \leftarrow \text{length of each tile} \\ \times 3.5 \leftarrow \text{number of tiles} \\ \hline 15\ 30 \\ 91\ 8 \\ \hline 107.10 \leftarrow \text{total length} \end{array}$$

*Answer:* The total length is 107.10 or 107.1 cm.

## CLASS EXERCISES

**1.** Multiply.

**a.** $.6 \times .3$ **b.** $.28 \times .09$ **c.** $5.22 \times 1.6$ **d.** $17.01 \times .32$ **e.** $.03 \times .7$ **f.** $.24 \times 1.9$

**g.** $.043 \times .5$ **h.** $25.5 \times 0.08$

**2.** Mr. O'Keefe's car averages 18.3 miles per gallon. How many miles can he travel using 12.8 gallons of gasoline?

**3.** Mark can lose 1.2 pounds per week on his diet. At that rate, how many pounds will Mark lose in 10.5 weeks?

## MULTIPLYING A DECIMAL BY A POWER OF 10

### THE MAIN IDEA

To multiply a decimal by a *power of 10* (numbers such as 10, 100, 1,000, 10,000, etc.), there are two methods:

1. Use the method you have learned to multiply a decimal by a whole number.

or

2. a. Count the number of zeros in the power of 10.
   b. Move the decimal point to the *right* the same number of places as there are zeros in the power of 10.
   c. Write extra zeros to the right if there are not enough decimal places for moving the decimal point.

**EXAMPLE 8** Multiply: 4.35 × 100

*Method 1:*

Multiply as with whole numbers.

```
   4.35
  ×100
    000
   000
  435
  43500
```

Since 4.35 has 2 decimal places, the product must have 2 decimal places.

Begin at the right and count off 2 decimal places. 4 3 5 . 0 0 (← ←)

*Answer:* 435.00 or 435

*Method 2:*

Count the number of zeros in 100. There are 2 zeros in 100.

Move the decimal point in 4.35 2 places to the right. 4 . 3 5 . (→ →)

*Answer:* 435

**EXAMPLE 9** Multiply: 17.34 × 10,000

Count the number of zeros in 10,000. There are 4 zeros in 10,000.

To move the decimal point in 17.34 4 places to the right, insert 2 zeros. 1 7 . 3 4 0 0 . (→ → → →)

*Answer:* 173,400

## CLASS EXERCISES

1. Multiply.

   **a.** 25.34 × 10 **b.** 18.005 × 100 **c.** 10,000 × 5.11756 **d.** 7.006 × 1,000

   **e.** 100 × 4.001 **f.** 100.71 × 100 **g.** 4.08 × 1,000 **h.** 8.4 × 100 **i.** .156 × 10

   **j.** 17.6 × 1,000 **k.** .7 × 10,000 **l.** 100 × .100

2. There are 2.54 centimeters in an inch. Find the number of centimeters in 100 inches.

3. There are 1.6 kilometers in a mile. Find the number of kilometers in 1,000 miles.

## HOMEWORK EXERCISES

1. Multiply.

   **a.** 23 × .7 **b.** 5 × .28 **c.** 18 × 3.14 **d.** 5.42 × 6

   **e.** 18.6 × 12 **f.** 14.62 × 45 **g.** 90.07 × 21 **h.** 47 × .026

2. Multiply.

   a. $\begin{array}{r} 4.8 \\ \times 3.2 \\ \hline \end{array}$ b. $\begin{array}{r} 5.19 \\ \times 4.6 \\ \hline \end{array}$ c. $\begin{array}{r} 9.2 \\ \times 0.09 \\ \hline \end{array}$ d. $\begin{array}{r} 15.6 \\ \times 2.5 \\ \hline \end{array}$

   e. $8.73 \times 4.2$ f. $9.07 \times .003$ g. $5.95 \times 1.07$ h. $.07 \times 2.005$

3. Multiply.

   a. $3.75 \times 10$ b. $19.84 \times 100$ c. $8.9 \times 1{,}000$ d. $0.4 \times 100$ e. $0.3 \times 100$

   f. $94.25 \times 1{,}000$ g. $8.4 \times 10$ h. $.91 \times 10{,}000$

4. Find the product of 209 and 0.08.

5. What is the total cost of 6 pounds of ham if the price is $2.55 per pound?

6. There are about .914 meters in a yard. The number of meters in 10,000 yards is
   (a) 9.14 (b) 91.4 (c) 914 (d) 9,140

7. Mr. Alcalde's car averages 31.7 miles per gallon. How many miles can he drive using 11.8 gallons of gasoline?

8. A crate weighs 37.8 pounds. What is the weight of 1,000 crates?

9. What is the cost of 8.9 gallons of gasoline at $1.40 per gallon?

## SPIRAL REVIEW EXERCISES

1. Find the sum of 12.75, 8.02, and 3.98.

2. Mary had $125.83 in her checking account and then wrote a check for $87.90. What was her new balance?

3. George bought a case of soda that cost $6.95. He paid with a $10 bill. How much change did he get?

4. Replace ? with $<$ or $>$ to make a true comparison.

   a. 0.8 ? 0.08 b. .903 ? .93

   c. 0.512 ? 0.513 d. 2.5 ? 5.2

5. Round 25.976 to the nearest hundredth.

6. How much smaller is 19.87 than 21.6?

7. Dr. Volpe flew 2,853 miles to New City, drove another 389 miles, and then took a train for 450 miles. What was the total number of miles that she traveled?

8. $\frac{32}{7}$ is equal to

   (a) $5\frac{4}{7}$ (b) $2\frac{5}{7}$ (c) $4\frac{5}{7}$ (d) $4\frac{4}{7}$

9. $\frac{1}{9}$ is equal to

   (a) $.\overline{1}$ (b) $.\overline{9}$ (c) $.\overline{19}$ (d) $.\overline{91}$

10. Find the product of $-34$ and $-7$.

11. Replace each ? with $+$, $-$, $\times$, or $\div$ to make each a true statement.

    a. $-8 \ ? -8 = -16$ b. $0 \ ? \ 5 = 0$

12. Which number is represented by point $A$ on the number line?

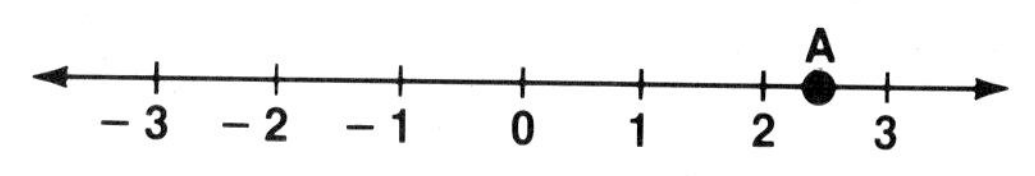


    (a) $\frac{12}{30}$ (b) $\frac{22}{18}$ (c) $\frac{30}{9}$ (d) $\frac{29}{12}$

### CHALLENGE QUESTION

How many lengths of 4.5 centimeters each can be marked off on a ribbon that is 450 centimeters long?

# UNIT 8–4 Dividing Decimals

## DIVIDING A DECIMAL BY A WHOLE NUMBER

### THE MAIN IDEA

To divide a decimal by a whole number:

1. Place a decimal point in the quotient just above the decimal point of the dividend.
2. Ignore the decimal points and divide as with whole numbers. If necessary, use zeros as placeholders in the quotient.

**EXAMPLE 1** Divide: 7.25 ÷ 5

Set up the division.

divisor → 5)7.25 ← dividend; quotient above

Place a decimal point in the quotient just above the decimal point of the dividend.

$$5\overline{)7.25}\ \text{with } . \text{ above}$$

Ignore the decimal points and divide as with whole numbers.

```
   1.45
5)7.25
  5
  2 2
  2 0
    25
    25
     0
```

*Answer:* 1.45

**EXAMPLE 2** Divide: .144 ÷ 12

Place a decimal point in the quotient just above the decimal point of the dividend.

```
     .
12).144
```

Ignore the decimal point and divide as with whole numbers. Since 12 is too large to divide into 1, place a zero above the 1.

```
    .0
12).144
```

Then divide 12 into 14.

```
    .012
12).144
     0
     14
     12
      24
      24
       0
```

*Answer:* .012

**EXAMPLE 3** Divide: 72.3 ÷ 5

Sometimes, to complete the division of a decimal, we must write zeros to the right in the dividend.

```
   14.4
5)72.3
  5
  22
  20
   2 3
   2 0
     3
```

Since there is still a remainder of 3, write a zero to the right, and continue dividing. Since the remainder is now 0, the division is complete.

```
   14.46
5)72.30
  5
  22
  20
   2 3
   2 0
     30
     30
      0
```

*Answer:* 14.46

**EXAMPLE 4** Five persons wanted to share equally the 30.7 pounds of fish they caught. How many pounds of fish will each person receive?

### THINKING ABOUT THE PROBLEM

Since the shares are to be equal, divide the total by the number of persons.

```
    6.14
 5)30.70
   30
   --
    0 7
      5
    ---
      20
      20
      --
       0
```

*Answer:* Each share is 6.14 pounds.

**EXAMPLE 5** If 4 bars of soap sell for \$.97, the amount the cashier will charge for one bar is
(a) \$.20 (b) \$.23 (c) \$.24 (d) \$.25

### THINKING ABOUT THE PROBLEM

To find the price of one bar when the total cost of several bars is given, divide the total cost by the number of bars. If there is a remainder, the store increases the price to the next cent.

Divide .97 by 4.

```
    .24
 4).97
    8
    --
    17
    16
    --
     1 ← remainder
```

Increase the price to the next cent. $.24\frac{1}{4} \rightarrow \$.25$

*Answer:* (d)

**EXAMPLE 6** If 5 oranges cost \$.80, what is the cost of 7 oranges?

### THINKING ABOUT THE PROBLEM

Use the strategy of breaking the problem into smaller, separate problems.

First, find the cost of one orange.

To find the cost of one orange, divide the total cost by the number of oranges.

```
   .16
 5).80
```

To find the cost of 7 oranges, multiply the cost of one orange by 7.

```
  $.16
   ×7
 -----
 $1.12
```

*Answer:* The cost of 7 oranges is \$1.12.

## CLASS EXERCISES

1. Divide.

   a. 3.125 ÷ 5 b. 28.8 ÷ 12 c. 12.58 ÷ 5 d. 2.87 ÷ 14 e. 227.7 ÷ 9

   f. 0.54 ÷ 12 g. 21.48 ÷ 24 h. 4,847.55 ÷ 85

2. Twelve boxes of candy weigh 5.4 kilograms. Find the weight of each box.
3. The height of a stack of 8 cans is 99.2 cm. Find the height of each can.
4. If 5 packages of paper towels cost $2.45, what is the cost of one package of paper towels?
5. If 3 bottles of apple juice cost $2.79, then the cost of one bottle is
   (a) $.83 (b) $.90 (c) $.93 (d) $.94
6. If 4 bagels cost $.76, what is the cost of 9 bagels?
7. If 3 slices of pizza cost $2.25, what is the cost of 5 slices?
8. If 2 quarts of orange juice cost $1.80, then 5 quarts cost
   (a) $9.00 (b) $5.50 (c) $4.50 (d) $3.00

## DIVIDING BY A DECIMAL

### THE MAIN IDEA

To divide by a decimal:

1. Move the decimal point of the divisor all the way to the right, after the last digit.
2. Count the number of places that the decimal point was moved.
3. Move the decimal point of the dividend the same number of places to the right. If necessary, write extra zeros as placeholders after the last digit.
4. Now divide as if by a whole number.

**EXAMPLE 7** Divide: 4.824 ÷ .12

Since the decimal point of the divisor must be moved 2 places to the right, the decimal point of the dividend must also be moved 2 places to the right.

$$.12.\overline{)4.82.4}$$

Divide by the whole number 12.

$$\begin{array}{r} 40.2 \\ 12\overline{)482.4} \\ \underline{48}\phantom{2.4} \\ 02\,4 \\ \underline{2\,4} \\ 0 \end{array}$$

*Answer:* 40.2

**EXAMPLE 8** Divide: 175.5 ÷ .25

Since the decimal point of the divisor must be moved 2 places to the right, the decimal point of the dividend must also be moved 2 places to the right. Insert 1 zero in the dividend.

.25.)175.50.

Divide by the whole number 25.

```
    702.
25)17550.
   175
     50
     50
      0
```

*Answer:* 702

**EXAMPLE 9** Divide: 124 ÷ .248

The decimal point in the dividend is at the end of the whole number 124.

.248)124.

Move the decimal points of the divisor and dividend 3 places to the right. Insert 3 zeros in the dividend.

.248.)124.000.

Complete the division.

```
       500.
248)124000.
    1240
       000
```

*Answer:* 500

## CLASS EXERCISES

1. Divide.

   a. .525 ÷ .05 b. 86.48 ÷ .2 c. 244 ÷ .08 d. 3.6048 ÷ .012 e. .968 ÷ .008

   f. 1,575 ÷ 2.5 g. .0297 ÷ .11 h. .01512 ÷ .0006

2. If 9.5 gallons of a liquid weigh 79.23 pounds, what is the weight of one gallon?

3. In 2.25 hours, a printing press uses 348.75 pounds of paper. How many pounds does it use each hour?

## DIVIDING A DECIMAL BY A POWER OF 10

### THE MAIN IDEA

To divide a decimal by a power of 10, there are two methods:

1. Use the method you have learned to divide a decimal by a whole number.

or

2. a. Count the number of zeros in the power of 10.
   b. Move the decimal point to the *left* the same number of places as there are zeros in the power of 10.
   c. Write extra zeros to the left if there are not enough decimal places for moving the decimal point.

**EXAMPLE 10** Divide: .046 ÷ 100

*Method 1:*

Set up in the usual way. $100\overline{).046}$

Insert 3 zeros in the quotient because 100 is too large to divide into 0, 4, or 46. Add a zero in the dividend in order to begin the division.

$$\begin{array}{r} .000 \\ 100\overline{).0460} \end{array}$$

Begin the division. Add another zero in the dividend to complete the division.

$$\begin{array}{r} .00046 \\ 100\overline{).04600} \\ \underline{400} \\ 600 \\ \underline{600} \\ 0 \end{array}$$

*Answer:* .00046

*Method 2:*

Count the number of zeros in the divisor, 100. There are 2 zeros in 100.

To move the decimal point 2 places to the left in .046, insert 2 zeros. . 0 0 . 0 4 6

*Answer:* .00046

**EXAMPLE 11** One thousand bolts cost $97. At the same rate, what do 450 bolts cost?

**THINKING ABOUT THE PROBLEM**

First find the item price, the cost of one bolt. Do not round costs until all calculations are completed.

To find the cost of one bolt, divide the total cost by the number of bolts. $97 ÷ 1,000 = . 0 9 7

To find the cost of 450 bolts, multiply the cost of one bolt by 450.

```
  $.097
  ×450
    000
   4 85
  38 8
$43.650
```

*Answer:* The cost of 450 bolts is $43.65.

## CLASS EXERCISES

**1.** Divide.

**a.** 14.732 ÷ 10 **b.** 9.683 ÷ 100 **c.** 148.9 ÷ 1,000 **d.** 87.1 ÷ 1,000

**e.** 42.6 ÷ 10,000 **f.** .017 ÷ 100 **g.** .917 ÷ 10 **h.** .005 ÷ 100

**2.** If 100 paper containers cost $24.75, what is the cost of one container?

**3.** If 10,000 key chains cost $850, what is the cost of one key chain?

**4.** Ten signs cost $465. At the same rate, what is the cost of 6 signs?

**5.** One thousand envelopes cost $120. At the same rate, what is the cost of 750 envelopes?

## HOMEWORK EXERCISES

**1.** Divide.

**a.** 42.85 ÷ 5 **b.** 3.79 ÷ 2 **c.** .081 ÷ 3 **d.** 9.18 ÷ 9 **e.** 13.28 ÷ 16

**f.** 69.72 ÷ 12 **g.** 159.12 ÷ 17 **h.** 644.8 ÷ 26

**2.** Divide.

**a.** 37.5 ÷ .5 **b.** 806 ÷ .2 **c.** 24.12 ÷ .04 **d.** 8,127 ÷ .009 **e.** 3.278 ÷ .11

**f.** 10.03 ÷ 1.7 **g.** 155 ÷ .25 **h.** .576 ÷ .018

**3.** Divide.

**a.** 25.9 ÷ 10 **b.** 37.1 ÷ 100 **c.** 936 ÷ 1,000 **d.** 4.8 ÷ 100 **e.** 92 ÷ 10

**f.** 103 ÷ 100 **g.** 15.7 ÷ 10 **h.** 97.09 ÷ 1,000

**4.** How many pieces measuring .6 meter each can be cut from a metal rod 7.2 meters long?

**5.** Fifteen boxes of nails weigh 6 pounds. How much does one box weigh?

**6.** If 5 cans of peas cost $1.40, what is the cost of one can?

**7.** If 3 packages of looseleaf paper cost $3.57, what is the cost of one package?

**8.** If 4 cans of soda cost $.92, then the cost of one can is
(a) $.20 (b) $.23 (c) $.30 (d) $.31

**9.** If 3 drinks cost $.84, what is the cost of 5 drinks?

**10.** If 5 boxes of raisins cost $1.05, what is the cost of 9 boxes?

## SPIRAL REVIEW EXERCISES

1. Find each of the products.

   a. $9.35 \times 12$  b. $0.8 \times 2.65$

   c. $12.45 \times 0.08$  d. $2.9 \times 12.5$

2. How much greater is 13.65 than 9.2?

3. Mr. Smith spent $16.48 at the supermarket. He gave the cashier a $20 bill. How much change did he receive?

4. Find the sum of 0.092 and 0.83.

5. Ms. Marker had $248.52 in her checking account. She wrote a check for $95.80. What was her new balance?

6. The largest decimal of the group 0.8, 0.82, 0.28, and 0.802 is
   (a) 0.8  (c) 0.28
   (b) 0.82  (d) 0.802

7. Which is largest, $\frac{1}{2}$, $\frac{3}{8}$, $\frac{3}{4}$, or $\frac{9}{16}$?

8. $\frac{4}{9}$ is equivalent to
   (a) $.\overline{4}$  (b) $.\overline{9}$  (c) $.\overline{94}$  (d) $.\overline{49}$

9. The sum of two numbers is 12. If one of the numbers is $8\frac{3}{8}$, what is the other number?

10. What is the largest prime number less than 100?

11. Divide 2,688 by 48.

12. A toll bridge token costs one dollar. Jessica's coins are shown in the photograph. Does she have enough money to buy a token?

**CHALLENGE QUESTION**

Find the value of $5.13 \div 3 + 22.55 \div 5$.

## UNIT 8–5 Using Decimals in Computing Wages

### THE MAIN IDEA

1. The amount of money earned by a person is called *salary* or *wage*. Wages are often paid by the hour.
2. To find a person's total wages when the hourly wage is known, multiply the number of hours worked by the hourly wage.

   Total Wages = Number of Hours Worked × Hourly Wage
3. To find a person's hourly wage when the total wages are known, divide the total wages by the number of hours worked.

   Hourly Wage = Total Wages ÷ Number of Hours Worked
4. To find the number of hours worked when the hourly wage and the total wages are known, divide the total wages by the hourly wage.

   Number of Hours Worked = Total Wages ÷ Hourly Wage
5. Sometimes, for working extra hours, a person is paid *overtime*. Generally, the number of extra hours worked is counted as $1\frac{1}{2}$ times or 2 times the actual number of overtime hours.

**EXAMPLE 1** Janice earns $6.80 per hour. If she works for 7 hours, how much money will she earn?

Multiply the hourly wage by the number of hours worked.

$$\begin{array}{rl} \$6.80 & \leftarrow \text{hourly wage} \\ \times 7 & \leftarrow \text{number of hours} \\ \hline \$47.60 & \leftarrow \text{total wages} \end{array}$$

*Answer:* Janice will earn $47.60 in 7 hours.

**EXAMPLE 2** Mr. Harris worked 40 hours last week and earned $260. What is Mr. Harris' hourly wage?

Divide the total wages by the number of hours worked.

$$\begin{array}{r} 6.5 \\ 40\overline{)260.0} \\ \underline{240\phantom{.0}} \\ 20\,0 \\ \underline{20\,0} \\ 0 \end{array}$$

*Answer:* Mr. Harris earns $6.50 per hour.

**EXAMPLE 3** Ralph worked 6 hours on Monday, 7 hours on Tuesday, and 4 hours on Friday. If Ralph earns $5.25 an hour, how much did he earn?

### THINKING ABOUT THE PROBLEM

Use the strategy of breaking a problem into smaller, separate problems.

Find the total number of hours worked.

$$\begin{array}{r} 6 \\ 7 \\ +4 \\ \hline 17 \end{array}$$

Multiply the total number of hours by the hourly wage.

$$\begin{array}{rl} \$5.25 & \leftarrow \text{hourly wage} \\ \times 17 & \leftarrow \text{total hours} \\ \hline 36\ 75 & \\ 52\ 5\ \ & \\ \hline \$89.25 & \leftarrow \text{total wages} \end{array}$$

*Answer:* Ralph earned $89.25.

**EXAMPLE 4** Mr. Sweeney earned $40.25 one day, working at the rate of $5.75 per hour. How many hours did he work?

Divide the total wages by the hourly rate.

$$\begin{array}{r} 7. \\ 5.75\overline{)40.25.} \\ 40\ 25 \\ \hline \end{array}$$

*Answer:* Mr. Sweeney worked 7 hours.

**EXAMPLE 5** Teresa earns $5.20 per hour and time and one-half for each hour that she works overtime. Last week, Teresa worked 35 hours and 6 hours overtime. How much money did she earn?

The 6 overtime hours count extra.

$$\begin{array}{rl} 6 & \leftarrow \text{overtime hours} \\ \times\ 1\frac{1}{2} & \leftarrow \text{time and one-half} \\ \hline 9 & \leftarrow \text{extra hours} \end{array}$$

Find the total number of hours counted.

$$\begin{array}{rl} 35 & \leftarrow \text{regular hours} \\ +9 & \leftarrow \text{extra hours} \\ \hline 44 & \leftarrow \text{total hours} \end{array}$$

Multiply the total number of hours by the hourly wage.

$$\begin{array}{rl} \$5.20 & \leftarrow \text{hourly wage} \\ \times 44 & \leftarrow \text{total hours} \\ \hline 20\ 80 & \\ 208\ 0\ \ & \\ \hline \$228.80 & \leftarrow \text{total wages} \end{array}$$

*Answer:* Teresa earned $228.80.

**EXAMPLE 6** Mrs. Franklin is paid double time for any hours worked above 40 hours per week. Last week, she worked 49 hours. How much money did she earn if her hourly wage is $8.40?

To find the number of hours worked overtime, subtract 40 hours from the total number of hours worked.

$$\begin{array}{rl} 49 & \leftarrow \text{total hours} \\ -40 & \leftarrow \text{regular hours} \\ \hline 9 & \leftarrow \text{overtime hours} \end{array}$$

Multiply the number of overtime hours by 2.

$$\begin{array}{rl} 9 & \leftarrow \text{overtime hours} \\ \times 2 & \leftarrow \text{double time} \\ \hline 18 & \leftarrow \text{extra hours} \end{array}$$

Find the sum of the regular hours and the extra hours.

$$\begin{array}{rl} 40 & \leftarrow \text{regular hours} \\ +18 & \leftarrow \text{extra hours} \\ \hline 58 & \leftarrow \text{sum of hours} \end{array}$$

Multiply the sum of the hours by the hourly wage.

$$\begin{array}{rl} \$8.40 & \leftarrow \text{hourly wage} \\ \times 58 & \leftarrow \text{sum of hours} \\ \hline 67\ 20 & \\ 420\ 0\phantom{0} & \\ \hline \$487.20 & \leftarrow \text{total wages} \end{array}$$

*Answer:* Mrs. Franklin earned $487.20.

## CLASS EXERCISES

1. Jane earns $8.25 per hour. If she works for 12 hours, how much money will she earn?
2. Mr. Roberts earns $15.70 per hour. How much will he earn if he works 35 hours?
3. Mr. Smith earns $7.50 an hour as a typist. What are his wages if he works 40 hours?
4. The Perfect Crating Company pays its workers $4.90 per hour for unloading its trucks. Find the amount of money earned for each of the following numbers of hours worked:

   **a.** 20 hours  **b.** 18 hours  **c.** 30 hours  **d.** 35 hours
5. Ms. Hope worked 35 hours last week and earned $332.50. What is Ms. Hope's hourly wage?
6. Bob worked 25 hours packing groceries and earned $87.50. What is Bob's hourly wage?
7. Jessica worked 18 hours and earned $104.40. What is Jessica's hourly wage?
8. An accountant billed a client $250 for 10 hours work. What is the accountant's hourly wage?
9. Bill worked 7 hours on Monday, 4 hours on Tuesday, 8 hours on Wednesday, 5 hours on Thursday, and 6 hours on Friday. If Bill earns $4.10 per hour, how much did he earn?
10. Julie earns $3.90 per hour working as a carpenter's assistant. Last week, she worked 3 hours on Monday, 5 hours on Wednesday, and 2 hours on Friday. How much did Julie earn last week?
11. Mr. James earns $16.80 per hour and time and one-half for each hour that he works overtime. Last week, Mr. James worked 35 hours plus 10 hours overtime. How much did Mr. James earn?
12. Mrs. Fox earned $39 working at the rate of $6.50 per hour. How many hours did she work?
13. Working at the rate of $8.40 per hour, Mr. Jones earned $63. How many hours did he work?
14. Mr. Cousins is paid time and one-half for any hours worked above 35 hours per week. Last week, he worked 41 hours. How much money did he earn if his hourly wage is $7.20?
15. Mrs. Robins is paid double time for any hours worked above 40 hours per week. Last week, she worked 50 hours. If her hourly wage is $8.50, how much did she earn?

## HOMEWORK EXERCISES

1. Phil earns $5.40 an hour. If he worked 15 hours this week, how much did he earn?
2. How much did Mrs. Osawa earn last week if she worked 40 hours at the rate of $6.50 an hour?

3. Mr. Davis worked 30 hours last week. If he earned a total of $450, what was his hourly wage?

4. What is Chuck's hourly wage if he earned $157.50 for 25 hours of work?

5. Helen worked 10 hours on Monday, 12 hours on Wednesday, and 8 hours on Friday. If she earns $5.80 an hour, how much money did she earn?

6. Mike earns $6.20 an hour and time and a half for any hours above 40. If he worked 43 hours, how much did he earn?

7. Flora kept a record, as shown, of the hours she worked last week. If she makes $4.50 an hour, how much did she earn last week?

| Day | Hours Worked |
|---|---|
| Monday | 3 |
| Tuesday | $2\frac{1}{2}$ |
| Wednesday | 2 |
| Thursday | 3 |
| Friday | $3\frac{1}{2}$ |

8. Mrs. Chan earned $90.40 working at a rate of $11.30 an hour. How many hours did she work?

9. Mr. Torre earns $10.50 an hour. One day, he earned a total of $57.75. How many hours did he work?

10. Sue earns double time for any hours that she works over 35. Her regular hourly wage is $4.50. Last week, she worked 39 hours. How much did she earn?

11. Copy and complete the table shown. Overtime is counted as time and one-half.

| Number of Regular Hours | Number of Overtime Hours | Hourly Wage | Total Wage |
|---|---|---|---|
| 20 | 10 | $3.60 | |
| 25 | 12 | $4.20 | |
| 35 | 6 | $4.80 | |
| 40 | 10 | $5.90 | |

12. Mr. Allison worked 40 hours last week and earned $380. Find his hourly wage.

## SPIRAL REVIEW EXERCISES

1. Write .14 as a fraction in simplest form.

2. Write $\frac{3}{7}$ as a decimal correct to the nearest hundredth.

3. Add: 27.02 + 93.8 + 18.75

4. Divide: 27.063 ÷ 0.03

5. Multiply: 184.72 × 1,000

6. Find the product of $\frac{3}{16}$ and $\frac{8}{9}$.

7. Rewrite the numbers 1.8, .81, .18, and .801 in order from the largest to the smallest.

8. Subtract: $-12 - (+7)$

9. Which number is between 5.2 and 5.3?
   (a) 5.023 (c) 5.230
   (b) 5.032 (d) 5.320

### CHALLENGE QUESTION

Debbie worked 35 hours at the rate of $4.75 an hour and then received a raise of 50 cents an hour. If she worked another 35 hours on her new salary, how much money did she earn altogether?

## UNIT 8–6 Using Decimals in Stepped Rates

### THE MAIN IDEA

1. Some services are priced so that the rate charged changes for different quantities purchased.
2. The rates are ***stepped*** so that the quantities in different steps are priced at different levels.
3. To find the total cost of a rate that has two steps:
   a. Find the cost of the first step.
   b. If more than the number of units allowed in the first step have been used, subtract the number of units allowed in the first step from the total number of units. Multiply the difference by the second step rate.
   c. Add the costs of the two steps to find the total cost.

**EXAMPLE 1** The rates for a long-distance telephone call are:

$.70 for the first five minutes or less
$.40 for each additional minute

What is the cost of a 3-minute call?

Since the call is shorter than the time allowed in the first step, only the first-step rate is charged.

*Answer:* A 3-minute call costs $.70.

**EXAMPLE 2** The rates for a long-distance telephone call are:

$.50 for the first three minutes
$.20 for each additional minute

What is the cost of a 10-minute call?

Since the call is longer than the time allowed in the first step, we must find the cost for each of two steps.

To find the number of additional minutes, subtract the time allowed in the first step from the length of the call.

10 ← total number of minutes
−3 ← number of minutes in first step
7 ← number of minutes over first-step allowance

To find the additional charge, multiply the additional number of minutes, 7, by $.20.

$.20 ← charge per minute
×7 ← number of minutes
$1.40 ← cost of second step

To find the total cost of the telephone call, add the costs of both steps.

$.50 ← cost of first step
+1.40 ← cost of second step
$1.90 ← total cost

*Answer:* The cost of a 10-minute call is $1.90.

**EXAMPLE 3** Julie works at the Quick Copy Center. The rates for making copies are:

$.15 for each of the first five copies
$.10 for each additional copy

How much should Julie charge a customer for 17 copies?

To find the cost of the first step, multiply $.15 by 5.

$.15 ← charge per copy
×5 ← number of copies
$.75 ← cost of first step

To find the number of copies in the second step, subtract 5 from 17.

17 ← total number of copies
−5 ← number of copies in first step
12 ← number of copies over first-step allowance

To find the cost of copies in the second step, multiply $.10, the rate for the second step, by 12.

$.10 ← charge per copy
×12 ← number of copies
20
10
$1.20 ← cost of second step

To find the total cost, add the costs of both steps.

$.75 ← cost of first step
+1.20 ← cost of second step
$1.95 ← total cost

*Answer:* The cost of 17 copies is $1.95.

## CLASS EXERCISES

1. A telephone call costs: $.80 for the first 5 minutes
   $.15 for each additional minute
   Find the cost of a 5-minute call.
2. The rates for a long-distance telephone call are: $.50 for the first 3 minutes
   $.10 for each additional minute
   What is the cost of a call lasting 12 minutes?
3. The rates for a long-distance telephone call are: $1.80 for the first 5 minutes
   $.30 for each additional minute
   How much does a call lasting 15 minutes cost?
4. A mailing service charges: $2.20 for the first pound
   $.90 for each additional pound up to 5 pounds
   Find the cost of mailing a package that weighs:
   **a.** 1 pound **b.** 3 pounds **c.** 5 pounds
5. The cost for a classified ad in a newspaper is: $2.50 for the first 15 words
   $.25 for each additional word
   How much does a classified ad of 21 words cost?
6. The rates at the Speedy Duplicating Shop are: $.10 for each of the first 10 copies
   $.05 for each additional copy
   How much will 25 copies cost?
7. The Cheap Pen Company sells pens at the following rate:
   $.50 for each of the first 20 pens bought
   $.30 for each additional pen
   Find the cost of: **a.** 15 pens **b.** 20 pens **c.** 30 pens **d.** 50 pens

## HOMEWORK EXERCISES

1. A telephone call costs: \$.40 for the first 3 minutes
   \$.10 for each additional minute

   Find the cost of a call that lasts:

   **a.** 2 minutes **b.** 5 minutes **c.** 20 minutes

2. The rates for a long-distance telephone call are: \$.95 for the first 5 minutes
   \$.20 for each additional minute

   What is the cost of a call that lasts 11 minutes?

3. The rates for a long-distance telephone call are: \$1.50 for the first 3 minutes
   \$.25 for each additional minute

   Find the cost of a call that lasts:

   **a.** 5 minutes **b.** 15 minutes **c.** 3 minutes

4. The Speedy Cab Co. charges \$1.00 for the first $\frac{1}{10}$ of a mile and \$.08 for each additional $\frac{1}{10}$ of a mile. What is the charge for a 2-mile ride?

5. Bloodhound Bus Co. charges \$12 for the first 50 miles and \$.15 for each additional mile. What is the price of a ticket for a 75-mile ride?

6. The rates at Greenlawn Tennis Club are \$10 for the first visit and \$8 for each additional visit. If Jodie visited the tennis club 5 times, how much did she pay?

7. An ad in the school newspaper costs \$5 for the first line and \$3 for each additional line. Find the cost of each ad:

   **a.** 1 line **b.** 2 lines **c.** 5 lines **d.** 10 lines

## SPIRAL REVIEW EXERCISES

1. Sandra worked a 35-hour week at an hourly wage of \$6.20, with time and a half for overtime. One week she worked 47 hours. What were her total earnings for that week?

2. How much smaller is 12.96 than 13?

3. **a.** If 12 candles cost \$18.60, what is the price of one candle?

   **b.** Using your answer for part **a,** find the price of 5 candles.

4. Mrs. Strong opened a new checking account with a balance of \$500. After writing checks for \$117.85, \$36.50, and \$210, what was the new balance?

5. Round 52,489 correct to the nearest thousand.

6. Which fraction is equal to $\frac{5}{6}$?

   (a) $\frac{10}{18}$ (b) $\frac{20}{24}$ (c) $\frac{15}{30}$ (d) $\frac{20}{30}$

### CHALLENGE QUESTION

Mr. Li called his brother in Tacoma and paid \$9.60 for a 12-minute call. If the rate for the first 3 minutes was \$2.40, how much was charged for each additional minute?

# THEME 5

# Percents

## MODULE 9



## MODULE 10

# UNIT 9–1 The Meaning of Percent; Changing Percents to Decimals and Decimals to Percents

## THE MEANING OF PERCENT

### THE MAIN IDEA

1. *Percent* means *hundredths*. (The symbol for percent is %.)
2. 100% means 100 hundredths or $\frac{100}{100}$ or 1, which represents a whole quantity.
3. Percents less than 100% are numbers less than 1, and represent less than a whole quantity.
4. Percents greater than 100% are numbers greater than 1, and represent more than a whole quantity.

**EXAMPLE 1** For each percent given in numbers, write the percent in words. Tell, in words and in numbers, how many hundredths there are in each percent.

| | *Percent in Numbers* | *Percent in Words* | *Hundredths in Words* | *Hundredths in Numbers* |
|---|---|---|---|---|
| **a.** | 1% | one percent | one hundredth | $\frac{1}{100}$ |
| **b.** | 95% | ninety-five percent | ninety-five hundredths | $\frac{95}{100}$ |
| **c.** | 100% | one hundred percent | one hundred hundredths | $\frac{100}{100}$ |
| **d.** | 250% | two hundred fifty percent | two hundred fifty hundredths | $\frac{250}{100}$ |

**EXAMPLE 2** Write 57 hundredths as a percent.

Since "percent" means "hundredths," 57 hundredths is the same as 57 percent.

*Answer:* 57%

**EXAMPLE 3** Replace ? by <, =, or > to make each comparison a true statement.

| | *Comparison* | *True Statement* |
|---|---|---|
| **a.** | 12% ? 1 | 12% < 1 |
| **b.** | 99% ? 1 | 99% < 1 |
| **c.** | 100% ? 1 | 100% = 1 |
| **d.** | 101% ? 1 | 101% > 1 |
| **e.** | 1,000% ? 1 | 1,000% > 1 |

**EXAMPLE 4** Jonathan has finished 68% of his homework. What percent of his homework remains to be done?

**THINKING ABOUT THE PROBLEM**

The key word "remains" tells you to subtract.

$$\begin{array}{rl} 100\% & \leftarrow \text{the whole quantity} \\ -\ 68\% & \leftarrow \text{the percent already done} \\ \hline 32\% & \leftarrow \text{the percent remaining to be done} \end{array}$$

*Answer:* Jonathan must do the remaining 32% of his homework.

## CLASS EXERCISES

1. For each percent given in numbers, write the percent in words. Tell, in words and in numbers, how many hundredths there are in each percent.

   **a.** 35% **b.** 42% **c.** 10% **d.** 100% **e.** 87% **f.** 110% **g.** 2% **h.** 200%

2. Write as a percent.

   **a.** $\frac{37}{100}$ **b.** 42 hundredths **c.** 3 hundredths **d.** $\frac{175}{100}$ **e.** 5 hundred hundredths

3. Replace ? by $<$, $=$, or $>$ to make each comparison a true statement.

   **a.** 300% ? 1 **b.** 30% ? 1 **c.** 90% ? 1 **d.** 4% ? 1 **e.** 120% ? 1

   **f.** 100% ? 1 **g.** 50% ? 1 **h.** 150% ? 1

4. Every morning, Samantha does her exercise routine.

   **a.** If she has completed 26% of the routine, what percent remains to be done?

   **b.** If she has completed 111% of the routine, by what percent has she exceeded her routine?

## CHANGING PERCENTS TO DECIMALS AND DECIMALS TO PERCENTS

### THE MAIN IDEA

1. To change a percent to a decimal:
   a. Drop the % sign.
   b. Move the decimal point two places to the *left,* adding zeros if necessary.
2. To change a decimal to a percent:
   a. Move the decimal point two places to the *right,* adding zeros if necessary.
   b. Write the resulting number followed by a % sign.

**EXAMPLE 5** Write each percent as a decimal.

| | Percent | Moving the Decimal Point | Decimal |
|---|---|---|---|
| a. | 52% | .52. | .52 |
| b. | 8% | .08. | .08 |
| c. | 4.3% | .04.3 | .043 |
| d. | 31.5% | .31.5 | .315 |
| e. | .6% | .00.6 | .006 |
| f. | 200% | 2.00. | 2 |

**EXAMPLE 6** Write each decimal as a percent.

| | Decimal | Moving the Decimal Point | Percent |
|---|---|---|---|
| a. | .34 | .34. | 34% |
| b. | .253 | .25.3 | 25.3% |
| c. | .047 | .04.7 | 4.7% |
| d. | 7 | 7.00. | 700% |
| e. | 8.5 | 8.50. | 850% |
| f. | 29.3 | 29.30. | 2,930% |

## CLASS EXERCISES

1. Write each percent as a decimal.

   **a.** 17% **b.** 9% **c.** 50% **d.** 6.2% **e.** 48.6% **f.** .8% **g.** 300%

   **h.** 120% **i.** 250% **j.** 115% **k.** 410% **l.** 500%

2. A certain fruit drink is 32.5% fruit juice. What is this percent written as a decimal?
3. The inflation rate in a certain country was recorded at 160%. What is this percent written as a decimal?
4. Of the eggs packed by Red Hen Dairy, 1.5% break before they reach the market. Write this percent as a decimal.
5. Sue is 120.5% of her normal weight. Write this percent as a decimal.
6. Write each decimal as a percent.

   **a.** .90 **b.** .06 **c.** 2 **d.** .2 **e.** .02 **f.** 1.5 **g.** .003

   **h.** .486 **i.** .038 **j.** 1.08 **k.** .0043 **l.** .184

7. The sales tax rate in a certain city is .075, written as a decimal. What is this tax rate written as a percent?
8. Carmen's family spends .33 of their income for rent. What percent of their income is spent for rent?

## HOMEWORK EXERCISES

1. For each percent given in numbers, write the percent in words. Tell, in words and in numbers, how many hundredths there are in each percent.

   **a.** 1% **b.** 100% **c.** 24% **d.** 400% **e.** 95%

2. Write as a percent.

   **a.** $\frac{5}{100}$ **b.** 12 hundredths **c.** one hundred hundredths **d.** $\frac{43}{100}$ **e.** 82 hundredths

3. Replace ? by $<$, $=$, or $>$ to make each comparison a true statement.

   **a.** 24% ? 1 **b.** 100% ? 1 **c.** 5% ? 1 **d.** 125% ? 1 **e.** 48% ? 1 **f.** 1% ? 1

4. On his diet, Mr. Watson is allowed no more than a certain number of calories daily.

   a. If he has consumed 82% of the allowed number of daily calories, what percent remains to be consumed?

   b. If he has consumed 136% of the allowed number of daily calories, by what percent has he exceeded his allowed number of calories?

5. Write each percent as a decimal.

   a. 98% b. 2% c. .7% d. 31.4% e. 8.5% f. 65% g. 99.9%

   h. 1.1% i. .08% j. 140% k. 500% l. 250% m. 101% n. 1,000%

6. A certain credit card company charges 11.5% as a service fee. Write this percent as a decimal.

7. Five percent of every dollar spent in Townville goes for sales tax. Write this percent as a decimal.

8. 1% written as a decimal is (a) .001 (b) .01 (c) .1 (d) 1

9. A soap company claims that its soap is 99.44% pure. Write this percent as a decimal.

10. Write each decimal as a percent.

    a. .6 b. .48 c. .09 d. 3.2 e. 3 f. .89 g. 1.1

    h. .70 i. .025 j. .512 k. .008 l. 3.04

11. In a certain high school, .258 of the senior class was accepted by Central University. What percent of the class was accepted by Central?

12. Thirteen thousandths of a certain yogurt is fat. Write this decimal as a percent.

13. Which is equal to 200%? (a) 200 (b) .200 (c) 2% (d) 2

## SPIRAL REVIEW EXERCISES

1. Multiply: $63.1 \times 4.6$

2. Mrs. Farland's checking account had a balance of \$117.50. Then she wrote checks for \$17.86, \$25, and \$35.98. What was her new balance?

3. B & G Clothing Store reduced the price of a \$35 bathing suit by \$7.50. What was the new price?

4. Three-eighths of the cars in a dealer's lot were black. If there were 160 cars, how many were black?

5. To make the statement $\frac{7}{8} = \frac{35}{?}$ true, ? should be replaced by
   (a) 48 (b) 8 (c) 40 (d) 56

6. What is the cost of 7 oranges if 12 cost \$1.32?

7. $\frac{5}{6}$ is equivalent to
   (a) $.8\overline{3}$ (b) $.\overline{83}$ (c) $.\overline{38}$ (d) $.\overline{8}$

8. The sum of $+7$ and $-7$ is
   (a) $-49$ (b) $-14$ (c) $+14$ (d) 0

### CHALLENGE QUESTION

Rearrange in order, from least to greatest:

13% $\frac{2}{100}$ .08 .8 3% 5 .5

# UNIT 9–2 Changing Percents to Fractions and Fractions to Percents; Fraction, Decimal, and Percent Equivalents

## CHANGING PERCENTS TO FRACTIONS

### THE MAIN IDEA

1. To change a percent to a fraction:
   a. Write a fraction whose numerator is the amount of the percent and whose denominator is 100.
   b. If possible, simplify the fraction.
2. To change a percent that contains a decimal (such as 3.8%) to a fraction:
   a. Change the percent to a decimal.
   b. Change the decimal to a fraction.
   c. If possible, simplify the fraction.

**EXAMPLE 1** Change 24% to a fraction.

Write a fraction whose numerator is 24 and whose denominator is 100. $24\% = \frac{24}{100}$

Simplify the fraction. $\frac{24 \div 4}{100 \div 4} = \frac{6}{25}$ *Ans.*

**EXAMPLE 2** Change 175% to a fraction.

Write a fraction whose numerator is 175 and whose denominator is 100. $175\% = \frac{175}{100}$

Simplify the fraction. $\frac{175 \div 25}{100 \div 25} = \frac{7}{4}$ or $1\frac{3}{4}$ *Ans.*

**EXAMPLE 3** Change $66\frac{2}{3}\%$ to a fraction.

Write a fraction whose numerator is $66\frac{2}{3}$ and whose denominator is 100. $66\frac{2}{3}\% = \frac{66\frac{2}{3}}{100}$

Since dividing by 100 is the same as multiplying by $\frac{1}{100}$, rewrite the fraction as a product. $66\frac{2}{3} \times \frac{1}{100}$

Write $66\frac{2}{3}$ as an improper fraction. $\frac{200}{3} \times \frac{1}{100}$

Multiply. $\frac{200}{300}$

Simplify. $\frac{200 \div 100}{300 \div 100} = \frac{2}{3}$ *Ans.*

**EXAMPLE 4** Change 6.5% to a fraction.

Change the percent to a decimal. $6.5\% = .065$

Write the decimal as a fraction. $.065 = \frac{65}{1000}$

Simplify. $\frac{65 \div 5}{1000 \div 5} = \frac{13}{200}$ *Ans.*

## CLASS EXERCISES

**1.** Change each percent to a fraction in lowest terms.

**a.** 9% **b.** 30% **c.** $33\frac{1}{3}\%$ **d.** 7.5% **e.** 47% **f.** 80% **g.** $12\frac{1}{2}\%$ **h.** 24.8%

**i.** 75% **j.** 495% **k.** $8\frac{1}{4}\%$ **l.** .04% **m.** 60% **n.** 1,000% **o.** $\frac{3}{4}\%$

**2.** A soft drink contains 15% sugar. Write as a fraction in simplest form the part of the drink that is sugar.

**3.** A quarterback completed 72% of the passes that he attempted. Write as a fraction in simplest form the part of the passes that he completed.

**4.** Marvelous Bond Trust Company offers a $9\frac{1}{4}\%$ rate of interest on all investments. Write this percent as a fraction in simplest form.

## CHANGING FRACTIONS TO PERCENTS

### THE MAIN IDEA

To change a fraction to a percent:

1. Change the fraction to a decimal by dividing the denominator into the numerator.
2. Change the resulting decimal to a percent by moving the decimal point two places to the right and writing a % sign.

**EXAMPLE 5** Change $\frac{3}{4}$ to a percent.

Change $\frac{3}{4}$ to a decimal by dividing the denominator 4 into the numerator 3.

$$\begin{array}{r} .75 \\ 4\overline{)3.00} \\ \underline{2\,8}\phantom{0} \\ 20 \\ \underline{20} \\ 0 \end{array}$$

Change .75 to a percent by moving the decimal point two places to the right and writing a % sign.

$.75. = 75\%$ *Ans.*

**EXAMPLE 6** Write $\frac{3}{8}$ as a percent.

Change $\frac{3}{8}$ to a decimal by dividing the denominator 8 into the numerator 3.

$$8\overline{)3.000} = .375$$
$$\begin{array}{r} 3.000 \\ \underline{2\,4\phantom{00}} \\ 60\phantom{0} \\ \underline{56\phantom{0}} \\ 40 \\ \underline{40} \\ 0 \end{array}$$

Change .375 to a percent by moving the decimal point two places to the right and writing a % sign.

$.37.5 = 37.5\%$ *Ans.*

**EXAMPLE 7** Change $\frac{1}{3}$ to a percent.

Change $\frac{1}{3}$ to a decimal by dividing the denominator 3 into the numerator 1. Since the decimal repeats, stop the division at the hundredths place.

$$3\overline{)1.00} = .33\ldots$$
$$\begin{array}{r} 1.00 \\ \underline{9\phantom{0}} \\ 10 \\ \underline{9} \\ 1 \end{array}$$

Change .33 to a percent.

$.33. = 33\%$

Use the remainder 1 and the divisor 3 to write the fraction $\frac{1}{3}$. Add this fraction to the percent.

$33\frac{1}{3}\%$ *Ans.*

**EXAMPLE 8** Change $3\frac{1}{2}$ to a percent.

Change $3\frac{1}{2}$ to an improper fraction.

$3\frac{1}{2} = \frac{7}{2}$

Change $\frac{7}{2}$ to a decimal.

$$2\overline{)7.0} = 3.5$$
$$\begin{array}{r} 7.0 \\ \underline{6\phantom{.0}} \\ 1\,0 \\ \underline{1\,0} \\ 0 \end{array}$$

Change 3.5 to a percent.

$3.50. = 350\%$ *Ans.*

## CLASS EXERCISES

1. Change to a percent:

a. $\frac{1}{2}$ b. $\frac{2}{3}$ c. $3\frac{1}{4}$ d. $\frac{9}{10}$ e. $\frac{3}{5}$ f. $\frac{1}{6}$ g. $5\frac{2}{5}$ h. $\frac{17}{20}$

i. $1\frac{5}{8}$ j. $\frac{4}{7}$ k. $7\frac{1}{12}$ l. $4\frac{3}{25}$ m. $\frac{5}{16}$ n. $\frac{7}{9}$ o. 9

2. Mr. Wise saves $\frac{13}{50}$ of his salary. What percent of his salary does he save?

3. A basketball team lost $\frac{7}{20}$ of the games that it played. What percent of the games did it lose?

4. Seven-eighths of the students in Ms. Jones's class passed the mathematics test. What percent of the class passed the test?

## FRACTION, DECIMAL, AND PERCENT EQUIVALENTS

### THE MAIN IDEA

Some often-used fractions and their decimal and percent equivalents are:

| Fraction | Decimal | Percent | Fraction | Decimal | Percent |
|---|---|---|---|---|---|
| $\frac{1}{2}$ | .5 | 50% | $\frac{1}{8}$ | .125 | $12\frac{1}{2}\%$ |
| $\frac{1}{3}$ | $.\overline{33}$ | $33\frac{1}{3}\%$ | $\frac{3}{8}$ | .375 | $37\frac{1}{2}\%$ |
| $\frac{2}{3}$ | $.\overline{66}$ | $66\frac{2}{3}\%$ | $\frac{5}{8}$ | .625 | $62\frac{1}{2}\%$ |
| $\frac{1}{4}$ | .25 | 25% | $\frac{7}{8}$ | .875 | $87\frac{1}{2}\%$ |
| $\frac{3}{4}$ | .75 | 75% | $\frac{1}{10}$ | .1 | 10% |
| $\frac{1}{5}$ | .2 | 20% | $\frac{3}{10}$ | .3 | 30% |
| $\frac{2}{5}$ | .4 | 40% | $\frac{7}{10}$ | .7 | 70% |
| $\frac{3}{5}$ | .6 | 60% | $\frac{9}{10}$ | .9 | 90% |
| $\frac{4}{5}$ | .8 | 80% | | | |

**EXAMPLE 9** Half of Marie's paycheck went for food this week. What percent of her money did she spend on food?

Without calculating, change $\frac{1}{2}$ to a percent. $\frac{1}{2} = 50\%$

*Answer:* 50% was spent on food.

**EXAMPLE 10** 75% of the class voted for Barry for class president. What fractional part of the class voted for Barry?

Without calculating, change 75% to a fraction. $75\% = \frac{3}{4}$

*Answer:* $\frac{3}{4}$ of the class voted for Barry.

## CLASS EXERCISES

1. Without calculating, write each fraction as a decimal and as a percent.

   a. $\frac{1}{10}$ b. $\frac{1}{4}$ c. $\frac{3}{8}$ d. $\frac{2}{5}$ e. $\frac{3}{4}$ f. $\frac{1}{3}$ g. $\frac{7}{8}$ h. $\frac{7}{10}$

2. Without calculating, write each decimal as a fraction and as a percent.

   a. .125 b. .75 c. $.\overline{6}$ d. .375 e. .1 f. .25 g. .5 b. .9

3. Without calculating, write each percent as a fraction and as a decimal.

   a. 25% b. 60% c. $87\frac{1}{2}\%$ d. 50% e. 10% f. $33\frac{1}{3}\%$ g. 90% h. $12\frac{1}{2}\%$

4. Ten percent of Sam's allowance goes into the bank. What fractional part of his money does Sam save?

5. A fourth of the puppies in a litter were tan. What percent of the litter was tan?

## HOMEWORK EXERCISES

1. Change each percent to a fraction in lowest terms.

   a. 7% b. 5% c. 51% d. 73% e. 97% f. 40% g. 70%

   h. 65% i. 25% j. 50% k. 130% l. 300% m. $66\frac{2}{3}\%$ n. $87\frac{1}{2}\%$

   o. $\frac{1}{4}\%$ p. $83\frac{1}{3}\%$ q. 12.5% r. 8.25% s. 0.8% t. .06%

2. A football team won 85% of the games it played. Write as a fraction in simplest form the part of the games that the team won.

3. Mr. Baker sold 63% of the shirts that he had on sale. Write as a fraction in simplest form the part of the shirts that he sold.

4. Money Savings Bank offers a $5\frac{1}{2}\%$ rate of interest. Write the interest rate as a fraction in simplest form.

5. Change to a percent:

   a. $\frac{3}{4}$ b. $\frac{7}{10}$ c. $\frac{5}{8}$ d. $\frac{7}{20}$ e. $\frac{11}{50}$ f. $\frac{8}{9}$ g. $\frac{13}{25}$ h. $\frac{2}{3}$

   i. $\frac{3}{16}$ j. $\frac{5}{12}$ k. 4 l. $2\frac{1}{8}$ m. $3\frac{4}{5}$ n. $\frac{9}{32}$ o. $2\frac{5}{6}$

6. Ms. Donland's children ate $\frac{6}{8}$ of a pizza pie. What percent of the pie did they eat?

7. Steve's height is $\frac{7}{8}$ of his father's height. What percent of his father's height is this?

8. Without calculating, write each fraction as a decimal and as a percent.

   a. $\frac{3}{10}$ b. $\frac{7}{10}$ c. $\frac{1}{2}$ d. $\frac{5}{8}$ e. $\frac{2}{3}$ f. $\frac{4}{5}$ g. $\frac{3}{4}$ h. $\frac{7}{8}$

**9.** Without calculating, write each decimal as a fraction and as a percent.

**a.** .875 **b.** .8 **c.** .25 **d.** .375 **e.** .5 **f.** .75 **g.** .9 **h.** .4

**10.** Without calculating, write each percent as a fraction and as a decimal.

**a.** $66\frac{2}{3}\%$ **b.** $12\frac{1}{2}\%$ **c.** 80% **d.** 62.5% **e.** 30% **f.** 75% **g.** 50% **h.** 40%

**11.** A carpenter completed 40% of a job. Without calculating, tell what fractional part of the job he completed.

**12.** A bolt is $\frac{3}{8}$ of an inch long. Without calculating, write this length as a decimal.

## *SPIRAL REVIEW EXERCISES*

**1.** Evaluate: $12 + 3 \times 8 \div 6$

**2.** The greatest common factor of 36, 54, 72, and 108 is
(a) 6 (b) 9 (c) 16 (d) 18

**3.** What is the total thickness of 100 sheets of paper if each sheet is .04 inch thick?

**4.** Divide $-360$ by 8.

**5.** Find $\frac{5}{9}$ of 81.

**6.** Bill opened a checking account with a balance of $750. He wrote two checks for $12.78 and $550, and deposited $200. What was his new balance?

**7.** The first rental of a videotape in the TV-Mania Video Club costs $35. Each rental after that costs $2. If the Carson family rented 10 tapes, what was the total cost?

## CHALLENGE QUESTION

Which of the following is not equal to the others?

$.5 \quad 50\% \quad \frac{1}{2} \quad \frac{50}{100} \quad \frac{5}{10} \quad .05$

# UNIT 9–3 Finding a Percent of a Number; Using Percents to Find Commissions

## FINDING A PERCENT OF A NUMBER

### THE MAIN IDEA

To find a percent of a number:

1. Write the percent as a decimal.
2. Multiply the number by the decimal.

**EXAMPLE 1** Find 25% of 84.

Write 25% as a decimal. .25

Multiply 84 by this decimal.

$$\begin{array}{r} 84 \\ \times .25 \\ \hline 4\ 20 \\ 16\ 8\phantom{0} \\ \hline 21.00 \end{array}$$

21.00 or 21 *Ans.*

**EXAMPLE 2** Find 22.4% of 72.

Write 22.4% as a decimal. .224

Multiply.

$$\begin{array}{r} .224 \\ \times 72 \\ \hline 448 \\ 15\ 68\phantom{0} \\ \hline 16.128 \end{array}$$

16.128 *Ans.*

**EXAMPLE 3** Mark spent 52% of his allowance for a record. If his allowance was \$9, find how much he spent for the record.

Write 52% as a decimal. 52% = .52

Multiply.

$$\begin{array}{r} .52 \\ \times 9 \\ \hline 4.68 \end{array}$$

*Answer:* Mark spent \$4.68 for the record.

**EXAMPLE 4** This year the Russo family spent 150% of what they spent last year for clothes. If they spent \$1,000 last year for clothes, what did they spend this year?

Write 150% as a decimal. 1.50

Multiply.

$$\begin{array}{r} 100\ 0 \\ \times 1.5 \\ \hline 500\ 0 \\ 1000\phantom{\ 0} \\ \hline 1500.0 \end{array}$$

*Answer:* The Russos spent \$1,500 for clothes.

## CLASS EXERCISES

**1.** Find the value of each expression.

**a.** 6% of 120 **b.** 42% of 80 **c.** 1% of 445 **d.** 15.4% of 200 **e.** 4.2% of .5

**2.** The Johnson family spent 34% of its weekly income for food. If the Johnson's weekly income was $300, how much did they spend for food?

**3.** Carl saved 25% of the money that he earned. If Carl earned $140, how much did he save?

**4.** The basketball team won 60% of the games that it played. If the team played 40 games, how many games did it win?

**5.** Membership this year in the Student Organization is 220% of last year's membership. There were 300 members last year. How many members are there this year?

## USING PERCENTS TO FIND COMMISSIONS

### THE MAIN IDEA

1. A *commission* is the amount of money that a salesperson earns when making a sale.
2. A percent is used to show the rate of commission.
3. To find the amount of commission:
   a. Change the rate of commission to a decimal.
   b. Multiply the dollar amount of the sales by this decimal.

   Amount of Commission = Amount of Sales × Rate of Commission

**EXAMPLE 5** Ms. Holt is a salesperson for the Zippy Used Car Company. She earns a commission, 9% of each sale. How much commission did she earn on the sale of a $2,400 car?

To find the commission, find 9% of $2,400.

$2400 ← amount of sale
× .09 ← rate of commission
$216.00 ← amount of commission

*Answer:* Ms. Holt earned $216 in commission.

**EXAMPLE 6** Mr. James earns $250 a week plus a 10% commission on his sales. Last week his sales totaled $1,490. Find his total earnings for last week.

Find the amount of commission by finding 10% of $1,490.

$1490 ← amount of sales
× .10 ← rate of commission
$149.00 ← amount of commission

Add the commission to the weekly salary.

$250 ← weekly salary
+ 149 ← commission
$399 ← total earnings

*Answer:* Mr. James' total earnings for last week were $399.

**EXAMPLE 7** Mrs. Baxter earns a 7% commission on her sales. She sold $240 worth of merchandise on Monday, $300 on Tuesday, and $190 on Wednesday. Find the commission for her total sales on these three days.

Find the amount of the total sales.

$$\begin{array}{r} \$240 \\ 300 \\ +190 \\ \hline \$730 \end{array} \leftarrow \text{total sales}$$

Find the commission.

$$\begin{array}{rl} \$730 & \leftarrow \text{amount of sales} \\ \times .07 & \leftarrow \text{rate of commission} \\ \hline \$51.10 & \leftarrow \text{amount of commission} \end{array}$$

*Answer:* Mrs. Baxter's commission was $51.10.

## CLASS EXERCISES

1. For each of the given sales and rates of commission, find the amount of commission.

   **a.** $180, 10%   **b.** $235, 5%   **c.** $12,500, 3%   **d.** $6,480, $4\frac{1}{2}$%   **e.** $3,000, 4.8%

2. Bill is paid a 5% commission on his magazine sales. He sold $180 worth of magazines. Find his commission.
3. Wanda earns an 8% commission. Find Wanda's commission on total sales of $450.
4. Betty's weekly salary is $320, plus a 9% commission on her sales. Betty's sales for the week totaled $800. Find Betty's total earnings for that week.
5. Jim's sales for four months were $1,500, $2,900, $960, and $3,400. Jim is paid a 10% commission. Find his commission for the four months.

## HOMEWORK EXERCISES

1. Find the value of each expression.

   **a.** 15% of 182   **b.** 4% of 98   **c.** 1% of 1,000   **d.** 98% of 200   **e.** 200% of 138

   **f.** 110% of 42.5   **g.** 52.4% of 800   **h.** 1.1% of 57   **i.** .3% of 125

2. Of Mr. Franklin's income, 32% is spent for rent. Mr. Franklin's monthly income is $1,250. How much does he spend for rent?
3. Of Mike's books, 58% are paperbacks. Mike has 50 books. How many are paperbacks?
4. Thirty-six percent of the members of the French Club voted to have their party at Chez Luis Restaurant. The club has 25 members. How many voted for Chez Luis?
5. Mary's weight is 82% of Jim's weight. Jim weighs 150 pounds. How much does Mary weigh?
6. This year, the price of a subscription to Sports Fair Magazine is 115% of what it was last year. Last year, a subscription cost $18. What is the price this year?
7. If there are 240 cubic centimeters of a solution and 3.2% of the solution is acid, how many cubic centimeters of acid are there?

8. Mr. Jacobs is paid 6% commission on his carpet sales. If he sold $450 worth of carpet, what was his commission?

9. How much commission did Mrs. Florio make on the sale of a $50,000 house if her rate of commission is 7%.

10. Valerie earns a salary of $175 a week, plus a 3% commission on her sales of encyclopedias. One week, she sold $550 worth of encyclopedias. What were Valerie's total earnings for the week?

11. Mr. Wolf is paid 11% commission on his sales of appliances. One day, he sold appliances for $348, $175.90, $244.50, and $79.60. What was his commission for the day?

## SPIRAL REVIEW EXERCISES

1. Which is equal to 20%?

   (a) $\frac{1}{5}$ (b) $\frac{1}{20}$ (c) $\frac{2}{100}$ (d) $\frac{1}{2}$

2. In a group of 9 children, 6 are wearing blue. The percent of children wearing blue is

   (a) 6% (c) $33\frac{1}{3}\%$

   (b) 60% (d) $66\frac{2}{3}\%$

3. Multiply: $185 \times .12$

4. Find the least common denominator for the fractions $\frac{3}{20}$, $\frac{12}{25}$, and $\frac{38}{100}$.

5. Find $\frac{32}{100}$ of 485.

6. Change $\frac{13}{25}$ to a decimal.

7. Round 15.645 to the nearest hundredth.

8. Divide $\frac{20}{39}$ by $\frac{2}{3}$.

9. Which digit in 748,325 is in the ten-thousands place?

10. What percent of the figure is shaded?

    (a) 3% (c) 50%

    (b) 30% (d) 75%

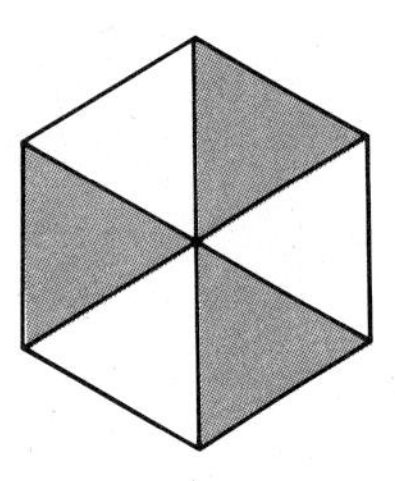

### CHALLENGE QUESTION

Replace ? by <, =, or > to make a true statement.

48% of 72 ? 72% of 48

# UNIT 9–4 Finding What Percent One Number Is of Another; Finding a Number When a Percent of It Is Known

## FINDING WHAT PERCENT ONE NUMBER IS OF ANOTHER

### THE MAIN IDEA

To find what percent one number is of another:

1. Write a fraction with the "is" number in the numerator and the "of" number in the denominator: $\frac{\text{is}}{\text{of}}$ ("is-over-of")
2. Simplify the fraction.
3. Change the fraction to a percent.

**EXAMPLE 1** What % is 8 of 40?

Write a fraction with the "is" number in the numerator and the "of" number in the denominator. $\frac{8}{40}$

Simplify the fraction. $\frac{8 \div 8}{40 \div 8} = \frac{1}{5}$

Change $\frac{1}{5}$ to a decimal.

$$\begin{array}{r} .20 \\ 5\overline{)1.00} \\ \underline{1\,0\phantom{0}} \\ 00 \end{array}$$

Change .20 to a percent. $.20. = 20\%$ *Ans.*

**EXAMPLE 2** What percent of 1,500 is 120?

Use $\frac{\text{is}}{\text{of}}$. $\frac{120}{1500}$

Simplify the fraction. $\frac{120 \div 60}{1500 \div 60} = \frac{2}{25}$

Change $\frac{2}{25}$ to a decimal.

$$\begin{array}{r} .08 \\ 25\overline{)2.00} \\ \underline{2\,00} \\ 0 \end{array}$$

Change .08 to a percent. $.08. = 8\%$ *Ans.*

**EXAMPLE 3** Mr. Rosen saved \$300 and spent \$42 of it on some new clothes. What percent of his savings did he spend?

## THINKING ABOUT THE PROBLEM

"What percent of his savings did he spend?" can be read as "What percent *of* the \$300 he saved *is* the \$42 he spent?"

Use $\frac{\text{is}}{\text{of}}$. $\frac{42}{300}$

Simplify the fraction. $\frac{42 \div 6}{300 \div 6} = \frac{7}{50}$

Change $\frac{7}{50}$ to a decimal.

$$\begin{array}{r} .14 \\ 50\overline{)7.00} \\ \underline{5\,0\phantom{0}} \\ 2\,00 \\ \underline{2\,00} \end{array}$$

Change .14 to a percent. $.14. = 14\%$

*Answer:* Mr. Rosen spent 14% of his savings.

## CLASS EXERCISES

1. What percent is 36 of 60?
2. What percent of 140 is 28?
3. What percent is 120 of 300?
4. What percent of 40 is 100?
5. In each case, find what percent the first number is of the second number.

   **a.** 5; 75 **b.** 100; 200 **c.** 3; 120 **d.** 12; 48 **e.** 15; 60 **f.** 22.8; 200
6. Alyse spent \$6 of her \$50 paycheck on a gift for her brother. What percent of her paycheck did she spend?
7. Of the 600 lenses that Mrs. McKinley inspected, 9 were defective. What percent of the lenses were defective?
8. One hundred seventy-five of the 1,400 students at Greenbay College have registered to vote. What percent of the students have registered?
9. Phil has 40 albums. Twenty-four of these are stereo albums. What percent of Phil's albums are stereo?
10. In an NBA playoff game with the Boston Celtics, a player for the New York Knicks made 17 of 25 attempted shots from the floor. On what percent of his shots from the floor did he score?

## FINDING A NUMBER WHEN A PERCENT OF IT IS KNOWN

### THE MAIN IDEA

To find a number when a percent of it is known:

1. Write a fraction with the known part in the numerator and the percent in the denominator.
2. Write the percent that is in the denominator as a decimal.
3. Divide this decimal into the numerator to find the answer.

**EXAMPLE 4** 16 is 8% of what number?

Use $\frac{\text{known part}}{\text{percent}}$. $\quad \frac{16}{8\%}$

Change 8% to a decimal. $\quad \frac{16}{.08}$

Divide the denominator .08 into the numerator 16. $\quad .08.\overline{)16.00.}$

$$\begin{array}{r} 200 \\ 8\overline{)1600} \\ \underline{16\downarrow\downarrow} \\ 000 \end{array}$$

*Answer:* 16 is 8% of 200.

**EXAMPLE 5** Thirty-five percent of Mrs. Chin's grocery bill was spent for meat. If she spent $14.70 for meat, how much was her grocery bill?

**THINKING ABOUT THE PROBLEM**

You must answer the question: $14.70 is 35% of what number?

Use $\frac{\text{known part}}{\text{percent}}$. $\quad \frac{14.70}{35\%}$

Write 35% as a decimal. $\quad \frac{14.70}{.35}$

Divide .35 into 14.70.

$$\begin{array}{r} 42 \\ .35.\overline{)14.70.} \\ \underline{14\ 0} \\ 70 \\ \underline{70} \\ 0 \end{array}$$

*Answer:* The total grocery bill was $42.

### CLASS EXERCISES

1. In each case, find the missing number.
   **a.** 30 is 25% of what number? **b.** 10.2 is 17% of what number?
   **c.** 57 is 100% of what number? **d.** 34 is 200% of what number?
   **e.** 58.016 is 11.2% of what number?
2. Eight percent of the price of a camera is $20. Find the price of the camera.
3. Forty-five percent of the senior class went on the senior class trip. If 135 seniors went on the senior class trip, how many students are there in the senior class?

4. The 420 freshmen at North Central High School are 28% of the entire student body. How many students are there at North Central?

5. The 8% service charge on Mrs. Spyros' bill came to \$2.60. What was the amount of the bill?

6. Four of Mr. White's mathematics students got 100% on their exam. This was 16% of the class. How many students were there in the class?

## HOMEWORK EXERCISES

1. What percent is 15 of 75?
2. What percent of 640 is 32?
3. What percent is 7 of 140?
4. What percent of 110 is 11?
5. What percent is 21 of 420?
6. What percent of 92 is 4.6?
7. What percent of 11.5 is 4.6?
8. Johnny saved \$8 of the \$32 he received as birthday gifts. What percent did he save?
9. Four of the 12 members of the Paulson family have red hair. What percent of the family has red hair?
10. What percent of Mr. Rawlings' bicycles are red if 18 of his 60 bicycles are red?
11. In each case, find the missing number.

    **a.** 50 is 75% of what number? **b.** 26 is 25% of what number?

    **c.** 15.8 is 5% of what number? **d.** 36.4 is 150% of what number?
12. Twelve of the city's buses were taken out of service for repairs. This represented 5% of all the city's buses. How many buses did the city have?
13. Fifteen percent of the pizzas made by J & W Pizzeria were made with mushrooms. If 48 pizzas were made with mushrooms, what was the total number of pizzas made?

## SPIRAL REVIEW EXERCISES

1. Change 48% to a fraction in simplest form.
2. Find 12% of \$6.50.
3. What is the commission on a \$450 sale if the rate of commission is 4%?
4. The price of a typewriter is \$248.98, and the East Townsend School District needs to buy 18 typewriters. What will be the total cost of the typewriters?
5. What is the total length of 12 wood blocks if each block is $4\frac{3}{4}$ inches long?
6. From $\frac{5}{12}$ subtract $\frac{1}{10}$.
7. Find the product of $-8$ and $-12$.
8. Evaluate: $10 \div 2 + 3$

## CHALLENGE QUESTION

Mr. Holt saves 20% of his total earnings. Last week, he earned a salary of \$280, plus a 6% commission on sales totaling \$3,000. How much did Mr. Holt save?

# UNIT 9–5 Finding the Percent of Increase or Decrease

## THE MAIN IDEA

To find the percent of increase or decrease:

1. Find the amount of increase or decrease by subtracting the smaller number from the larger.
2. Write a fraction in which the numerator is the amount of increase or decrease and the denominator is the original amount.
3. If possible, simplify the fraction.
4. Change the fraction to a percent.

**EXAMPLE 1** The temperature increased from 60°F to 75°F. Find the percent of increase.

Find the amount of increase by subtraction.

$$\begin{array}{rl} 75 & \leftarrow \text{new temperature} \\ -60 & \leftarrow \text{original temperature} \\ \hline 15 & \leftarrow \text{amount of increase} \end{array}$$

Write a fraction in which 15, the amount of increase, is the numerator and 60, the original amount, is the denominator.

$$\frac{15}{60} \begin{array}{l} \leftarrow \text{amount of increase} \\ \leftarrow \text{original amount} \end{array}$$

Simplify $\frac{15}{60}$.

$$\frac{15 \div 15}{60 \div 15} = \frac{1}{4}$$

Change $\frac{1}{4}$ to a percent.

$$4\overline{)1.00} = .25 = 25\%$$

*Answer:* The percent of increase is 25%.

**EXAMPLE 2** Before dieting, Walter weighed 180 pounds. He now weighs 162 pounds. Find the percent of decrease in his weight.

Find the amount of decrease by subtraction.

$$\begin{array}{rl} 180 & \leftarrow \text{original weight} \\ -162 & \leftarrow \text{new weight} \\ \hline 18 & \leftarrow \text{amount of decrease} \end{array}$$

Write a fraction.

$$\frac{18}{180} \begin{array}{l} \leftarrow \text{amount of decrease} \\ \leftarrow \text{original amount} \end{array}$$

Simplify $\frac{18}{180}$.

$$\frac{18 \div 18}{180 \div 18} = \frac{1}{10}$$

Change $\frac{1}{10}$ to a percent.

10%

*Answer:* The percent of decrease is 10%.

**EXAMPLE 3** The student organization had 400 members. The membership was increased to 1,000 students. Find the percent of increase.

Find the amount of increase.

$$\begin{array}{r} 1000 \\ -400 \\ \hline 600 \end{array}$$

Write a fraction and simplify.

$$\frac{600 \div 200}{400 \div 200} = \frac{3}{2}$$

Change $\frac{3}{2}$ to a percent.

$$2\overline{)3.0} = 1.5 = 150\%$$

*Answer:* The percent of increase is 150%.

## CLASS EXERCISES

**1.** Find the percent of increase.

**a.** 50 to 70 **b.** 25 to 40 **c.** 45 to 72

**2.** Find the percent of decrease.

**a.** 24 to 12 **b.** 64 to 16 **c.** 75 to 50

**3.** In January, Martha sold 20 cars. The next month, she sold 35 cars. Find the percent of increase in the number of cars that Martha sold.

**4.** During a sale, the price of a shirt was changed from \$20 to \$12. Find the percent of decrease in the price of the shirt.

**5.** In one hour, the temperature went from 25° to 27°. Find the percent of increase in temperature.

**6.** Before going on a diet, Mr. Coleman's weight was 200 pounds. After dieting, he reached a weight of 170 pounds. What was the percent of decrease in his weight?

## HOMEWORK EXERCISES

**1.** Find the percent of increase.

**a.** 25 to 38 **b.** 50 to 52 **c.** 20 to 20.2 **d.** 10 to 67

**2.** Find the percent of decrease.

**a.** 100 to 90 **b.** 75 to 25 **c.** 50 to 49 **d.** 200 to 148

**3.** The price of a dress went from \$35 to \$42. Find the percent of increase.

**4.** The number of people in a chorus went from 27 to 18. Find the percent of decrease.

**5.** In one year, Mark's height went from 60 inches to 63 inches. What was the percent of increase in Mark's height?

**6.** The number of student organization members increased from 400 to 500. Find the percent of increase.

**7.** Mr. James sold 40 cars in June and 25 cars in July. Find the percent of decrease in sales.

8. In one year, the number of cases of measles reported in Southside High School dropped from 24 to 8. What was the percent of decrease?

9. Before exercising, Martha's pulse rate was 75. After exercising, her pulse rate was 90. Find the percent of increase.

10. The price of a share of stock went from $24 to $20. What was the percent of decrease?

11. The price of a computer dropped from $300 to $240. Find the percent of decrease.

## SPIRAL REVIEW EXERCISES

1. A player for the Knicks missed 16 of the 24 shots he took in playoff games in New York. On what percent of his shots did he score?

2. At the end of the regular season, a member of the Boston Celtics team had made 348 of the 870 field goals he had attempted. What percent of the shots did he make?

3. Of the 1,600 pages in Jim's encyclopedia, 55.5% have illustrations. How many pages have illustrations?

4. Twenty-four percent of Mrs. Reed's employees work part-time. If 18 of the employees work part-time, how many employees does Mrs. Reed have?

5. Find $\frac{4}{5}$ of $\frac{15}{16}$.

6. Round 12,498 to the nearest thousand.

7. The number of tickets sold at the Bijou theater each day of one week is as shown. What was the total number of tickets sold that week?

| Sunday | 505 |
|---|---|
| Monday | 68 |
| Tuesday | 47 |
| Wednesday | 59 |
| Thursday | 103 |
| Friday | 452 |
| Saturday | 651 |

8. Subtract 17.38 from 20.2.

9. Which number is not a multiple of 7?
(a) 7 (b) 17 (c) 42 (d) 70

### CHALLENGE QUESTION

Charles can lift 1.5 more pounds of weight this month than he could last month. If this represents a 5% increase, how many pounds can he lift this month?

## UNIT 10–1 Using Percents to Find Discounts

### THE MAIN IDEA

1. The amount by which an original price, or *list price*, is lowered is called the *discount*. The new price is called the *sale price*.
2. Percents are used to show the *rate of discount*.
3. To find the amount of discount:
   a. Change the rate of discount to a decimal.
   b. Multiply the list price by this decimal.

   Amount of Discount = List Price × Rate of Discount
4. To find the sale price:
   a. Find the amount of discount.
   b. Subtract the amount of discount from the list price.

   Sale Price = List Price − Amount of Discount

**EXAMPLE 1** A television set is on sale at 15% off the list price of $429. Find the amount of discount.

Write the percent as a decimal. 15% = .15

Multiply the list price by the decimal.

```
  $429   ← list price
  ×.15   ← rate of discount
 21 45
 42 9
$64.35   ← amount of discount
```

*Answer:* The amount of discount is $64.35.

**EXAMPLE 2** A book that has a list price of $19.50 is on sale at a 20% discount. Find the sale price.

To find the amount of discount, multiply the list price by the rate of discount written as a decimal.

```
  $19.50   ← list price
    ×.20   ← rate of discount
 $3.9000   ← amount of discount
= $3.90
```

To find the sale price, subtract the amount of discount from the list price.

```
$19.50   ← list price
 −3.90   ← amount of discount
$15.60   ← sale price
```

*Answer:* The sale price is $15.60.

**EXAMPLE 3** Find the sale price of a camera that has a list price of $148.70 and is being sold at a 35% discount.

To find the amount of discount, multiply the list price by the rate of discount written as a decimal.

```
 $148.70 ← list price
   ×.35  ← rate of discount
  7 4350
 44 610
$52.0450 ← amount of discount
```

Round the amount of discount to the nearest cent.

$52.04(5)0 or $52.05

To find the sale price, subtract the amount of discount from the list price.

```
$148.70 ← list price
 −52.05 ← amount of discount
 $96.65 ← sale price
```

*Answer:* The sale price is $96.65.

**EXAMPLE 4** For the George Washington's Birthday Sale, everything at the Best Department Store was discounted 25%. Mr. Billings bought a shirt listed at $15.90, a tie listed at $8.75, and a sweater listed at $22.95. What was the total sale price for these items?

**THINKING ABOUT THE PROBLEM**

This problem should be broken down into three smaller problems.

To find the total list price, add the list price of each of the items.

```
 $15.90
   8.75
 +22.95
 $47.60 ← total list price
```

To find the amount of discount, multiply the total list price by the rate of discount expressed as a decimal.

```
  $47.60 ← total list price
    ×.25 ← rate of discount
  2 3800
  9 520
$11.9000 ← amount of discount
= $11.90
```

To find the total sale price, subtract the amount of discount from the total list price.

```
$47.60 ← total list price
−11.90 ← amount of discount
$35.70 ← total sale price
```

*Answer:* The total sale price was $35.70.

## CLASS EXERCISES

1. A tie that has a list price of $12 is on sale at 10% off. Find the amount of discount.
2. A cassette player that regularly sells for $240 is on sale at a 25% discount. Find the amount of discount.
3. A shirt that regularly sells for $18.99 is on sale at 35% off. What is the amount of discount?
4. Find the sale price of a baseball glove that has a list price of $35 and is being sold at a 15% discount.
5. Find the sale price of an $8,000 car that is being sold at a 12% discount.

6. Find the sale price of a bicycle that has a list price of $190.80 when it is discounted at each of the following rates.

   a. 10% b. 25% c. 30% d. 45%

7. A $35.75 dress is being sold at a 15% discount. What is the sale price?

8. Virginia bought a pen listed at $3.90, a notebook listed at $1.40, and a desk calendar listed at $5.95 at a store that discounts all items at 20%. What was the total amount that she paid?

9. During a 40% discount sale, Mr. Waters bought a toaster oven regularly priced at $35.90, a clock regularly priced at $29, and a radio regularly priced at $45.50.

   a. How much did Mr. Waters save?

   b. What was the total amount that he paid?

10. Louise bought 3 books originally priced at $3.49 each at a 15%-off book sale. What was the total amount that she spent?

## HOMEWORK EXERCISES

1. A jacket that has a list price of $60 is on sale at 20% off. Find the amount of discount.

2. A radio that regularly sells for $39 is on sale at a 15% discount. What is the amount of discount?

3. A pair of shoes that usually costs $23.99 is on sale at 25% off. How much money is saved by buying a pair of shoes on sale?

4. How much will the discount be on a suit listed at $185 and sold at a discount of 30%?

5. A game listed at $19.50 was sold at a discount of 15%. What was the sale price?

6. Find the sale price of a $12,000 station wagon that is being sold at a 10% discount.

7. A $499 video recorder is being sold at a 20% discount. What is the sale price?

8. Find the sale price of a television set that has a list price of $369 when it is discounted at the rate of:

   a. 10% b. 15% c. 25% d. 50%

9. A shirt listed at $25.99 is being sold at a 25% discount. What is the sale price?

10. Ms. Jewel bought a blouse that is regularly $15.75, a skirt regularly $24.50, and a handbag regularly $18.95 at a 30%-off sale.

    a. What was the total amount that she saved?

    b. What was the total amount that she spent?

11. Benjamin bought 4 records listed at $7.99 each at a 15%-off sale. How much did he spend?

12. Roberta bought a stereo listed at $319 and a cassette deck listed at $119.80 at a 20%-off sale. What was the total amount that she spent?

13. Albert bought 2 pens regularly priced at $1.99 each and 3 pens regularly priced at $2.50 each, at a store that discounts all items at 15%. What was the total amount that he spent?

14. Ron bought 2 shirts listed at $18.70 each, 4 ties listed at $10.50 each, and a pair of slacks listed at $31, all at a 20%-off sale. What was the total amount that he spent?

## SPIRAL REVIEW EXERCISES

**1.** Of the 600 trees on the Dyson property, 210 are apple trees. What percent are apple trees?

**2.** The temperature decreased from 90°F to 75°F. What was the percent of decrease?

**3.** Mary earns $5.50 an hour. How much will she earn in 8 hours?
(a) $45 (b) $44 (c) $40 (d) $38

**4.** Which decimal has the largest value?
(a) .356 (b) .298 (c) .9 (d) .87

**5.** If 3 gallons of gasoline cost $4.20, the cost of 7 gallons is
(a) $29.40 (c) $8.40
(b) $9.80 (d) $7.70

**6.** The value of $5 \times 4 + 2 \times 9$ is
(a) 198 (b) 48 (c) 38 (d) 270

**7.** Three stores are featuring the same folding chairs, with prices as shown in the display signs. What is the lowest price for four chairs?

| ALLEN's Best Price only $14.25 each | BONDIO's Special 4 for $59 | 2 for $27! Buy at CORMAN's |
|---|---|---|

### CHALLENGE QUESTION

A video game that has a list price of $24 is first discounted by 20% and then by an additional 10%. Find the final sale price.

## UNIT 10–2 Using Percents to Find Sales Tax

### THE MAIN IDEA

1. *Taxes* are monies collected by city, state, and federal governments to help pay for schools, highways, buildings, and other public services.
2. A tax on something bought in a store is called a *sales tax*. The rate of sales tax is given as a percent.
3. To find the amount of sales tax:
   a. Write the percent as a decimal.
   b. Multiply the price by this decimal.

   Amount of Sales Tax = Price × Rate of Sales Tax

4. To find the total cost when a sales tax is collected:
   a. Find the sales tax.
   b. Add the sales tax to the price.

   Total Cost = Price + Amount of Sales Tax

**EXAMPLE 1** Iris bought a jacket for $70. If the sales tax is 8%, how much tax must she pay?

Write the percent as a decimal. 8% = .08

Multiply the price by this decimal.

$$\begin{array}{rl} \$70 & \leftarrow \text{price} \\ \underline{\times .08} & \leftarrow \text{rate of sales tax} \\ \$5.60 & \leftarrow \text{amount of sales tax} \end{array}$$

*Answer:* The sales tax is $5.60.

**EXAMPLE 2** Saleem bought a pair of slacks for $29.95. If the sales tax is $7\frac{1}{2}\%$, how much tax must he pay?

Write the percent as a decimal. $7\frac{1}{2}\% = 7.5\% = .075$

Multiply the price by this decimal.

$$\begin{array}{rl} \$29.95 & \leftarrow \text{price} \\ \underline{\times .075} & \leftarrow \text{rate of sales tax} \\ 14975 & \\ \underline{2\ 0965\phantom{0}} & \\ \$2.24625 & \leftarrow \text{amount of sales tax} \end{array}$$

Round the amount of sales tax to the nearest cent. $2.24⑥25 or $2.25

*Answer:* The sales tax is $2.25.

**EXAMPLE 3** Marvin bought a camera for \$390 and film for \$12. If the sales tax is 7%, how much tax must he pay?

To find the total price, add the individual prices.

$$\begin{array}{r} \$390 \\ +12 \\ \hline \$402 \end{array} \leftarrow \text{total price}$$

Write the percent as a decimal.

$7\% = .07$

Multiply the total price by this decimal.

$$\begin{array}{rl} \$402 & \leftarrow \text{total price} \\ \times .07 & \leftarrow \text{rate of sales tax} \\ \hline \$28.14 & \leftarrow \text{amount of sales tax} \end{array}$$

*Answer:* The sales tax is \$28.14.

**EXAMPLE 4** Ray bought 4 pairs of socks at \$2.49 for each pair. If the sales tax is 8%, how much tax did he pay?

To find the total price, multiply the item price, \$2.49, by the number of items, 4.

$$\begin{array}{r} \$2.49 \\ \times 4 \\ \hline \$9.96 \end{array} \leftarrow \text{total price}$$

Write the percent as a decimal.

$8\% = .08$

Multiply the total price by the decimal.

$$\begin{array}{rl} \$9.96 & \leftarrow \text{total price} \\ \times .08 & \leftarrow \text{rate of sales tax} \\ \hline \$.7968 & \leftarrow \text{amount of sales tax} \end{array}$$

Round the sales tax to the nearest cent.

\$.79⑥8 or \$.80

*Answer:* The sales tax is \$.80.

**EXAMPLE 5** A computer sells for \$499. If the rate of sales tax is $8\frac{1}{4}\%$, find the total cost.

Write the percent as a decimal.

$8\frac{1}{4}\% = 8.25\% = .0825$

Multiply the price by this decimal.

$$\begin{array}{rl} \$499 & \leftarrow \text{price} \\ \times .0825 & \leftarrow \text{rate of sales tax} \\ \hline 2495 & \\ 998 & \\ 39\ 92\phantom{00} & \\ \hline \$41.1675 & \leftarrow \text{amount of sales tax} \end{array}$$

Round the sales tax to the nearest cent.

\$41.16⑦5 or \$41.17

To find the total cost, add the sales tax to the price.

$$\begin{array}{rl} \$499.00 & \leftarrow \text{price} \\ +41.17 & \leftarrow \text{sales tax} \\ \hline \$540.17 & \leftarrow \text{total cost} \end{array}$$

*Answer:* The total cost is \$540.17.

**EXAMPLE 6** The Salengo Department Store charges its customers the county sales tax rate of 7%. Shown here is part of the table used by store clerks to determine how much tax must be charged.

**Sales Tax Table: 7% Tax**

| Amount of Sale | | Tax | Amount of Sale | | Tax |
|---|---|---|---|---|---|
| *From* | *To* | | *From* | *To* | |
| .08 | .21 | .01 | 2.22 | 2.35 | .16 |
| .22 | .35 | .02 | 2.36 | 2.49 | .17 |
| .36 | .49 | .03 | 2.50 | 2.64 | .18 |
| .50 | .64 | .04 | 2.65 | 2.78 | .19 |
| .65 | .78 | .05 | 2.79 | 2.92 | .20 |
| .79 | .92 | .06 | 2.93 | 3.07 | .21 |
| .93 | 1.07 | .07 | 3.08 | 3.21 | .22 |
| 1.08 | 1.21 | .08 | 3.22 | 3.35 | .23 |
| 1.22 | 1.35 | .09 | 3.36 | 3.49 | .24 |
| 1.36 | 1.49 | .10 | 3.50 | 3.64 | .25 |
| 1.50 | 1.64 | .11 | 3.65 | 3.78 | .26 |
| 1.65 | 1.78 | .12 | 3.79 | 3.92 | .27 |
| 1.79 | 1.92 | .13 | 3.93 | 4.07 | .28 |
| 1.93 | 2.07 | .14 | 4.08 | 4.21 | .29 |
| 2.08 | 2.21 | .15 | 4.22 | 4.35 | .30 |

Sally Sims bought a bracelet for $2.39. What was the total cost of the purchase, including sales tax?

Refer to the tax table to find the amount of the tax. The amount of the sale, $2.39, is between the table entries of 2.36 and 2.49. The tax shown is .17.

Add the sales tax to the price.

$$\begin{array}{rl} \$2.39 & \leftarrow \text{price} \\ +.17 & \leftarrow \text{sales tax} \\ \hline \$2.56 & \leftarrow \text{total cost} \end{array}$$

*Answer:* The total cost was $2.56.

## CLASS EXERCISES

1. If the rate of sales tax is 5%, find the amount of tax for each price.

   **a.** $15  **b.** $38  **c.** $105  **d.** $240.50  **e.** $19.95

2. Frank bought a suit for $95. If the rate of sales tax is 7%, how much tax must he pay?

3. What is the tax on a cassette recorder costing $120.50 if the rate of sales tax is 8%?

4. The rate of sales tax is 6%. How much tax will Roberta pay if she buys a book for $12 and a pen for $5?

5. Fred spent $50.95 on records and $38 for stereo headphones. If the rate of sales tax is 8%, how much tax did Fred pay?
   (a) $7.11  (b) $7.12  (c) $.71  (d) $.72

6. Rosa bought 3 cassettes at $2.99 each. If the rate of sales tax is 8%, how much tax did she pay?

7. Find the total cost of each purchase if the rate of sales tax is 6% and the price is:

   **a.** $120  **b.** $210  **c.** $325  **d.** $590.25  **e.** $14.99

**8.** A color television set sells for $380. If the rate of sales tax is 8%, find the total cost.

**9.** Jason bought a clock radio for $38.50. What is the total amount that he paid if the rate of sales tax is 7%?

**10.** Carla bought a dress for $48.99. How much did she spend if the rate of sales tax is 6%?

**11.** How much would you pay for a jacket that costs $90 if the rate of sales tax is 8%?
(a) $5.40 (b) $95.40 (c) $7.20 (d) $97.20

**12.** Ben bought two pieces of luggage costing $54 each. How much did he spend if the rate of sales tax is 5%?

**13.** Jane bought a bowling ball for $35.49 and a $22.99 bag to carry the ball. How much did she spend if the rate of sales tax is 7%?

In 14 and 15, use the tax table on page 226.

**14.** If the rate of sales tax is 7%, find the amount of tax on a purchase of

**a.** 57¢ **b.** $4 **c.** $2.50 **d.** $1.75

**15.** Find the total cost of each purchase if the rate of sales tax is 7%.

**a.** $1.25 **b.** $3.98 **c.** $.49 **d.** $2

## *HOMEWORK EXERCISES*

**1.** Find the amount of sales tax on each price if the rate of sales tax is 6%.

**a.** $18 **b.** $25 **c.** $52 **d.** $80.50 **e.** $19.99

**2.** Find the amount of sales tax on each price if the rate of sales tax is $8\frac{1}{2}$%.

**a.** $5 **b.** $17 **c.** $39 **d.** $47.50 **e.** $24.95

**3.** Philip bought a jacket that sold for $50. If the rate of sales tax is 7%, how much sales tax did he pay?

**4.** Willis bought a $22 shirt and a $9 tie. If the rate of sales tax is 7%, how much tax did he pay?

**5.** Mrs. Williams bought a coat costing $120 and a pair of shoes costing $50. How much tax did she pay if the rate of sales tax is 8%?

**6.** Elizabeth bought a handbag for $24.99, a pair of gloves for $9.95, and a scarf for $7.49. If the rate of sales tax is 6%, how much tax did she pay?

**7.** Ricardo bought 3 books for $3.75 each. If the rate of sales tax is 7%, how much tax did he pay?

**8.** Find the total cost of each purchase if the rate of sales tax is 7% and the price is:

**a.** $20 **b.** $60 **c.** $19 **d.** $86 **e.** $14.50 **f.** $36.25 **g.** $42.99 **h.** $112.68

**9.** Records are on sale for $7.99 each. Don bought 4 records. If the rate of sales tax is 6%, find the total cost.

**10.** Mr. Manners bought 2 suits on sale at $79.95 each. If the rate of sales tax is 8%, how much did he spend?

**11.** Jill bought a baseball glove costing $28.49 and a bat costing $19.99. If the rate of sales tax is 6%, how much did she spend?

**12.** Mary added \$700 in options to a car that sells for \$8,000. If the rate of sales tax is 5%, how much must Mary pay for the car?

**13.** Luis bought 5 pairs of socks that cost \$2.79 for each pair and 2 belts that cost \$8.99 each. If the rate of sales tax is 7%, how much did Luis pay?

**14.** A music teacher bought 12 records for the class library. Three of the records cost \$3.99 each, 3 cost \$4.99 each, 4 cost \$5.49 each, and 2 cost \$7 each. If the rate of sales tax is 8%, how much was the total cost?

In 15 and 16, use the tax table on page 226.

**15.** If the rate of sales tax is 7%, find the amount of tax on a purchase of

**a.** 89¢ **b.** \$3.25 **c.** \$4.19 **d.** \$1

**16.** Find the total cost of each purchase if the rate of sales tax is 7%.

**a.** \$1.85 **b.** \$3 **c.** 50¢ **d.** \$4.29

## SPIRAL REVIEW EXERCISES

**1.** Expressed as a percent, the fraction $\frac{9}{10}$ is

(a) 9% (c) 90%
(b) 29% (d) .09%

**2.** 85% written as a fraction is

(a) $\frac{85}{1000}$ (b) $\frac{4}{5}$ (c) $\frac{15}{17}$ (d) $\frac{17}{20}$

**3.** $\frac{3}{5} + \frac{8}{5}$ equals

(a) $\frac{11}{10}$ (b) $2\frac{1}{5}$ (c) $\frac{24}{25}$ (d) $\frac{3}{8}$

**4.** $\frac{1}{3} + \frac{2}{9}$ equals

(a) $\frac{3}{12}$ (b) $\frac{1}{2}$ (c) $\frac{5}{9}$ (d) $\frac{15}{18}$

**5.** $\frac{22}{4}$ is equal to

(a) $4\frac{1}{2}$ (b) $5\frac{1}{2}$ (c) $6\frac{3}{4}$ (d) $5\frac{3}{4}$

**6.** $\frac{3}{4}$ of 80 is

(a) 60 (b) 6 (c) 48 (d) $20\frac{3}{4}$

**7.** Find the sale price of a sofa bed that has a list price of \$499 and is being sold at a 15% discount.

**8.** From the sum of 5 and −7, subtract −2.

**CHALLENGE QUESTION**

If the total cost for a coat that sells for \$80 is \$84.80, then the rate of sales tax is
(a) 5% (b) 6% (c) 7% (d) 8%

# UNIT 10–3 Using Percents to Find Interest

## FINDING THE AMOUNT OF INTEREST

### THE MAIN IDEA

1. *Interest* is a charge for money that is borrowed. For example, when you deposit money in a savings account, the bank pays you interest because it borrows your money to make investments. When you take a loan from a bank, you pay the bank interest for borrowing the bank's money.
2. Interest rates are given per year and are written as percents.
3. The amount of money that is invested or borrowed is called the ***principal.***
4. To find the amount of interest for one year:
   a. Change the rate of interest to a decimal.
   b. Multiply the principal by this decimal.

$$\text{Interest} = \text{Principal} \times \text{Rate}$$

5. To find the total amount that a borrower must repay or to find the total amount in a savings account, add the interest to the principal.

$$\text{Amount} = \text{Principal} + \text{Interest}$$

**EXAMPLE 1** Mr. Smith deposited $300 into his savings account. How much interest will he receive at the end of one year if the interest rate is 5%?

Write the interest rate as a decimal. $5\% = .05$

Multiply the principal, $300, by this decimal (the rate).

```
  $300 ← principal
  ×.05 ← rate of interest
$15.00 ← amount of interest
```

*Answer:* The interest is $15.

**EXAMPLE 2** Jerry has $250 in his savings account. If the bank pays $6\frac{1}{2}\%$ interest, how much money will Jerry have in his account a year from now?

Write the rate as a decimal. $6\frac{1}{2}\% = 6.5\% = .065$

To find the amount of interest, multiply the principal, $250, by the rate, .065.

```
   $250 ← principal
  ×.065 ← rate of interest
  1 250
  15 00
$16.250
= $16.25 ← amount of interest
```

Add the amount of interest, $16.25, to the principal.

```
 $250.00 ← principal
  +16.25 ← amount of interest
 $266.25 ← amount at end of one year
```

*Answer:* At the end of one year, Jerry will have $266.25 in his savings account.

## CLASS EXERCISES

1. Find the interest for one year for each of the following:

   **a.** \$80 at 6%   **b.** \$200 at 10%   **c.** \$420 at 9%   **d.** \$1,500 at 12%

   **e.** \$5,000 at $8\frac{1}{2}$%   **f.** \$250,000 at $9\frac{1}{4}$%

2. What is the interest on \$500 paid by a bank at the end of one year if the rate of interest is 6%?
3. What is the yearly interest on \$2,000 invested at $5\frac{1}{2}$%?
4. What is the interest on a one-year loan of \$800 if the interest rate is 12%?
5. Carol borrowed \$120 for one year at 10% interest. What is the amount of interest that Carol must pay?
6. Jason invested \$250 at 16%. How much interest will Jason earn at the end of one year?
7. Roberto borrowed \$300 for one year at 15%. What is the total amount of money that Roberto must repay?
8. Della deposits \$1,000 in a savings account. If the interest rate is 8%, how much money will be in Della's account at the end of one year?

## FINDING INTEREST FOR TIME OTHER THAN ONE YEAR

### THE MAIN IDEA

To find the amount of interest when the amount of time is other than a year:

1. Change the rate of interest to a decimal.
2. Multiply the principal by this decimal.
3. Multiply the product by the time expressed in years.

Interest = Principal × Rate × Time

**EXAMPLE 3** Don invested \$500 at 10% for 3 years. How much interest will Don earn?

Multiply the principal by the rate of interest written as a decimal.

```
  $500 ← principal
  ×.10 ← rate of interest
$50.00 ← amount of interest for one year
```

Multiply the amount of interest for 1 year, \$50, by the time in years.

```
 $50
  ×3 ← time in years
$150
```

*Answer:* Don will earn \$150 in 3 years.

**EXAMPLE 4** Mr. Walsh invested $8,000 at an 11% annual (yearly) interest. What will be the value of his investment 4 years later?

Multiply the principal by the rate of interest written as a decimal.

$$\begin{array}{rl} \$8000 & \leftarrow \text{principal} \\ \times .11 & \leftarrow \text{rate of interest} \\ \hline 80\ 00 & \\ 800\ 0 & \\ \hline \$880.00 & \leftarrow \text{amount of interest for 1 year} \end{array}$$

Multiply by the time in years.

$$\begin{array}{rl} \$880 & \\ \times 4 & \leftarrow \text{time in years} \\ \hline \$3520 & \leftarrow \text{amount of interest for 4 years} \end{array}$$

Add the amount of interest to the principal.

$$\begin{array}{rl} \$8000 & \leftarrow \text{principal} \\ +3520 & \leftarrow \text{amount of interest} \\ \hline \$11520 & \leftarrow \text{value of investment 4 years later} \end{array}$$

*Answer:* In 4 years, the value of the investment will be $11,520.

**EXAMPLE 5** Adele invested $3,000 at 8% for 3 months. How much interest did she earn?

**THINKING ABOUT THE PROBLEM**

This problem should be broken down into three smaller problems.

To find the amount of interest for one full year, multiply $3,000 by the rate of interest expressed as a decimal.

$$\begin{array}{rl} \$3000 & \leftarrow \text{principal} \\ \times .08 & \leftarrow \text{rate of interest} \\ \hline \$240.00 & \leftarrow \text{amount of interest for 1 full year} \end{array}$$

Express the part of the year represented by 3 months as a fraction.

$$\frac{3 \text{ months}}{12 \text{ months}} = \frac{1}{4} \text{ year}$$

Multiply $240 by the fractional part of the year.

$$\frac{1}{\cancel{4}_1} \times \cancel{240}^{60} = \$60$$

*Answer:* Adele earned $60 interest in 3 months.

## CLASS EXERCISES

1. Sam invested $1,000 at 9% for 2 years. How much interest will he earn?
2. Sandra deposited $700 in a savings account. If the interest rate is 6%, how much interest will she earn after 5 years?
3. Mr. McCarthy invested $10,000 at 12% for 4 years. How much interest will he earn?
4. What is the value of a $5,000 investment after 10 years if the interest rate is 8%?

5. Ms. Santiago invested $4,000 at 10% interest. How much is her investment worth 5 years later?

6. How much interest is earned on a $2,000 investment for 6 months at 8%?

7. If $1,000 is invested at 10%, the amount of interest earned after 9 months is
(a) $100 (b) $900 (c) $75 (d) $90

## HOMEWORK EXERCISES

1. Find the interest for one year for each investment.

**a.** $100 at 7% **b.** $350 at 5% **c.** $500 at 10% **d.** $4,000 at $6\frac{1}{2}$%

**e.** $650 at $12\frac{1}{4}$% **f.** $2,000 at $9\frac{3}{4}$%

2. What is the interest paid at the end of one year on a $1,000 investment if the interest rate is 8%?

3. What is the yearly interest on $10,000 invested at $9\frac{1}{2}$%?

4. How much interest is paid on a one-year loan of $250 if the interest rate is 20%?

5. Ben borrowed $500 for one year at 12% interest. What is the amount of interest that he must pay?

6. Karen invests $200 at a yearly interest rate of 9%. What will be the total value of Karen's investment at the end of one year?

7. Mr. Woods borrowed $5,000 for one year at an interest rate of 14%. What is the total amount of money that he must repay?

8. Find the total value of each investment at the end of one year.

**a.** $1,000 at 9.5% **b.** $500 at 5.5% **c.** $10,000 at 8% **d.** $600 at 11%

**e.** $850 at 7% **f.** $20,000 at 12%

9. Melanie invested $500 at 8% for 3 years. How much interest will she earn?

10. David invested $1,000 at $9\frac{1}{2}$% interest for 5 years. How much interest will he earn?

11. Find the amount of interest earned for each investment.

**a.** $100 at 7% for 2 years **b.** $350 at 8% for 5 years **c.** $10,000 at 10% for 3 years

**d.** $800 at 5% for 9 months **e.** $4,500 at $8\frac{1}{2}$% for 6 months

**f.** $100,000 at 9% for 3 months

12. Ms. Keith invested $500 at 11% interest for 8 years. How much is her investment worth at the end of the 8 years?

13. What is the value of a $3,000 investment after 7 years if the interest rate is 7%?

14. Mr. Williams invested $2,500 at 8% interest. What is the value of his investment 6 months later?

## SPIRAL REVIEW EXERCISES

1. If sales tax is charged at the rate of 4%, what is the total cost of a book priced at $15?

2. The balance in Fred's checking account was $133.78 before he made a deposit of $75. What was his new balance?

3. Mario's Pizza charges $6 for a whole pie of 8 slices and 90¢ for each slice. If Barbara bought a pie and 3 slices, how much did she spend?

4. If 3 pairs of sneakers cost $33, what is the cost of 5 pairs of sneakers?

5. How much greater is 8.39 than 7.86?

6. Find the product of $\frac{5}{8}$ and $\frac{24}{35}$.

7. Round 2,563,419 to the nearest hundred thousand.

8. Evaluate: $-3 + (-8) + (+10)$

**CHALLENGE QUESTION**

A stock valued at $50 a share had a 12% increase in value followed by a 5% decrease. What was the final value of a share of the stock?

## UNIT 10–4 Using Percents in a Budget

### THE MAIN IDEA

1. A *budget* is a plan for spending money. Each item in a budget is allowed a certain part of the total income.
2. The parts of the income that are allowed for individual items are written as percents.
3. To find the amount of money that a budget allows for a particular item:
   a. Write the percent budgeted for that item as a decimal.
   b. Multiply the total income by this decimal.

   Amount Budgeted = Total Income × Percent Budgeted
4. The sum of the percents budgeted for all the items is 100%.
5. The sum of the individual amounts of money budgeted for all the items equals the total income.

**EXAMPLE 1** The Clark family budget allows 10% of the weekly income for entertainment. How much money can be spent on entertainment if the total weekly income is $520?

Write the percent as a decimal. 10% = .10

Multiply the total income by this decimal.

$520 ← total income
×.10 ← percent budgeted
$52.00 ← amount budgeted

*Answer:* The Clarks can spend $52 weekly on entertainment.

**EXAMPLE 2** The Vivaldi family budget allows 35% for rent and utilities, 30% for food, 23% for clothing and entertainment, and 12% for savings. If their monthly income is $1,500, how much money does their budget allow for each item?

Find the amount of money allowed for each item by changing the percent budgeted to a decimal and multiplying $1,500 by this decimal.

| *Item* | *Percent Budgeted* | *Calculation* | *Amount Budgeted* |
|---|---|---|---|
| Rent and Utilities | 35% | $1500<br>×.35<br>75 00<br>450 0<br>$525.00 | $525 |
| Food | 30% | $1500<br>×.30<br>$450.00 | $450 |

| Item | Percent Budgeted | Calculation | Amount Budgeted |
|---|---|---|---|
| Clothing and Entertainment | 23% | $1500<br>×.23<br>45 00<br>300 0<br>$345.00 | $345 |
| Savings | 12% | $1500<br>×.12<br>30 00<br>150 0<br>$180.00 | $180 |
| Total | 100% | | $1,500 |

Since the sum of the monies allowed for each budgeted item is equal to the total income, the work checks.

**EXAMPLE 3** The chart shown gives the Hunter family weekly budget. Calculate the percent of the budget that was planned for clothing.

| Item | Percent Budgeted |
|---|---|
| Food | 26 |
| Housing | 34 |
| Entertainment | 7 |
| Savings | 9 |
| Clothing | |
| Other | 12 |

**THINKING ABOUT THE PROBLEM**

The sum of all the individual percents budgeted is 100%. To find the missing percent, subtract the sum of all the other percents from 100%.

Add the percents planned for food, housing, entertainment, savings, and other.

$$\begin{array}{r} 26\% \\ 34\% \\ 7\% \\ 9\% \\ +12\% \\ \hline 88\% \end{array}$$

Subtract 88% from 100%, the whole amount.

$$\begin{array}{r} 100\% \\ -88\% \\ \hline 12\% \end{array}$$

*Answer:* 12% of the budget was planned for clothing.

**EXAMPLE 4** The Carter budget allows 35% for housing, 25% for food, 15% for clothing, 9% for transportation, 10% for entertainment, and 6% for savings. If the Carter monthly income is $1,350, how much should they plan to spend on both food and clothing?

Since 25% is budgeted for food and 15% is budgeted for clothing, 25% + 15% or 40% is budgeted for both food and clothing.

Find 40% of $1,350.

$$\begin{array}{r} \$1350 \\ \times .40 \\ \hline \$540.00 \end{array}$$

*Answer:* The Carters should plan to spend $540 on food and clothing.

**EXAMPLE 5** Mr. Jerome earns \$2,000 a month. His monthly expenses for housing are \$800. What percent of his income should he budget for housing?

Write as a fraction the part of Mr. Jerome's income that he spends for housing. $\frac{800}{2000}$

Simplify the fraction. $\frac{800 \div 400}{2000 \div 400} = \frac{2}{5}$

Change $\frac{2}{5}$ to a decimal. $5\overline{)2.00}$ = .4

$\frac{2}{5} = .4$

Change .4 to a percent. $.4 = 40\%$

*Answer:* Mr. Jerome should budget 40% of his income for housing.

## CLASS EXERCISES

1. Mrs. Johnson earns \$1,200 a month. In her budget, she allows 30% for the rental of her apartment. How much money has she budgeted for rent?
2. Ethan plans to save 40% of the money that he earns baby-sitting. If Ethan earned \$50, how much should he save to keep to his budget?
3. The Smith family budget allows 15% of the weekly income for clothing. How much money can be spent on clothing if the total weekly income is \$440?
4. John's budget allows 20% for clothing, 40% for food, and 26% for entertainment. If savings is the only other item on the budget, what percent is allowed for savings?
5. Copy and complete the table shown for the Solomon family budget if the monthly income is \$1,900.

**Monthly Budget for the Solomon Family**

| Item | Percent Budgeted | Amount Budgeted |
|---|---|---|
| Food | 20% | |
| Housing | 25% | |
| Clothing | 22% | |
| Savings | 6% | |
| Entertainment | 12% | |
| Other | 15% | |

6. The chart shown gives Jane's weekly budget.
   a. What percent of the budget was planned for transportation?
   b. If Jane's weekly income was \$580, how much money did she budget for clothing and savings?

| Item | Percent Budgeted |
|---|---|
| Rent | 28% |
| Food | 22% |
| Utilities | 20% |
| Transportation | ? |
| Clothing | 15% |
| Savings | 8% |

7. The Dawson family budget allows 28% for housing, 22% for food, 18% for clothing, 11% for transportation, 12% for entertainment, and 9% for savings. If the Dawsons' weekly income is \$650, how much money is budgeted for entertainment and savings?

8. Mr. Reader earns $480 a week. If his weekly expenses for food are $120, what percent of his income should he budget for food?

9. Ms. Rose earns $800 a month. What percent of her income should she budget for rent if she pays $320 a month for rent?

## HOMEWORK EXERCISES

1. Jill earns $80 a week as a cashier. If she budgets 35% for savings, how much should she save each week?

2. Mr. George earns $950 a month. His budget allows 15% for entertainment. How much money has he budgeted for entertainment?

3. The Jones family budget allows 20% of the weekly income for food. How much money can be spent on food if the total weekly income is $720?

4. A guide for a family budget recommends that 30% of the monthly income be budgeted for housing. How much money should be budgeted for housing if the monthly income is $1,800?

5. Copy and complete the weekly budget shown if the weekly income is $600.

| Item | Percent Budgeted | Amount Budgeted |
|---|---|---|
| Housing | 28% | |
| Food | 18% | |
| Savings | 10% | |
| Clothing | 15% | |
| Entertainment | 10% | |
| Transportation | 12% | |
| Miscellaneous | 7% | |

6. The Johnson family budget is shown in the chart.
   a. What percent of the budget was planned for food?
   b. If the Johnson monthly income was $2,000, how much money did they budget for transportation and miscellaneous?

| Item | Percent Budgeted |
|---|---|
| Housing | 29% |
| Food | ? |
| Clothing | 18% |
| Savings | 10% |
| Transportation | 12% |
| Miscellaneous | 8% |

7. Roberta's weekly budget allows 30% for food, 25% for clothing, 12% for savings, and 20% for entertainment. If transportation is the only other item on her budget, what percent is allowed for transportation?

8. The Reliable Delivery Company budgets 30% for transportation, 25% for repairs, and 45% for the workers' salaries. If the total weekly expenses are $3,500, how much is budgeted for repairs and salaries?

9. The utility bills for the Levine family are $180 a month. If the Levine family's monthly income is $1,800, what percent of the income should be budgeted for utility expenses?

10. Mary earns $40 a week on her part-time job. What percent of her income should she budget for savings if she wants to save $10 a week?

11. Mr. Rome earns $500 a week. If his weekly expenses for transportation are $25, what percent of his weekly income should he budget for transportation?

## SPIRAL REVIEW EXERCISES

1. How much tax is paid on a $20 shirt if the sales tax is 6%?

2. If 4 pounds of tomatoes cost $2.80, what is the cost of 5 pounds of tomatoes?

3. Joe invested $500 at 7% interest. What is the value of his investment one year later?

4. Ms. Joyce had a balance of $785.52 in her checking account. What is her new balance after she wrote a check for $119.80?

5. Mr. Mizer spent $17.20 on groceries. How much change will he receive from a $20 bill?
   (a) $37.20 (c) $2.80
   (b) $3.80 (d) $1.80

6. Doris bought a photograph frame regularly priced at $7.99 on sale at 15% off. How much did she pay for the frame?

7. The rates for a telephone call are:
   $.50 for the first three minutes
   .15 for each additional minute
   The cost of a 10-minute call is
   (a) $1.85 (c) $.95
   (b) $1.55 (d) $.65

8. The value of $-7 + (-2) + (-9)$ is
   (a) $-81$ (b) $-18$ (c) $-14$ (d) 0

### CHALLENGE QUESTION

From Martin's $250 weekly paycheck, he budgets 20% for food. Last week he spent $60 for food. By how much did he increase the percent budgeted for food?

# UNIT 10–5 Using Percents in Installment Buying

## THE MAIN IDEA

1. ***Installment buying*** is sometimes used instead of paying the entire price at the time of purchase. In an installment plan, a ***down payment*** is usually made at the time of purchase, followed by equal monthly installments. The down payment is often a percent of the purchase price.
2. To find the cost of an item bought on an installment plan:
   a. Multiply the amount of each installment by the number of installments.
   b. Add the amount of the down payment.

$$\text{Cost of Item on Installment Plan} = \text{Amount of Each Installment} \times \text{Number of Installments} + \text{Down Payment}$$

3. The cost of an item bought on an installment plan is more than the original price. This increase in cost is called the ***carrying charge*** of the installment plan.
4. To find the carrying charge, subtract the original price from the installment plan cost.

$$\text{Carrying Charge} = \text{Installment Plan Cost} - \text{Original Price}$$

**EXAMPLE 1** Find the cost of a television set that is bought on an installment plan by making a $50 down payment and paying 10 monthly installments of $40 each.

Multiply the amount of each installment by the number of installments.

```
 $40 ← amount of each installment
 ×10 ← number of installments
$400
```

Add the amount of the down payment.

```
$400
 +50 ← down payment
$450
```

*Answer:* The cost of the television set is $450.

**EXAMPLE 2** A vacation that costs $780 is paid for on an installment plan by making a 10% down payment and paying 15 monthly installments of $60 each. Find the total cost of the vacation.

To find the amount of the down payment, multiply $780 by 10% written as a decimal.

```
  $780
  ×.10
$78.00 ← down payment
```

Multiply the amount of each installment by the number of installments.

```
 $60 ← amount of each installment
 ×15 ← number of installments
 300
 60
$900
```

Add the down payment.

```
$900
 +78 ← down payment
$978
```

*Answer:* The total cost of the vacation is $978.

**EXAMPLE 3** A $110 coat is bought on an installment plan by making a $20 down payment and paying 8 installments of $15 each. Find the carrying charge.

Multiply the amount of each installment by the number of installments.

$$\begin{array}{r l} \$15 & \leftarrow \text{amount of each installment} \\ \times 8 & \leftarrow \text{number of installments} \\ \hline \$120 & \end{array}$$

Add the $20 down payment.

$$\begin{array}{r l} \$120 & \\ +20 & \leftarrow \text{down payment} \\ \hline \$140 & \leftarrow \text{installment plan cost} \end{array}$$

To find the carrying charge, subtract the original price from the installment plan cost.

$$\begin{array}{r l} \$140 & \leftarrow \text{installment plan cost} \\ -110 & \leftarrow \text{original price} \\ \hline \$30 & \leftarrow \text{carrying charge} \end{array}$$

*Answer:* The carrying charge is $30.

## CLASS EXERCISES

1. A cassette recorder is purchased by making a $30 down payment and paying 6 monthly installments of $15. How much does the cassette recorder cost?
2. A refrigerator is bought by making a $400 down payment and paying 12 monthly installments of $25 each. How much does the refrigerator cost?
3. Find the installment price of a stereo if a down payment of $200 and 20 payments of $40 were made.
4. Mr. James bought a used car by making a down payment of $500 and 12 monthly payments of $300. How much did Mr. James pay for the car?
5. A $400 ring is bought on installment by making a 15% down payment and paying 10 installments of $38 each. Find the cost of the ring.
6. A $650 washing machine is bought on an installment plan by giving a 20% down payment. What is the cost of the washing machine if 24 installments of $25 each will be paid?
7. A $480 moped was bought by paying 25% down and 15 installments of $30 each. What is the installment price of the moped?
8. A $450 television set was purchased by making a down payment of $\frac{1}{3}$ of the price and arranging for 12 installments of $34.95 each. What is the installment price of the television set?
9. You can buy a $300 color television set by making a $60 down payment and paying 12 installments of $25 each. What is the carrying charge?
10. A $540 stereo set can be bought by paying $150 down and 15 installments of $30 each. What is the carrying charge?
11. Mr. Jackson bought a $400 video recorder by making a $60 down payment and paying 20 installments of $22 each. How much more did Mr. Jackson pay for the video recorder by using the installment plan than if he had paid cash?
12. You can buy a $1,500 fur coat by paying $300 down and 20 installments of $80 each. The carrying charge is
(a) $400 (b) $500 (c) $1,600 (d) $1,900

## HOMEWORK EXERCISES

1. A television set is bought by paying $50 down and 10 monthly installments of $38. How much did the television set cost?
2. Find the installment price of an air conditioner if a down payment of $45 and 12 payments of $30 were made.
3. Mr. Reynolds bought a new car by making a down payment of $3,500 and 24 monthly payments of $200.95. What is the total cost of the car?
4. Elizabeth bought a moped by paying $95 down and 15 installments of $20.49 each. What is the installment plan cost?
5. Copy and complete the table shown.

| Item | Down Payment | Amount of Each Installment | Number of Installments | Installment Plan Cost |
|---|---|---|---|---|
| Camera | $20 | $10 | 15 | |
| Stereo | $55 | $22 | 10 | |
| Tape Recorder | $35 | $18 | 12 | |
| Moped | $28 | $20 | 14 | |
| Used Car | $1,200 | $150 | 24 | |
| Boat | $4,000 | $280 | 48 | |

6. A $500 video recorder is bought on an installment plan by giving a 10% down payment. What is the cost of the video recorder if 20 installments of $28 are paid?
7. Albert bought a $180 bicycle by paying 20% down and 10 installments of $19.50 each. What was the installment plan cost of the bicycle?
8. A $425 dishwasher is bought by paying $50 down and 12 installments of $35.99 each. What is the carrying charge?
9. Marcia bought a $1,200 personal computer by making a $100 down payment and 20 payments of $65 each. What was the carrying charge?
10. Mr. Cobb bought a $5,000 used car by paying $1,800 down and 36 installments of $120.25 each. How much more than the $5,000 price did he pay for the car by using the installment plan?
11. A $625 sofa is bought by paying $125 down and 12 installments of $50 each. The carrying charge is
(a) $200 (b) $100 (c) $725 (d) $600
12. A trumpet that costs $380 is bought by making a $50 down payment and 8 payments of $50 each. The installment plan cost is
(a) $450 (b) $400 (c) $70 (d) $20

**13.** Copy and complete the chart shown.

| Item | Original Price | Down Payment | Amount of Each Installment | Number of Installments | Installment Plan Cost | Carrying Charge |
|---|---|---|---|---|---|---|
| Guitar | \$250 | \$25 | \$22 | 12 | | |
| Kitchen Set | \$528 | \$100 | \$30 | 18 | | |
| Typewriter | \$198 | \$50 | \$28 | 6 | | |
| New Car | \$9,800 | \$2,900 | \$190 | 42 | | |

## *SPIRAL REVIEW EXERCISES*

**1.** If sales tax is charged at a rate of 7%, what is the tax on a jacket priced at \$65?

**2.** If Mrs. Jolson bought two items priced at \$7.50 and \$3.10 and sales tax is charged at the rate of 6%, what was the total cost of her purchases?

**3.** 18% of 400 is
(a) 720 (b) 72 (c) 7.2 (d) .72

**4.** $9\frac{3}{8}$ is equal to
(a) $\frac{75}{8}$ (b) $\frac{66}{8}$ (c) $\frac{35}{8}$ (d) $\frac{12}{8}$

**5.** The fraction $\frac{2}{5}$ is equal to
(a) 20% (b) .2 (c) .4 (d) 25%

**6.** 125.862 rounded to the nearest hundred is
(a) 100 (c) 125.86
(b) 200 (d) 125.87

**7.** Thirteen thousand, forty-eight written as a numeral is
(a) 130,048 (c) 13,480
(b) 1,348 (d) 13,048

## CHALLENGE QUESTION

Mr. Crane bought a \$600 refrigerator by making a 10% down payment and 12 payments of \$48 each. What percent of the price of the refrigerator is he paying in carrying charges?

THEME 6

# Measurement

## MODULE 11



## MODULE 12

# UNIT 11–1 Customary Measures of Length

## THE MAIN IDEA

1. The most common *Customary (English) units of length* are the *inch* (in.), the *foot* (ft.), the *yard* (yd.), and the *mile* (mi.).

   An inch is about the length across a bottle cap.
   A foot is about the length of a regular loaf of bread.
   A yard is about the distance from a doorknob to the floor.
   A mile is about as long as 18 football fields.

2. These units have the following relationships to each other:

   12 inches = 1 foot
   (12″ = 1′)
   3 feet = 1 yard
   5,280 feet = 1 mile

3. In general:

   To change from a smaller unit to a larger unit, divide.

   To change from a larger unit to a smaller unit, multiply.

**EXAMPLE** Make each of the required unit changes.

| | *Required Unit Change* | *Conversion Information* | *Arithmetic* |
|---|---|---|---|
| **1.** | Change 66 inches to feet. | To change from the smaller unit (inches) to the larger unit (feet), divide by 12. | $66 \div 12 = 5\frac{6}{12} = 5\frac{1}{2}$<br>$12\overline{)66}$ = 5; 60; 6 |
| **2.** | Change 7 feet to inches. | To change from the larger unit (feet) to the smaller unit (inches), multiply by 12. | $7 \times 12 = 84$ |

*Answer:* 66 inches $= 5\frac{1}{2}$ feet

*Answer:* 7 feet = 84 inches

| *Required Unit Change* | *Conversion Information* | *Arithmetic* |
|---|---|---|
| 3. Change 14 feet to yards. | To change from the smaller unit (feet) to the larger unit (yards), divide by 3. | $14 \div 3 = 4\frac{2}{3}$ <br> $3\overline{)14}$ = 4, 12, remainder 2 |

*Answer:* 14 feet $= 4\frac{2}{3}$ yards

| | | |
|---|---|---|
| 4. Change 8 yards to feet. | To change from the larger unit (yards) to the smaller unit (feet), multiply by 3. | $8 \times 3 = 24$ |

*Answer:* 8 yards = 24 feet

| | | |
|---|---|---|
| 5. Change 4 miles to feet. | To change from the larger unit (miles) to the smaller unit (feet), multiply by 5,280. | $4 \times 5{,}280 = 21{,}120$ |

*Answer:* 4 miles = 21,120 feet

## CLASS EXERCISES

1. Change the given number of inches to feet.

   **a.** 72 inches **b.** 144 inches **c.** 30 inches **d.** 54 inches

2. Change the given number of feet to inches.

   **a.** 12 feet **b.** 100 feet **c.** $8\frac{1}{4}$ feet **d.** $18\frac{3}{4}$ feet

3. Helen needed 48 inches of trim for her skirt. How many feet of trim did she need?
4. Hank threw a baseball 140 feet. How many inches did he throw it?
5. Annie is 42 inches tall. What is her height in feet?
6. A golf ball landed $5\frac{3}{4}$ feet from the flag. How many inches away from the flag was it?
7. Change the given number of feet to yards.

   **a.** 36 feet **b.** 300 feet **c.** 144 feet **d.** 13 feet

8. Change the given number of yards to feet.

   **a.** 4 yards **b.** 100 yards **c.** $10\frac{2}{3}$ yards **d.** $\frac{1}{6}$ yard

9. Joe ran 17 yards for a touchdown. How many feet did he run?
10. Mrs. Rivera measured her windows for curtains and found that she needed 45 feet of material. How many yards of material did she need?
11. Charlene ran $\frac{1}{2}$ mile. How many feet did she run?

12. Mrs. Roth's vegetable garden was 48 feet long. How many yards long was it?

13. Change the given number of miles to feet.

    a. 1 mile  b. 3 miles  c. 10 miles  d. $2\frac{1}{2}$ miles

14. Marcia and Bill walked 4 miles to get help for their disabled car. How many feet did they walk?

## HOMEWORK EXERCISES

1. Change the given number of inches to feet.

   a. 156 inches  b. 1,440 inches  c. 111 inches  d. 9 inches

2. Change the given number of feet to inches.

   a. 120 feet  b. 45 feet  c. $9\frac{1}{2}$ feet  d. $19\frac{3}{4}$ feet

3. A shelf is 48 inches above the floor. How many feet above the floor is it?

4. A basketball player is 7 feet tall. What is his height in inches?

5. Nancy is 63 inches tall. What is her height in feet?

6. Roberta broad-jumped $7\frac{3}{4}$ feet. How many inches did she jump?

7. Change the given number of feet to yards.

   a. 300 feet  b. 480 feet  c. 10 feet  d. 101 feet

8. Change the given number of yards to feet.

   a. 9 yards  b. 42 yards  c. 100 yards  d. $8\frac{2}{3}$ yards

9. A living room is 5 yards long. How many feet long is it?

10. A telephone installer found that she needed 75 feet of wire. How many yards of wire did she need?

11. A racetrack is 440 yards long. What is its length in feet?

12. The distance from third base to home plate is 90 feet. How many yards is this?

13. Change the given number of miles to feet.

    a. 2 miles  b. 5 miles  c. 20 miles  d. $7\frac{1}{2}$ miles

14. Joe biked for 12 miles in the Bike-a-thon. How many feet did he travel?

15. Change to the indicated unit.

    a. 6 feet = ? inches  b. 48 inches = ? feet  c. 17 yards = ? feet

    d. 35 feet = ? yards  e. 10 miles = ? feet  f. $9\frac{1}{3}$ yards = ? feet

    g. 20 inches = ? feet  h. 48 feet = ? yards  i. 19 feet = ? inches

## SPIRAL REVIEW EXERCISES

1. How much would a $1\frac{1}{2}\%$ city sales tax add to the cost of a $399.99 stereo? Round the answer to the nearest cent.

2. A plumber charges $36 for the first hour and $12.50 for each half-hour after the first hour. If the plumber worked for $2\frac{1}{2}$ hours to fix a sink, how much was his bill?

3. On a map, 1 inch represents 5 miles. How long a line segment would be used to represent 35 miles?

4. 10% of $256 is
   (a) $256 (c) $2.56
   (b) $25.60 (d) $221.40

5. 32.39 rounded to the nearest tenth is
   (a) 30 (b) 32 (c) 32.3 (d) 32.4

### CHALLENGE QUESTION

A telephone repair team has 10 miles of cable. How many 100-yard lengths of cable can they cut?

# UNIT 11–2 Adding and Subtracting Measures

## THE MAIN IDEA

To add or subtract measures that are given in two or more related units, such as feet and inches:

1. Write the measures in line with each other so that feet are in one column and inches are in another column.
2. Add or subtract the feet and inches separately, renaming units as needed.

**EXAMPLE 1** Add 7 ft. 3 in. to 5 ft. 4 in.

Write the measures in line with each other. Add feet and inches separately.

```
  7 ft. 3 in.
 +5 ft. 4 in.
 ------------
 12 ft. 7 in.   Ans.
```

**EXAMPLE 2** Add 4 ft. 9 in. to 8 ft. 5 in.

Write the measures in line with each other. Add feet and inches separately.

```
  4 ft.  9 in.
 +8 ft.  5 in.
 -------------
 12 ft. 14 in.
```

Since the number of inches is greater than 12, divide by 12.

```
     1
 12)14
    12
    --
     2
```

The quotient 1 gives one more foot, and the remainder 2 stays as 2 inches.

```
   12 ft. 14 in.
 = 12 ft. + 1 ft. 2 in.
 = 13 ft. 2 in.   Ans.
```

**EXAMPLE 3** Add 10 yd. 2 ft. 8 in. to 4 yd. 2 ft. 9 in.

Add yards, feet, and inches separately.

```
  10 yd. 2 ft.  8 in.
  +4 yd. 2 ft.  9 in.
 --------------------
  14 yd. 4 ft. 17 in.
```

Since the number of inches is greater than 12, divide 17 by 12.

```
     1
 12)17
    12
    --
     5
```

The quotient 1 gives one more foot, and the remainder 5 stays as 5 inches.

```
   14 yd. 4 ft. 17 in.
 = 14 yd. 4 ft + 1 ft. 5 in.
 = 14 yd. 5 ft. 5 in.
```

Since the number of feet is greater than 3, divide 5 by 3.

```
   1
 3)5
   3
   -
   2
```

The quotient 1 gives one more yard, and the remainder 2 stays as 2 feet.

```
   14 yd. 5 ft. 5 in.
 = 14 yd. + 1 yd. 2 ft. 5 in.
 = 15 yd. 2 ft. 5 in.   Ans.
```

**EXAMPLE 4** Subtract 4 ft. 7 in. from 10 ft. 11 in.

Write the measures in line with each other. Subtract feet and inches separately.

```
  10 ft. 11 in.
 − 4 ft.  7 in.
   6 ft.  4 in.  Ans.
```

**EXAMPLE 5** Subtract 8 ft. 6 in. from 14 ft. 3 in.

Write the measures in line with each other. We cannot subtract 6 in. from 3 in.

```
  14 ft. 3 in.
 − 8 ft. 6 in.
        X
```

In the minuend, rename 1 foot as 12 inches.

```
  14 ft. 3 in.
= 13 ft. 12 in. + 3 in.
= 13 ft. 15 in.
```

Now we can subtract.

```
  13 ft. 15 in.
 − 8 ft.  6 in.
   5 ft.  9 in.  Ans.
```

## CLASS EXERCISES

1. Add the given measures.

   a. 6 ft. 8 in. and 5 ft. 3 in.  b. 11 ft. 6 in. and 1 ft. 8 in.  c. 10 ft. and 3 ft. 9 in.

   d. 17 ft. 8 in. and 11 in.  e. 4 yd. 1 ft. 10 in. and 5 yd. 1 ft. 1 in.

   f. 7 yd. 1 ft. 9 in. and 4 yd. 2 ft. 6 in.  g. 4 ft. 11 in. and 6 yd. 2 ft. 5 in.

   h. 2 yd. 2 in. and 10 ft. 10 in.  i. 6 mi. 4,176 ft. and 10 mi. 943 ft.

   j. 4 mi. 5,100 ft. 4 in. and 3 mi. 1,500 ft. 10 in.

2. Sarah wishes to fence in her vegetable garden, which is shaped in the form of a triangle. If the measures of the sides of the triangle are 6 ft. 3 in., 5 ft. 4 in., and 7 ft. 5 in., how many feet of fencing must Sarah buy?

3. If Sarah were to enlarge her garden so that the measurements were 11 ft. 11 in., 10 ft. 10 in., and 9 ft. 9 in., how many feet of fencing would she have to buy?

   (a) 30 ft.  (b) 31 ft.  (c) 32 ft.  (d) $32\frac{1}{2}$ ft.

4. If Sarah were to rearrange her garden into the shape of a quadrilateral (4-sided figure) with sides measuring 11 ft. 11 in., 10 ft. 10 in., 9 ft. 9 in., and 8 ft. 8 in., how many feet of fencing would she have to buy?
   (a) 39 ft.  (b) 41 ft.  (c) 42 ft.  (d) 43 ft.

5. Subtract the given measures.

   a. 11 ft. 7 in. from 19 ft. 9 in.  b. 7 ft. 11 in. from 9 ft. 9 in.  c. 9 ft. from 16 ft. 6 in.

   d. 5 ft. 4 in. from 8 ft.  e. 10 yd. 1 ft. 4 in. from 14 yd. 2 ft. 7 in.

   f. 8 yd. 1 ft. 3 in. from 15 yd. 2 ft. 1 in.  g. 2 yd. 2 ft. 7 in. from 8 yd. 1 ft. 4 in.

   h. 12 mi. 4,444 ft. from 15 mi. 4,700 ft.  i. 3 mi. 3,300 ft. from 7 mi. 2,000 ft.

   j. $10\frac{1}{2}$ mi. from 20 mi. 2,800 ft.

**6.** A tailor has $8\frac{1}{3}$ yd. of material from which to make:

**a.** two dresses. If the first dress requires 3 yd. 2 ft. 7 in., how much material is left for the second dress?

**b.** a three-piece suit. If the jacket requires 2 yd. 1 ft. 9 in. and the vest requires 1 yd. 4 in., how much material is left for the skirt?

## HOMEWORK EXERCISES

**1.** Add the given measures.

**a.** 8 ft. 1 in. and 3 ft. 9 in. **b.** 10 ft. 4 in. and 7 ft. 11 in. **c.** 16 ft. and 4 ft. 8 in.

**d.** 7 ft. 3 in. and 9 in. **e.** 3 yd. 1 ft. 7 in. and 5 yd. 1 ft. 4 in.

**f.** 9 yd. 2 ft. 4 in. and 3 yd. 1 ft. 8 in. **g.** 4 mi. 2,000 ft. and 20 mi. 1,000 ft.

**h.** 19 mi. 3,300 ft. and 9 mi. 2,200 ft.

**2.** Mr. Thatcher wishes to frame a tapestry to hang on the wall. If two sides of the tapestry measure 4 ft. 2 in. each and the other two sides measure 5 ft. 8 in. each, how much framing wood does Mr. Thatcher need?

**3.** Mrs. Elkins wishes to trim a blouse with ribbon. If each sleeve requires 2 ft. 4 in. of ribbon and the front of the blouse requires 4 ft. 4 in., how many yards of ribbon must Mrs. Elkins buy?
(a) 3 yd. (b) 4 yd. (c) 6 yd. (d) 9 yd.

**4.** In an athletic contest, Jane runs 2,500 feet, and bikes for $1\frac{1}{2}$ miles. She qualifies for the finals if she has covered a distance of 10,000 feet. Does Jane qualify?

**5.** Subtract the given measures.

**a.** 7 ft. 4 in. from 10 ft. 8 in. **b.** 11 ft. 9 in. from 15 ft. 1 in. **c.** 5 ft. from 12 ft. 8 in.

**d.** 6 ft. 7 in. from 10 ft. **e.** 6 yd. 1 ft. 8 in. from 10 yd. 2 ft. 11 in.

**f.** 8 yd. 2 ft. 6 in. from 12 yd. 1 ft. 4 in. **g.** 16 mi. 100 ft. from 20 mi. 1,000 ft.

**h.** 4 mi. 1,000 ft. from 10 mi. 500 ft.

**6.** A tailor has $4\frac{2}{3}$ yd. of material from which to make:

**a.** a skirt and a blouse. If the blouse requires 1 yd. 2 ft. 7 in., how much material is left for the skirt?

**b.** a three-piece outfit. If the blouse requires 1 yd. 1 ft. 11 in. and the skirt requires 2 yd. 2 ft. 6 in., how much material is left for the sash?

## SPIRAL REVIEW EXERCISES

**1.** Sally ran a 100-yard dash. How many feet did she run?

**2.** Mr. Thompson needed 84 feet of webbing to reweb a chair. How many yards did he need?

**3.** $6\frac{2}{3}$ yards is equivalent to
(a) 21 feet (c) 19 feet
(b) 20 feet (d) 18 feet

**4.** Change $1\frac{1}{4}$ miles to feet.

5. A library lends records at the rate of \$1.50 for the first day and \$.25 for each day thereafter. If Camille borrowed a record for 7 days, how much must she pay?

6. If Susan can type 50 words per minute, how many minutes would it take her to type 750 words?

7. Change $\frac{7}{8}$ to a decimal.

8. Which is the larger value, $\frac{1}{4}$ or $\frac{2}{9}$?

9. Which is the smaller value, .1 or .01?

10. Add $+7$ and $-11$.

11. 42.7 rounded to the nearest ten is
    (a) 40 (b) 42 (c) 43 (d) 50

12. 20% of \$10,000 is
    (a) \$2 (c) \$200
    (b) \$20 (d) \$2,000

13. Multiply $-5$ by $-3$.

14. Evaluate: $11 + 4 \times 2 \times 3$

**CHALLENGE QUESTION**

Mrs. Veley has a piece of ribbon 10 ft. 3 in. long. She wants to cut it into three pieces of equal length. How long should each piece be?

# UNIT 11–3 Customary Measures of Weight

## THE MAIN IDEA

1. The most common *Customary units of weight* are the *ounce* (oz.), the *pound* (lb.), and the *ton* (T.).

   An ounce is about the weight of a spoon.

   A pound is about the weight of a loaf of bread.

   A ton is about the weight of a small car.

2. These units have the following relationships to each other:

   16 ounces = 1 pound

   2,000 pounds = 1 ton

**EXAMPLE 1** Make each of the required unit changes.

| *Required Unit Change* | *Conversion Information* | *Arithmetic* |
|---|---|---|
| **a.** Change 48 ounces to pounds. | To change from the smaller unit (ounces) to the larger unit (pounds), divide by 16. | $48 \div 16 = 3$ |

*Answer:* 48 ounces = 3 pounds

| *Required Unit Change* | *Conversion Information* | *Arithmetic* |
|---|---|---|
| **b.** Change $2\frac{3}{4}$ pounds to ounces. | To change from the larger unit (pounds) to the smaller unit (ounces), multiply by 16. | $2\frac{3}{4} \times 16 = \frac{11}{\cancel{4}_1} \times \frac{\cancel{16}^4}{1} = 44$ |

*Answer:* $2\frac{3}{4}$ pounds = 44 ounces

| *Required Unit Change* | *Conversion Information* | *Arithmetic* |
|---|---|---|
| **c.** Change 3,500 pounds to tons. | To change from the smaller unit (pounds) to the larger unit (tons), divide by 2,000. | $3500 \div 2000 = 1\frac{1500}{2000} = 1\frac{3}{4}$; $2000\overline{)3500}$ = 1, $-2000$, remainder 1500 |

*Answer:* 3,500 pounds = $1\frac{3}{4}$ tons

| *Required Unit Change* | *Conversion Information* | *Arithmetic* |
|---|---|---|
| **d.** Change $2\frac{1}{2}$ tons to pounds. | To change from the larger unit (tons) to the smaller unit (pounds), multiply by 2,000. | $2\frac{1}{2} \times 2000 = \frac{5}{\cancel{2}_1} \times \frac{\cancel{2000}^{1000}}{1} = 5000$ |

*Answer:* $2\frac{1}{2}$ tons = 5,000 pounds

**EXAMPLE 2** Combine the measures as indicated.

**a.**
```
  10 lb.  5 oz.
 +7 lb.  8 oz.
 17 lb. 13 oz.   Ans.
```

**b.**
```
  16 T. 500 lb.
 −4 T. 200 lb.
  12 T. 300 lb.   Ans.
```

No renaming of units is necessary here.

**EXAMPLE 3** Combine the measures as indicated.

**a.**
```
   14 lb. 11 oz.
  +3 lb.  7 oz.
   17 lb. 18 oz.
 = 17 lb. + 1 lb. 2 oz.
 = 18 lb. 2 oz.   Ans.
```

Since the number of ounces is greater than 16:

```
      1 lb.
 16)18
    16
     2 oz.
```

**b.**
```
  18 T.  700 lb.
 −2 T. 1000 lb.
       X
```

We cannot subtract 1,000 lb. from 700 lb.

```
  17 T. 2700 lb.
 −2 T. 1000 lb.
  15 T. 1700 lb.   Ans.
```

In the minuend, rename 1 T. as 2,000 lb.

## CLASS EXERCISES

1. Change the given number of ounces to pounds.

   **a.** 160 oz. **b.** 48 oz. **c.** 68 oz. **d.** 8 oz.

2. Change the given number of pounds to ounces.

   **a.** 5 lb. **b.** 20 lb. **c.** $\frac{1}{4}$ lb. **d.** $2\frac{1}{2}$ lb.

3. Change the given number of pounds to tons.

   **a.** 4,000 lb. **b.** 10,000 lb. **c.** 5,000 lb. **d.** 1,000 lb.

4. Change the given number of tons to pounds.

   **a.** 3 tons **b.** 10 tons **c.** $8\frac{3}{4}$ tons **d.** $4\frac{1}{2}$ tons

5. Joseph collected 196 ounces of scrap aluminum. How many pounds of aluminum did he collect?

6. Mrs. Green used $2\frac{3}{4}$ pounds of butter in her baking. How many ounces of butter did she use?

7. If a loaded shipping crate weighs 7,500 pounds, what is its weight in tons?

8. What is the weight, in pounds, of a $4\frac{1}{4}$-ton truck?

**9.** Combine the measures as indicated.

**a.** $\begin{array}{r} 8 \text{ lb. } 1 \text{ oz.} \\ +3 \text{ lb. } 9 \text{ oz.} \\ \hline \end{array}$ **b.** $\begin{array}{r} 18 \text{ T. } 1000 \text{ lb.} \\ -12 \text{ T. } 100 \text{ lb.} \\ \hline \end{array}$ **c.** $\begin{array}{r} 17 \text{ lb. } 14 \text{ oz.} \\ +3 \text{ lb. } 8 \text{ oz.} \\ \hline \end{array}$ **d.** $\begin{array}{r} 100 \text{ T. } 900 \text{ lb.} \\ -27 \text{ T. } 1200 \text{ lb.} \\ \hline \end{array}$

**10.** 14 oz. + 22 oz. + 10 oz. is equal to
(a) 1 lb. 12 oz. (b) 2 lb. 6 oz. (c) 2 lb. 14 oz. (d) 3 lb. 2 oz.

## *HOMEWORK EXERCISES*

**1.** Change the given number of ounces to pounds.

**a.** 480 oz. **b.** 176 oz. **c.** 40 oz. **d.** 20 oz.

**2.** Change the given number of pounds to ounces.

**a.** 4 lb. **b.** 15 lb. **c.** $3\frac{1}{2}$ lb. **d.** $\frac{3}{4}$ lb.

**3.** Change the given number of pounds to tons.

**a.** 8,000 lb. **b.** 120,000 lb. **c.** 9,000 lb. **d.** 10,200 lb.

**4.** Change the given number of tons to pounds.

**a.** 2 tons **b.** 25 tons **c.** $\frac{1}{4}$ ton **d.** $9\frac{3}{10}$ tons

**5.** How many pounds of turkey are there in 240 ounces?

**6.** Helen lifted 228 ounces of canned goods. How many pounds did she lift?

**7.** Jack caught a 3-pound flounder. How many ounces did the fish weigh?

**8.** A recipe calls for $1\frac{1}{4}$ pounds of chicken. How many ounces of chicken is this?

**9.** An airplane had a 6,000-pound cargo. How many tons was this?

**10.** What is the weight, in tons, of a truck that weighs 5,500 pounds?

**11.** There is a 3-ton limit on a certain stretch of road. What is the limit in pounds?

**12.** A truck dumped $\frac{3}{4}$ of a ton of gravel. How many pounds of gravel was this?

**13.** Combine the measures as indicated.

**a.** $\begin{array}{r} 11 \text{ lb. } 4 \text{ oz.} \\ +10 \text{ lb. } 10 \text{ oz.} \\ \hline \end{array}$ **b.** $\begin{array}{r} 19 \text{ T. } 900 \text{ lb.} \\ -7 \text{ T. } 100 \text{ lb.} \\ \hline \end{array}$ **c.** $\begin{array}{r} 1 \text{ lb. } 1 \text{ oz.} \\ +15 \text{ lb. } 15 \text{ oz.} \\ \hline \end{array}$ **d.** $\begin{array}{r} 20 \text{ T. } 100 \text{ lb.} \\ -10 \text{ T. } 500 \text{ lb.} \\ \hline \end{array}$

**14. a.** At birth, the Johnson twins weighed 5 lb. 14 oz. and 6 lb. 3 oz. What was the combined birth weight?

**b.** During the first month, the smaller twin gained 1 lb. 10 oz. and the larger twin gained 11 oz. What was the new combined weight?

**c.** At a checkup, the doctor put both twins on the scale at the same time and the recorded weight was 20 lb. 3 oz. If one twin weighed 9 lb. 8 oz., what was the weight of the other twin?

**15.** 15 oz. + 21 oz. + 28 oz. is equal to

(a) 2 lb. (b) 3 lb. (c) $3\frac{1}{2}$ lb. (d) 4 lb.

## SPIRAL REVIEW EXERCISES

1. Change the given measures of length to the indicated units.

   a. 16 feet = __?__ inches

   b. 10 miles = __?__ feet

   c. 15 yards = __?__ feet

2. The quarterback of the Clinton High School football team completed 60% of his passes. How many passes did he complete if he threw a total of 225 passes?

3. If Cheryl buys 3 bottles of nail polish at \$1.75 each, how much change should she receive from a \$10 bill?

4. If 5 pounds of ham cost \$12.75, what is the cost of 3 pounds?

5. For a long-distance call from a pay telephone, Wanda has \$4.75 in quarters. How many quarters does she have?

### CHALLENGE QUESTION

If a 1-foot section of a beam weighs 42 pounds, what is the weight, in tons, of an 80-foot section of the same beam?

# UNIT 11–4 Customary Measures of Liquid Volume

## THE MAIN IDEA

1. *Volume* is a measure of how much room something takes up.
2. The most common *Customary units of liquid volume* are the *fluid ounce* (fl. oz.), the *cup* (c.), the *pint* (pt.), the *quart* (qt.), and the *gallon* (gal.).

   The fluid ounce, which is a measure of volume, is not the same as the ounce that is a measure of weight.

   A fluid ounce is the size of the contents of a bottle of iodine.
   A cup is about as much as a coffee mug holds.
   A pint is about as much as a tall spray-paint can holds.
   A quart is the size of the contents of a tall milk container.
   A gallon is about as much as a small sink holds.
3. These units have the following relationships to each other:

   16 fluid ounces = 1 pint
   8 fluid ounces = 1 cup
   2 cups = 1 pint
   2 pints = 1 quart
   4 quarts = 1 gallon

**EXAMPLE 1** Make each of the required unit changes.

| | *Required Unit Change* | *Conversion Information* | *Arithmetic* |
|---|---|---|---|
| **a.** | Change 3 pints to fluid ounces. | To change from the larger unit (pints) to the smaller unit (fluid ounces), multiply by 16. | $3 \times 16 = 48$ |
| | *Answer:* 3 pints = 48 fluid ounces | | |
| **b.** | Change 80 fluid ounces to pints. | To change from the smaller unit (fluid ounces) to the larger unit (pints), divide by 16. | $80 \div 16 = 5$ |
| | *Answer:* 80 fluid ounces = 5 pints | | |
| **c.** | Change 5 pints to quarts. | To change from the smaller unit (pints) to the larger unit (quarts), divide by 2. | $5 \div 2 = 2\frac{1}{2}$ |
| | *Answer:* 5 pints = $2\frac{1}{2}$ quarts | | |
| **d.** | Change $5\frac{1}{2}$ quarts to pints. | To change from the larger unit (quarts) to the smaller unit (pints), multiply by 2. | $5\frac{1}{2} \times 2 = \frac{11}{\cancel{2}_1} \times \frac{\cancel{2}^1}{1} = 11$ |
| | *Answer:* $5\frac{1}{2}$ quarts = 11 pints | | |

**EXAMPLE 2** Combine the measures as indicated.

**a.**

```
   10 gal. 3 qt. 1 pt.
 +  9 gal. 2 qt. 1 pt.
 ---------------------
   19 gal. 5 qt. 2 pt.
```

= 19 gal. 6 qt. — Change 2 pints to 1 quart.

= 19 gal. + 1 gal. 2 qt. — Since the number of quarts is greater than 4:

= 20 gal. 2 qt. *Ans.*

$$\begin{array}{r} 1 \text{ gal.} \\ 4\overline{)6} \\ \underline{4} \\ 2 \text{ qt.} \end{array}$$

**b.**

```
   3 pt.  8 fl. oz.
 - 1 pt. 12 fl. oz.
 ------------------
         X
```

We cannot subtract 12 fl. oz. from 8 fl. oz.

```
   2 pt. 24 fl. oz.
 - 1 pt. 12 fl. oz.
 ------------------
   1 pt. 12 fl. oz.   Ans.
```

In the minuend, rename 1 pt. as 16 fl. oz.

## CLASS EXERCISES

**1.** Change the given number of pints to fluid ounces.

**a.** 11 pints **b.** 6 pints **c.** $3\frac{1}{2}$ pints

**2.** Change the given number of fluid ounces to pints.

**a.** 32 fl. oz. **b.** 56 fl. oz. **c.** 172 fl. oz.

**3.** Change the given number of quarts to pints.

**a.** 5 quarts **b.** 2 quarts **c.** $3\frac{1}{2}$ quarts

**4.** Change the given number of pints to quarts.

**a.** 2 pints **b.** 1 pint **c.** 14 pints

**5.** Change each given quantity to four other equivalent quantities.

| Gallons | Quarts | Pints | Cups | Fluid Ounces |
|---|---|---|---|---|
| 8 | | | | |
| | 20 | | | |
| | | 40 | | |
| | | | 104 | |
| | | | | 320 |

**6.** Mrs. Meyers' recipe for cookies called for 3 cups of milk. How many fluid ounces of milk were needed?

**7.** At the service station, Mr. Jackson discovered that his car needed 2 quarts of oil. How many pints of oil did his car need?

8. The Russo family wanted a half-gallon of ice cream. They decided to get separate pints instead of one large container, so that they could order a different flavor for each pint. How many different flavors could they order?

9. A bottle of milk contained 2 quarts and a bottle of juice contained 72 fluid ounces. Which bottle contained more?

10. Mr. Rose's recipe for pizza dough called for $3\frac{1}{2}$ fluid ounces of olive oil. He wanted to make 10 times the usual amount of dough. How many cups of oil should he have used?

11. Combine the measures as indicated.

| a. | 15 gal. 2 qt. 1 pt. | b. | 7 pt. 9 fl. oz. |
|---|---|---|---|
| | + 5 gal. 1 qt. 1 pt. | | − 2 pt. 14 fl. oz. |

## HOMEWORK EXERCISES

1. Change the given number of pints to fluid ounces.

   a. 3 pints  b. 8 pints  c. $4\frac{1}{2}$ pints

2. Change the given number of fluid ounces to pints.

   a. 64 fl. oz.  b. 40 fl. oz.  c. 108 fl. oz.

3. Change the given number of quarts to pints.

   a. 3 quarts  b. 10 quarts  c. $1\frac{1}{2}$ quarts

4. Change the given number of pints to quarts.

   a. 4 pints  b. 12 pints  c. 27 pints

5. Change each given quantity to four other equivalent quantities.

| Gallons | Quarts | Pints | Cups | Fluid Ounces |
|---|---|---|---|---|
| $1\frac{1}{2}$ | | | | |
| | 16 | | | |
| | | 60 | | |
| | | | 160 | |
| | | | | 288 |

6. A batch of bread dough requires 4 cups of water. How many fluid ounces is this?

7. At lunch, a group of children drank 3 quarts of milk. How many pints of milk did they drink?

8. For the school fair, 4 gallons of ice cream were bought. If the ice cream was packed into 1-pint containers, how many containers were needed?

9. Which contains more, a 2-gallon picnic jug or a container that holds 260 fluid ounces?

**10.** If a jar contains 12 fluid ounces, how many cups do 16 of these jars hold?

**11.** Combine the measures as indicated.

**a.** $\begin{array}{r} 12 \text{ gal. } 3 \text{ qt.} \\ +\ 5 \text{ gal. } 2 \text{ qt.} \\ \hline \end{array}$ **b.** $\begin{array}{r} 8 \text{ pt. } 3 \text{ fl. oz.} \\ -1 \text{ pt. } 9 \text{ fl. oz.} \\ \hline \end{array}$

**12.** A gourmet grocery had a total of 14 gal. 2 qt. 1 pt. of its special salad dressing. The grocer sold the following amounts: 3 gal. 3 qt., 1 qt. 1 pt., and 1 pt. How much salad dressing remained in the store?

## SPIRAL REVIEW EXERCISES

**1.** Change each of the given weights to the indicated unit.

**a.** 80 oz. = ? lb. **b.** 10,000 lb. = ? T.

**c.** 9 lb. = ? oz. **d.** 10 oz. = ? lb.

**e.** $3\frac{1}{4}$ T. = ? lb. **f.** $\frac{1}{2}$ lb. = ? oz.

**2.** A bowling alley is 60 feet long. What is its length in yards?

**3.** The distance from the floor to the ceiling is 8 feet. How many inches is this?

**4.** A truck weighs 11,000 lb. How many tons does it weigh?

**5.** Ms. Fine, a real estate agent, earns a 7% commission on her home sales. How much does she earn on selling a home that costs $85,000?

**6.** Multiply $1\frac{2}{3}$ by 18.

**7.** What is the total cost of an $8 book if the sales tax rate is 5%?

**8.** Find the sum of 12.8, 17.05, 26, and 2.11.

**9.** Calculate: $2 \times 8 + 5 \times 3$

**10.** Subtract $-7$ from $-9$.

### CHALLENGE QUESTION

Get Well Pharmacy pays $7.99 for a gallon of cough syrup. It repackages the syrup in 8-ounce bottles that sell for $1.39 each. How much profit does the pharmacy make?

# UNIT 11–5 Measuring Time

## THE MAIN IDEA

1. The *units of time* have the following relationships to each other:

   60 seconds (sec.) = 1 minute (min.)
   60 minutes = 1 hour (hr.)
   24 hours = 1 day (da.)

2. The time of day is divided into 2 parts:

   A.M. before noon
   P.M. after noon

3. To compute the *elapsed* time (time gone by) between two given times:
   a. If both times are A.M. or both times are P.M. in the same day, subtract the earlier time from the later time.
   b. If the first time is A.M. and the second is P.M. within the same day, subtract the A.M. time from 12:00 noon and add the result to the P.M. time.
   c. If the first time is P.M. and the second is A.M. within a 24-hour period, subtract the P.M. time from 12:00 midnight and add the result to the A.M. time.

**EXAMPLE 1** How much time has elapsed between 3:15 P.M. and 7:40 P.M.?

Since both times are P.M., subtract directly.

$$\begin{array}{r} 7{:}40 \\ -\,3{:}15 \\ \hline 4{:}25 \end{array}$$

*Answer:* 4 hr. 25 min.

**EXAMPLE 2** How much time has elapsed from the first clock to the second?

A.M.

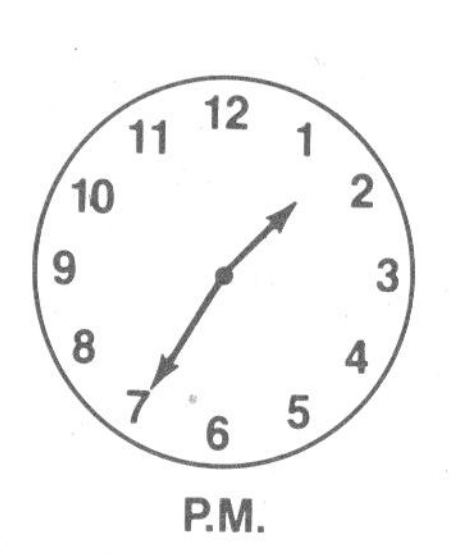

P.M.

First, read the clocks to tell the times they show. 11:15 A.M. to 1:35 P.M.

Subtract 11:15 from 12:00 noon.

$$\begin{array}{r} 12{:}00 \\ -\,11{:}15 \\ \hline \end{array}$$

To do the subtraction, rename 12:00 noon as 11:60.

$$\begin{array}{r} 11{:}60 \\ -\,11{:}15 \\ \hline 00{:}45 \end{array}$$

Add this result to the P.M. time.

$$\begin{array}{r} 1{:}35 \\ +\,00{:}45 \\ \hline 1{:}80 \end{array}$$

Rename 80 minutes as 1 hour 20 minutes.

1:80 = 1 hour + 1 hour 20 minutes
= 2 hours 20 minutes *Ans.*

**EXAMPLE 3** Rose must get to work by 8:10. If her trip from home takes 35 minutes, by what time must she leave home in order to get to work on time?

Subtract 35 minutes from 8:10.

```
  8:10
 - :35
```

Change 1 hour to 60 minutes.

```
  7:70
 - :35
  7:35
```

*Answer:* She must leave by 7:35.

**EXAMPLE 4** Combine the measures as indicated.

**a.**

```
   4 da. 17 hr. 25 min.
 + 8 da. 19 hr. 45 min.
  12 da. 36 hr. 70 min.
```

Since the number of minutes is greater than 60:

```
     1 hr.
 60)70
    60
    10 min.
```

= 12 da. 36 hr. + 1 hr. 10 min.
= 12 da. 37 hr. 10 min.

Since the number of hours is greater than 24:

```
     1 da.
 24)37
    24
    13 hr.
```

= 12 da. + 1 da. 13 hr. 10 min.
= 13 da. 13 hr. 10 min. *Ans.*

**b.**

```
   15 hr. 20 min. 10 sec.
 -  2 hr. 30 min. 15 sec.
```

We cannot subtract 15 sec. from 10 sec.

```
   15 hr. 19 min. 70 sec.
 -  2 hr. 30 min. 15 sec.
```

In the minuend, rename 1 min. as 60 sec.
Since we cannot subtract 30 min. from 19 min., rename 1 hr. as 60 min.

```
   14 hr. 79 min. 70 sec.
 -  2 hr. 30 min. 15 sec.
   12 hr. 49 min. 55 sec.   Ans.
```

## CLASS EXERCISES

1. How much time elapsed between the first time and the second time?

**a.** 5:00 A.M. to 10:25 A.M. **b.** 2:15 P.M. to 4:25 P.M. **c.** 3:45 A.M. to 5:15 A.M.

**d.** 8:52 P.M. to 11:12 P.M. **e.** 3:37 A.M. to 5:00 A.M. **f.** 9:30 A.M. to 12:00 noon

**g.** 10:30 A.M. to 1:15 P.M. **h.** 6:45 A.M. to 2:40 P.M. **i.** 10:15 A.M. to 3:45 P.M.

**j.** 9:52 P.M. to 2:55 A.M. **k.** 10:45 A.M. to 12:00 midnight

2. How much time has elapsed from the first clock to the second?

a.

A.M. A.M.

b.

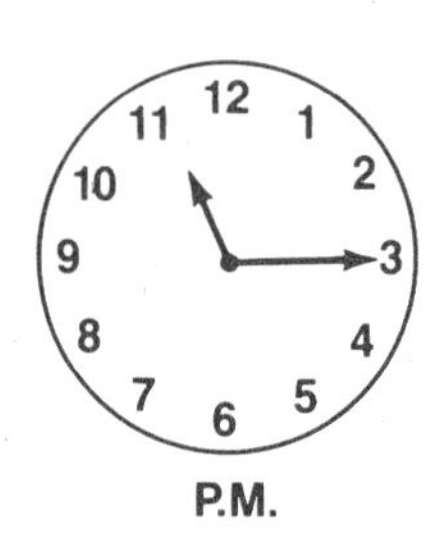


P.M. P.M.

c.

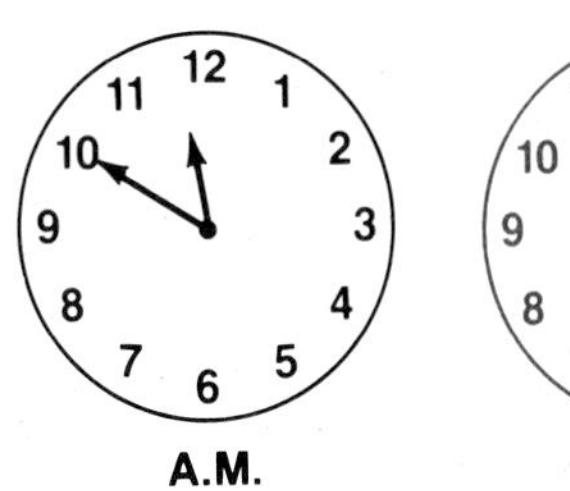


A.M. P.M.

d.

P.M. A.M.

3. A train started at 8:12 A.M. and arrived at its destination at 11:31 A.M. How long was the trip?

4. Jennifer worked from 2:35 P.M. to 4:00 P.M. How long did she work?

5. How much time elapsed between sunrise at 6:01 A.M. and sunset at 9:29 P.M.?

6. Workers at Acme, Inc. start the day at 8:25 A.M. and leave at 4:05 P.M. How long is their workday?

7. Mike slept from 11:17 P.M. to 6.15 A.M. How long did he sleep?

8. Combine the measures as indicated.

| a. | 7 da. 19 hr. 30 min. | b. | 17 hr. 10 min. 12 sec. |
|---|---|---|---|
| | + 1 da. 15 hr. 35 min. | | − 5 hr. 15 min. 26 sec. |

9. Mr. Ford worked 8 hours and 10 minutes on Monday and 7 hours and 35 minutes on Tuesday. What was the total amount of time that he worked?

10. A turkey cooked for 3 hours and 15 minutes and a roast cooked for 1 hour and 55 minutes. How much longer than the roast did the turkey cook?

## *HOMEWORK EXERCISES*

1. How much time elapsed between the first time and the second time?

| | | |
|---|---|---|
| **a.** 3:00 A.M. to 6:25 A.M. | **b.** 5:10 P.M. to 7:35 P.M. | **c.** 8:50 A.M. to 11:15 A.M. |
| **d.** 7:47 P.M. to 10:22 P.M. | **e.** 6:39 A.M. to 11:00 A.M. | **f.** 8:42 A.M. to 12:05 P.M. |
| **g.** 2:42 P.M. to 4:50 P.M. | **h.** 11:12 A.M. to 12:17 P.M. | **i.** 10:39 A.M. to 2:15 P.M. |
| **j.** 10:40 P.M. to 5:35 A.M. | **k.** 12:00 noon to 2:35 A.M. | |

**2.** How much time elapsed from the first clock to the second?

**a.**

**b.**

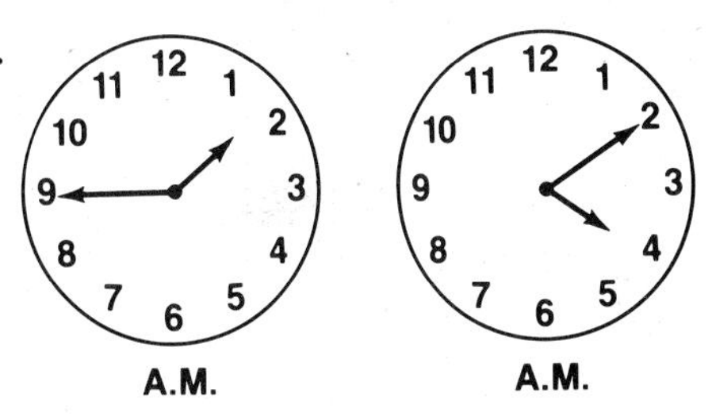


**c.**

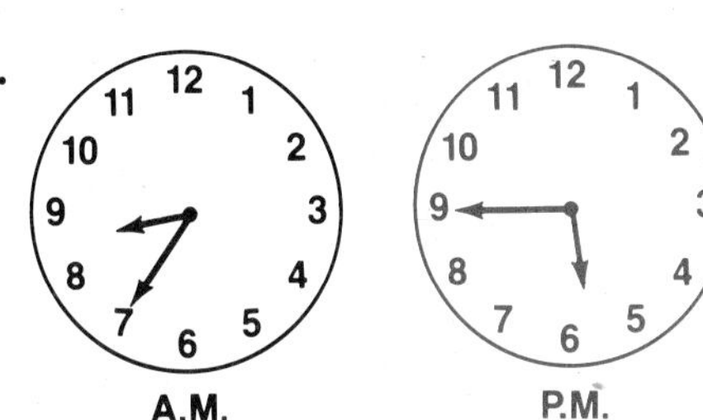


**d.**

**3.** Mrs. Miller wants to set the timer on her stove to ring 25 minutes before 3:15 P.M. For what time should she set the timer?

**4.** The Carlson family left their home at 5:50 P.M. and drove for 47 minutes to pay a visit. At what time did they arrive?

**5.** Harry left for work at 7:20 A.M. and Joan left an hour and 45 minutes later. At what time did Joan leave?

**6.** The #6 bus arrived at the bus terminal at 5:39 P.M. If the trip took 2 hours and 45 minutes, at what time did the bus start the trip?

**7.** Combine the measures as indicated.

**a.** 15 da. 18 hr. 47 min.
+11 da. 10 hr. 20 min.

**b.** 16 hr. 4 min. 1 sec.
− 5 hr. 10 min. 5 sec.

**8.** Peter practiced the trumpet for 1 hour and 40 minutes in the afternoon and 2 hours and 20 minutes in the evening.

**a.** What was the total amount of time that Peter practiced?

**b.** How much longer did Peter practice in the evening?

## SPIRAL REVIEW EXERCISES

**1.** Change each measure of volume to the indicated unit.

**a.** 160 fl. oz. = ? quarts

**b.** 11 gallons = ? quarts

**c.** 6 pints = ? fl. oz.

**d.** $1\frac{1}{2}$ cups = ? fl. oz.

**2.** How many ounces does a 2-pound steak weigh?

**3.** What is the length in inches of a 9-foot rope?

**4.** A loaded cargo crate weighed 9,500 pounds. How many tons did it weigh?

5. George purchased items for \$3.75, \$1.98, and \$1.57. How much change did he receive from a \$10 bill?

6. The product of .02 and 50 is
   (a) .1 (b) 1.0 (c) 10 (d) 50.02

7. The quotient of 25 divided by .05 is
   (a) 5 (b) 50 (c) 500 (d) 5,000

8. Mr. Blank had \$2,475 in his savings account and made a withdrawal of \$298. What is the new balance?

9. Mrs. Chase weighs 120 pounds and her daughter weighs 80 pounds. The ratio of Mrs. Chase's weight to her daughter's weight is
   (a) 3:2 (b) 5:2 (c) 2:5 (d) 2:3

10. The selling price of a \$30 sweater that is sold at a 15% discount is
    (a) \$34.50 (c) \$15.00
    (b) \$26.50 (d) \$25.50

## CHALLENGE QUESTION

Robert worked for 3 hr. 20 min. on Monday, 2 hr. 45 min. on Tuesday, 4 hr. 50 min. on Thursday, and 1 hr. 35 min. on Friday. What are Robert's total earnings if his salary is \$4.50 an hour?

# UNIT 12–1 Metric Measures of Length

## THE MAIN IDEA

1. The most common *metric units of length* are the *millimeter* (mm), the *centimeter* (cm), the *meter* (m), and the *kilometer* (km).

   A millimeter is about the thickness of a piece of cardboard.

   A centimeter is about the width of a telephone push button.

   A meter is about the length of a baseball bat.

   A kilometer could be the distance between two towns.

2. In metric measurement, the *prefix* in the name of a unit tells you how that unit compares to a basic unit. Some of the most common prefixes are:

| *Prefix* | *Meaning* | *Name of Unit* | *Relationship to a Meter* |
|---|---|---|---|
| milli | $\frac{1}{1,000}$ | *milli*meter | $\frac{1}{1,000}$ of a meter |
| centi | $\frac{1}{100}$ | *centi*meter | $\frac{1}{100}$ of a meter |
| kilo | 1,000 | *kilo*meter | 1,000 meters |

3. The Customary and metric units of length compare in the following way:

   A meter is slightly longer than a yard.

   $$1 \text{ m} = 1.1 \text{ yd.}$$

   A centimeter is slightly less than half an inch.

   $$1 \text{ cm} = .4 \text{ in.}$$

   A kilometer is slightly more than half a mile.

   $$1 \text{ km} = .6 \text{ mi.}$$

**EXAMPLE 1** Which would be the most convenient metric unit of length to use in measuring each item? What would be the corresponding Customary unit of length?

| | *Item to be Measured* | *The Most Convenient Metric Unit of Length* | *The Corresponding Customary Unit* |
|---|---|---|---|
| **a.** | a length of carpeting | the meter | the yard |
| **b.** | the length of an auto trip | the kilometer | the mile |
| **c.** | the length of a book | the centimeter | the inch |
| **d.** | the thickness of a coin | the millimeter | the inch |

**EXAMPLE 2** Which is the most likely length of a home television screen?
(a) 40 mm (b) 40 cm (c) 40 m (d) 40 km

A centimeter is roughly the thickness of a finger. Forty of these units would be a reasonable size for a television screen.

40 mm would be only about the thickness of 40 pieces of cardboard—too small for a television screen. 40 m would be longer than most rooms in a house, and 40 km could be the distance between two cities.

*Answer:* (b)

**EXAMPLE 3** Make each of the required unit changes.

| *Required Unit Change* | *Conversion Information* | *Arithmetic* |
|---|---|---|
| **a.** Change 47 millimeters to centimeters. | To change from the smaller unit (millimeters) to the larger unit (centimeters), divide by 10. The easy way to divide by 10 is to move the decimal point one place to the left. | $47 \div 10 = 4.7$<br>$\begin{array}{r} 4.7 \\ 10\overline{)47.0} \\ \underline{40\phantom{.0}} \\ 7\,0 \\ \underline{7\,0} \\ 0 \end{array}$<br>4.7. |

*Answer:* 47 millimeters = 4.7 centimeters

| *Required Unit Change* | *Conversion Information* | *Arithmetic* |
|---|---|---|
| **b.** Change 11.3 centimeters to millimeters. | To change from the larger unit (centimeters) to the smaller unit (millimeters), multiply by 10 (move the decimal point one place to the right). | $11.3 \times 10 = 113$<br>$\begin{array}{r} 11.3 \\ \underline{\times 10} \\ 00\,0 \\ \underline{113\phantom{\,0}} \\ 113.0 \end{array}$<br>11.3. |

*Answer:* 11.3 centimeters = 113 millimeters

| *Required Unit Change* | *Conversion Information* | *Arithmetic* |
|---|---|---|
| **c.** Change 25 centimeters to meters. | To change from the smaller unit (centimeters) to the larger unit (meters), divide by 100 (move the decimal point two places to the left). | $25 \div 100 = .25$<br>.25. |

*Answer:* 25 centimeters = .25 meter

| *Required Unit Change* | *Conversion Information* | *Arithmetic* |
|---|---|---|
| **d.** Change 4.6 meters to centimeters. | To change from the larger unit (meters) to the smaller unit (centimeters), multiply by 100 (move the decimal point two places to the right). | $4.6 \times 100 = 460$<br>4.60. |

*Answer:* 4.6 meters = 460 centimeters

| *Required Unit Change* | *Conversion Information* | *Arithmetic* |
|---|---|---|
| **e.** Change 480 meters to kilometers. | To change from the smaller unit (meters) to the larger unit (kilometers), divide by 1,000 (move the decimal point three places to the left). | $480 \div 1000 = .48$<br>.480. |

*Answer:* 480 meters = .48 kilometer

| *Required Unit Change* | *Conversion Information* | *Arithmetic* |
| --- | --- | --- |
| **f.** Change 17.6 kilometers to meters. | To change from the larger unit (kilometers) to the smaller unit (meters), multiply by 1,000 (move the decimal point three places to the right). | $17.6 \times 1000 = 17{,}600$<br>17.600. |

*Answer:* 17.6 kilometers = 17,600 meters

## CLASS EXERCISES

**1.** Which would be the most convenient metric unit of length to use in measuring each item? What would be the corresponding Customary unit of length?

**a.** the length of window curtains **b.** the distance from the earth to the moon

**c.** the length of a swimming pool **d.** the thickness of a piece of sheet metal

**e.** the height of a tree **f.** the length of a train track **g.** the width of a tooth

**h.** the width of a piece of movie film **i.** the width of a record album cover

**2.** Which is the most likely height of a basketball player?
(a) 2 mm (b) 2 cm (c) 2 m (d) 2 km

**3.** Which is the most likely length of a playing card?
(a) 85 mm (b) 85 cm (c) 85 m (d) 85 km

**4.** Change the given number of millimeters to centimeters.

**a.** 68 mm **b.** 10 mm **c.** 300 mm

**5.** Change the given number of centimeters to millimeters.

**a.** 8 cm **b.** 48.5 cm **c.** 13 cm

**6.** Change each measurement to an equivalent measurement, using the given units.

**a.** 32 cm to m **b.** 7.2 m to cm **c.** 2,420 m to km **d.** 9.6 km to m

## HOMEWORK EXERCISES

**1.** Which would be the most convenient metric unit of length to use in measuring each item? What would be the corresponding Customary unit of length?

**a.** the width of a doorway **b.** the thickness of a magazine **c.** the height of a house

**d.** the length of a garden hose **e.** the length of a dollar bill

**f.** the space between two teeth **g.** the width of a hair ribbon

**h.** the distance from Los Angeles to San Francisco **i.** a person's neck size

**j.** the length of the equator

2. Choose the most likely measurement.
   a. the length of a person's arm
      (a) 1 mm (b) 1 cm (c) 1 m (d) 1 km
   b. the diameter of a nickel
      (a) 20 mm (b) 20 cm (c) 20 m (d) 20 km
   c. the thickness of a nickel
      (a) .2 mm (b) .2 cm (c) .2 m (d) .2 km
   d. the height of the Empire State Building
      (a) .446 mm (b) .446 cm (c) .446 m (d) .446 km

3. Change the given number of millimeters to centimeters.
   a. 54 mm b. 1,482 mm c. 97 mm

4. Change the given number of centimeters to millimeters.
   a. 37 cm b. 2.05 cm c. .03 cm

5. Change each measurement to an equivalent measurement using the given units.
   a. 620 cm to m b. 8.04 m to cm c. 6,540 m to km d. .48 km to m

## SPIRAL REVIEW EXERCISES

1. Add: 5 hr. 18 min.
   2 hr. 47 min.

2. Subtract: 8 hr. 21 min.
   3 hr. 38 min.

3. Philip left for Chicago at 7:35 A.M. and arrived on the same day at 2:05 P.M. (same time zone). How long was the trip?

4. Mike found that 4 boxes of canned goods weighed 28 pounds. What is the weight of 7 boxes?

5. Last year, Mark's height was 50 inches. If his height increased by 10%, what is his new height?

6. 50 is 20% of what number?

7. $\frac{9}{12}$ written as a decimal is
   (a) .25 (b) .50 (c) $.66\frac{2}{3}$ (d) .75

8. Helen purchased items totaling $7.83. The amount of change that she should get from a $20 bill is
   (a) $2.17 (c) $3.17
   (b) $12.17 (d) $13.17

9. The product of .9 and .07 is
   (a) 6.3 (b) .63 (c) .063 (d) .97

10. Which of the following has the largest value?
    (a) $\frac{1}{4}$ (b) $\frac{2}{5}$ (c) .35 (d) 30%

### CHALLENGE QUESTION

Combine the measures as indicated.

a.
```
  15 km 600 m 30 cm
+50 km 800 m 90 cm
```

b.
```
  40 m 10 cm 1 mm
−30 m 20 cm 5 mm
```

# UNIT 12–2 Using a Ruler

## THE INCH SCALE

### THE MAIN IDEA

1. On a ruler, an inch scale is marked in the following way:
   The inches are numbered next to the longest marks.
   The half inches are shown by the second longest marks, which divide each inch into two equal parts.

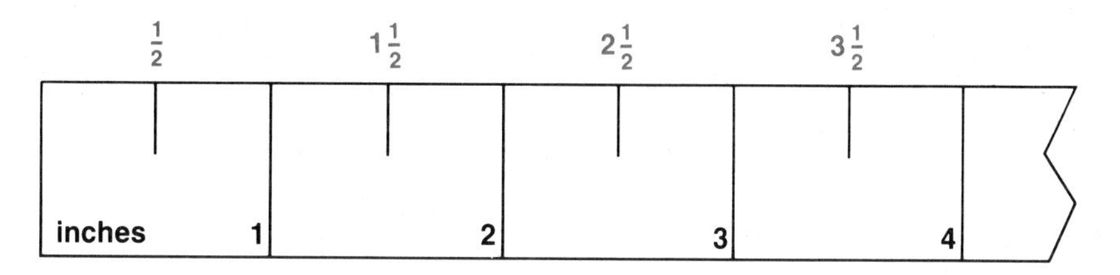


As the marks decrease in length, each inch is broken into a larger number of equal parts.

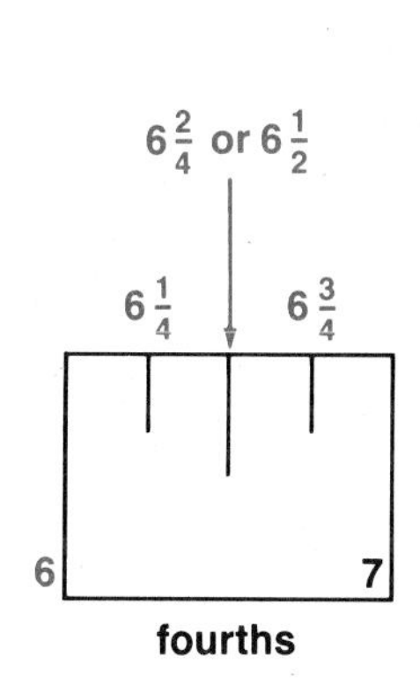


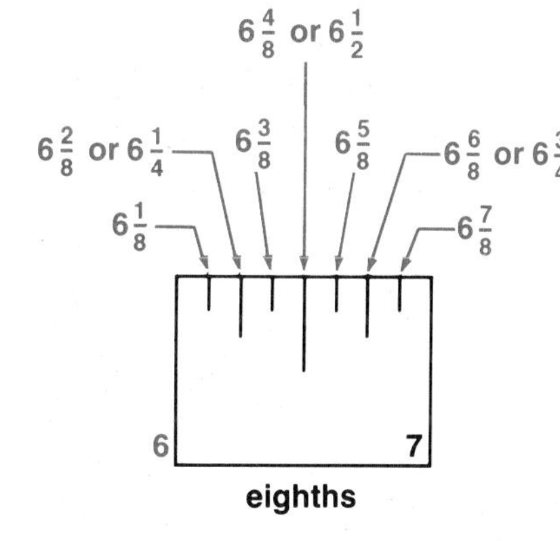


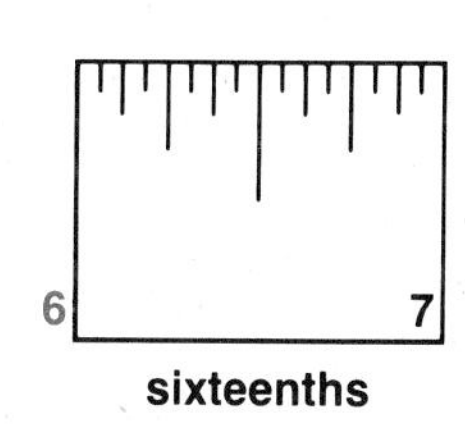


2. To measure a line segment with the inch scale on a ruler, line up one end of the line segment with the beginning of the ruler. Then read:
   a. the number of the inch just to the left of the other end of the line segment, and
   b. the fraction mark closest to that end of the line segment.

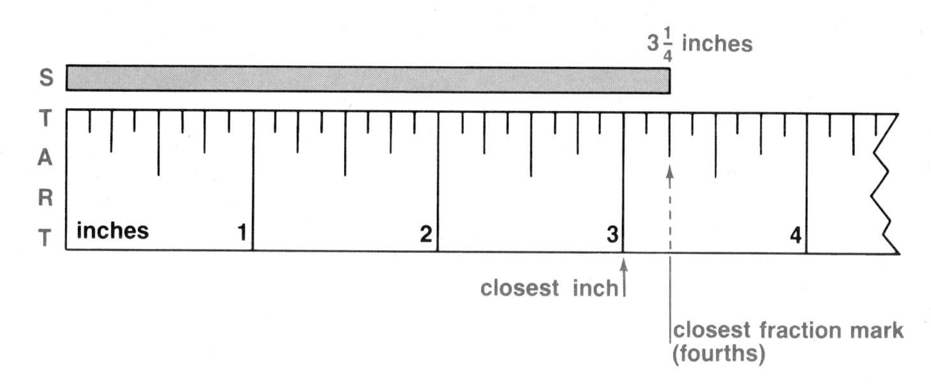

**EXAMPLE 1** Tell how the inch is divided.

a. Since the marks divide the inch into 8 equal parts, they show eighths. *Ans.*

b. 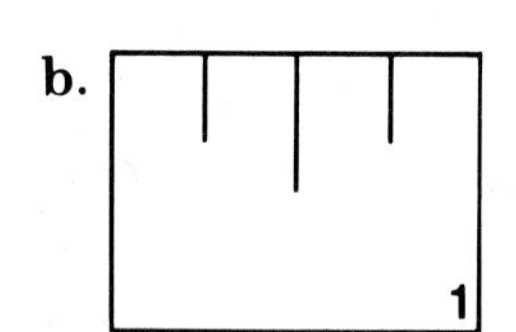

Since the marks divide the inch into 4 equal parts, they show fourths. *Ans.*

**EXAMPLE 2** Draw a line segment $4\frac{3}{4}$ inches long.

Count 3 fourths of an inch past 4 inches. Draw a line segment with the ruler, starting at the beginning and going up to $4\frac{3}{4}$ inches.

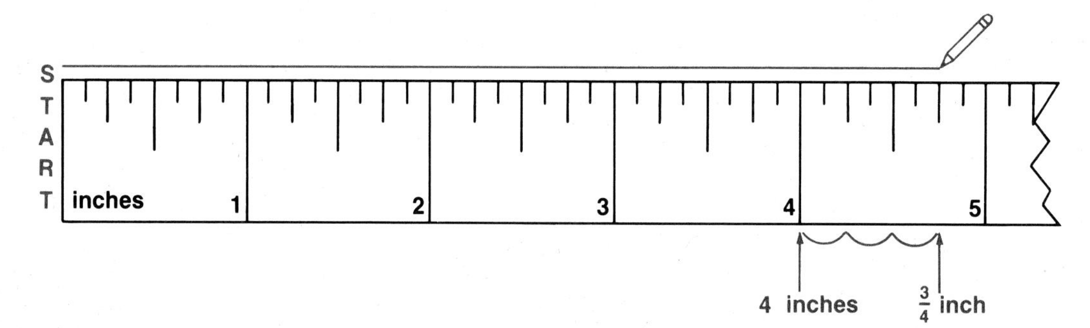


## CLASS EXERCISES

1. Tell how each inch is divided.

a. 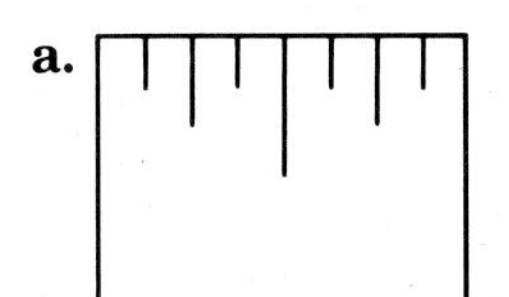 b. c. 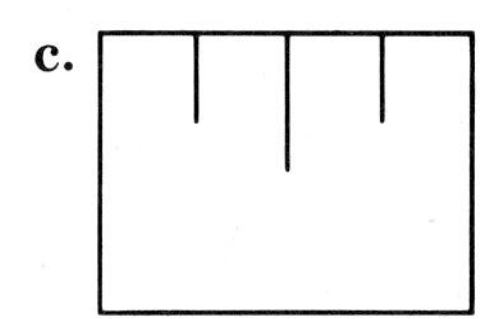 d. 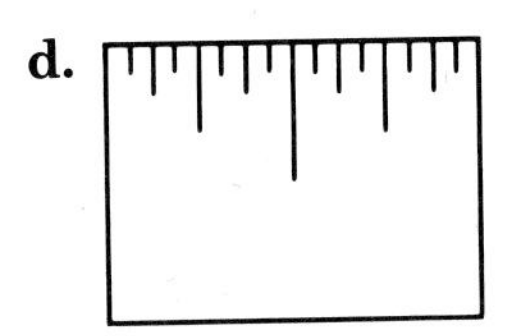

2. Draw a line segment having each length.

a. 5 inches b. $5\frac{1}{4}$ inches c. $4\frac{1}{2}$ inches d. $3\frac{3}{4}$ inches e. $5\frac{1}{8}$ inches

f. $3\frac{5}{8}$ inches g. $5\frac{6}{8}$ inches h. $5\frac{3}{4}$ inches i. $4\frac{7}{8}$ inches

3. Measure the length, in inches, of each line segment.

a. ______________________________

b. ______________________________________________________

c. ____________ d. ______________________________

e. ______________________________________________

f. ________________ g. ______

h. ____________________________

i. _______ j. _______________________

k. _ l. ________

## THE METRIC SCALE

### THE MAIN IDEA

1. On a ruler, a metric scale is marked in the following way:

   The centimeters are numbered next to the longest marks.

   Each centimeter is divided into 10 equal parts, 10 millimeters. The second longest marks show 5-millimeter intervals (half centimeters).

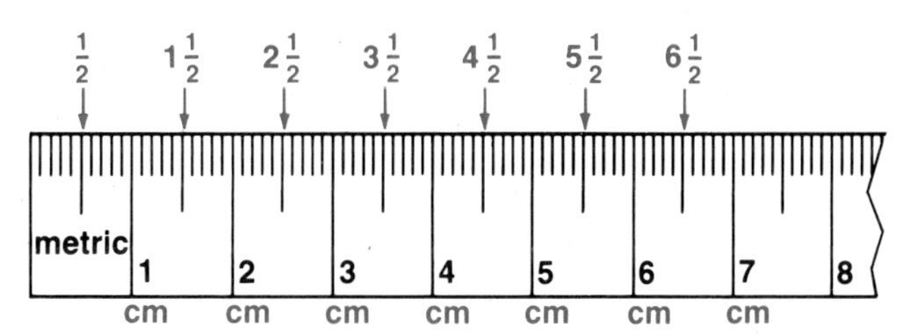


2. To measure a line segment with the metric scale on a ruler, line up one end of the line segment with the beginning of the ruler. Then read:
   a. the number of the centimeter just to the left of the other end of the line segment, and
   b. the millimeter mark closest to that end of the line segment. Since $1 \text{ mm} = \frac{1}{10}$ cm, the millimeters may be written as a decimal.

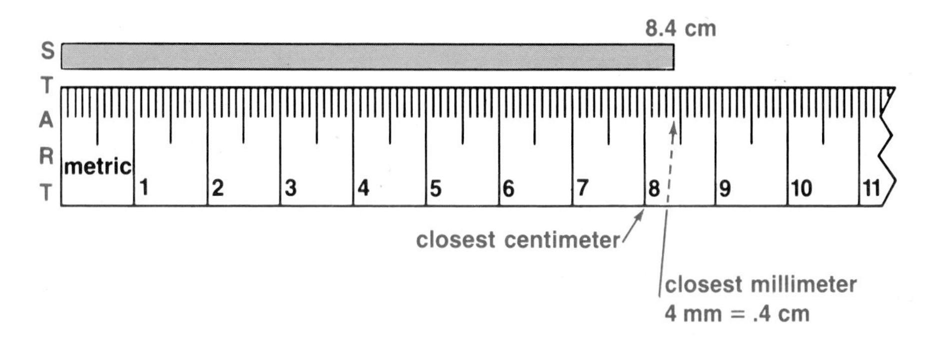

**EXAMPLE 3** Write the number of millimeters as a decimal fraction of a centimeter.

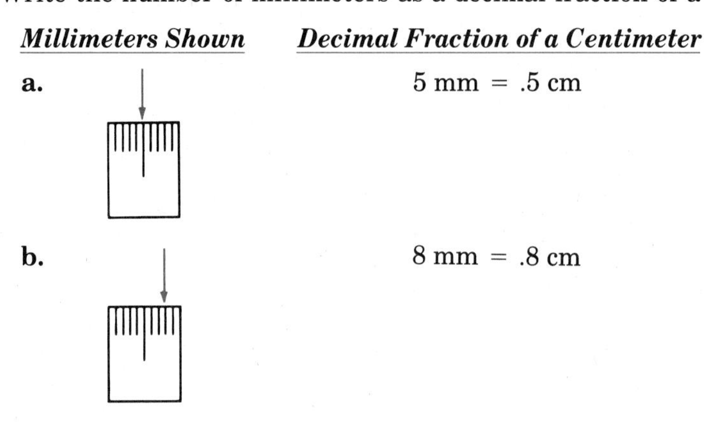


**EXAMPLE 4** Draw a line segment 8.6 cm long.

Since 6 mm = .6 cm, count 6 mm past 8 cm. Draw a line segment from the start of the ruler up to 8.6 cm.

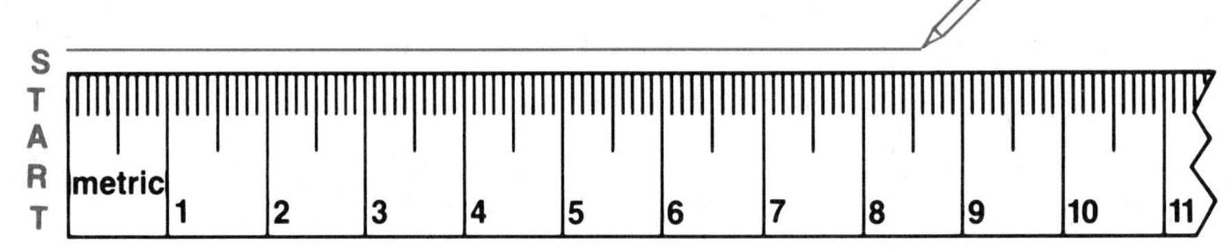


## CLASS EXERCISES

1. Write the number of millimeters as a decimal fraction of a centimeter.

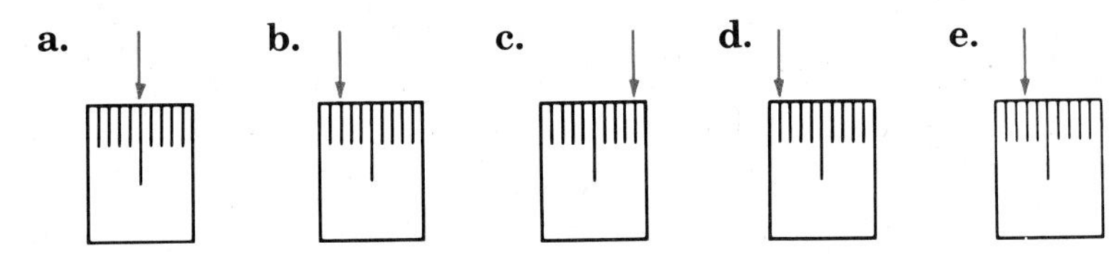


2. Draw a line segment having each length.

**a.** 4.5 cm **b.** 3.2 cm **c.** 6.8 cm **d.** 10.1 cm **e.** 1.1 cm

**f.** .7 cm **g.** .9 cm **h.** 9.9 cm **i.** 6.0 cm

3. Measure the length, in centimeters, of each line segment.

**a.** ______________________________ **b.** ________

**c.** ________________________ **d.** ___ **e.** ______________

**f.** ________________________________ **g.** ____________________

**h.** ______________________________________________________

**i.** ____ **j.** _______________

## HOMEWORK EXERCISES

1. Tell how each inch is divided.

   a. b. c. d.

2. Draw a line segment having each length.

   a. 4 inches  b. $4\frac{1}{8}$ inches  c. $4\frac{1}{4}$ inches  d. $6\frac{1}{2}$ inches  e. $4\frac{3}{8}$ inches

   f. $2\frac{5}{8}$ inches  g. $1\frac{6}{8}$ inches  h. $1\frac{3}{4}$ inches  i. $5\frac{7}{8}$ inches

3. Measure the length of each line segment in inches.

   a.

   b.

   c.

   d.

   e.

   f.

   g.

   h.

   i.

   j.

   k.

   l.

4. Write the numbers of millimeters as a decimal fraction of a centimeter.

   a. b. c. 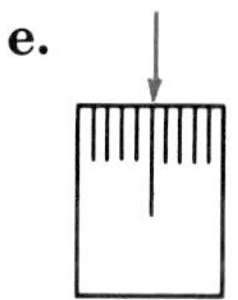 d. e.

5. Draw a line segment having each length.

   a. 3.8 cm  b. 5.2 cm  c. 1.4 cm  d. 4.1 cm  e. 1.9 cm

   f. .8 cm  g. .2 cm  h. 8.8 cm  i. 3.0 cm

**6.** Measure the length of each line segment in centimeters.

**a.** ______

**b.** ________________________________________

**c.** ___________________

**d.** ___________________

**e.** ______________________________________________________

**f.** ____

**g.** ___________________________________

**h.** _____________

**i.** ____________________________

**j.** __________________________________________________________

## SPIRAL REVIEW EXERCISES

**1.** Change each given measure of length to the indicated unit.

**a.** 11 km = ? m  **b.** .6 cm = ? mm

**c.** 1.8 m = ? cm  **d.** 250 cm = ? m

**2.** The width of a record album cover is closest to
(a) 30 mm (c) 3 m
(b) 30 cm (d) 30 m

**3.** The length of a football field is closest to
(a) 92 mm (c) 92 m
(b) 92 cm (d) 92 km

**4.** Combine the measures as indicated.

**a.**
 4 yd. 2 ft. 10 in.
+6 yd. 1 ft. 8 in.

**b.**
 7 gal. 1 qt.
−4 gal. 2 qt. 1 pt.

**5.** If it takes Mary 25 minutes to get to school, at what time should she leave in order to get there at 8:10 A.M.?

**6.** Mr. Johnson earns $5.40 an hour. One week, he worked 40 hours and 2 hours overtime. If he is paid time and one-half for overtime, how much did he earn?

**7.** Ms. Byron invested $2,000 at 12% annual interest. How much interest did she earn the first year?

**8.** Round 1,475,892 to the nearest ten thousand.

**9.** Add: $\frac{3}{8} + \frac{3}{4}$

**10.** Subtract 19.48 from 28.07.

**11.** $\frac{23}{4}$ written as a mixed number is

(a) $4\frac{3}{4}$ (b) $5\frac{1}{4}$ (c) $5\frac{3}{4}$ (d) $7\frac{1}{2}$

**12.** Divide .27 by 1,000.

**13.** 16 is 20% of what number?

**14.** The greatest common factor of 24 and 48 is
(a) 6 (b) 12 (c) 24 (d) 48

## CHALLENGE QUESTION

In this drawing of a field, each centimeter represents 3 meters. If fencing costs $21 per meter, what would be the total cost of fencing the four sides of the field?

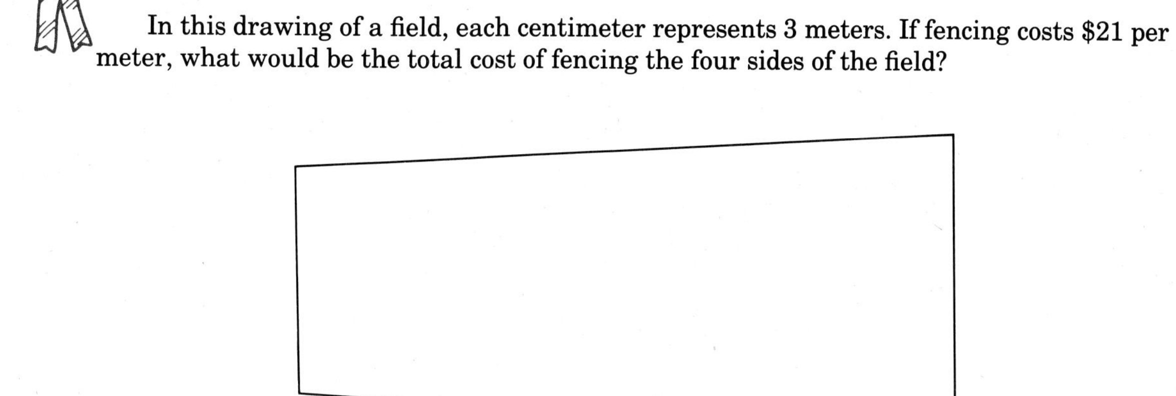

# UNIT 12–3 Metric Measures of Weight

## THE MAIN IDEA

1. The most common *metric units of weight* are the *milligram* (mg), the *gram* (g), the *kilogram* (kg), and the *metric ton* (t).

   A milligram is about the weight of a few grains of salt.

   A gram is about the weight of a pinch of salt.

   A kilogram could be the weight of a dozen tomatoes.

   A metric ton might be the weight of a small car.

2. The prefixes in the names of the most common metric units of weight have the following meanings:

| *Prefix* | *Meaning* | *Name of Unit* | *Relationship to a Gram* |
|---|---|---|---|
| milli | $\frac{1}{1,000}$ | *milli*gram | $\frac{1}{1,000}$ of a gram |
| kilo | 1,000 | *kilo*gram | 1,000 grams |

3. A metric ton is 1,000 times a kilogram. (A metric ton is the only commonly used metric unit of weight that is not identified by a prefix.)

4. The Customary and metric units of weight compare in the following way:

   A kilogram is slightly more than two pounds.

   $$1 \text{ kg} = 2.2 \text{ lb.}$$

   A metric ton is slightly more than a Customary ton.

   $$1 \text{ t} = 1.1 \text{ T.}$$

   A gram is a very small part of an ounce.

   $$1 \text{ g} = \frac{1}{28} \text{ oz.}$$

**EXAMPLE 1** Which would be the most convenient metric unit of weight to use in measuring each item? What would be the corresponding Customary unit of weight?

| | *Item to be Measured* | *The Most Convenient Metric Unit of Weight* | *The Corresponding Customary Unit* |
|---|---|---|---|
| **a.** | the weight of a bicycle | the kilogram | the pound |
| **b.** | the weight of a wristwatch | the gram | the ounce |
| **c.** | the amount of fat in a serving of cornflakes | the milligram | the ounce |
| **d.** | the weight of a truckload of sand | the metric ton | the ton |

**EXAMPLE 2** Make each of the required unit changes.

| *Required Unit Change* | *Conversion Information* | *Arithmetic* |
|---|---|---|
| **a.** Change 1,450 milligrams to grams. | To change from the smaller unit (milligrams) to the larger unit (grams), divide by 1,000 (move the decimal point three places to the left). | 1450 ÷ 1000 = 1.45<br>1.450. |
| *Answer:* 1,450 milligrams = 1.45 grams | | |
| **b.** Change 3.2 grams to milligrams. | To change from the larger unit (grams) to the smaller unit (milligrams), multiply by 1,000 (move the decimal point three places to the right). | 3.2 × 1000 = 3,200<br>3.200. |
| *Answer:* 3.2 grams = 3,200 milligrams | | |
| **c.** Change 3,200 grams to kilograms. | To change from the smaller unit (grams) to the larger unit (kilograms), divide by 1,000. | 3200 ÷ 1000 = 3.2<br>3.200. |
| *Answer:* 3,200 grams = 3.2 kilograms | | |
| **d.** Change 12.8 kilograms to grams. | To change from the larger unit (kilograms) to the smaller unit (grams), multiply by 1,000. | 12.8 × 1000 = 12,800<br>12.800. |
| *Answer:* 12.8 kilograms = 12,800 grams | | |
| **e.** Change 3,758 kilograms to metric tons. | To change from the smaller unit (kilograms) to the larger unit (metric tons), divide by 1,000. | 3758 ÷ 1000 = 3.758<br>3.758. |
| *Answer:* 3,758 kilograms = 3.758 metric tons | | |
| **f.** Change 1.75 metric tons to kilograms. | To change from the larger unit (metric tons) to the smaller unit (kilograms), multiply by 1,000. | 1.75 × 1000 = 1,750<br>1.750. |
| *Answer:* 1.75 metric tons = 1,750 kilograms | | |

## CLASS EXERCISES

**1.** Which would be the most convenient metric unit of weight to use in measuring each item? What would be the corresponding Customary unit of weight?

**a.** the weight of a person **b.** the weight of a truck **c.** the weight of a suitcase

**d.** the weight of a chair **e.** the amount of chlorine in a gallon of drinking water

**f.** the weight of 2 slices of bread **g.** the weight of a truckload of bricks

**h.** the amount of Vitamin C in an orange **i.** the weight of a heavy gold ring

**j.** the amount of salt in a can of peas

2. How many grams are there in each of the following weights?

   a. 3,200 mg  b. 7,050 mg  c. 850 mg

3. How many milligrams are there in each of the following weights?

   a. 5 gm  b. 7.4 gm  c. 13.64 gm

4. Change each weight to kilograms.

   a. 4,800 gm  b. 5,675 gm  c. 80 gm

5. Change each weight to grams.

   a. 8.9 kg  b. 502.7 kg  c. .002 kg

6. Change each weight to metric tons.

   a. 10,500 kg  b. 650 kg  c. 91.8 kg

7. Change each weight to kilograms.

   a. 3 t  b. 4.6 t  c. .75 t

8. 5.6 gm is equivalent to
   (a) 5,600 mg  (b) .056 kg  (c) 560 mg  (d) 5.6 kg

9. Mrs. Smith's diamond earrings weigh 3.6 grams. How many milligrams do they weigh?

10. ABC Contractors dumped 3.6 metric tons of concrete. How many kilograms of concrete was this?

11. A loaf of bread contains 28,000 mg of protein. What is this weight in grams?

## HOMEWORK EXERCISES

1. Which would be the most convenient metric unit of weight to use in measuring each item? What would be the corresponding Customary unit of weight?

   a. the weight of a television set  b. the weight of a pen

   c. the weight of a briefcase  d. the weight of a stack of books

   e. the weight of a postage stamp  f. the weight of a full cement mixer

   g. the weight of a deck of playing cards  h. the amount of iron in a vitamin pill

   i. a baby's weight  j. the weight of a small insect

2. How many grams are there in each of the following weights?

   a. 4,800 mg  b. 998 mg  c. 34 mg

3. How many milligrams are there in each of the following weights?

   a. 4.1 gm  b. 29.02 gm  c. .007 gm

4. Change each weight to kilograms.

   a. 6,720 gm  b. 205 gm  c. 24 gm

5. Change each weight to grams.

   a. 12 kg  b. 3.01 kg  c. .5 kg

6. Change each weight to metric tons.

   a. 12,000 kg  b. 725 kg  c. 683.4 kg

7. Change each weight to kilograms.

   a. 7 t  b. 8.3 t  c. .403 t

8. 22.2 gm is equivalent to
   (a) 2.22 mg  (b) 222 mg  (c) 22.2 mg  (d) .0222 kg

9. A loaf of bread contained 5 grams of fat. How many kilograms of fat were in the loaf?

10. Mr. James used 2.4 metric tons of sand to mix concrete for his driveway. How many kilograms of sand did he use?

## SPIRAL REVIEW EXERCISES

1. Change each given measure of length to the indicated unit.

   a. 30 cm = ? mm  b. 5 m = ? cm

   c. 720 mm = ? cm  d. 8 km = ? m

2. Find the measure of each line segment, using the indicated unit of measure.

   a. ____________
   centimeters

   b. ______________________________
   inches

3. If 5 cans of tomato sauce cost $1.65, what is the cost of 7 cans?

4. Add: 5 hr. 29 min.
   2 hr. 37 min.

5. What time is it 2 hours and 15 minutes after 11:15 A.M.?

6. If there is a 5% sales tax, what will be the total cost of a $38 radio?

7. Evaluate: $12 + 18 \times 5 \div 6$

8. Philip got 18 hits in 40 times at bat. What percent of his times at bat did he get a hit?

9. To the product of $-7$ and $-3$, add $-6$.

## CHALLENGE QUESTION

The Ace Refrigerated Truck Company charges the following rates to ship dairy products:

$1.95 for the first 50 kilograms
$.35 for each additional 50 kilograms or portion thereof

Farmer Bill shipped 450 kg of cottage cheese, 280 kg of butter, and 610 kg of milk. What was the total shipping cost?

## UNIT 12–4 Metric Measures of Volume

### THE MAIN IDEA

1. The most common *metric units of volume* are the *milliliter* (mL), the *liter* (L), and the *kiloliter* (kL).

   A milliliter is about the size of a small sugar cube.

   It is the same as a cubic centimeter, which is the volume contained in a small box all of whose sides are 1 cm long.

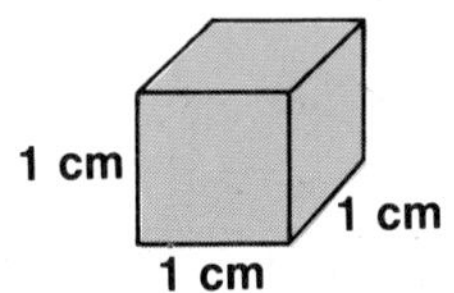


   A liter is just a little larger than a tall milk container.

   A kiloliter could be the size of a large carton.

   It is the same as a cubic meter, which is the volume contained in a box all of whose sides are 1 m long.

2. The prefixes in the names of the most common metric units of volume have the following meanings:

| *Prefix* | *Meaning* | *Name of Unit* | *Relationship to a Liter* |
|---|---|---|---|
| milli | $\frac{1}{1{,}000}$ | *milli*liter | $\frac{1}{1{,}000}$ of a liter |
| kilo | 1,000 | *kilo*liter | 1,000 liters |

3. The Customary and metric units of volume compare in the following way:

   A liter is slightly more than a quart.

$$1 \text{ L} = 1.1 \text{ qt.}$$

**EXAMPLE 1** Name the metric unit of volume that would be most convenient to use in measuring each item.

| *Item to be Measured* | *The Most Convenient Metric Unit of Volume* |
|---|---|
| **a.** the capacity of a bathtub | the kiloliter |
| **b.** the amount of juice in a large picnic cooler | the liter |
| **c.** the amount of medicine in a bottle | the milliliter |

EXAMPLE 2 Make each of the required unit changes.

| *Required Unit Change* | *Conversion Information* | *Arithmetic* |
|---|---|---|
| **a.** Change 7,520 milliliters to liters. | To change from the smaller unit (milliliters) to the larger unit (liters), divide by 1,000 (move the decimal point three places to the left). | 7520 ÷ 1000 = 7.52<br>7.520. |
| *Answer:* 7,520 milliliters = 7.52 liters | | |
| **b.** Change 18.6 liters to milliliters. | To change from the larger unit (liters) to the smaller unit (milliliters), multiply by 1,000 (move the decimal point three places to the right). | 18.6 × 1000 = 18,600<br>18.600. |
| *Answer:* 18.6 liters = 18,600 milliliters | | |
| **c.** Change 86.3 liters to kiloliters. | To change from the smaller unit (liters) to the larger unit (kiloliters), divide by 1,000. | 86.3 ÷ 1000 = .0863<br>.086.3 |
| *Answer:* 86.3 liters = .0863 kiloliter | | |
| **d.** Change .124 kiloliter to liters. | To change from the larger unit (kiloliters) to the smaller unit (liters), multiply by 1,000. | .124 × 1000 = 124<br>.124. |
| *Answer:* .124 kiloliter = 124 liters | | |

## CLASS EXERCISES

1. Name the metric unit of volume that would be most convenient to use in measuring each item.

   **a.** the amount of cider in a large jug **b.** the capacity of a refrigerator

   **c.** the capacity of a car trunk **d.** the volume of a room

2. Change each volume to liters.

   **a.** 3,450 mL **b.** 785 mL **c.** 124.3 mL

3. Change each volume to milliliters.

   **a.** 9 L **b.** 32.5 L **c.** .0385 L

4. Find the number of kiloliters in the given number of liters.

   **a.** 8,000 L **b.** 660 L **c.** 8.25 L

5. Find the number of liters in the given number of kiloliters.

   **a.** 17 kL **b.** 8.5 kL **c.** .95 kL

6. The number of cubic centimeters in 5 milliliters is (a) .05 (b) .5 (c) 5 (d) 50

7. Dr. Gomez gave Anita a 2-cubic-centimeter injection of vaccine. How many milliliters of vaccine did he give her?

8. Joe and Barbara drank a half liter of soda with their pizza. How many milliliters of soda was this?

9. How many cubic meters of sand are there in 750 liters?

10. Mr. Gallo bought 4 liters of gasoline. How many milliliters was this?

## HOMEWORK EXERCISES

1. Name the metric unit of volume that would be most convenient to use in measuring each item.

   a. the amount of water in a small sink b. the contents of a tea bag

   c. the contents of a spoon d. the amount of cement needed to make a section of sidewalk

   e. the contents of a swimming pool f. the contents of an automobile gas tank

2. Change each volume to liters.

   a. 2,005 mL b. 583 mL c. 326.25 mL

3. Change each volume to milliliters.

   a. 20 L b. 49.4 L c. .0036 L

4. Find the number of kiloliters in the given number of liters.

   a. 7,750 L b. 11 L c. 95.4 L

5. Find the number of liters in the given number of kiloliters.

   a. 48 kL b. 1.04 kL c. .008 kL

6. How many cubic meters are there in 500 liters?
   (a) .5 cubic meter (b) 5 cubic meters (c) 50 cubic meters (d) 500 cubic meters

7. Rosalie bought a bottle of perfume containing 3.5 cubic centimeters. How many milliliters of perfume did it contain?

8. Mr. Charles bought 200 cubic meters of topsoil for his lawn. How many kiloliters was this?

9. At the Browns' New Year's Eve party, the guests drank 8 liters of club soda. How many milliliters was this?

10. A chemist measured out 2 liters of distilled water. How many cubic centimeters was this?

## SPIRAL REVIEW EXERCISES

1. Change each of the given measures of weight to the indicated unit.

   a. 540 g = ? kg b. 72 g = ? mg

   c. 4,000 g = ? t d. 25 mg = ? g

2. A cake weighs 1.5 kilograms. What is its weight in grams?

3. If it costs $.90 to ship one kilogram of freight, what is the cost of shipping a package weighing 3.5 kilograms?

4. If a bus ride takes 3 hr. 35 min. and the bus arrives at its destination at 8:15 P.M., at what time did it start out?

5. Subtract $3\frac{5}{8}$ from $6\frac{1}{2}$.

6. Find 250% of 64.

7. Round 241.7 to the nearest ten.

8. Carol bowled 3 games. If each game cost $1.35, how much change did she receive from a $5 bill?

9. Jim buys a watch by making a $25 down payment and paying 15 installments of $12 each. What is the total cost of the watch?

10. Subtract −10 from −40.

**CHALLENGE QUESTION**

A syrup concentrate is mixed with carbonated water, using one liter of syrup concentrate to 5 liters of carbonated water, to make a soft drink that sells for $1.10 a liter. When 8 liters of syrup concentrate are used in this way, what is the total value of the resulting soft drink?

## UNIT 12–5 Temperature

### THE MAIN IDEA

1. Two different scales for measuring temperature are the *Celsius (C)* scale and the *Fahrenheit (F)* scale.
2. On both scales, the unit of measure is called the *degree* (°), but a Celsius degree is almost twice as large as a Fahrenheit degree.

   Because Celsius degrees are larger units than Fahrenheit degrees, it takes fewer of them to measure the same amount of heat. A Celsius temperature above zero will always be a smaller number than the equivalent Fahrenheit temperature.
3. Here is a comparison of some familiar temperatures on both scales.

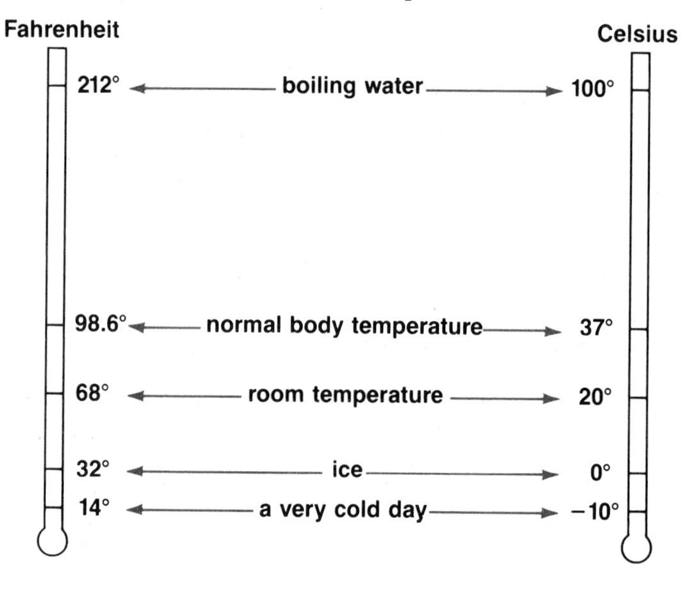


**EXAMPLE 1** On a very hot day, Jerry measured the temperature with a Celsius thermometer and a Fahrenheit thermometer at the same time. He jotted down the two temperatures, 35° and 95°, but forgot which was which. Which temperature was measured with a Celsius thermometer and which with a Fahrenheit thermometer?

For the same amount of heat above zero, the Celsius temperature will be a smaller number than the Fahrenheit temperature.

*Answer:* The temperature is 35°*C* or 95°*F*.

**EXAMPLE 2** State whether each temperature is probably Fahrenheit or Celsius.

**a.** Cooking a roast beef in an oven: 300°.

This is probably 300°*F*, higher than the temperature of boiling water. *Ans.* Fahrenheit

**b.** Swimming in the ocean on a summer day: 25°.
This is probably 25°*C*, since 25°*F* is colder than the temperature of ice. *Ans.* Celsius

**c.** A person's temperature during a fever: 102°.
This is 102°*F*, since 102°*C* would be hotter than boiling water. *Ans.* Fahrenheit

**EXAMPLE 3** Choose the more reasonable temperature.

**a.** A snowy day: 30°*C* or 30°*F*.
30°*F* is around the temperature of ice.
30°*C* is more than normal room temperature.

*Answer:* 30°*F*

**b.** A cup of hot soup: 90°*C* or 90°*F*.
90°*C* is a bit lower than the temperature of boiling water.
90°*F* is cooler than normal body temperature.

*Answer:* 90°*C*

**EXAMPLE 4** Determine whether the temperature change represents an increase or a decrease, and state the amount of change.

To determine the change in temperature, subtract the first value from the second.

| *Temperature Change* | *Subtraction* | *Result* |
|---|---|---|
| **a.** 40°*F* to 31°*F* | $31 - 40$<br>$= -9$ | decrease of 9° |
| **b.** 54.8°*C* to 56.7°*C* | $56.7 - 54.8$<br>$= 1.9$ | increase of 1.9° |
| **c.** −10°*F* to −3°*F* | $-3 - (-10)$<br>$= -3 + 10$<br>$= +7$ | increase of 7° |
| **d.** −4°*C* to 15°*C* | $15 - (-4)$<br>$= 15 + 4$<br>$= 19$ | increase of 19° |
| **e.** 20°*F* to −5°*F* | $-5 - 20$<br>$= -25$ | decrease of 25° |

## CLASS EXERCISES

**1.** Two scientists measured the temperature of the same body of water at the same time, one with a Celsius thermometer and one with a Fahrenheit thermometer. If their readings were 25° and 77°, which was the Fahrenheit temperature?

**2.** Mrs. White gave Mrs. Ross her recipe for apple pie. The recipe said, "Heat the oven to 150°," but it didn't tell whether this was 150° Celsius or Fahrenheit. Which was it, probably?

**3.** Ada's pen pal in France wrote to her and said that the temperature in her town was 30°*C*. Was it a hot day or a cold day?

**4.** In Montreal one day the temperature had dropped 10 Celsius degrees and in New York it had dropped 10 Fahrenheit degrees. In which city did the temperature drop more?

**5.** State whether each temperature is probably Fahrenheit or Celsius.

**a.** a warm spring day: 25° **b.** an iced drink: 35° **c.** a person's body temperature: 37°

**d.** room temperature in a school: 70° **e.** a cup of warm coffee: 150°

6. In each case, choose the more reasonable temperature.
   a. a cup of hot chocolate: 180°*C* or 180°*F*
   b. swimming pool water: 25°*C* or 25°*F*
   c. a person's forehead: 99°*C* or 99°*F*
   d. a chilly autumn day: 5°*C* or 5°*F*

7. In each case, find the change in temperature.
   a. from 15°*C* to 27°*C*
   b. from 85°*F* to 73°*F*
   c. from 48.6°*C* to 37.8°*C*
   d. from 101.5°*F* to 102.6°*F*
   e. from 72.5°*F* to 70.4°*F*
   f. from 10.1°*C* to 20°*C*

8. Find each change in temperature.
   a. from +18°*C* to +24°*C*
   b. from −32°*F* to +8°*F*
   c. from +11°*C* to −8°*C*
   d. from −21°*F* to −18°*F*
   e. from −5°*C* to −12°*C*
   f. from −14°*F* to 0°*F*

9. Find the change in temperature from the first thermometer reading to the second.

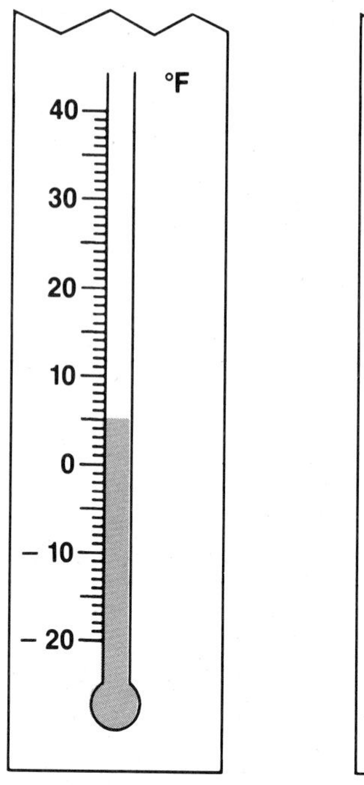


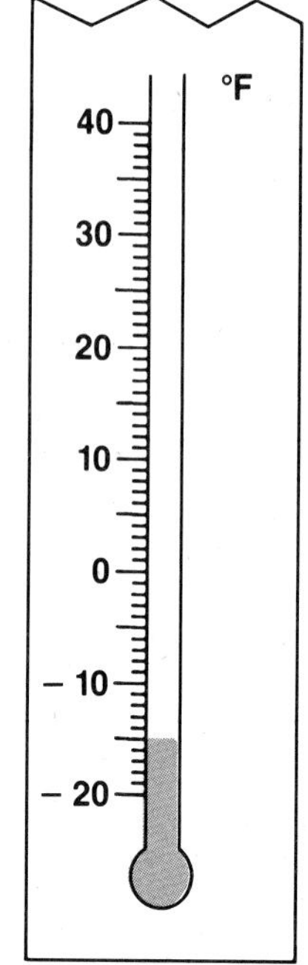


## HOMEWORK EXERCISES

1. A display sign outside a bank gave the temperature in both Celsius and Fahrenheit. If the sign showed 30° and 86°, which was the Celsius temperature?
2. The instruction manual for Mr. Henry's new oil burner told him to set his thermometer to 68°. Is it more likely that this meant 68° Celsius or 68° Fahrenheit?
3. The temperature of Joe's coffee was 40°*C*. Was the coffee warm or cold?
4. Which is a greater increase of heat, a rise of 5 Celsius degrees or a rise of 5 Fahrenheit degrees?
5. State whether each temperature is probably Fahrenheit or Celsius.
   a. the inside of a freezer: 10°
   b. the inside of a refrigerator: 40°
   c. a glass of iced tea: 38°
   d. a hot summer day: 35°
6. In each case, choose the more reasonable temperature.
   a. a pizza oven: 150°*C* or 150°*F*
   b. bath water: 30°*C* or 30°*F*
   c. aquarium water: 75°*C* or 75°*F*
   d. a hot meal: 65°*C* or 65°*F*
7. The temperature of Mrs. Sweeney's car engine changed from 75°*F* to 150°*F*. Find the change in temperature.
8. In each case, find the change in temperature.
   a. from 32°*F* to 15°*F*
   b. from 100°*C* to 117°*C*
   c. from 0°*C* to 15°*C*
   d. from 20.5°*C* to 11.4°*C*
   e. from 101.5°*F* to 102.6°*F*
   f. from 98°*F* to 89°*F*

**9.** In each case, find the change in temperature.

**a.** from $+15°C$ to $+7°C$ **b.** from $-12°F$ to $+8°F$ **c.** from $-34°F$ to $-27°F$

**d.** from $+14°C$ to $-26°C$ **e.** from $-28°C$ to $-37°C$ **f.** from $0°F$ to $-16°F$

**10.** Find the change in temperature from the first thermometer reading to the second.

**a.**

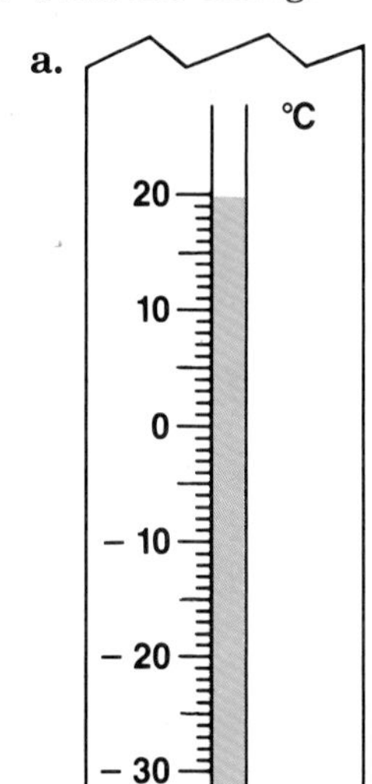


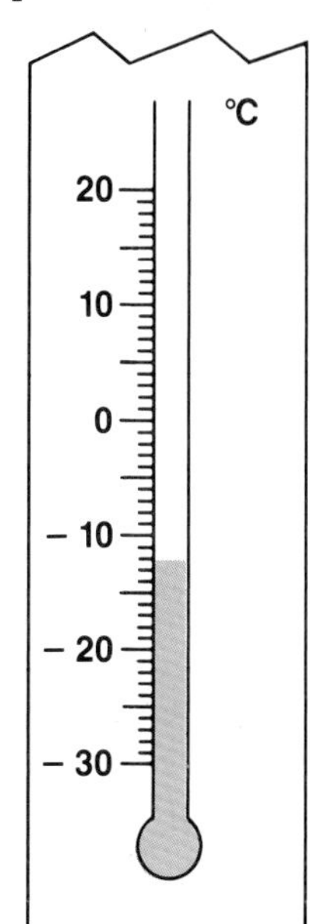


**b.**

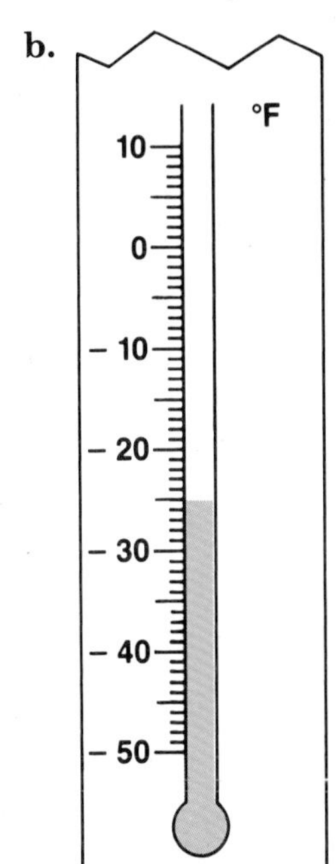


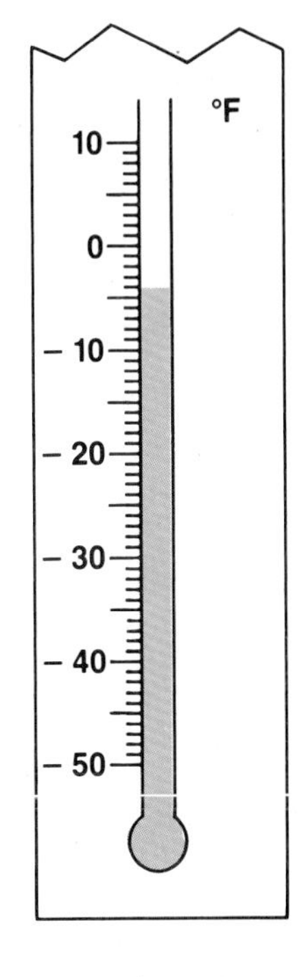


## SPIRAL REVIEW EXERCISES

**1.** Change each of the given measures of volume to the indicated unit.

**a.** 75 mL = ? L **b.** 7.2 kL = ? L

**c.** 14 L = ? mL **d.** 2,800 L = ? kL

**2.** Find 1% of 500. **3.** Add: $5\frac{3}{4} + 2\frac{3}{16}$

**4.** Subtract 11.7 from 19.2.

**5.** Divide: $\frac{7}{8} \div \frac{1}{2}$

**6.** The cost of a home computer went from $400 to $300. What was the percent decrease in price?

**7.** Write .93 as a percent.

**8.** $\frac{3}{8}$ written as a decimal is

(a) .125 (b) .25 (c) .375 (d) .5

**9.** Marla earns $5\frac{1}{2}\%$ annual interest on $500 savings. How much interest did she earn at the end of one year?

**10.** On a map, 4 cm represent 1 km. How many cm are used to represent 275 km?

**11.** Find the product of $-7$ and $-9$.

### CHALLENGE QUESTION

The temperature of a chemical solution drops 2 Celsius degrees every 30 minutes. The temperature at 11:25 A.M. was $55°C$. What will the temperature of the solution be at 2:55 P.M.?

THEME 7

# Ratio, Proportion, and Probability

## MODULE 13



## MODULE 14

# UNIT 13–1 The Meaning of Ratio; Using Ratios to Express Rates

## THE MEANING OF RATIO

### THE MAIN IDEA

1. A *ratio* is a comparison between two numbers.
2. A ratio can be written in three ways:
   a. by using the word "to" 2 to 3
   b. by using a colon, :, instead of the word "to" 2:3
   c. by writing a fraction in which the fraction bar replaces the word "to" $\frac{2}{3}$
3. A ratio is read using the word "to."

2 to 3, 2:3, and $\frac{2}{3}$ are all read "2 to 3."

**EXAMPLE 1** In an English class, there are 16 girls and 14 boys. In three different ways, write:

**a.** the ratio of girls to boys

**b.** the ratio of boys to girls

**THINKING ABOUT THE PROBLEM**

In writing a ratio, write the number of the item mentioned first as the first number of the ratio.

| *Ways to Write a Ratio* | *Ratio of Girls to Boys* | *Ratio of Boys to Girls* |
|---|---|---|
| Using the word "to" | 16 to 14 | 14 to 16 |
| Using a colon | 16:14 | 14:16 |
| Using a fraction | $\frac{16}{14}$ | $\frac{14}{16}$ |

**EXAMPLE 2** In a litter of brown puppies and black puppies, 3 were born brown and 5 were born black. What is the ratio of brown puppies to the total number of puppies?

To find the total number of puppies, add.

$$\begin{array}{rl} 3 & \leftarrow \text{brown} \\ +5 & \leftarrow \text{black} \\ \hline 8 & \leftarrow \text{total} \end{array}$$

Since the question "What is the ratio of . . .?" mentions brown puppies first, 3 is the first number in the ratio.

$$\frac{3}{8}$$

*Answer:* $\frac{3}{8}$ or 3:8 or 3 to 8

**EXAMPLE 3** In a bag containing 4 red marbles and 8 blue marbles, find:

**a.** the ratio of red marbles to the total number of marbles

**b.** the ratio of blue marbles to the total number of marbles

**c.** the ratio of red marbles to blue marbles

To find the total number of marbles, add.

$$\begin{array}{rl} 4 & \leftarrow \text{red} \\ +8 & \leftarrow \text{blue} \\ \hline 12 & \leftarrow \text{total} \end{array}$$

*Answers:* **a.** The ratio of red marbles to the total number of marbles is $\frac{4}{12}$ or $\frac{1}{3}$.

**b.** The ratio of blue marbles to the total number of marbles is $\frac{8}{12}$ or $\frac{2}{3}$.

**c.** The ratio of red marbles to blue marbles is $\frac{4}{8}$ or $\frac{1}{2}$.

## CLASS EXERCISES

1. Copy the chart shown and write each ratio in two other ways.

| Using "to" | Using ":" | As a Fraction |
|---|---|---|
| 5 to 8 | | |
| | 3:7 | |
| | | $\frac{4}{11}$ |

2. At Carla's party, there were 8 girls and 6 boys. What was the ratio of boys to girls?
3. Fieldston High School's football team won 7 games and lost 5 games. What was the ratio of the number of games won to the number of games lost?
4. In Bill's aquarium, there are 7 angelfish and 6 guppies. Find:

   **a.** the ratio of angelfish to the total number of fish

   **b.** the ratio of angelfish to guppies

   **c.** the ratio of guppies to angelfish

5. In Professor Wilson's library, there are 30 history books and 40 mathematics books. Find:
   a. the ratio of the number of history books to the number of mathematics books
   b. the ratio of the number of mathematics books to the total number of history books and mathematics books

6. In a box of candy, there are 5 chocolates, 3 marshmallows, and 4 caramels. Find:
   a. the ratio of the number of chocolates to the number of marshmallows
   b. the ratio of the number of chocolates to the number of caramels
   c. the ratio of the number of chocolates to the total number of candies

7. Linda bought 9 yellow balloons, 12 red balloons, and 20 white balloons to decorate her home for her party. Write a ratio for:
   a. the number of yellow balloons to the number of red balloons
   b. the number of red balloons to the number of white balloons
   c. the number of yellow balloons to the total number of balloons

## USING RATIOS TO EXPRESS RATES

### THE MAIN IDEA

A ratio can be used to compare two different kinds of quantities. For example:

A car traveling at a *rate* of 55 miles *per* hour compares the number of miles traveled, 55, to the number of hours traveled, 1.

**EXAMPLE 4** Express each ratio as a rate in simplest form.

| *Ratio* | *Rate* | *Rate in Simplest Form* |
|---|---|---|
| a. 300 miles in 6 hours | $\frac{300 \text{ miles}}{6 \text{ hours}}$ | $\frac{50}{1}$ or 50 miles per hour |
| b. 135 words in 3 minutes | $\frac{135 \text{ words}}{3 \text{ minutes}}$ | $\frac{45}{1}$ or 45 words per minute |
| c. \$2.80 for 14 pencils | $\frac{\$2.80}{14 \text{ pencils}}$ | $\frac{20}{1}$ or 20 cents per pencil |

## CLASS EXERCISES

1. Express each ratio as a rate in simplest form.

   a. 234 kilometers in 6 hours
   b. 400 words in 10 minutes
   c. $2.70 for 9 candy bars
   d. $23.75 for 5 hours
   e. 275 students for 11 teachers

2. If Nancy can do 30 push-ups in 6 minutes, what is Nancy's exercise rate?
3. If Ernesto can read 350 words in 7 minutes, what is Ernesto's reading rate?
4. If Evan earns $25.20 for 6 hours of work, what is Evan's rate of pay?
5. Mr. Hall rode 208 miles using 16 gallons of gasoline. At what rate did he use gasoline?
6. In 120 minutes of television viewing, Marcie counted 40 commercials. At what rate did the commercials appear?

## HOMEWORK EXERCISES

1. Copy the chart shown and write each ratio in two other ways.

| Using "to" | Using ":" | As a Fraction |
|---|---|---|
| 9 to 11 | | |
| | | $\frac{5}{12}$ |
| | 3:4 | |
| | | $\frac{15}{11}$ |

2. In Mr. Paulson's class, there are 17 boys and 15 girls. What is the ratio of the number of boys to the number of girls?
3. Newton's soccer team won 12 games and lost 7 games. What was the ratio of the number of games lost to the number of games won?
4. On a class test, 19 students passed and 7 students failed. Find:

   a. the ratio of the number of students who passed to the number of students who failed
   b. the ratio of the number of students who failed to the total number of students who took the test

5. In a pet store, there are 15 dogs and 10 cats. Find:

   a. the ratio of the number of cats to the number of dogs
   b. the ratio of the number of dogs to the total number of dogs and cats

6. In a basket of fruit, there were 7 apples, 5 pears, and 9 oranges. Write a ratio for:

   a. the number of apples to the number of oranges
   b. the number of oranges to the number of pears
   c. the number of pears to the total number of pieces of fruit

7. Mary sold 17 student tickets, 24 adult tickets, and 11 senior citizen tickets for the school play. Write a ratio for:

   a. the number of student tickets to the number of adult tickets

   b. the number of student tickets to the number of senior citizen tickets

   c. the number of senior citizen tickets to the total number of tickets

8. Express each ratio as a rate in simplest form.

   a. 249 miles in 5 hours
   b. 2 defective tires in every 100 tires produced
   c. 5 teachers for 60 students
   d. 1,000 miles using 40 gallons of gasoline
   e. 50 strikeouts in 35 innings
   f. 85 baskets in 100 attempts

9. John can do 300 sit-ups in 5 minutes. What is his exercise rate?

10. Milton can type 240 words in 8 minutes. What is his typing rate?

11. If Florence earns $48.60 for 12 hours of work, what is her rate of pay?

12. An accountant takes 48 hours to complete 8 tax forms. At what rate did the accountant complete the tax forms?

## SPIRAL REVIEW EXERCISES

1. Write each fraction in simplest form.

   a. $\frac{25}{35}$ b. $\frac{14}{32}$ c. $\frac{42}{28}$

   d. $\frac{10}{6}$ e. $\frac{55}{33}$ f. $\frac{62}{62}$

2. Mr. Homer's bank pays him 11% interest per year on his special savings account. If he has invested $320 in this account, how much interest will he earn in one year?

3. The Diaz family budget allows 75% for rent and food. If their monthly income is $800, how much does the budget allow for rent and food?

4. At its white sale, Mancy's department store is offering a 20% discount on designer sheets. If the original selling price of a queen size sheet is $16.99, how much will a customer pay after the discount?

5. Janice plans to save $\frac{1}{4}$ of the money she earns baby-sitting. This week Janice earned $56. How much money should she save?

6. Mr. Math bought a calculator by making a down payment of $10 and paying 6 installments of $9 each. The total cost of the calculator was
   (a) $60 (b) $54 (c) $64 (d) $96

7. Which is equal to $\frac{1}{2}$?
   (a) .5 (b) 5% (c) .05 (d) .5%

8. If one pair of shoes costs $21, then the cost of 5 pairs of shoes is
   (a) $26 (c) $110
   (b) $100 (d) $105

9. To the product of 8 and −3, add −9.

### CHALLENGE QUESTION

At Ana's party, there were 8 girls and 6 boys. At Maria's party, there were 12 girls and 9 boys. How did the ratio of the number of boys to the number of girls at Ana's party compare to the ratio of the number of boys to the number of girls at Maria's party?

# UNIT 13–2 The Meaning of Proportion; Determining If a Proportion Is True

## THE MEANING OF PROPORTION

### THE MAIN IDEA

1. A ***proportion*** is a statement that two ratios are equal.
2. A proportion tells that, in two different ratios, the numbers compare to each other in the same way.

   The proportion $\frac{3}{4} = \frac{9}{12}$ is read "3 *is to* 4 *as* 9 *is to* 12." This means
   3 compares to 4 in the same way that 9 compares to 12.
3. There are four numbers in a proportion. The first and last numbers are called the ***extremes***. The second and third numbers are called the ***means***.

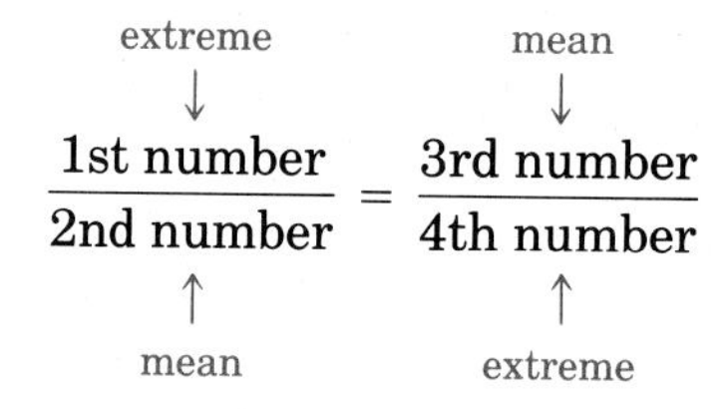


**EXAMPLE 1** Write each proportion in words.

| | *Proportion in Numbers* | *Proportion in Words* |
|---|---|---|
| **a.** | $\frac{3}{7} = \frac{9}{21}$ | 3 is to 7 as 9 is to 21. |
| **b.** | 5:4 = 20:16 | 5 is to 4 as 20 is to 16. |

**EXAMPLE 2** Write each statement as a proportion.

| *Statement* | *Proportion* |
|---|---|
| **a.** 8 is to 16 as 12 is to 24. | $\frac{8}{16} = \frac{12}{24}$ or 8:16 = 12:24 |
| **b.** 9 is to 3 as 15 is to 5. | $\frac{9}{3} = \frac{15}{5}$ or 9:3 = 15:5 |

**EXAMPLE 3** In each proportion, name the means and the extremes.

| | *Proportion* | *Means* | *Extremes* |
|---|---|---|---|
| **a.** | $\frac{9}{18} = \frac{1}{2}$ | 18 and 1 | 9 and 2 |
| **b.** | 20:4 = 5:1 | 4 and 5 | 20 and 1 |

## CLASS EXERCISES

**1.** Write each proportion in words.

**a.** $\frac{7}{28} = \frac{2}{8}$ **b.** $1:3 = 6:18$ **c.** $\frac{9}{72} = \frac{1}{8}$ **d.** $9:81 = 4:36$ **e.** $\frac{10}{2} = \frac{25}{5}$ **f.** $\frac{.5}{2} = \frac{1}{4}$

**2.** In each proportion, name the means and the extremes.

**a.** $\frac{4}{20} = \frac{2}{10}$ **b.** $9:27 = 2:6$ **c.** $\frac{40}{90} = \frac{8}{18}$ **d.** $27:54 = 50:100$

**e.** $\frac{25}{75} = \frac{50}{150}$ **f.** $81:18 = 18:4$

**3.** Write each statement as a proportion. Name the means and the extremes.

**a.** 4 is to 24 as 2 is to 12 **b.** 50 is to 6 as 25 is to 3

**c.** 4 is to 8 as 8 is to 16 **d.** 100 is to 25 as 8 is to 2

## DETERMINING IF A PROPORTION IS TRUE

### THE MAIN IDEA

A proportion is true if:

1. the two fractions are equivalent.
2. the ***cross products*** are equal.

**EXAMPLE 4** Tell if each proportion is *true* or *false*.

| | *Proportion* | *Are the Fractions Equivalent?* | *Are the Cross Products Equal?* | *Answer* |
|---|---|---|---|---|
| **a.** | $\frac{3}{12} = \frac{8}{32}$ | $\frac{3 \div 3}{12 \div 3} = \frac{1}{4}$ | $\frac{3}{12} \times \frac{8}{32}$ | true |
| | | $\frac{8 \div 8}{32 \div 8} = \frac{1}{4}$ | $12 \times 8 \stackrel{?}{=} 3 \times 32$ | |
| | | | $96 = 96$ ✓ | |
| | | yes | yes | |
| **b.** | $\frac{2}{5} = \frac{12}{30}$ | $\frac{2}{5}$ | $\frac{2}{5} \times \frac{12}{30}$ | true |
| | | $\frac{12 \div 6}{30 \div 6} = \frac{2}{5}$ | $5 \times 12 \stackrel{?}{=} 2 \times 30$ | |
| | | | $60 = 60$ ✓ | |
| | | yes | yes | |

| Proportion | Are the Fractions Equivalent? | Are the Cross Products Equal? | Answer |
|---|---|---|---|
| c. $\frac{3}{7} = \frac{2}{5}$ | $\frac{3}{7}$ <br> $\frac{2}{5}$ <br> no | $\frac{3}{7} \times \frac{2}{5}$ <br> $7 \times 2 \stackrel{?}{=} 3 \times 5$ <br> $14 = 15$ ✗ <br> no | false |

**EXAMPLE 5** Use the numbers 1, 3, 5, and 15 to write a true proportion.

Try the numbers in the arrangement shown. $\frac{1}{3} \stackrel{?}{=} \frac{5}{15}$

Since both ratios are equal to $\frac{1}{3}$, the proportion is true. $\frac{5 \div 5}{15 \div 5} = \frac{1}{3}$

or

Since the cross products are equal, the proportion is true.

$$\frac{1}{3} \times \frac{5}{15}$$

$$3 \times 5 \stackrel{?}{=} 1 \times 15$$

$$15 = 15 \checkmark$$

There are other arrangements of the four numbers that also form true proportions:

$$\frac{1}{5} = \frac{3}{15}, \frac{3}{1} = \frac{15}{5}, \text{ and } \frac{15}{3} = \frac{5}{1}$$

**EXAMPLE 6** Mona and Amy were comparing their monthly food budgets. Mona spent \$15 for snacks and \$125 for basic groceries. Amy spent \$12 for snacks and \$100 for basic groceries. Do these amounts form a true proportion?

Write the "skeleton" of a proportion that follows the word pattern.

$$\frac{\text{Mona's snacks}}{\text{Mona's groceries}} \stackrel{?}{=} \frac{\text{Amy's snacks}}{\text{Amy's groceries}}$$

Substitute the given numbers, following the pattern.

$$\frac{15}{125} \stackrel{?}{=} \frac{12}{100}$$

Test the cross products.

$$125 \times 12 \stackrel{?}{=} 15 \times 100$$

$$1500 = 1500 \checkmark$$

*Answer:* The comparison of Mona's and Amy's food budgets results in a true proportion.

## CLASS EXERCISES

1. Tell whether each proportion is *true* or *false*.

**a.** $\frac{7}{5} = \frac{28}{20}$ **b.** $100:5 = 50:2$ **c.** $\frac{40}{30} = \frac{8}{6}$ **d.** $11:66 = 4:24$ **e.** $\frac{2}{8} = \frac{8}{34}$

**f.** $\frac{4}{9} = \frac{8}{16}$ **g.** $\frac{27}{36} = \frac{24}{32}$ **h.** $15:10 = 10:6$ **i.** $90:60 = 2:3$

2. Use the numbers 2, 7, 10, and 35 to write four true proportions.

3. Use the numbers 45, 39, 30, and 26 to write four true proportions.

4. Joe and Cheryl made photographs of their families. Joe's photograph was 8 inches wide and 10 inches long. Cheryl's photograph was 11 inches wide and 14 inches long. Were the dimensions of the photographs in proportion?

5. Mr. Rand rode 200 miles using 16 gallons of gas. Mr. Burton rode 150 miles using 14 gallons of gas. Are these rates in proportion?

6. Herb is 60 inches tall and weighs 120 pounds. Ralph is 72 inches tall and weighs 144 pounds. Are these heights and weights in proportion?

7. The height of a building is 300 feet and its width is 200 feet. In Martha's model of the building, the height is 6 inches and the width is 5 inches. Did Martha make her model in proportion to the building?

## HOMEWORK EXERCISES

1. Write each proportion in words.

   **a.** $\frac{3}{8} = \frac{9}{24}$ **b.** $2:5 = 14:35$ **c.** $\frac{12}{48} = \frac{3}{12}$ **d.** $\frac{16}{48} = \frac{2}{6}$

   **e.** $50:10 = 5:1$ **f.** $\frac{40}{25} = \frac{16}{10}$

2. In each proportion, name the means and the extremes.

   **a.** $19:38 = 1:2$ **b.** $8:5 = 24:15$ **c.** $12:20 = 15:25$ **d.** $\frac{20}{16} = \frac{10}{8}$

   **e.** $\frac{8}{16} = \frac{16}{32}$ **f.** $\frac{55}{33} = \frac{10}{6}$

3. Write each statement as a proportion. Name the means and the extremes.

   **a.** 2 is to 3 as 8 is to 12 **b.** 22 is to 16 as 33 is to 24

   **c.** 9 is to 54 as 12 is to 72 **d.** 90 is to 36 as 135 is to 54

4. Tell whether each proportion is *true* or *false*.

   **a.** $\frac{3}{9} = \frac{6}{18}$ **b.** $28:7 = 64:16$ **c.** $95:100 = 17:20$ **d.** $\frac{60}{80} = \frac{14}{16}$ **e.** $\frac{200}{300} = \frac{24}{36}$

   **f.** $98:100 = 49:50$ **g.** $\frac{18}{24} = \frac{27}{34}$ **h.** $800:500 = 16:10$ **i.** $20:30 = 3:2$

5. Use the numbers 3, 5, 15, and 25 to write four true proportions.

6. Use the numbers 81, 63, 36, and 28 to write four true proportions.

7. Use the numbers 18, 24, 27, and 36 to write four true proportions.

8. Carol walked 8 miles in 3 hours. John walked 15 miles in 6 hours. Are these rates in proportion?

9. Bill bought 6 apples for \$.90. Jane bought 5 apples for \$.75. Are these rates in proportion?

10. In a basketball game, Maxine made 12 shots out of 25 attempts, and Rosa made 6 shots out of 13 attempts. Are these rates in proportion?

11. Elizabeth is 64 inches tall and weighs 120 pounds. Joan is 60 inches tall and weighs 100 pounds. Are these heights and weights in proportion?

12. On a map, 3 inches represent 90 miles. On a second map, 2 inches represent 60 miles. Are the two maps in proportion?

## SPIRAL REVIEW EXERCISES

1. In Mr. Benson's class, 5 students have blond hair, 8 students have brown hair, and 13 students have black hair. Write a ratio for:

   a. the number of students who have brown hair to the number of students who have black hair

   b. the number of students who have black hair to the number of students who have blond hair

   c. the number of students who have blond hair to the total number of students

2. Ms. Phelps drove 320 miles using 25 gallons of gasoline. What was the rate of gasoline use?

3. Carl bought 2 records at $4.99 each. How much change did Carl receive from a $20 bill?

4. Mr. Stein bought a dishwasher having a list price of $380 on an installment plan. If he made a 20% down payment and 15 installment payments of $25 each, how much did he pay for the dishwasher?

5. If sales tax is charged at the rate of 8%, what is the total cost of a suit priced at $84?

6. 36 is what percent of 48?

7. Bananas are 75.7% water. If a bunch of bananas weighs about 4 pounds, about how many pounds of water does the bunch of bananas contain?

8. The sum of 5.8 and .23 is
   (a) 81 (b) 8.1 (c) 6.03 (d) 5.93

9. Evaluate: $12 + \frac{1}{2} \times 10 \times 40$

## CHALLENGE QUESTION

Harry wants to enlarge a photograph that is now 5 inches wide and 7 inches long so that it will be 21 inches long. How wide will the enlargement be?

# UNIT 13–3 Solving Proportions

## THE MAIN IDEA

1. If one of the numbers in a proportion is unknown, represent it by using a letter. For example: $\frac{5}{20} = \frac{n}{48}$

2. To find an unknown number in a proportion:

   a. Write the cross products and set them equal to each other. (In a proportion, the product of the means is equal to the product of the extremes.)

   $$20 \times n = 5 \times 48$$
   $$20 \times n = 240$$

   b. Divide the multiplier of $n$ into the other cross product.

   $$n = \frac{240}{20}$$

   The result is the unknown number.

   $$n = 12$$

   c. Check this value by substituting it into the original proportion. The two fractions should be equivalent.

   $$\frac{5}{20} = \frac{n}{48}$$
   $$\frac{5}{20} \stackrel{?}{=} \frac{12}{48}$$
   $$\frac{5 \div 5}{20 \div 5} \stackrel{?}{=} \frac{12 \div 12}{48 \div 12}$$
   $$\frac{1}{4} = \frac{1}{4} \checkmark$$

**EXAMPLE 1** Find the value of $n$.

$$\frac{5}{7} = \frac{n}{42}$$

Write the cross products and set them equal to each other.

$$7 \times n = 5 \times 42$$
$$7 \times n = 210$$

Divide 7 into 210.

$$n = \frac{210}{7}$$
$$n = 30$$

Check the value of $n$ in the original proportion.

$$\frac{5}{7} = \frac{n}{42}$$
$$\frac{5}{7} \stackrel{?}{=} \frac{30}{42}$$
$$\frac{5}{7} \stackrel{?}{=} \frac{30 \div 6}{42 \div 6}$$

The two fractions are equivalent.

$$\frac{5}{7} = \frac{5}{7} \checkmark$$

*Answer:* 30

**EXAMPLE 2** From a class of 30 girls and 18 boys, a committee will be formed. The number of boys and girls on the committee should be proportional to the number of boys and girls in the class. If the committee will contain 3 boys, how many girls should be on the committee?

Use the words of the problem to think out a pattern.

$$\frac{\text{number of girls in class}}{\text{number of boys in class}} = \frac{\text{number of girls on committee}}{\text{number of boys on committee}}$$

Three of the four numbers are given. Let $n$ represent the missing number, the number of girls on the committee.

$$\frac{30}{18} = \frac{n}{3}$$

Set the cross products equal.

$$18 \times n = 30 \times 3$$
$$18 \times n = 90$$

Divide by the multiplier of $n$.

$$n = \frac{90}{18}$$
$$n = 5$$

Check the value of $n$ in the original proportion.

$$\frac{30}{18} = \frac{n}{3}$$
$$\frac{30}{18} \stackrel{?}{=} \frac{5}{3}$$
$$\frac{30 \div 6}{18 \div 6} \stackrel{?}{=} \frac{5}{3}$$

The two fractions are equivalent.

$$\frac{5}{3} = \frac{5}{3} \checkmark$$

*Answer:* There should be 5 girls on the committee.

**EXAMPLE 3** If Julie can type 200 words in 4 minutes, how many words can she type in 10 minutes?

Use the words of the problem to think out a pattern, letting $n$ represent the number of words in 10 minutes.

$$\frac{200 \text{ words}}{4 \text{ minutes}} = \frac{n \text{ words}}{10 \text{ minutes}}$$

Solve the proportion.

$$4 \times n = 2000$$
$$n = \frac{2000}{4}$$
$$n = 500$$

Check the value of $n$ in the original proportion.

$$\frac{200}{4} = \frac{n}{10}$$
$$\frac{200}{4} \stackrel{?}{=} \frac{500}{10}$$
$$\frac{200 \div 4}{4 \div 4} \stackrel{?}{=} \frac{500 \div 10}{10 \div 10}$$

The two fractions are equivalent.

$$\frac{50}{1} = \frac{50}{1} \checkmark$$

Note that the check shows the rate at which Julie types, 50 words per minute.

*Answer:* Julie can type 500 words in 10 minutes.

## CLASS EXERCISES

1. In each proportion, find the value of $n$.

   a. $\frac{6}{9} = \frac{n}{72}$  b. $8:3 = n:15$  c. $\frac{n}{55} = \frac{3}{5}$  d. $\frac{16}{n} = \frac{28}{35}$  e. $n:12 = 5:4$

   f. $33:n = 11:5$  g. $7:13 = 21:n$  h. $\frac{8}{3} = \frac{56}{n}$  i. $.7 : 1.4 = 7 : n$

2. A recipe calls for 4 cups of flour and 2 tablespoons of sugar. Mrs. James wants to increase the recipe so that it will use 28 cups of flour. How much sugar should she use?

3. In a certain country, the banks exchange 5 buckos for every 4 U.S. dollars. How many buckos will they exchange for 52 U.S. dollars?

4. Jenny thought that 3 boys for every 2 girls was a good ratio for her party. She invited 8 girls. How many boys should she invite?

5. Mr. Harris knew that his car could go 14 miles for every gallon of gasoline. How far could he travel using 12 gallons?

6. If 100 children eat 20 pounds of cheese for lunch in the school cafeteria, how much cheese should the dietician order for 250 children?

## HOMEWORK EXERCISES

1. In each proportion, find the value of $n$.

   a. $\frac{n}{12} = \frac{21}{36}$  b. $\frac{20}{n} = \frac{4}{5}$  c. $\frac{45}{50} = \frac{n}{20}$  d. $\frac{8}{12} = \frac{14}{n}$  e. $\frac{5}{8} = \frac{n}{72}$

   f. $12:15 = 30:n$  g. $5:4 = n:28$  h. $\frac{n}{1.5} = \frac{36}{27}$  i. $\frac{30}{n} = \frac{3}{.5}$

   j. $18:12 = 24:n$  k. $2.4 : n = 1.2 : 3.6$  l. $\frac{200}{120} = \frac{n}{150}$

2. If a recipe for 4 loaves of bread uses 3 tablespoons of honey, how much honey will be needed for 6 loaves?

3. A psychologist formed 2 groups of subjects in which the numbers of men and women were proportional. In one group, she placed 8 men and 5 women. In the second group, she placed 40 men. How many women were in the second group?

4. A nurse usually prepares a certain medicine by mixing 3 cc of water with 7 cc of liquid. How much water should he use with 98 cc of liquid?

5. If Paul can read 6 pages in 24 minutes, how many pages can he read in 60 minutes?

6. It has been raining at the rate of .14 inch per hour. At this rate, how much will it rain in 24 hours?

## *SPIRAL REVIEW EXERCISES*

1. On a car lot, there were 24 American-made cars and 14 imported cars. What was the ratio of imported cars to the total number of cars?

2. Use the numbers 28, 5, 4, and 35 to write four true proportions.

3. What is the cost of 5 records if each record costs $5.99?

4. $5\frac{1}{2}\%$ written as a decimal is
   (a) .55 (c) .055
   (b) .05 (d) .0055

5. $\frac{3}{20}$ written as a percent is
   (a) 35% (c) 15%
   (b) 25% (d) 5%

6. Ms. Jones bought 3 items priced at $4.80, $11.20, and $6.40. If sales tax is charged at the rate of 6%, what was the total cost of her purchases?

7. What is the total cost of a car if the down payment is $600 and there are 24 installments of $300 each?

8. What is the price of each pen if a dozen pens cost $8.28?

9. The first word printed on a T-shirt costs $3 and each additional word costs $.75. What is the cost of a 5-word message?

10. Which is a composite number?
    (a) 2 (b) 7 (c) 11 (d) 15

11. Find the sum of $11\frac{2}{3}$ and $\frac{1}{6}$.

12. How many 18-inch-long pieces can be cut from 6,300 inches of cable?

13. 28 is 50% of
    (a) 14 (b) 56 (c) 2.8 (d) 50

14. The reciprocal of $-1$ is
    (a) $-1$ (b) 1 (c) 0 (d) $-\frac{1}{2}$

## CHALLENGE QUESTION

Explain why you cannot enlarge a 4-inch by 5-inch photograph to an 11-inch by 14-inch size.

# UNIT 13–4 Using Proportions in Scale Drawings

## THE MAIN IDEA

1. A *scale drawing* is a drawing in which all the dimensions of the actual objects are reduced or enlarged proportionally. For example, a map is a scale drawing.
2. The ratio between any actual length and the corresponding length in the drawing is always the same ratio.
3. To find an unknown length in a scale drawing, or an actual distance, write and solve a proportion.

**EXAMPLE 1** The 30-kilometer distance between Avon and Boyle is represented on a map by a line segment 2 centimeters long. The distance between Boyle and Carson is represented by a line segment 3 centimeters long. What is the actual distance between Boyle and Carson?

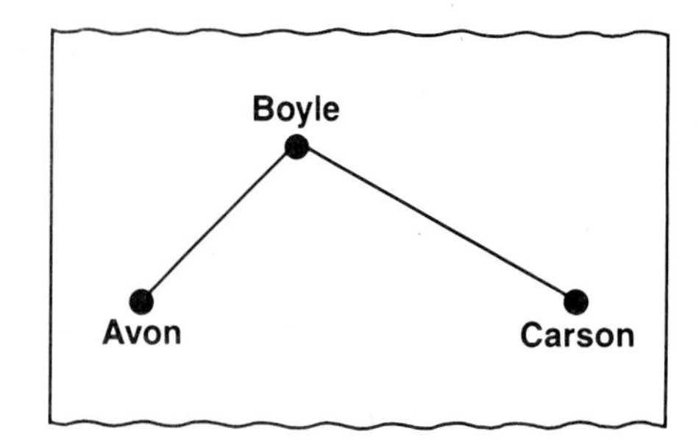


Use the words of the problem to think out a pattern.

$$\frac{\text{1st actual distance}}{\text{1st map length}} = \frac{\text{2nd actual distance}}{\text{2nd map length}}$$

Three of the four numbers are given. Let $n$ represent the missing number, the 2nd actual distance.

$$\frac{30 \text{ km}}{2 \text{ cm}} = \frac{n \text{ km}}{3 \text{ cm}}$$

Solve the proportion: set the cross products equal; divide by 2.

$$2 \times n = 90$$

$$n = \frac{90}{2}$$

$$n = 45$$

*Answer:* The actual distance between Boyle and Carson is 45 miles.

**EXAMPLE 2** What is the approximate distance between Acton and Bream?

(a) 40 km (c) 80 km
(b) 60 km (d) 100 km

According to the scale shown on the bottom of the map, 1 centimeter represents 24 kilometers.

Write a proportion.

$$\frac{\text{scale length}}{\text{actual distance}} = \frac{\text{map length}}{\text{unknown distance}}$$

Use the centimeter scale shown on the map to estimate the map length between Acton and Bream.

The length is about 2.5 cm.

Write a proportion using the three known numbers, and using $n$ to represent the unknown distance.

$$\frac{1 \text{ cm}}{24 \text{ km}} = \frac{2.5 \text{ cm}}{n \text{ km}}$$

Solve the proportion by cross-multiplying.

$$1 \times n = 24 \times 2.5$$
$$n = 60$$

*Answer:* (b)

**EXAMPLE 3** On a certain scale drawing, all the dimensions are shown $\frac{1}{16}$ actual size. What is the actual length of a desk that is 3 inches on the drawing?

Use the words of the problem to think out a pattern.

$$\frac{\text{drawing length}}{\text{actual length}} = \frac{1}{16}$$

Let $n$ represent the missing length.

$$\frac{3 \text{ in.}}{n \text{ in.}} = \frac{1}{16}$$

Solve the proportion.

$$1 \times n = 48$$
$$n = 48$$

*Answer:* The actual length is 48 in.

**EXAMPLE 4** On a diagram, a doorway is represented by a rectangle that is 2 inches wide and $4\frac{1}{2}$ inches high. If the width of the actual doorway is 4 feet, what is the height of the actual doorway?

Write a proportion.

$$\frac{\text{diagram width}}{\text{diagram height}} = \frac{\text{actual width}}{\text{actual height}}$$

Let $n$ represent the actual height.

$$\frac{2 \text{ in.}}{4\frac{1}{2} \text{ in.}} = \frac{4 \text{ ft.}}{n \text{ ft.}}$$

Solve the proportion.

$$2 \times n = 4\frac{1}{2} \times 4$$

$$2 \times n = \frac{9}{\cancel{2}_1} \times \cancel{4}^2$$

$$2 \times n = 18$$

$$n = \frac{18}{2}$$

$$n = 9$$

*Answer:* The height of the actual doorway is 9 ft.

**EXAMPLE 5** A stamp of length 20 mm and width 30 mm is enlarged to be reproduced in a book. If the length of the enlargement is $1\frac{1}{3}$ in., what is the width of the enlargement?

Write a proportion.

$$\frac{\text{stamp length}}{\text{stamp width}} = \frac{\text{enlarged length}}{\text{enlarged width}}$$

Let $n$ represent the enlarged width.

$$\frac{20 \text{ mm}}{30 \text{ mm}} = \frac{1\frac{1}{3} \text{ in.}}{n \text{ in.}}$$

Solve the proportion.

$$20 \times n = 30 \times 1\frac{1}{3}$$

$$20 \times n = \overset{10}{\cancel{30}} \times \frac{4}{\underset{1}{\cancel{3}}}$$

$$20 \times n = 40$$

$$n = \frac{40}{20}$$

$$n = 2$$

*Answer:* The width of the enlargement is 2 in.

## CLASS EXERCISES

1. A length of 15 miles is represented on a scale drawing by a line segment 3 centimeters long. What length is represented by a line segment 3.5 centimeters long?
2. The width of an island that is actually 100 mi. wide is represented on a map by a line segment 2 in. long. If the length of the island is represented by a line segment 3 in. long, how long is the island?
3. How would you represent a distance of 75 kilometers on a map that represents 25 kilometers by a line segment 4 centimeters long?
4. A scale drawing shows all dimensions $\frac{1}{20}$ actual size. How long is a steel bar that is represented by a line segment $2\frac{1}{2}$ inches long?
5. A truck is 32 feet long and 8 feet high. In a photograph, the length of the truck is 5 inches. How high is the truck in the photograph?
6. Stephanie made an accurate model of an airplane. In her model, the length of the airplane is 11 inches and the wingspread is 9 inches. If the wingspread of the actual airplane is 27 feet, what is its length?
7. For his science project, Mark made a giant model of a bacterium. The actual length of the bacterium is .05 inch and its width is .02 inch. Mark wants his model to be 24 centimeters long. How wide should he make it?
8. A computer chip is $\frac{1}{4}$ inch wide and $\frac{1}{2}$ inch long. The diagram for the chip is 36 inches long. How wide is the diagram?

## HOMEWORK EXERCISES

1. On a map, a distance of 200 miles is represented by a line segment 5 inches long. What distance is represented by a line segment 4 inches long on this map?
2. A length of 35 cm is shown as a line segment 2 cm long on a scale drawing. If a length is shown as a line segment 7 cm long on this drawing, what is its actual size?
3. On an architect's blueprint, a room that is 12 feet wide and 20 feet long is represented by a rectangle that is 5 cm long. How wide is the rectangle?
4. On a scale drawing, every length is shown $\frac{1}{12}$ actual size. What is the length of a line segment that represents a length of 30 feet?
5. How would you represent a 42-foot length on a scale drawing that represents a 20-foot length by a line segment that is 2 centimeters long?
6. A sheet of paper that is $8\frac{1}{2}$ inches wide and 11 inches long is to be photographed and reduced so that the width of the photograph will be $4\frac{1}{4}$ inches. How long will the photograph be?
7. A man is 6 feet tall and his daughter is 4 feet tall. In a photograph, the man's height is $2\frac{1}{2}$ inches. What is the height of his daughter in the photograph?
8. On a sign, a bottle that is actually 15 inches tall is shown 10 feet tall. If the bottle is actually 6 inches wide, how wide is it on the sign?

## SPIRAL REVIEW EXERCISES

1. In each proportion, find the value of $n$.

   **a.** $\frac{n}{32} = \frac{33}{24}$ **b.** $6:n = 9:3$

   **c.** $\frac{27}{9} = \frac{9}{n}$ **d.** $1.5:15 = n:40$

2. Ms. Philrup makes a salary of $165 a week and a 2% commission on her gasoline sales. Last week, she sold $2,580 worth of gasoline. Find her total earnings.
3. The baby-sitter at the health club charges $1.00 for the first hour and $.25 for each additional hour or part of an hour. Joy left her daughter with the baby-sitter for $2\frac{1}{2}$ hours. How much did this cost her?
4. If the population of Trenton, New Jersey, was about 105,000 one year and about 94,500 ten years later, by what percent did the population decrease in those 10 years?
5. Joel got 15 strikes in 60 frames of bowling. What percent of his frames bowled were strikes?
6. Mr. Lopez buys a diamond ring by making a $300 down payment and 24 installments of $50 each. The total cost of the ring is
   (a) $1,200 (c) $900
   (b) $1,500 (d) $1,600
7. If the rate of sales tax is 4%, the amount of tax charged on a $15 purchase is
   (a) $6.00 (c) $15.60
   (b) $.42 (d) $.60

**8.** 24% written as a fraction is
(a) $\frac{12}{100}$ (b) $\frac{13}{50}$ (c) $\frac{6}{25}$ (d) $\frac{24}{50}$

**9.** 16 is 25% of
(a) 4 (b) 8 (c) 32 (d) 64

**10.** Find the product of $\frac{28}{39}$ and $\frac{26}{35}$.

**11.** Find the greatest common factor of 60 and 45.

**12.** When 58 is divided by 7, the remainder is
(a) 2 (b) 3 (c) 5 (d) 6

## CHALLENGE QUESTION

In the past 5 years, John's weight increased from 130 pounds to 170 pounds and his height increased from 64 inches to 70 inches. Which showed the greater increase, his height or his weight?

# UNIT 14–1 The Meaning of Probability

## THE MAIN IDEA

1. A *probability* is a number that describes how certain it is that a particular event will happen.
2. A probability is expressed as the *ratio* of the number of successful ways that an event can happen to the total number of ways that the event can happen:

$$\textit{probability} = \frac{\textit{number of successful ways}}{\textit{total number of ways}}$$

3. To find a probability:
   a. Count the number of successful ways that the event can happen.
   b. Count the total number of ways that the event can happen.
   c. Form a ratio that compares the number of successful ways to the total number of ways.

**EXAMPLE 1** What is the probability that a card picked at random from a standard deck will be a king?

### THINKING ABOUT THE PROBLEM

Some of the information that you need to know in order to solve probability problems involving playing cards is not stated. "At random" means to pick without looking. In a standard deck, there is a total of 52 cards divided so that there are 4 suits (hearts, diamonds, clubs, and spades). There are 13 cards in each suit (2, 3, 4, 5, 6, 7, 8, 9, 10, jack, queen, king, and ace). Two suits are red: hearts and diamonds; two suits are black: clubs and spades.

$$\text{probability} = \frac{\text{number of successful ways}}{\text{total number of ways}}$$

Write the number of kings in a standard deck. 4

Write the total number of cards. 52

Write the probability as a ratio. $\frac{4}{52}$ $\begin{matrix}\leftarrow \text{number of kings} \\ \leftarrow \text{total number of cards}\end{matrix}$

Simplify the fraction. $\frac{4 \div 4}{52 \div 4} = \frac{1}{13}$

*Answer:* The probability of picking a king at random from a standard deck of cards is $\frac{4}{52}$ or $\frac{1}{13}$.

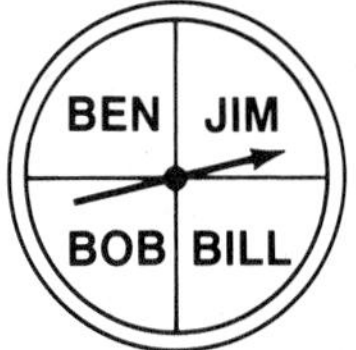


**EXAMPLE 2** What is the probability that the arrow on the spinner shown will stop on Jim's space?

$$\text{probability} = \frac{\text{number of successful ways}}{\text{total number of ways}}$$

Count the number of spaces that are labeled "Jim." 1

Count the total number of spaces. 4

Write the probability as a ratio. $\frac{1}{4}$ ← number of spaces labeled "Jim" / ← total number of spaces

*Answer:* The probability that the arrow will stop on Jim's space is $\frac{1}{4}$.

**EXAMPLE 3** What is the probability that the arrow on the spinner in Example 2 will stop on a space with a name that begins with the letter "B"?

$$\text{probability} = \frac{\text{number of successful ways}}{\text{total number of ways}}$$

Count the number of spaces with a name that begins with the letter "B." 3

Count the total number of spaces. 4

Write the probability as a ratio. $\frac{3}{4}$ ← number of spaces beginning with "B" / ← total number of spaces

*Answer:* The probability that the arrow will stop on a space with a name that begins with the letter "B" is $\frac{3}{4}$.

**EXAMPLE 4** A box contains 3 red marbles and 2 yellow marbles. What is the probability that a marble chosen at random from that box will be red?

$$\text{probability} = \frac{\text{number of successful ways}}{\text{total number of ways}}$$

Write the number of red marbles. 3

Find the total number of marbles.

3 ← red marbles
+2 ← yellow marbles
5 ← total number of marbles

Write the probability as a ratio. $\frac{3}{5}$ ← number of red marbles / ← total number of marbles

*Answer:* The probability that a red marble will be chosen is $\frac{3}{5}$.

**EXAMPLE 5** What is the probability that a month of the year chosen at random will have a name beginning with the letter "J"? Write the answer as a fraction, as a decimal, and as a percent.

**THINKING ABOUT THE PROBLEM**

It would be helpful to list the names of the months of the year:

January, February, March, April, May, June, July,
August, September, October, November, December

Write the ratio of the number of successful ways to the total number of ways.

$\frac{3}{12}$ ← January, June, July / ← total number of months

Simplify the fraction.

$\frac{3 \div 3}{12 \div 3} = \frac{1}{4}$

Change $\frac{1}{4}$ to a decimal by dividing 4 into 1, or remember the decimal value of $\frac{1}{4}$.

$$\begin{array}{r} .25 \\ 4\overline{)1.00} \\ \underline{8}\downarrow \\ 20 \\ \underline{20} \\ 0 \end{array}$$

Change .25 to a percent by moving the decimal point 2 places to the right and adding a percent sign.

$.25. = 25\%$

*Answer:* The probability of randomly picking a month with a name that begins with "J" is $\frac{1}{4}$ or .25 or 25%.

**EXAMPLE 6** The ratio of the number of boys to the total number of students in a class is 4:5. If a student is chosen at random from that class, what is the probability that the student will be a boy?

The probability of choosing a boy is the same as the ratio of the number of boys to the total number of students.

4:5

Write 4:5 as a fraction.

$\frac{4}{5}$

*Answer:* The probability of choosing a boy is $\frac{4}{5}$.

**EXAMPLE 7** In a lot of 50 cars, 30 are used cars. If a car is chosen at random from that lot, what is the probability that it will be a new car?

**THINKING ABOUT THE PROBLEM**

Use the strategy of breaking a problem into smaller, separate problems.

Find the number of new cars.

$$\begin{array}{rl} 50 & \leftarrow \text{total number of cars} \\ -30 & \leftarrow \text{number of used cars} \\ \hline 20 & \leftarrow \text{number of new cars} \end{array}$$

Write the probability of choosing a new car.

$\frac{20}{50}$ ← number of new cars / ← total number of cars

Simplify the fraction.

$\frac{20 \div 10}{50 \div 10} = \frac{2}{5}$

*Answer:* The probability of choosing a new car is $\frac{2}{5}$.

## CLASS EXERCISES

1. A box contains 25 coins, of which 7 are nickels. If a coin is chosen at random from that box, what is the probability of choosing a nickel?

2. Find the probability that the arrow on the spinner shown will stop on the space marked with:

   a. "A" b. "B" c. "C"

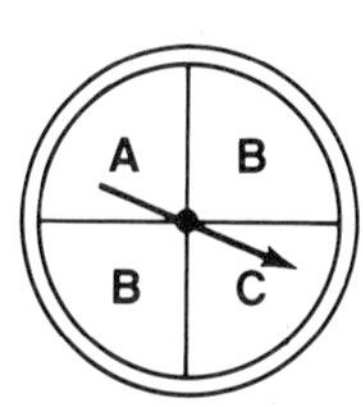


3. In a room containing 32 people, 11 persons are wearing blue shirts. What is the probability that the first person leaving the room will be wearing a blue shirt?

4. In our alphabet, there are 26 letters, of which 5 are vowels, and the rest are consonants. Each letter is written on a separate card and placed in a box. If a card is chosen at random from the box, what is the probability that a vowel is chosen?

5. There are 4 aces in a standard deck of 52 cards. If the cards have been shuffled, what is the probability that an ace will be the top card?

6. A box contains only 3 red marbles, 5 yellow marbles, and 4 green marbles. If a marble is chosen at random from the box, what is the probability of choosing:

   a. a red marble b. a green marble c. a yellow marble

7. In **a–e,** a probability is given as a fraction. If possible, simplify the given fraction and write each given probability as a decimal and as a percent.

   a. The probability of choosing a red card from a standard deck is $\frac{26}{52}$.

   b. The probability of choosing a 3 from the digits 0, 1, 2, 3, 4, 5, 6, 7, 8, 9 is $\frac{1}{10}$.

   c. The probability of choosing an "m" from the letters of the word "mummy" is $\frac{3}{5}$.

   d. The probability of choosing a girl from a class of 12 girls and 24 boys is $\frac{12}{36}$.

   e. The probability of choosing an even number from the integers 1 through 25 is $\frac{12}{25}$.

8. What is the probability that a card chosen at random from a standard deck of cards will be:

   a. a black card b. a heart

   c. a picture card (count only jacks, queens, and kings as picture cards)

   d. a number card (count an ace as the number 1)

9. What is the probability that a pet chosen at random from among all the pets in a certain animal shelter will be a dog if the ratio of the number of dogs to the total number of pets is 3:5? Write the answer as a fraction, as a decimal, and as a percent.

10. In a bakery window, there are 16 cakes. Twelve of these are chocolate cakes. What is the probability that a cake chosen at random from the window will not be a chocolate cake?

## HOMEWORK EXERCISES

1. There are 13 hearts in a standard deck of 52 cards. If a card is chosen at random from the deck, find the probability of choosing a heart.

2. A box contains 14 marbles, of which 7 are red. If a marble is chosen at random from the box, what is the probability of choosing a red marble?

3. A drawer contains 10 pairs of socks, of which 6 pairs are brown. If a pair of socks is drawn at random from the drawer, what is the probability of drawing a pair of brown socks?

4. Find the probability that the arrow on the spinner shown will stop on the red space.

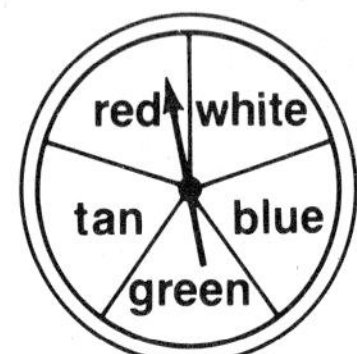


5. Find the probability that the arrow on the spinner shown will stop on an even number.

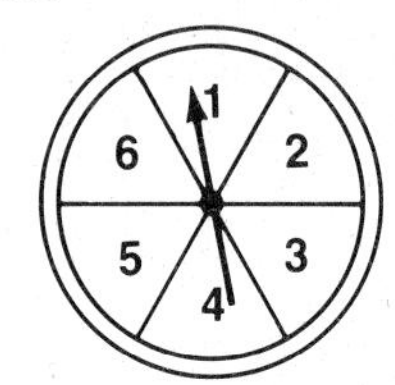


6. Find the probability that the arrow on the spinner shown in Exercise 5 will stop on a multiple of 3.

7. What is the probability that a month of the year chosen at random will have a name containing 4 letters?

8. What is the probability that a day of the week chosen at random will have a name that begins with the letter "S"?

9. The ratio of the number of boys to the total number of guests at Judd's party is 3:8. If a person is chosen at random from these guests, what is the probability that the person will be a boy?

10. The ratio of the number of chocolate cakes to the total number of cakes in a bakeshop window is 4:9. If a cake is chosen at random from this window, what is the probability that it will be a chocolate cake?

11. If a marble is chosen at random from a bag containing 8 red marbles and 10 green marbles, what is the probability of choosing a green marble?

12. Mrs. Jones has 5 nickels and 10 quarters in her purse. If she picks a coin at random from her purse, the probability that it will be a nickel is

    (a) $\frac{5}{10}$ (b) $\frac{1}{3}$ (c) $\frac{10}{15}$ (d) $\frac{10}{5}$

13. Write the probability of each event as a fraction in simplest form, as a decimal, and as a percent.

    **a.** selecting a 5 from the numbers 1, 5, 10, 15, and 20

    **b.** choosing a "T" from the letters of the word "TURTLE"

    **c.** choosing a boy from a group of 14 girls and 6 boys

    **d.** choosing a black card from a standard deck of cards

14. A record collection contains 8 jazz records and 24 classical records. The probability that a record chosen at random from this collection is a classical record is

    (a) 75% (b) 50% (c) $33\frac{1}{3}\%$ (d) 25%

**15.** If a letter is chosen at random from the letters of the word "SLEEP," the probability of choosing an "E" is
(a) .2 (b) .4 (c) .5 (d) .6

**16.** There is a 15% chance that it will rain. Express this probability as a fraction in simplest form.

**17.** Change each of the given probabilities, which are expressed as percents, into fractions in simplest form.

**a.** 30% **b.** 80% **c.** 50% **d.** 41% **e.** 32%

**18.** The probability of Carl's getting a base hit is .250. Express this probability as a fraction in simplest form.

**19.** Change each of the given probabilities, which are expressed as decimals, into fractions in simplest form.

**a.** .9 **b.** .75 **c.** .37 **d.** .005 **e.** .85

## SPIRAL REVIEW EXERCISES

**1.** 2.4 m is the same as
(a) 24 cm (c) 2,400 cm
(b) 240 cm (d) .24 cm

**2.** Mr. Fein began work at 8:25 A.M. and finished at 4:55 P.M. How many hours did he work?

**3.** From a group of 40 students, 12 of them went to summer school. What percent of them went to summer school?

**4.** Round 135.967 to the nearest:

**a.** hundred **b.** hundredth

**5.** Find the product of $2\frac{3}{5}$ and 20.

**6.** Find the sum of 8.9, .06, and 12.39.

**7.** The largest value of the choices below is
(a) $\frac{1}{7}$ (b) 7 (c) 7.7 (d) 77%

## CHALLENGE QUESTION

If one card is drawn at random from a standard deck, what is the probability that it is an ace or a red card?

## UNIT 14–2 Probabilities of 0 and 1

### THE MAIN IDEA

1. If there are *no successful ways* for an event to happen, then the probability is 0.

$$\text{probability} = \frac{\text{number of successful ways}}{\text{total number of ways}}$$

$$= \frac{0}{\text{total}} = 0$$

This means that the event is *impossible.*

2. If *every way is a successful way* for an event to happen, then the probability is 1.

$$\text{probability} = \frac{\text{number of successful ways}}{\text{total number of ways}}$$

$$= \frac{\text{total}}{\text{total}} = 1$$

This means that the event is *certain.*

3. The smallest possible probability is 0. The largest possible probability is 1. All other probabilities are numbers between 0 and 1.

**EXAMPLE 1** What is the probability that the arrow on the spinner shown will stop on a space marked "John"?

$$\text{probability} = \frac{\text{number of successful ways}}{\text{total number of ways}}$$

There are no spaces marked "John," and there are 4 spaces in all.

$$\text{probability} = \frac{0}{4}$$

$$= 0$$

*Answer:* The probability that the arrow will stop on a space marked "John" is 0. (The event is impossible.)

**EXAMPLE 2** What is the probability that the arrow on the spinner shown will stop on an odd number?

$$\text{probability} = \frac{\text{number of successful ways}}{\text{total number of ways}}$$

All 4 of the numbers are odd.

$$\text{probability} = \frac{4}{4}$$

$$= 1$$

*Answer:* The probability that the arrow will stop on an odd number is 1. (The event is certain.)

**EXAMPLE 3** What is the probability that if a day of the week is chosen at random its name will have 4 letters?

### THINKING ABOUT THE PROBLEM

It would be helpful to list the names of the days of the week:

Monday, Tuesday, Wednesday, Thursday, Friday, Saturday, Sunday

$$\text{probability} = \frac{\text{number of successful ways}}{\text{total number of ways}}$$

There are no days of the week with a name having 4 letters.

$$\text{probability} = \frac{0}{7}$$
$$= 0$$

*Answer:* The probability of randomly choosing a day with a four-letter name is 0.

**EXAMPLE 4** What is the probability that a month of the year chosen at random will have fewer than 32 days?

### THINKING ABOUT THE PROBLEM

Some of the information that you need to know in order to solve this problem is not stated. The number of days in a month can only be 28, 29, 30, or 31.

$$\text{probability} = \frac{\text{number of successful ways}}{\text{total number of ways}}$$

All the months have fewer than 32 days.

$$\text{probability} = \frac{12}{12}$$
$$= 1$$

*Answer:* The probability of randomly choosing a month with fewer than 32 days is 1.

## CLASS EXERCISES

1. What is the probability that the arrow on the spinner shown will stop on an even number?

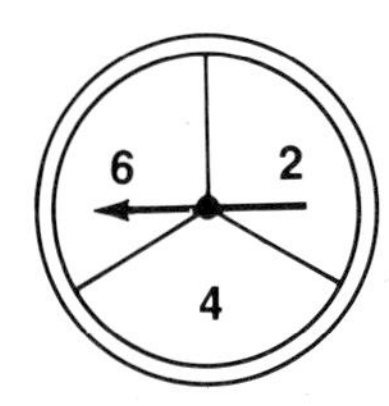


2. What is the probability of picking a green marble at random from a box that contains only 11 blue marbles?
3. What is the probability that a day of the week chosen at random will have a name that begins with the letter "X"?
4. If a bag contains only 5 nickels and 7 dimes, what is the probability of picking, at random, a quarter?

5. Using the spinner shown, find the probability that the arrow will stop on:
   a. a name that has 3 letters
   b. a name that begins with the letter "A"
   c. a name that begins with the letter "O"
   d. a name that contains the letter "N"
   e. a name that has more than 5 letters

6. A box contains 9 marbles, numbered from 1 to 9. Find the probability of picking at random:
   a. a marble having the number 9
   b. a marble having an even number
   c. a marble having an odd number
   d. a marble having a number less than 10
   e. a marble having a number greater than 20

## HOMEWORK EXERCISES

1. What is the probability that the arrow on the spinner shown will stop on a space marked "W"?

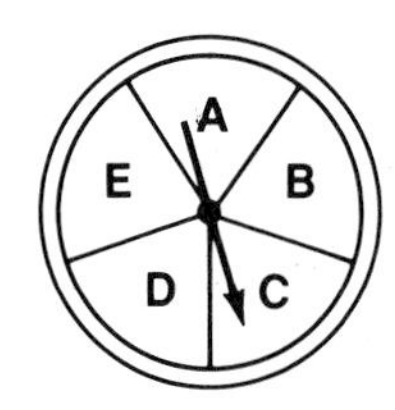


2. What is the probability that a day of the week chosen at random will have a name ending with the letter "y"?
3. What is the probability that a month of the year chosen at random will have a name beginning with the letter "B"?
4. What is the probability of picking, at random, a red marble from a box that contains only 15 red marbles?
5. A bag contains only 50 dimes. If a coin is chosen at random from that bag, what is the probability that the coin is a nickel?

## SPIRAL REVIEW EXERCISES

1. A box contains only 5 quarters, 3 dimes, and 8 nickels. If one coin is selected at random from that box, find the probability of choosing:
   a. a quarter b. a dime c. a nickel
   d. a penny e. a coin f. a dollar

2. Change each of the given probabilities, which are expressed as percents, into fractions in simplest form.
   a. 25% b. 40% c. 75%
   d. 17% e. 84% f. 100%

3. Find the probability that the arrow on the spinner shown will stop on:

   a. the number 6
   b. an odd number
   c. an even number
   d. a prime number
   e. a number less than 10
   f. a two-digit number

4. Change each of the given probabilities, which are expressed as decimals, into fractions in simplest form.

   **a.** .17  **b.** .05  **c.** .8  **d.** .019

5. If one letter is chosen at random from the word "PUPPIES," the probability of choosing a "P" is

   (a) $\frac{3}{4}$  (b) $\frac{4}{7}$  (c) $\frac{3}{7}$  (d) 3

6. In a stack of 20 playing cards, 6 are red. What is the probability that a card chosen at random from the stack will be black?

   (a) .3  (b) .6  (c) .7  (d) 1.0

7. Find 15% of 240.

8. Multiply: $\frac{5}{12} \times \frac{3}{10}$

9. $\frac{19}{7}$ is equal to

   (a) $2\frac{5}{7}$  (b) $2\frac{3}{7}$  (c) $1\frac{5}{7}$  (d) $\frac{7}{19}$

10. Round 2.739 to the nearest tenth.

11. Ernesto earns $8.40 an hour, and he makes time-and-a-half for any time over 35 hours a week. If he worked 41 hours during a week, how much did he earn?

12. Mr. Wilkins earns $2,000 a month. If he pays $800 for rent, what percent of his income should he budget for rent?

13. 842 mL has the same value as

    (a) .842 L  (c) 84.2 L
    (b) 8.42 L  (d) 842,000 L

14. Find the sum of $-20$ and $+42$.

15. Evaluate: $38 - 18 \div 2 + 7$

## CHALLENGE QUESTION

There are only 15 red marbles and 21 yellow marbles in a box. One marble is chosen at random, and it is yellow. The chosen yellow marble is not put back into the box, and a second marble is chosen.

**a.** What is the probability that the second marble is yellow?

**b.** What is the probability that the second marble is red?

# UNIT 14–3 Probability Involving Dice

## THE MAIN IDEA

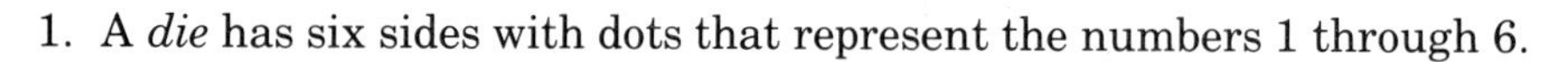

1. A *die* has six sides with dots that represent the numbers 1 through 6.

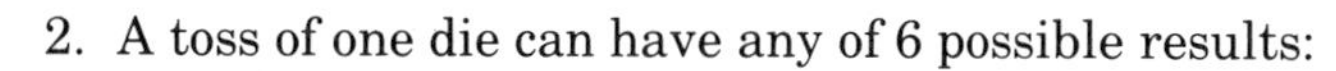

2. A toss of one die can have any of 6 possible results:

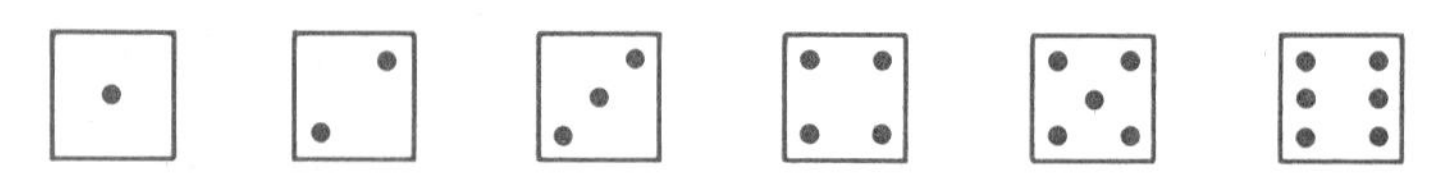

3. A toss of two *dice* can have any of 36 possible results:

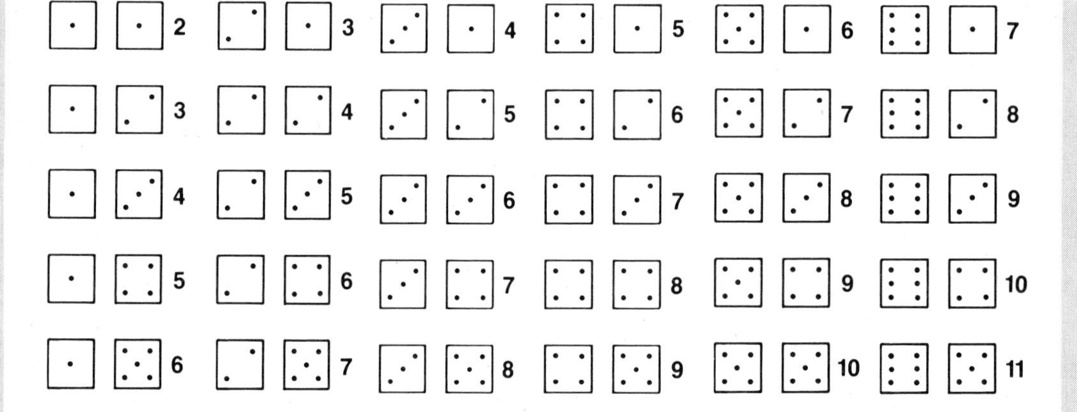


These results show sums from 2 through 12.

**EXAMPLE 1** If a die is tossed, what is the probability that it will show the value "5"?

There are 6 ways that a die can come up. Only 1 of these ways shows the value "5."

$$\text{probability} = \frac{1}{6}$$

*Answer:* The probability that a die will show the value "5" is $\frac{1}{6}$.

**EXAMPLE 2** If a die is tossed, what is the probability that it will show an even value?

Of the 6 ways the die can come up, 3 of them are even values: "2," "4," and "6."

$$\text{probability} = \frac{3}{6}$$

$$= \frac{1}{2}$$

*Answer:* The probability that a die will show an even value is $\frac{3}{6}$ or $\frac{1}{2}$.

**EXAMPLE 3** What is the probability of rolling a sum of 7 with a pair of dice?

There are 36 ways in which a pair of dice can fall. Six of these show a sum of 7.

$$\text{probability} = \frac{6}{36} = \frac{1}{6}$$

*Answer:* The probability that a pair of dice will show a sum of 7 is $\frac{6}{36}$ or $\frac{1}{6}$.

**EXAMPLE 4** Which sums from 2 through 12 have the lowest probabilities of coming up in a toss of a pair of dice?

### THINKING ABOUT THE PROBLEM

A key word in this problem is "lowest." This suggests comparing the probabilities of all the possible sums, from 2 through 12, and choosing the lowest.

There are 36 ways in which a pair of dice can fall. Only 1 of these ways shows a sum of 2: 1 + 1

probability of getting a sum of 2 $= \frac{1}{36}$

Only 1 way shows a sum of 12: 6 + 6

probability of getting a sum of 12 $= \frac{1}{36}$

All other sums can be formed in more than 1 way (for example, 5 = 2 + 3 or 4 + 1). Thus, all other sums have probabilities that are larger than $\frac{1}{36}$.

*Answer:* The sums with the lowest probabilities are 2 and 12.

## CLASS EXERCISES

**1.** In rolling a die, find the probability that it will show:

**a.** "1" **b.** "2" **c.** "3" **d.** "4" **e.** "5" **f.** "6"

**g.** an even number **h.** an odd number **i.** a number less than 4

**2.** In rolling a pair of dice, find the probability of getting a sum of:

**a.** 2 **b.** 3 **c.** 4 **d.** 5 **e.** 6 **f.** 7 **g.** 8 **h.** 9 **i.** 10

**j.** 11 **k.** 12 **l.** 1

**3.** When a pair of dice is rolled, what is the probability that the sum will be an odd number?

**4.** When a pair of dice is rolled, what is the probability that a sum greater than 9 will result?

**5.** When a pair of dice is rolled, what sum has the same probability of coming up as does a sum of 5?

## HOMEWORK EXERCISES

**1.** A die is rolled. What is the probability that it will show the value "4"?

**2.** In rolling a die, find the probability that it will show:

**a.** "2" **b.** "1" **c.** "9" **d.** a number less than 5 **e.** a number less than or equal to 6

3. The probability of rolling a die and having it show the value "3" is
(a) $\frac{1}{5}$ (b) $\frac{1}{6}$ (c) $\frac{5}{6}$ (d) $\frac{1}{2}$

4. The probability of rolling a number that is a multiple of 3 with a die is
(a) $\frac{1}{6}$ (b) $\frac{1}{3}$ (c) $\frac{1}{2}$ (d) $\frac{2}{3}$

5. When a pair of dice is rolled, what is the probability that the sum will be 7?

6. When a pair of dice is rolled, what is the probability that the sum will be 2?

7. In rolling a pair of dice, the probability that the sum will be 8 is
(a) $\frac{1}{8}$ (b) $\frac{5}{36}$ (c) $\frac{8}{36}$ (d) $\frac{1}{11}$

8. In rolling a pair of dice, the probability of getting a sum less than 15 is
(a) 0 (b) $\frac{15}{36}$ (c) $\frac{1}{15}$ (d) 1

## SPIRAL REVIEW EXERCISES

1. What is the probability that the arrow on the spinner shown will stop on a space:

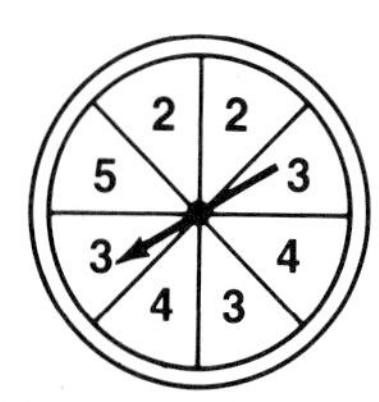


   a. marked "3"
   b. marked with a number less than 5
   c. marked with an even number
   d. marked with a number less than 7
   e. marked "7"

2. If a card is drawn at random from a standard deck of playing cards, what is the probability that the card chosen is:
   a. a club  b. a red card
   c. a picture card  d. a red queen
   e. a black picture card

3. Find the sum of $\frac{1}{2}$ and $\frac{2}{5}$.

4. Change $\frac{37}{24}$ to a mixed number.

5. Express $\frac{5}{8}$ as a decimal.

6. Find the product of $(-4)$ and $(+14)$.

7. If a temperature rises from $-3°$ to $+7°$, how many degrees has it risen?

8. A \$120 bike is on sale at 10% off. The sale price is
(a) \$108 (c) \$96
(b) \$100 (d) \$80

9. John spent \$4.38 on a record. How much change did he receive from a \$10 bill?
(a) \$15.62 (c) \$5.62
(b) \$5.72 (d) \$6.62

10. $6.2 \times .3 =$
(a) 1.86 (c) 18.6
(b) .186 (d) .0186

## CHALLENGE QUESTION

A coin is tossed at the same time that a die is rolled. What is the probability of getting a "head" on the coin and a "4" or higher on the die?

# UNIT 14–4 Using Probability to Predict Outcomes

## THE MAIN IDEA

If you know the probability of an event and you know the number of trials that will take place, you can predict the *number of times* that you can expect the event to happen:

1. Find the probability that the event will happen.
2. Multiply this probability by the number of trials.

expected number of successes = probability of success × number of trials

**EXAMPLE 1** How many times can you expect to get the value "4" if you roll a die 180 times in a row?

Find the probability of getting the value "4" on 1 roll of a die.

$$\text{probability} = \frac{1}{6}$$

Multiply the probability by the number of trials.

$$\frac{1}{\cancel{6}_{1}} \times \cancel{180}^{30} = \frac{30}{1}$$

$$= 30$$

*Answer:* You can expect to get the value "4" thirty times in 180 rolls of a die.

**EXAMPLE 2** A die is rolled 1,000 times. How many times can the value "7" be expected to appear?

Find the probability of getting the value "7" on 1 roll of a die.

$$\text{probability} = \frac{0}{6}$$

$$= 0$$

Multiply the probability by the number of trials.

$$0 \times 1{,}000 = 0$$

*Answer:* You cannot expect the value "7" to appear on a die no matter how many times you roll the die.

**EXAMPLE 3** In 144 rolls of a pair of dice, how many times can you expect to roll a sum of 8?

Find the probability of rolling a sum of 8 with a pair of dice:

There are 36 ways in which a pair of dice can fall. Five of these show a sum of 8.

$$\text{probability} = \frac{5}{36}$$

Multiply the probability by the number of trials.

$$\frac{5}{\cancel{36}_{1}} \times \cancel{144}^{4} = 20$$

*Answer:* You can expect to get a sum of 8 twenty times in 144 rolls of a pair of dice.

## CLASS EXERCISES

1. If the probability of getting a "head" is $\frac{2}{3}$ when a certain weighted coin is tossed, how many "heads" would you expect to get in 60 tosses of this coin?
2. If a die is rolled 120 times in a row, how many times would you expect to roll the value "1"?
3. If you rolled a pair of dice 180 times in a row, state how many times you would expect to roll:

   **a.** a sum of 10 **b.** a sum of 7 **c.** a sum of 11 **d.** a sum of 3
4. The probability that Warren will make a basket is $\frac{2}{3}$. How many baskets can Warren be expected to make in 330 attempts?
5. In 150 spins of the spinner shown, state how many times you can expect it to stop on:

   **a.** the letter "a" **b.** the letter "c" **c.** the letter "z"

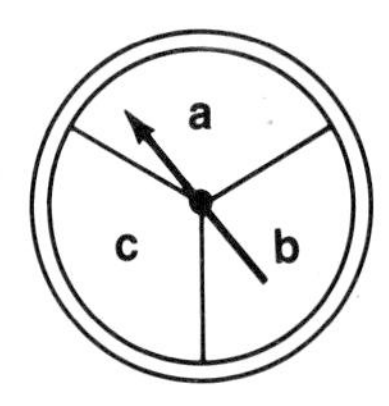


6. A magician shuffles a deck of cards, chooses a card at random, and then puts the chosen card back into the deck. If he does this 32 times in a row, how many hearts would you expect him to pick?
7. How many times would you expect to get each of the following in 104 individual random selections of a card from a standard deck, if the chosen card is returned to the deck each time?

   **a.** a king **b.** a picture card **c.** a red card

   **d.** a number card (count an ace as number 1)

## HOMEWORK EXERCISES

1. In 300 rolls of a die, how many times would you expect the die to show the value "4"?
2. In 500 tosses of a fair coin, how many times would you expect the coin to land heads up?
3. If you rolled a pair of dice 108 times in a row, state how many times you would expect to roll:

   **a.** a sum of 5 **b.** a sum of 2 **c.** a sum of 8 **d.** a sum of 14
4. In 3,000 rolls of a die, how many times would you expect the die to show a number less than 7?
5. The probability that it will rain on a given day is .2. In the next 100 days, how many rainy days can you expect?
6. The probability that Reggie will hit a home run is $\frac{1}{15}$. How many home runs can Reggie be expected to hit in 450 times at bat?
7. In 100 spins of the spinner shown, state how many times you can expect the arrow to stop on:

   **a.** the number "1" **b.** the number "5"

   **c.** the number "3" **d.** the number "7"

8. A marble is picked at random from a box containing 3 red marbles, 2 blue marbles, and 5 yellow marbles. The marble is then placed back into the box. If this is done 150 times, how many times can you expect a red marble to be picked?

9. In 180 rolls of a die, the number of times you would expect it to show a number less than 10 is
(a) 30 (b) 90 (c) 120 (d) 180

## SPIRAL REVIEW EXERCISES

1. A jar contains only 4 red marbles, 5 white marbles, and 6 blue marbles. If a marble is chosen at random from this jar, find the probability that the marble chosen is:

   **a.** red **b.** green **c.** not red

   **d.** not green **e.** red or blue

2. Given: the set of positive integers that are less than 20 and that are multiples of 3. If a number is chosen at random from this set, find the probability that the number chosen is:

   **a.** a two-digit number **b.** an odd number

   **c.** a prime number **d.** a multiple of 4

   **e.** a multiple of 7 **f.** less than 30

3. If a die is rolled, find the probability that it will show:

   **a.** the value "5"

   **b.** a value that is not "5"

   **c.** a value that is less than 5

   **d.** a value that is not less than 5

4. 80% is equivalent to
(a) $\frac{7}{10}$ (b) $\frac{4}{5}$ (c) $\frac{8}{100}$ (d) $\frac{20}{100}$

5. The sum of $\frac{1}{2}$ and $\frac{2}{3}$ is
(a) $\frac{3}{5}$ (b) $\frac{1}{3}$ (c) $\frac{3}{4}$ (d) $\frac{7}{6}$

6. The quotient of $+3$ and $-6$ is
(a) $-18$ (b) $+18$ (c) $+\frac{1}{2}$ (d) $-\frac{1}{2}$

7. Find, to the nearest cent, the amount of a 7% sales tax on an item that sells for $10.99.

8. If 3 apples cost $.66, then the cost of 5 apples is
(a) $.22 (c) $1.10
(b) $.88 (d) $1.98

## CHALLENGE QUESTION

Copy and complete the chart shown.

| Probability of Success | Number of Trials | Expected Number of Successes |
|---|---|---|
| .3 | 50 | |
| 20% | 90 | |
| | 200 | 100 |
| $\frac{3}{4}$ | | 75 |

# UNIT 14–5 Finding the Probability That an Event Will Not Happen

THE MAIN IDEA

1. The sum of the probability that an event will happen and the probability that it will not happen is 1.

   probability of happening + probability of not happening = 1

2. To find the probability that an event will not happen, subtract the probability that the event will happen from 1.

   *probability of not happening = 1 − probability of happening*

**EXAMPLE 1** The probability of tossing a "head" with a certain weighted coin is $\frac{3}{8}$. What is the probability of not tossing a "head" with this coin?

probability of not tossing a "head" = 1 − probability of tossing a "head"

$$= 1 - \frac{3}{8}$$

$$= \frac{8}{8} - \frac{3}{8} = \frac{5}{8}$$

*Answer:* The probability of not tossing a "head" with this coin is $\frac{5}{8}$.

**EXAMPLE 2** What is the probability of not getting a heart when choosing a card at random from a standard deck?

Find the probability of getting a heart.

$$\text{probability} = \frac{13}{52} \begin{array}{l}\leftarrow \text{number of hearts}\\ \leftarrow \text{total number of cards}\end{array}$$

Simplify the fraction.

$$\frac{13 \div 13}{52 \div 13} = \frac{1}{4}$$

probability of not getting a heart = 1 − probability of getting a heart

$$= 1 - \frac{1}{4}$$

$$= \frac{4}{4} - \frac{1}{4} = \frac{3}{4}$$

*Answer:* The probability of not getting a heart is $\frac{3}{4}$.

**EXAMPLE 3** If the probability of choosing a girl from a certain class is .45, what is the probability of not choosing a girl?

probability of not choosing a girl = 1 − probability of choosing a girl
= 1 − .45
= 1.00 − .45 = .55

*Answer:* The probability of not choosing a girl is .55.

**EXAMPLE 4** The probability that a train will be on time is 72%. What is the probability that the train will not be on time?

probability of not being on time = 1 − probability of being on time
= 1 − 72%
= 100% − 72% = 28%

*Answer:* The probability that the train will not be on time is 28%.

## CLASS EXERCISES

1. Each number represents the probability that an event will happen. Find the probability that the event will not happen.

   **a.** $\frac{5}{6}$ **b.** .36 **c.** 85% **d.** 1 **e.** .01 **f.** 0

2. The probability of tossing a "head" with a certain weighted coin is $\frac{3}{10}$. What is the probability of not tossing a "head" with this coin?

3. The probability of selecting a vowel from the letters of our alphabet is $\frac{5}{26}$. What is the probability of selecting a consonant?

4. The probability that it will not rain tomorrow is 60%. What is the probability that it will rain tomorrow?

5. If a card is picked at random from a standard deck, what is the probability that the card will not be a picture card?

## HOMEWORK EXERCISES

1. Each number represents the probability that an event will happen. Find the probability that the event will not happen.

   **a.** 37% **b.** $\frac{3}{8}$ **c.** 0 **d.** .45 **e.** $\frac{9}{16}$ **f.** 100%

2. The probability of tossing a "head" with a certain weighted coin is $\frac{4}{9}$. What is the probability of not tossing a "head" with this coin?

3. The probability of John's winning the race is .29. What is the probability of John's not winning the race?

4. The probability that a card chosen at random from a standard deck of playing cards will not be a heart is (a) $\frac{1}{4}$ (b) $\frac{1}{2}$ (c) $\frac{3}{4}$ (d) 1

5. The probability that it will rain today is 40%. What is the probability that it will not rain?

6. The probability of not picking a red ball at random from a bag is $\frac{7}{35}$. What is the probability of picking a red ball at random from the bag?

7. The probability of a candidate's winning an election is .52. What is the probability of the candidate's not winning that election?

## SPIRAL REVIEW EXERCISES

1. Jim has a $\frac{3}{10}$ probability of getting a hit. What is the number of hits Jim can be expected to get in 200 times at bat?

2. The names of the days of the week are written on separate slips of paper and placed in a hat. One slip of paper is chosen at random from the hat. Find the probability of:

   a. picking a day whose name begins with the letter "T"

   b. picking a day whose name ends with the letter "y"

   c. picking a day whose name has 7 letters

   d. picking a day whose name starts with the letter "Z"

3. Change each probability into a fraction in simplest form.

   a. 83%  b. .95  c. .125  d. 19%

   e. 58%  f. 25%  g. .007  h. .98

4. Find the probability that the arrow on the spinner shown will stop on:

   a. the number "5"

   b. an even number

   c. an odd number

   d. a prime number

   e. a number less than 3

5. A box contains 10 coins: 3 quarters, 2 dimes, and 5 nickels. If one coin is chosen at random from the box, find the probability of picking:

   a. a quarter  b. a dime  c. a nickel

   d. a half-dollar  e. a dime or a nickel

6. What percent of 40 is 8?

7. If the sales tax is 5%, what is the total amount paid for a shirt that is priced at $14?

8. The greatest common factor of 12 and 18 is
   (a) 2  (b) 3  (c) 6  (d) 9

9. Subtract $-15$ from 35.

10. Round 0.9827 to the nearest thousandth.

11. The largest value of the choices below is
   (a) $\frac{5}{8}$  (b) .67  (c) 68%  (d) $\frac{9}{16}$

12. The product of .05 and .9 is
   (a) 4.5  (c) .045
   (b) .45  (d) .0045

13. Divide: $3\frac{3}{4} \div \frac{5}{16}$

14. Evaluate: $3(17 + 5) - 2 \times 6$

### CHALLENGE QUESTION

James said that the probability of drawing a heart at random from a standard deck is $\frac{1}{2}$. He reasoned that there are only two possibilities: either the card will be a heart or it will not be a heart. Explain why James' reasoning is wrong.

# Statistics and Graphs

## MODULE 15



## MODULE 16

# UNIT 15–1 The Meaning of Statistics; Constructing a Frequency Table

## THE MAIN IDEA

1. ***Statistics*** is the branch of mathematics that deals with organizing numerical facts, or ***data,*** in order to study them more clearly.
2. One way we organize data is to record the number of times each value appears, the ***frequency.*** This count, or ***tally,*** produces a list of values that is called a ***frequency table.***

The model examples that follow will show you how to read a frequency table and how to make your own frequency table for given data.

**EXAMPLE 1** The frequency table shows the number of long-distance phone charges paid by the Jefferson Accounting Office during the month of March.

| Charge | Number of Calls (frequency) |
|---|---|
| $1.00–$1.99 | 5 |
| $2.00–$2.99 | 8 |
| $3.00–$3.99 | 6 |
| $4.00–$4.99 | 1 |
| $5.00–$5.99 | 2 |
| $6.00 and over | 3 |

**a.** How many long-distance calls were made by the office during the month of March?

The table shows all the calls for March. Add to find the total number of calls.

$$\begin{array}{r} 5 \\ 8 \\ 6 \\ 1 \\ 2 \\ +3 \\ \hline 25 \end{array}$$

*Answer:* There were 25 calls made in the month of March.

**b.** Which category had the smallest number of calls?

The smallest number of calls is 1, which is the number of calls costing $4.00–$4.99.

*Answer:* The category that had the smallest number of calls was $4.00–$4.99.

**c.** Which category had the greatest number of calls?

The greatest number of calls is 8, which is the number of calls costing $2.00–$2.99.

*Answer:* The category that had the greatest number of calls was $2.00–$2.99.

**EXAMPLE 2** Make a frequency table for these weights of 30 high school students.

| | | | | | | | | | |
|---|---|---|---|---|---|---|---|---|---|
| 140 | 130 | 120 | 110 | 115 | 100 | 120 | 115 | 140 | 120 |
| 110 | 120 | 115 | 100 | 95 | 95 | 130 | 115 | 120 | 95 |
| 100 | 115 | 95 | 140 | 115 | 110 | 120 | 100 | 110 | 95 |

First, list the various weights in order.

| Weight |
|---|
| 140 |
| 130 |
| 120 |
| 115 |
| 110 |
| 100 |
| 95 |

Second, tally (count, using stroke marks in groups of 5) the weights to show how many times each weight is found in the data.

| Weight | Tally |
|---|---|
| 140 | ||| |
| 130 | || |
| 120 | 卌 | |
| 115 | 卌 | |
| 110 | |||| |
| 100 | |||| |
| 95 | 卌 |

Finally, count the tally marks and write the totals, or frequencies.

| Weight | Tally | Frequency |
|---|---|---|
| 140 | ||| | 3 |
| 130 | || | 2 |
| 120 | 卌 | | 6 |
| 115 | 卌 | | 6 |
| 110 | |||| | 4 |
| 100 | |||| | 4 |
| 95 | 卌 | 5 |

To check, find the sum of all the frequencies and see if this equals the total number of weights.

$$\begin{array}{r} 3 \\ 2 \\ 6 \\ 6 \\ 4 \\ 4 \\ +5 \\ \hline 30 \end{array} \checkmark$$

**EXAMPLE 3** The numbers listed are the heights, in inches, of 50 schoolchildren. Make a frequency table of these heights, grouping the data in 1-inch intervals.

60 48 47 63 50 61 51 51 53 49
61 59 49 48 48 49 55 57 60 53
49 58 61 50 54 56 55 59 61 57
56 48 46 60 61 54 49 60 48 51
59 50 55 59 51 53 57 62 49 59

In order to label the intervals, begin with the smallest height (46 inches) and count off inches (46–47, 48–49, etc.) until you have included the largest height (63 inches).

Then, tally the heights according to these intervals and write the frequencies.

| Height (inches) | Tally | Frequency |
|---|---|---|
| 46–47 | \|\| | 2 |
| 48–49 | 卌 卌 \| | 11 |
| 50–51 | 卌 \|\| | 7 |
| 52–53 | \|\|\| | 3 |
| 54–55 | 卌 | 5 |
| 56–57 | 卌 | 5 |
| 58–59 | 卌 \| | 6 |
| 60–61 | 卌 \|\|\|\| | 9 |
| 62–63 | \|\| | 2 |
| | | Check: 50 ✓ |

## CLASS EXERCISES

**1.** This is a frequency table of the average number of repairs made on different brands of automobiles during the first two years after purchase.

**a.** Which brand had the worst record of repairs (the greatest number)?

**b.** Which brand had the best record of repairs (the least number)?

**c.** Which brands had equal frequencies of repair?

| Brand | Number of Repairs |
|---|---|
| A | 2 |
| B | 7 |
| C | 3 |
| D | 5 |
| E | 8 |
| F | 3 |
| G | 1 |

**2. a.** Make a frequency table of the given weights (in pounds) of pieces of airplane luggage. Use the intervals 14–16, 17–19, 20–22, 23–25, 26–28, and 29–31.

27 22 18 15 22 16 22 16 21 18
28 22 15 14 22 27 29 31 30 28
27 22 19 15 18 17 17 30 29 30

**b.** How many pieces of luggage are there?

**c.** Which interval has the greatest frequency?

**d.** Which interval has the smallest frequency?

**e.** What is the sum of all the frequencies?

3. a. Given are lengths of wire, to the nearest tenth of an inch. Copy and complete the frequency table for these lengths.

4.5 6.0 14.4 15.8 7.3 7.9 6.5
8.0 12.0 14.0 8.0 5.3 4.9 14.5
8.0 5.0 11.0 15.0 11.1 11.5

| Interval (inches) | Tally | Frequency |
|---|---|---|
| 4–5.9 | | |
| 6–7.9 | | |
| 8–9.9 | | |
| 10–11.9 | | |
| 12–13.9 | | |
| 14–15.9 | | |

b. Which interval has the greatest frequency?

c. Which interval has the smallest frequency?

d. What is the sum of all the frequencies?

## HOMEWORK EXERCISES

1. This is a frequency table showing numbers of patients and their blood pressures.

a. In which group was the number of patients the highest?

b. In which group was the number of patients the lowest?

c. How many patients had their blood pressure measured?

| Blood Pressure | Number of Patients |
|---|---|
| 110–119 | 10 |
| 120–129 | 27 |
| 130–139 | 23 |
| 140–149 | 12 |
| 150–159 | 3 |

2. This is a summary of the number of typing errors made one day by 6 secretaries.

a. Who had the best record (the least number)?

b. Who had the worst record (the greatest number)?

c. How many typing errors did this group of secretaries make that day?

| Secretary | Number of Errors |
|---|---|
| Miss A | 5 |
| Mr. B | 8 |
| Miss C | 3 |
| Mrs. D | 11 |
| Mr. E | 10 |
| Ms. F | 1 |

3. The highest temperatures recorded in a city on 20 consecutive days are as listed.

87 90 85 87 89 85 91 87 88 84
86 88 86 90 85 85 86 85 84 87

a. Make a frequency table for these data.

b. What is the temperature with the highest frequency?

c. Which temperatures have the lowest frequencies?

d. What is the sum of the frequencies?

4. The ages of 30 teachers at Adams High School are given.

| | | | | | | | | | |
|---|---|---|---|---|---|---|---|---|---|
| 40 | 53 | 50 | 30 | 53 | 39 | 45 | 48 | 30 | 27 |
| 45 | 29 | 28 | 41 | 32 | 25 | 41 | 32 | 31 | 25 |
| 43 | 46 | 33 | 48 | 33 | 26 | 29 | 51 | 32 | 50 |

**a.** Make a frequency table of these ages, grouping them in 4-year intervals, starting with 25–28.

**b.** Which age group has the highest frequency?

## SPIRAL REVIEW EXERCISES

1. What is the probability that a letter chosen at random from the letters of the word "GIGGLES" will be a G?
2. In 180 tosses of a fair coin, how many times could you expect to get "heads"?
3. Find 40% of 180.
4. If a $50 item is discounted 20%, what is its sale price?
5. A bag contains 5 red balls and 4 black balls. The probability of picking a red ball at random is
   (a) $\frac{4}{9}$ (b) $\frac{5}{9}$ (c) $\frac{2}{3}$ (d) 1
6. 25% written as a fraction is
   (a) $\frac{1}{4}$ (b) $\frac{3}{10}$ (c) $\frac{1}{25}$ (d) $\frac{4}{5}$
7. Divide: $1.2\overline{)12.36}$
8. Find the sum of $2\frac{3}{4}$ and $4\frac{7}{8}$.
9. What percent of 20 is 16?
10. Mr. Strong earns a weekly salary of $250 plus a 15% commission on all of his sales. Last week, his sales totaled $1,200. What were his total earnings?
11. Round 25,371,846 to the nearest hundred thousand.

## CHALLENGE QUESTION

Mary made a frequency table of the 100 books that she sold. How do you know that she counted wrong?

| Type of Book | Frequency |
|---|---|
| Fiction | 13 |
| Travel | 15 |
| Humor | 17 |
| History | 11 |
| Science | 14 |
| Cookbooks | 18 |
| Home Repair | 17 |

## UNIT 15–2 Finding the Mode

### THE MAIN IDEA

1. The *mode* of a set of data is the value that has the *greatest frequency.*
2. To find the mode of a set of data:
   a. Tally the values and write the frequencies.
   b. Choose the value with the greatest frequency.
3. A set of data may have two modes, if there are two values that both have the greatest frequency.

**EXAMPLE 1** What is the mode of this group of numbers? 72, 23, 35, 23, 34, 23

The number that has the greatest frequency is 23. Therefore, 23 is the mode.

*Answer:* The mode is 23.

**EXAMPLE 2** In the frequency table of weights (in pounds) of schoolchildren, which is the mode?

The weight that has the greatest frequency, 9, is 56.

| Weight | Frequency |
|---|---|
| 50 | 1 |
| 51 | 3 |
| 52 | 4 |
| 53 | 2 |
| 54 | 4 |
| 55 | 8 |
| 56 | 9 |
| 57 | 7 |
| 58 | 5 |
| 59 | 1 |
| 60 | 1 |

*Answer:* The mode is 56.

**EXAMPLE 3** For this list of 25 prices of grocery orders, find the mode.

| | | | | |
|---|---|---|---|---|
| $ 3.50 | 12.75 | 4.25 | 9.00 | 10.00 |
| 8.00 | 9.00 | 8.00 | 8.75 | 10.25 |
| 11.00 | 7.50 | 11.50 | 8.50 | 7.00 |
| 9.50 | 9.00 | 10.00 | 8.00 | 7.50 |
| 8.50 | 9.50 | 9.00 | 9.00 | 11.00 |

First, make a frequency table.

| Price | Tally | Frequency |
|---|---|---|
| $ 3.50 | \| | 1 |
| 4.25 | \| | 1 |
| 7.00 | \| | 1 |
| 7.50 | \|\| | 2 |
| 8.00 | \|\|\| | 3 |
| 8.50 | \|\| | 2 |
| 8.75 | \| | 1 |
| 9.00 | ~~\|\|\|\|~~ | 5 |
| 9.50 | \|\| | 2 |
| 10.00 | \|\| | 2 |
| 10.25 | \| | 1 |
| 11.00 | \|\| | 2 |
| 11.50 | \| | 1 |
| 12.75 | \| | 1 |

From the frequency table, read the value that has the highest frequency: The mode is $9.00. *Ans.*

**EXAMPLE 4** For this set of average temperatures (in degrees Fahrenheit) of 30 cities, find the mode.

54 48 31 46 40 72 32 54 40 48
31 46 70 31 46 34 46 72 32 40
68 31 46 32 31 46 31 45 46 31

Make a frequency table.

| Temperature (°F) | Tally | Frequency |
|---|---|---|
| 31 | ~~\|\|\|\|~~ \|\| | 7 |
| 32 | \|\|\| | 3 |
| 34 | \| | 1 |
| 40 | \|\|\| | 3 |
| 45 | \| | 1 |
| 46 | ~~\|\|\|\|~~ \|\| | 7 |
| 48 | \|\| | 2 |
| 54 | \|\| | 2 |
| 68 | \| | 1 |
| 70 | \| | 1 |
| 72 | \|\| | 2 |

From the table, you can see that there are two different temperatures with the same highest frequency, 7. Therefore, there are two modes: 31° and 46°.

*Answer:* The modes are 31° and 46°.

## CLASS EXERCISES

1. Find the mode of each group of numbers.

**a.** 12, 27, 12, 14, 21, 12

**b.** 13, 16, 31, 13, 13

**c.** 57, 75, 62, 61, 75, 56, 61, 75, 57

**d.** 102, 105, 205, 103, 105, 102, 105

**e.** 89, 98, 87, 79, 88, 97, 79, 87, 89, 79

**f.** $2\frac{3}{4}, 2\frac{1}{2}, 2\frac{3}{4}, 2, 2\frac{1}{8}, 2\frac{3}{4}, 2\frac{1}{2}$

**g.** 7.1, 7.4, 7.5, 7.4, 7.6, 7.4, 7.4, 7.9

**h.** 78.5, 114.6, 98.3, 89.3, 78.6, 89.3, 94.3, 89.3, 93.8

2. For each frequency table, find the mode.

**a.**

| Height (in.) | Frequency |
|---|---|
| 62 | 7 |
| 63 | 11 |
| 64 | 12 |
| 65 | 10 |
| 66 | 5 |

**b.**

| Price | Frequency |
|---|---|
| $10.00 | 7 |
| 11.00 | 6 |
| 12.00 | 6 |
| 13.00 | 5 |
| 14.00 | 7 |
| 15.00 | 6 |

**c.**

| Distance (mi.) | Frequency |
|---|---|
| 10 | 7 |
| 15 | 12 |
| 20 | 21 |
| 25 | 17 |
| 30 | 29 |
| 35 | 25 |
| 40 | 19 |
| 45 | 16 |

**3.** Make a frequency table and find the mode for this set of 20 shoe sizes.

$10\frac{1}{2}$ $8\frac{1}{2}$ $9\frac{1}{2}$ 9 8 8 11 8 $9\frac{1}{2}$ 10

9 $9\frac{1}{2}$ $9\frac{1}{2}$ $10\frac{1}{2}$ $8\frac{1}{2}$ $9\frac{1}{2}$ $8\frac{1}{2}$ $8\frac{1}{2}$ 11 $9\frac{1}{2}$

**4.** Make a frequency table and find the mode for this set of distances, in miles.

| | | | | | | | |
|---|---|---|---|---|---|---|---|
| 12.4 | 10.1 | 5.6 | 3.1 | 5.6 | 6.3 | 2.4 | 12.3 |
| 7.1 | 6.9 | 9.7 | 6.5 | 2.4 | 3.1 | 4.2 | 4.2 |
| 6.9 | 7.5 | 5.6 | 3.1 | 2.4 | 7.1 | 3.1 | 6.5 |

## *HOMEWORK EXERCISES*

**1.** What is the mode for this set of numbers? 72, 76, 89, 72, 76, 93, 86, 76

**2.** What is the mode for this set of numbers? 4, 7, 9, 11, 9, 4, 10, 8, 9, 3, 7

**3.** For the set of numbers 5, 15, 25, 5, 35, 50, 5, 15, 5, 25 the mode is
(a) 4 (b) 5 (c) 15 (d) 50

**4.** Find the mode for this set of temperatures. 45°, 54°, 40°, 45°, 55°, 54°, 45°, 45°, 55°

**5.** In each case, find the mode of the given set of numbers.

**a.** 11, 32, 27, 33, 32, 12, 32 **b.** 26, 49, 55, 38, 26, 19, 67

**c.** 59, 60, 61, 61, 61, 70, 72 **d.** 101, 117, 101, 132, 115, 119

**e.** 125, 156, 198, 165, 156, 120, 156 **f.** $3\frac{1}{2}$, 4, $3\frac{1}{2}$, $4\frac{1}{2}$, $3\frac{1}{2}$, 5, $3\frac{1}{2}$, $5\frac{1}{2}$, $4\frac{1}{2}$

**g.** 10.4, 10.5, 10.3, 10.2, 10.3, 10.4, 10.3, 10.3 **h.** 98.7, 97.8, 99.3, 93.7, 97.8, 98.3, 97.8, 98.7

**6.** For each frequency table, find the mode.

**a.**

| Age | Number of Students |
|---|---|
| 13 | 35 |
| 14 | 68 |
| 15 | 129 |
| 16 | 134 |
| 17 | 112 |
| 18 | 52 |

**b.**

| Exam Score | Number of Students |
|---|---|
| 70 | 12 |
| 75 | 34 |
| 80 | 20 |
| 85 | 15 |
| 90 | 34 |
| 95 | 14 |
| 100 | 8 |

**7.** Make a frequency table and find the mode for this group of lengths (in centimeters) of electrical wire.

| | | | | | | | | |
|---|---|---|---|---|---|---|---|---|
| 11.4 | 10.1 | 9.6 | 9.5 | 11.5 | 9.4 | 11.0 | 9.6 | 6.9 |
| 9.4 | 9.6 | 11.0 | 25.3 | 11.4 | 25.2 | 11.4 | 9.6 | 9.5 |
| 9.5 | 9.6 | 10.1 | 9.7 | 10.1 | 9.6 | 6.9 | 10.2 | 10.1 |

8. Make a frequency table and find the mode for this group of shirt sizes.

$16\frac{1}{2}$ 14 $16\frac{1}{2}$ 15 16 16 $14\frac{1}{2}$

$14\frac{1}{2}$ $16\frac{1}{2}$ 17 $15\frac{1}{2}$ $16\frac{1}{2}$ 16 $16\frac{1}{2}$

14 $16\frac{1}{2}$ $15\frac{1}{2}$ 15 16 $16\frac{1}{2}$ $15\frac{1}{2}$

## SPIRAL REVIEW EXERCISES

1. The table shows Mary's absences for 5 months. In which month did she have the greatest frequency of absence?

| Month | Number of Absences |
|---|---|
| Sept. | 0 |
| Oct. | 2 |
| Nov. | 5 |
| Dec. | 1 |
| Jan. | 4 |

2. What is the probability that a month chosen at random will have a name that ends with the letter "R"?

3. In 240 random selections of a card from a standard deck of cards, how often could you expect to get a diamond?

4. If the probability of tossing a "head" with a certain unbalanced coin is $\frac{5}{8}$, what is the probability of tossing a "tail"?

5. In choosing a day of the week at random, what is the probability that its name will have fewer than 10 letters?

6. Copy and complete the chart to express each number in two other ways.

| Fraction | Decimal | Percent |
|---|---|---|
| $\frac{3}{4}$ | | |
| | .7 | |
| | | 45% |
| $\frac{3}{5}$ | | |
| | | 72% |
| | .24 | |

7. 18 is 90% of what number?

8. Find the sum of 4.86, 11.9, and 28.09.

9. Find the product of $\frac{30}{39}$ and $\frac{78}{105}$.

### CHALLENGE QUESTION

The men and women in a college ski club are to be arranged in order of height for a group photograph. Their heights are listed.

| | | | | | | | |
|---|---|---|---|---|---|---|---|
| 6′0″ | 5′5″ | 5′11″ | 6′1″ | 5′11″ | 5′4″ | 5′7″ | 6′4″ |
| 5′1″ | 4′11″ | 5′11″ | 5′8″ | 6′1″ | 5′9″ | 5′2″ | 5′4″ |
| 5′6″ | 5′11″ | 5′5″ | 6′3″ | 5′9″ | 5′5″ | 5′4″ | 5′10″ |
| 5′8″ | 5′2″ | 5′3″ | 5′8″ | 5′6″ | 5′11″ | 5′4″ | 5′4″ |

a. How many modes are shown in the data?
(a) 0 (b) 1 (c) 2 (d) 3

b. How can you explain the number of modes?

## UNIT 15–3 Finding the Mean

**THE MAIN IDEA**

1. The *mean* of a set of data is the *average*.
2. To find the mean:
   a. Find the sum of the items.
   b. Divide the sum by the number of items.

**EXAMPLE 1** John took 4 exams in his mathematics class. His grades were 72, 76, 86, and 90. What is his mean grade?

John's mean grade is the average of his grades.

First, find the sum of his grades.

$$\begin{array}{r} 72 \\ 76 \\ 86 \\ +90 \\ \hline 324 \end{array}$$

Then, divide the sum by the number of grades. There were 4 exams.

$$\begin{array}{r} 81 \\ 4\overline{)324} \\ \underline{32}\downarrow \\ 04 \\ \underline{4} \\ 0 \end{array}$$

*Answer:* John's mean grade is 81.

**EXAMPLE 2** What is the mean of this group of numbers?

11, 14, 16, 15, 13, 10, 17, 12

First, find the sum of the numbers.

$$\begin{array}{r} 11 \\ 14 \\ 16 \\ 15 \\ 13 \\ 10 \\ 17 \\ +12 \\ \hline 108 \end{array}$$

Then, divide by the number of numbers.

$$13\tfrac{4}{8} = 13\tfrac{1}{2} \quad \textit{Ans.}$$

$$\begin{array}{r} 8\overline{)108} \\ \underline{8}\downarrow \\ 28 \\ \underline{24} \\ 4 \end{array}$$

**EXAMPLE 3** During one week, Carl did this many push-ups: 20, 24, 26, 26, 30, 30, and 32. Joe did this many push-ups during the same week: 20, 16, 18, 25, 25, 34, and 36. Who had the higher average number of push-ups?

### THINKING ABOUT THE PROBLEM

Use the strategy of breaking a problem into simpler problems. First, find Carl's average. Then, find Joe's average. Finally, compare the two averages to find the answer.

Add the numbers of push-ups for Carl.

$$\begin{array}{r} 20 \\ 24 \\ 26 \\ 26 \\ 30 \\ 30 \\ +32 \\ \hline 188 \end{array}$$

Divide the sum by the number of items.

$$\begin{array}{r} 26\frac{6}{7} \\ 7\overline{)188} \\ \underline{14}\phantom{8} \\ 48 \\ \underline{42} \\ 6 \end{array}$$

Carl's average is $26\frac{6}{7}$.

Add the numbers of push-ups for Joe.

$$\begin{array}{r} 20 \\ 16 \\ 18 \\ 25 \\ 25 \\ 34 \\ +36 \\ \hline 174 \end{array}$$

Divide the sum by the number of items.

$$\begin{array}{r} 24\frac{6}{7} \\ 7\overline{)174} \\ \underline{14}\phantom{4} \\ 34 \\ \underline{28} \\ 6 \end{array}$$

Joe's average is $24\frac{6}{7}$.

*Answer:* Carl had the higher average.

**EXAMPLE 4** Stephanie has grades of 78, 92, and 84 in 3 math tests. What grade must she get on a fourth test in order to have an average grade of 85?

### THINKING ABOUT THE PROBLEM

The mean is given and one of the grades must be found. This is a good place to guess and check. Guess a value for the missing grade and check by computing the mean. If the resulting mean is higher than 85, try a lower guess. If the resulting mean is lower than 85, try a higher guess.

Guess a grade and use it to compute the mean.

$$\begin{array}{r} 80 \leftarrow \text{guess} \\ 78 \\ 92 \\ \underline{84} \\ 334 \end{array} \qquad \begin{array}{r} 83.5 \\ 4\overline{)334.0} \end{array} \leftarrow \text{mean (too low)}$$

Try a higher guess.

$$\begin{array}{r} 85 \leftarrow \text{higher guess} \\ 78 \\ 92 \\ \underline{84} \\ 339 \end{array} \qquad \begin{array}{r} 84.75 \\ 4\overline{)339.00} \end{array} \leftarrow \text{mean (still too low, but closer)}$$

Try again.

$$\begin{array}{r} 86 \leftarrow \text{higher guess} \\ 78 \\ 92 \\ \underline{84} \\ 340 \end{array} \qquad \begin{array}{r} 85 \\ 4\overline{)340} \end{array} \leftarrow \text{mean (correct)}$$

The guess of 86 leads to a mean of 85.

*Answer:* Stephanie needs a grade of 86 on the fourth test.

## CLASS EXERCISES

1. Find the mean of each group of numbers.

   **a.** 27, 27, 27, 27, 27 **b.** 11, 17, 13, 16, 18 **c.** 52, 61, 59, 63, 57

   **d.** 36, 48, 52, 37, 50, 46, 38, 51 **e.** 79, 89, 87, 78, 88, 89

   **f.** 104, 212, 365, 248, 167, 190 **g.** 1.1, 1.3, 1.2, .4

   **h.** 26.1, 23.4, 28.6, 20.4, 18.0, 21.1 **i.** 63.02, 48.12, 52.35, 60.03 **j.** 3, $2\frac{3}{4}$, $2\frac{1}{4}$

2. Which set of numbers has the greatest mean?
   (a) 45, 46, 46, 47 (b) 45, 46, 46, 47, 45, 46 (c) 47, 46, 45, 44 (d) 48, 42, 46, 47

3. During one term, Tom's grades in mathematics were 72, 83, 68, 80, and 92. What was Tom's average (mean) grade?

4. The noon temperature (°F) in a certain city for 5 consecutive days was 82°, 78°, 85°, 84°, and 83°. What was the mean noon temperature for these 5 days?

5. Carla, Maria, and James each took 5 exams in mathematics. What was the average grade for each student?

| Carla's Grades | Maria's Grades | James' Grades |
|---|---|---|
| 68 | 100 | 77 |
| 74 | 83 | 76 |
| 95 | 68 | 78 |
| 55 | 54 | 79 |
| 75 | 43 | 74 |

6. Mrs. Harris, the English teacher, thought that the average number of students in her classes was greater than the average number of students in Mr. Smith's classes. Was she right?

| Number of Students in Mrs. Harris' Classes | Number of Students in Mr. Smith's Classes |
|---|---|
| 35 | 34 |
| 36 | 35 |
| 42 | 35 |
| 28 | 37 |
| 30 | 34 |

7. The average height of four students was 67 inches. The heights of three students were 64 inches, 68 inches, and 70 inches. What was the height of the fourth student?

## HOMEWORK EXERCISES

1. Find the mean of each group of numbers.

   **a.** 20, 30, 40 **b.** 15, 16, 14, 17, 18, 16 **c.** 28, 32, 47, 63 **d.** 38, 42, 56, 26

   **e.** 30, 50, 60, 80, 100 **f.** 28, 30, 26, 38, 22 **g.** 46, 39, 54, 18, 71, 92 **h.** 5, 5.5, 6.2, 5.3

   **i.** 9.6, 8.3, 2.4, 11.7 **j.** 2, 5, $5\frac{1}{2}$, 6, $6\frac{1}{2}$ **k.** 3, $3\frac{1}{2}$, $3\frac{1}{4}$, $3\frac{3}{4}$, 3, 4

2. Mark's grades on 4 exams were 78, 83, 85, and 90. What was the mean of his grades?

3. Mary's temperature (°F) for the past 5 mornings was 99°, 100°, 100°, 101°, 100°. What was her mean temperature?

4. Which set of numbers has the greatest mean?
(a) 23, 27, 28, 30 (b) 32, 22, 26, 22 (c) 22, 23, 22, 24, 23, 24 (d) 28, 27, 26, 25, 24, 23

5. These are the numbers of miles run by 3 joggers over a period of 5 days. Which jogger had the highest mean mileage? Which jogger had the lowest mean mileage?

| Jane | Carlo | Milly |
|---|---|---|
| 3 | 1 | 5 |
| 6 | 9 | 4 |
| 2 | 2 | 5 |
| 3 | 2 | 6 |
| 5 | 3 | 5 |

6. These are the tolls that Mr. Reed and Mr. Sweeney each paid over a period of 5 days. Which man had the higher mean toll?

| Mr. Reed | Mr. Sweeney |
|---|---|
| \$2.00 | \$2.00 |
| \$2.50 | \$3.00 |
| \$1.50 | \$2.50 |
| \$3.00 | \$1.50 |
| \$2.50 | \$3.00 |

7. Find the mean of this group of numbers. 63, 68, 64, 68, 68, 65, 62

8. The mean of three numbers is 120. If two of the numbers are 90 and 150, what is the third number?

9. The average yardage gained by four running backs in a football game was 42 yards. If three of the running backs gained 21 yards, 38 yards, and 50 yards, how many yards did the fourth running back gain?

## SPIRAL REVIEW EXERCISES

1. Find the mode for this group of numbers.
48, 44, 41, 47, 48

2. A shelf contains 4 Spanish books and 8 mathematics books. What is the probability that a book chosen at random will be a Spanish book?

3. When a pair of dice is tossed, what is the probability that the sum will be 14?

4. Kim's restaurant bill came to \$12. She wanted to leave the waiter a 15% tip. How much money should she have left for the tip?

5. The first 3 visits to the Health Club cost \$5 each. Each visit after that costs \$4. If Albert visited the Health Club 10 times, how much did he pay?

6. 95% written as a fraction is
(a) $\frac{19}{20}$ (b) $\frac{9}{10}$ (c) $\frac{4}{5}$ (d) $\frac{3}{4}$

7. The value of $\frac{3}{8} + \frac{1}{4} \times \frac{1}{2}$ is
(a) $\frac{1}{2}$ (b) $\frac{5}{16}$ (c) $\frac{1}{6}$ (d) $\frac{3}{64}$

8. $5\frac{1}{2}\%$ written as a decimal is
(a) 5.5 (b) .55 (c) .055 (d) .0505

9. The value of the digit 7 in 238.175 is
(a) 700 (b) 70 (c) $\frac{7}{10}$ (d) $\frac{7}{100}$

10. The greatest common factor of 16 and 24 is
(a) 4 (b) 8 (c) 12 (d) 24

### CHALLENGE QUESTION

The average noontime temperature for four days in December was 8°F. The noontime temperatures for the first three days were 9°F, 10°F, and 20°F. Find the noontime temperature for the fourth day.

# UNIT 15–4 Finding the Median

## THE MAIN IDEA

1. The *median* of a group of numbers is the middle value when the numbers are arranged in order.
2. There are as many values greater than the median as there are less than the median.
3. To find the median, count the number of items.
   a. If the number of items is odd, there is one middle value, which is the median.
   b. If the number of items is even, there are two middle values. The mean of these two middle values is the median. In this case, the median will be a value that is not one of the original values.

**EXAMPLE 1** Find the median of this group of numbers.

24, 26, 27, 29, 30, 34, 36

Since these numbers are already in order, count them. There are 7, which is an odd number. The middle number is the fourth number.

24, 26, 27, 29, 30, 34, 36

There are 3 numbers less than the median. | The middle number is the median. (29) | There are 3 numbers greater than the median.

*Answer:* The median is 29.

**EXAMPLE 2** Find the median of this group of numbers.

21, 23, 24, 26, 28, 34, 36, 39

Since the numbers are already in order, count them. There are 8, which is an even number. The median is halfway between the two middle numbers.

two middle numbers (26, 28)

21, 23, 24, 26, 28, 34, 36, 39

There are 3 numbers less than the median. | median | There are 3 numbers greater than the median.

Find the mean of the two middle numbers by adding them and dividing by 2.

$$\begin{array}{r} 26 \\ +28 \\ \hline 54 \end{array}$$

$$2\overline{)54} = 27$$

*Answer:* The median is 27 (not one of the original values).

**EXAMPLE 3** Find the median of this group of numbers. 26, 34, 14, 13, 27, 33, 32, 28, 29

First, put the numbers in order.

13, 14, 26, 27, 28, 29, 32, 33, 34

Then, locate the middle number by counting. Since there are 9 numbers, the middle number is the fifth number.

13, 14, 26, 27, 28, 29, 32, 33, 34

There are 4 numbers less than the median.

The middle number is the median.

There are 4 numbers greater than the median.

*Answer:* The median is 28.

**EXAMPLE 4** Find the median of this set of numbers. 3, 10, 5, 12, 4, 2

Put the numbers in order and locate the middle.

two middle numbers

2, 3, 4, 5, 10, 12

median

Find the mean of the two middle numbers.

$$\begin{array}{r} 4 \\ +5 \\ \hline 9 \end{array} \qquad 9.0 \div 2 = 4.5$$

*Answer:* The median is 4.5.

## CLASS EXERCISES

**1.** Find the median of each group of numbers.

**a.** 47, 53, 38, 21, 50 **b.** 102, 98, 112 **c.** 12, 3, 9, 15, 11, 8, 4 **d.** 27, 18, 14, 6

**e.** 22, 28, 20, 26 **f.** 75, 80, 67, 92 **g.** 97, 96, 95, 99, 92, 94, 93

**h.** 54, 55, 55, 54, 54, 55, 55, 54, 54, 54, 55 **i.** 18.6, 17.8, 18.7, 19 **j.** 32.4, 32.6, 31.8, 32.2

**2.** These are the prices of 5 automobiles: $8,300 $9,240 $8,210 $10,335 $8,877
What is the median price?

**3.** Find the median of the following group of prices of houses in Brickville:
$52,000 $48,500 $63,000 $71,350

4. The number of cars sold in each of the first 5 months of the year by the Acme Car Lot are shown in the table.
What was the median number of sales?

| Month | Number of Cars Sold |
|---|---|
| Jan. | 20 |
| Feb. | 24 |
| March | 18 |
| April | 40 |
| May | 52 |

## HOMEWORK EXERCISES

1. Find the median of each group of numbers.

   a. 37, 22, 54, 23, 27 b. 11, 9, 8, 12, 10 c. 126, 117, 152 d. 5, 7, 7, 12, 6, 6, 5

   e. 42, 48, 52, 50 f. 33, 34, 34, 33, 33, 34, 34, 33, 33 g. 66, 67, 68, 69, 68, 67

   h. 13.1, 12.6, 16.1, 14.5 i. 36.7, 37.5, 37.6 j. $8\frac{1}{8}, 6\frac{3}{8}, 9\frac{1}{2}, 3$ k. $3\frac{1}{3}, 3\frac{5}{6}, 3\frac{1}{6}, 3\frac{2}{3}, 3\frac{1}{2}, 3$

2. Find the median of this group of book prices. $15.25 $11.30 $10.00 $14.75 $13.25

3. In the group of annual salaries, find the median.

   $34,500 $27,000 $21,750 $30,000 $31,000 $32,500

4. The table shows the numbers of new subscriptions sold by Bobby on his paper route last week. What was the median number of new subscriptions?

| **Day** | Monday | Tuesday | Wednesday | Thursday | Friday |
|---|---|---|---|---|---|
| **Number of New Subscriptions** | 2 | 8 | 3 | 4 | 2 |

## SPIRAL REVIEW EXERCISES

1. For the group of numbers given:
   25, 27, 25, 26, 24, 25, 27

   a. Find the mode. b. Find the mean.

2. The rainfall, in inches, for 7 days in New City was:
   2, 0, 1, 2, 0, 1, 1
   Find the average rainfall for the 7 days.

3. In a bag there are 13 red marbles and 17 green marbles. The probability that a marble picked at random will be red is

   (a) $\frac{13}{30}$ (b) $\frac{17}{30}$ (c) $\frac{13}{17}$ (d) $\frac{17}{13}$

4. Add: $\frac{2}{5} + \frac{3}{20}$ 5. Multiply: $.18 \times .02$

6. If sales tax is charged at the rate of 8%, what is the final cost of a $20 dress?

7. The mean of the set of numbers 12, 12, 15, 20, and 36 is

   (a) 12 (b) 15 (c) 19 (d) 20

8. A shirt is on sale at 25% off the regular price of $30. The amount of discount is

   (a) $7.25 (c) $37.25
   (b) $7.50 (d) $37.50

9. What is the probability that the arrow on the spinner will stop on an even-numbered space?

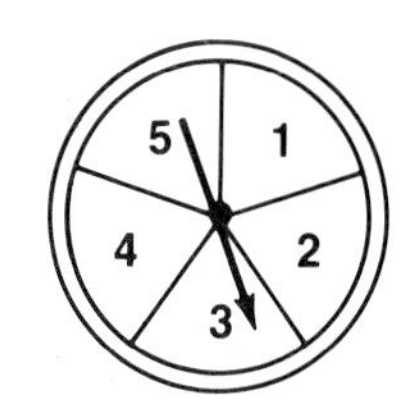


10. What is the mode of the group of numbers shown?

    $1\frac{1}{2}, 1\frac{2}{3}, 2\frac{1}{4}, 2\frac{1}{2}, 2\frac{1}{2}, 2\frac{1}{2}, 3\frac{1}{3}$

11. The smallest of the values shown is

    (a) 9% (b) $\frac{1}{10}$ (c) .08 (d) $7\frac{3}{4}\%$

12. 318.234 rounded to the nearest tenth is

    (a) 310 (c) 318.2
    (b) 320 (d) 318.3

### CHALLENGE QUESTION

In a contest, the 5 winners received cash prizes of $50, $50, $50, $500, and $50,000. After calculating the mean, median, and mode for these values, tell which value most fairly reflects the "average" winning.

## UNIT 16–1 Bar Graphs

### THE MAIN IDEA

1. A *graph* is a visual way to organize and present data.
2. In a *bar graph*, the data are shown as bars of different lengths, which correspond to the numbers that they represent.

**EXAMPLE 1** This bar graph shows the number of push-ups Mike did each day last week.

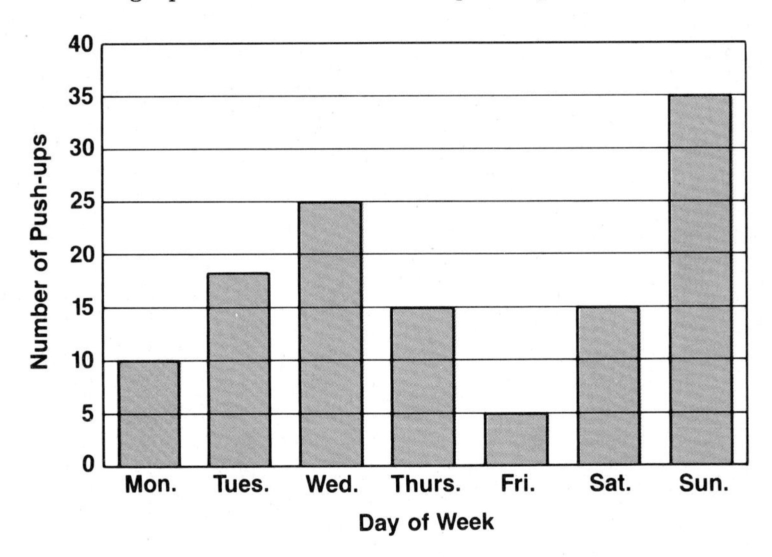


**a.** On which day did Mike do the greatest number of push-ups?

Look for the tallest bar.

*Answer:* Mike did the greatest number of push-ups on Sunday.

**b.** On which day did Mike do the least number of push-ups?

Look for the shortest bar.

*Answer:* Mike did the least number of push-ups on Friday.

**c.** On which two days did Mike do the same number of push-ups?

Look for the two bars that have the same height.

*Answer:* The two days on which Mike did the same number of push-ups are Thursday and Saturday.

**d.** Over which three-day period did the number of push-ups decrease?

Look for three consecutive bars that are decreasing in height.

*Answer:* The three days over which the number of push-ups decreased are Wednesday, Thursday, and Friday.

**e.** How many push-ups did Mike do on Wednesday?

The bar for Wednesday reaches up to the number 25 on the scale at the left.

*Answer:* On Wednesday, Mike did 25 push-ups.

**f.** On which day did Mike do twice as many push-ups as he did on Friday?

Mike did 5 push-ups on Friday. Twice as many is 10 push-ups. The day on which Mike did 10 push-ups is Monday.

*Answer:* The day on which Mike did twice as many push-ups as he did on Friday is Monday.

**g.** How many more push-ups did Mike do on Sunday than on Wednesday?

Mike did 35 push-ups on Sunday and 25 on Wednesday.

Subtract. $35 - 25 = 10$

*Answer:* Mike did 10 push-ups more on Sunday than on Wednesday.

**h.** On which day did Mike do between 15 and 20 push-ups?

Look for the bar that reaches a height that is greater than 15 and less than 20.

*Answer:* Mike did between 15 and 20 push-ups on Tuesday.

**i.** Which number is closest to the number of push-ups that Mike did on Tuesday?
(a) 20 (b) 15 (c) 22 (d) 18

The bar for Tuesday has a height between 15 and 20. The only answer that is between 15 and 20 is 18.

*Answer:* (d)

**EXAMPLE 2** This bar graph shows the average number of calories consumed at breakfast by 8 schoolchildren.

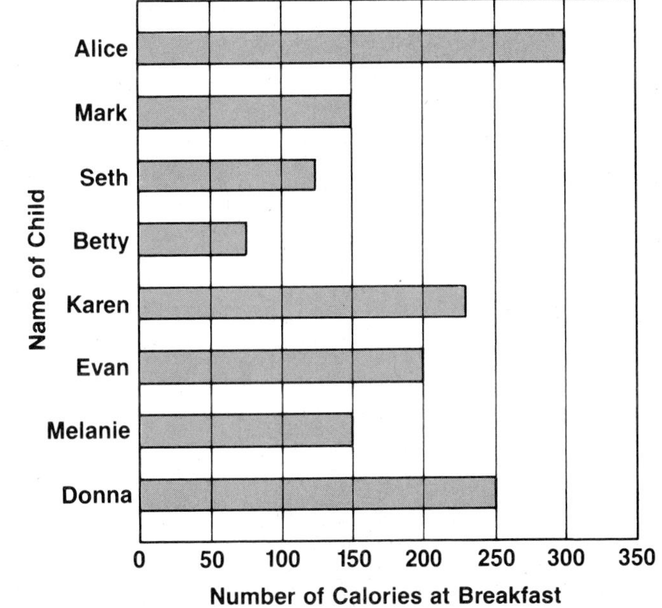


**a.** Which child consumes the largest number of calories at breakfast?

Look for the longest bar.

*Answer:* Alice

**b.** Who consumes the fewest calories at breakfast?

Look for the shortest bar.

*Answer:* Betty

**c.** Which children consume the same number of calories at breakfast?

Look for bars that are equal in length.

*Answer:* Mark and Melanie

**d.** About how many calories does Seth consume at breakfast?

The bar for Seth stops about halfway between 100 and 150.

*Answer:* 125 calories

**e.** Who consumes twice as many calories as Melanie?

Melanie consumes 150 calories. The child who consumes 300 calories is Alice.

*Answer:* Alice

**f.** How many more calories does Donna consume than Evan?

Donna consumes 250 calories and Evan consumes 200 calories. The difference is 50 calories.

*Answer:* 50 calories

**EXAMPLE 3** Make a bar graph to show the given information.

**Number of Pairs of Sneakers Sold at the Sneaker Den**

| | |
|---|---|
| Monday | 15 |
| Tuesday | 20 |
| Wednesday | 10 |
| Thursday | 18 |
| Friday | 24 |
| Saturday | 37 |

Choose to make either a horizontal or vertical graph.

Choose a scale that will fit the given numbers. Starting at 0, mark off the scale on graph paper. Use labels.

Draw the proper lengths for the bars. Be sure the bars all have the same width.

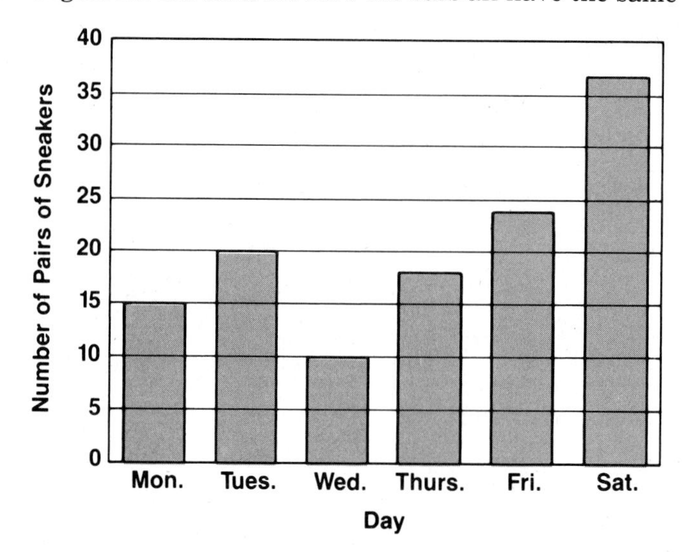

## CLASS EXERCISES

1. This bar graph shows the number of books read by some students during the month of October.

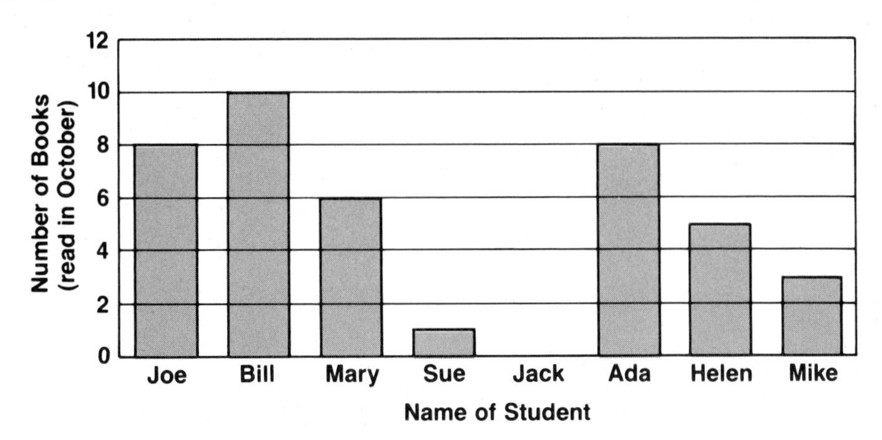


  a. Who read the greatest number of books in October?
  b. Who read no books in October?
  c. Which two students read the same number of books?
  d. Who read twice as many books as Mike?
  e. Who read half the number of books that Bill read?
  f. How many more books did Joe read than Helen?

2. This bar graph shows the expected enrollment for Plainview School during the years 1990 to 1996.

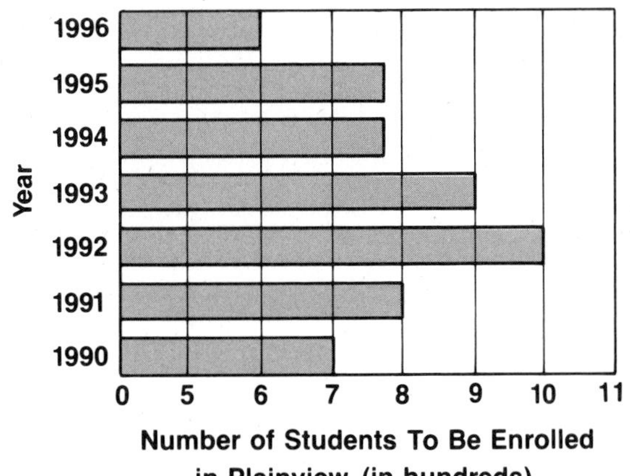


  a. During which year is enrollment expected to be the lowest?
  b. When is enrollment expected to reach its highest level?
  c. During which two years is it expected that enrollment will stay the same?
  d. How many more students are expected to enroll in 1991 than in 1996?
  e. In which year are 300 fewer students expected than in 1992?

3. Make a bar graph for the information in each table.

a.

| Salaries of Executives at Topper Company | |
|---|---|
| Ms. Jones | $40,000 |
| Mr. Brown | $25,000 |
| Miss White | $17,000 |
| Mr. Smith | $37,000 |
| Mr. Green | $30,000 |

b.

| Number of Trees Sold by Green Nursery | |
|---|---|
| March | 150 |
| April | 170 |
| May | 220 |
| June | 280 |
| July | 190 |
| August | 100 |

## *HOMEWORK EXERCISES*

1. This bar graph shows Mary's baby-sitting earnings for 6 months.

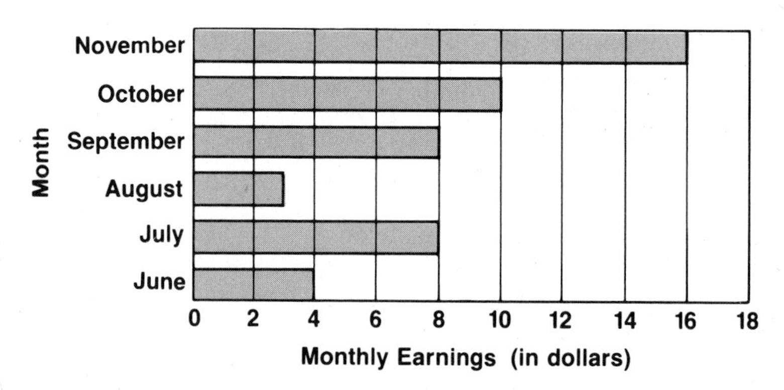


a. How much baby-sitting money did Mary earn in August?
(a) $2 (b) $3 (c) $4 (d) $6

b. In which month did Mary earn the least?

c. In which two months did Mary earn the same amount?

d. How much more did Mary earn in November than in October?

e. For which two months combined did Mary earn a total of $20?

2. This bar graph shows the number of miles traveled by a salesperson during each of five months.

a. What is the best estimate of the number of miles traveled in January?
(a) 400 (b) 375 (c) 350 (d) 300

b. The total number of miles traveled over the five-month period is
(a) under 1,000
(b) between 1,000 and 1,500
(c) between 1,500 and 2,000
(d) over 2,000

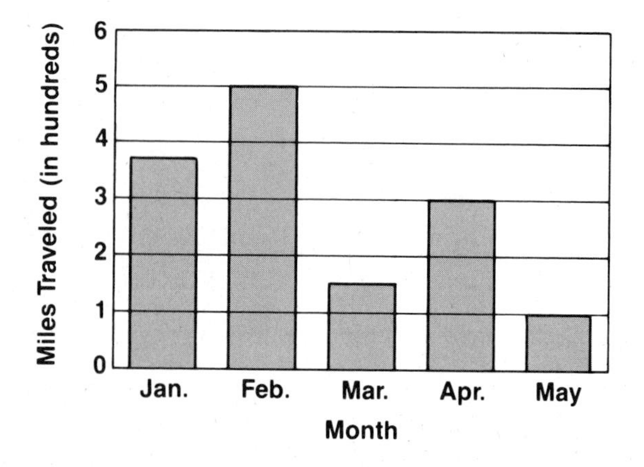

3. This bar graph shows the number of milligrams of cholesterol in an average loaf of 5 different brands of bread.

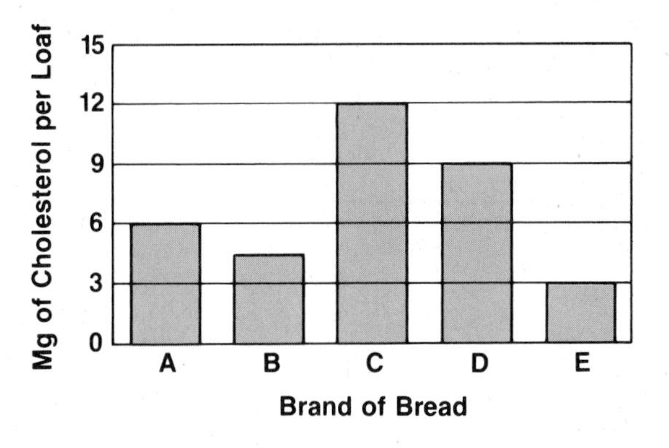


a. Which brand has about half the amount of cholesterol as Brand D?
(a) Brand A (c) Brand C
(b) Brand B (d) Brand E

b. Which two brands have as their sum the same amount of cholesterol as Brand D?

4. This bar graph shows the number of lunches served each month at Beefburg High School.

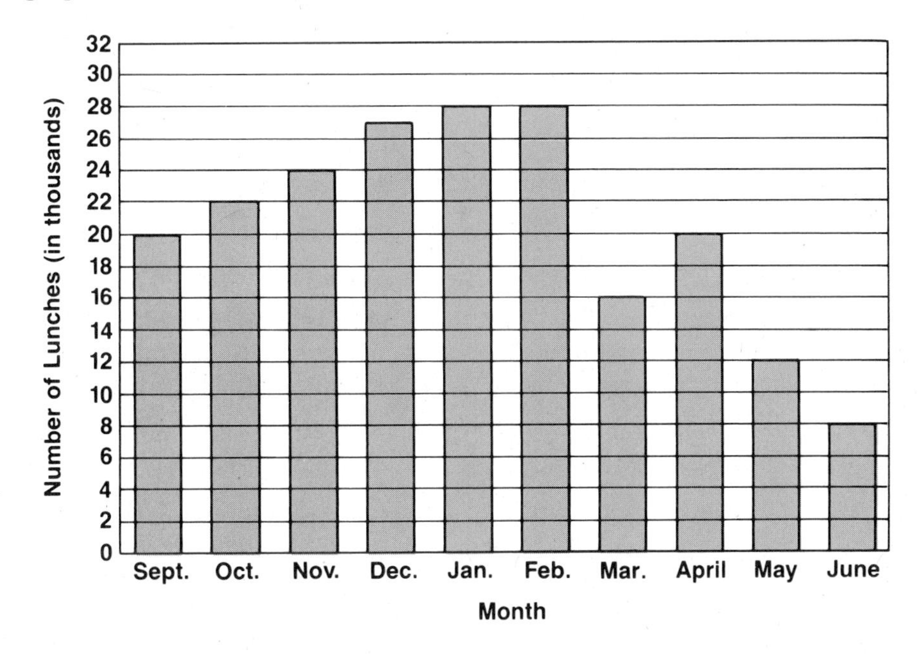


a. What was the greatest number of lunches served in one month?

b. In which months was the greatest number of lunches served?

c. What was the smallest number of lunches served in one month?

d. Estimate the number of lunches served in December.

e. In which month was the same number of lunches served as in September?

f. How many more lunches were served in April than in May?

g. In which month was the greatest drop from the previous month in the number of lunches served?

h. In which month was twice the number of lunches served as in May?

i. In which month was one-third the number of lunches served as in November?

j. During how many months were fewer than 20,000 lunches served?

**5.** Make a bar graph for the information in the table.

| Number of Students Taking Language Courses at Endwood High School | |
|---|---|
| French | 550 |
| Latin | 200 |
| Spanish | 700 |
| German | 300 |
| Italian | 250 |
| Chinese | 100 |

## SPIRAL REVIEW EXERCISES

**1.** If Debra's grades on her mathematics tests are 75, 85, 65, 90, and 80, find the average of her grades.

**2.** Find the median for the following test scores: 55, 60, 75, 75, 80, 80, 95

**3.** What is the mode of the given set of numbers?

23, 29, 29, 35, 46, 47, 47, 47, 50

**4.** What is the probability that the arrow on the spinner shown will stop on a prime number?

**5.** A $120 jacket is being sold at a 40% discount. What is the sale price?

**6.** How much change is received from a $20 bill on a $14.80 purchase?

**7.** At a practice game, Jeff made 20 out of 50 baskets he attempted. What percent of the attempted baskets did Jeff make?

**8.** Five quarts of milk cost $3.00. What is the cost of 2 quarts?

**9.** On a winter day, 7 trains arrived on time and 13 trains were late. What is the ratio of the number of on-time trains to the total number of trains?

**10.** Find the sum of $\frac{7}{16}$ and $\frac{5}{8}$.

**11.** Subtract 15.98 from 22.17.

**12.** Multiply: $1.9 \times .26$

**13.** The number of grams in 3.8 kilograms is

(a) .38 (c) 380

(b) .038 (d) 3,800

## CHALLENGE QUESTION

This bar graph shows the number of students attending basketball games at Harris High School.

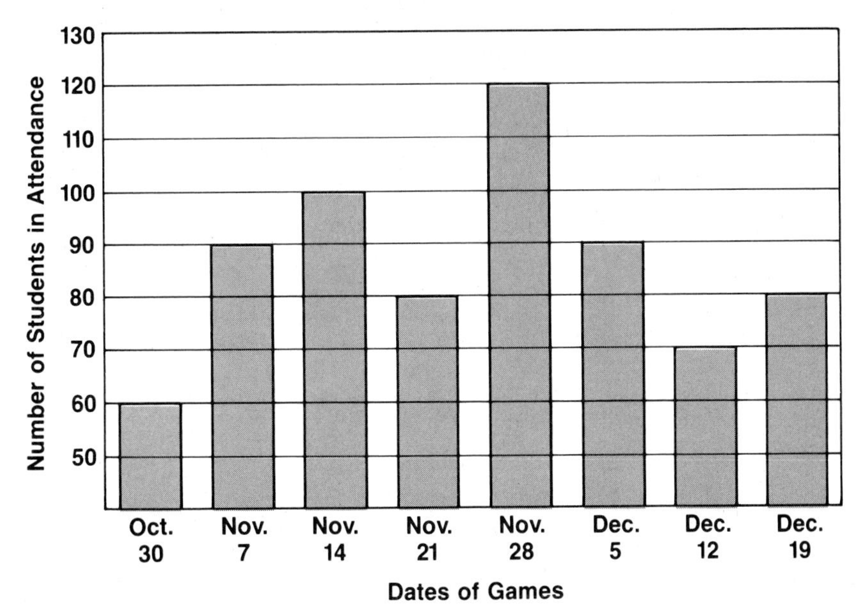


For how many games was the number of students in attendance above average?

# UNIT 16–2 Line Graphs

## THE MAIN IDEA

1. On a *line graph,* pairs of points representing numbers are connected to form a *broken line.*
2. The rise or fall of the line shows the increase or decrease between numbers.

**EXAMPLE 1** This line graph shows the 12:00 noon temperatures in Beefburg last week.

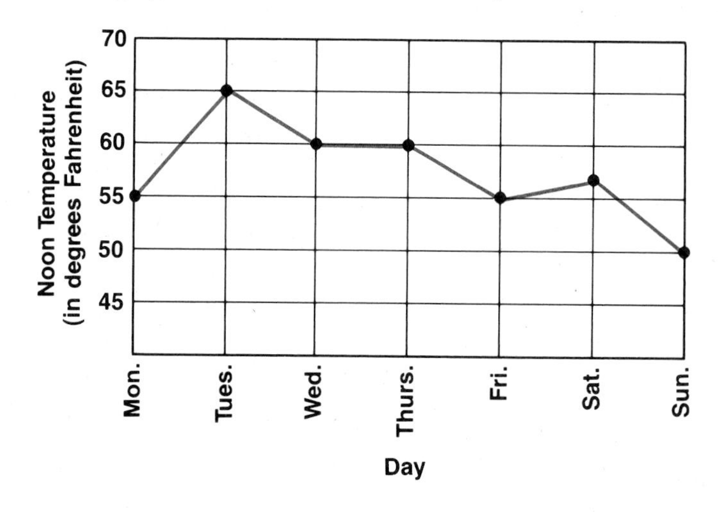


**a.** Which day had the highest noon temperature?

Look for the highest point on the graph.

*Answer:* Tuesday

**b.** Which day had the lowest noon temperature?

Look for the lowest point on the graph.

*Answer:* Sunday

**c.** For which two consecutive days did the noon temperature remain the same?

Look for a piece of the line *(line segment)* that is neither rising nor falling.

*Answer:* Wednesday and Thursday

**d.** On which day did the noon temperature increase the most from the previous day?

There are two places where the broken line rises: from Monday to Tuesday and from Friday to Saturday. The line segment rises more from Monday to Tuesday than from Friday to Saturday. Therefore, Tuesday's temperature shows the greatest increase from the previous day.

*Answer:* Tuesday

**e.** What was the noon temperature in Beefburg on Thursday?

From the point on the broken line above the name "Thurs." look left to the vertical scale and read the number "60."

*Answer:* 60°

**f.** On which day was the noon temperature 50°?

Find 50 on the vertical scale and look right for a point on the broken line that has the same height. The point is above the name "Sun."

*Answer:* Sunday

**g.** Which is the best estimate of the noon temperature on Saturday?
(a) 52° (b) 57° (c) 59° (d) 68°

The height of the point above "Sat." is between 55° and 60°, but closer to 55°. The closest choice is 57°.

*Answer:* (b)

**h.** On which day was the noon temperature the same as it was on Monday?

The noon temperature on Monday was 55°. It was 55° also on Friday.

*Answer:* Friday

**EXAMPLE 2** This line graph shows the population, in thousands, of Apex City for a period of 7 years.

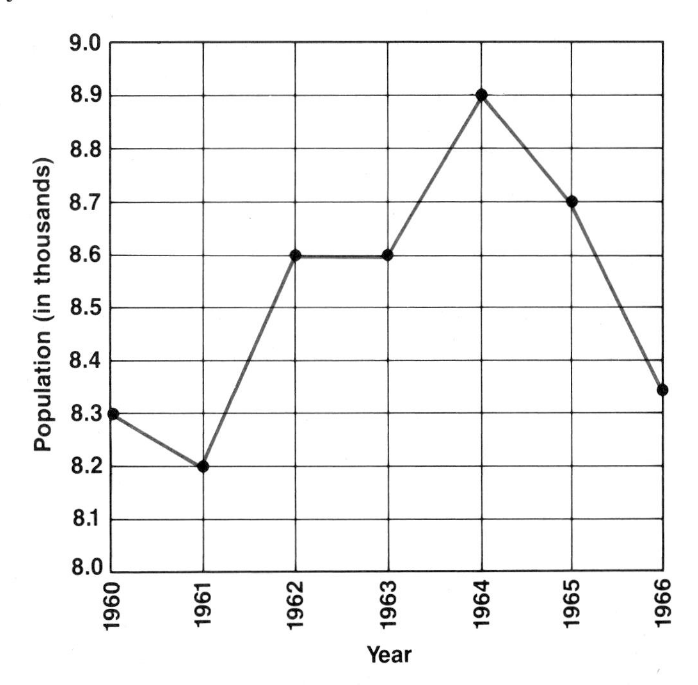


**a.** What was the population during the year when the population was the largest?

The year when the population was the largest was 1964. The point above 1964 corresponds to the number 8.9 at the left on the vertical scale.

Since the scale gives the population "in thousands," multiply by 1,000. Move the decimal point three places to the right.

1000 × 8.9

8.900.

*Answer:* 8,900

**b.** Estimate the population for 1966.

The point above 1966 lies halfway between 8.3 and 8.4 on the vertical scale. 8.3 is the same as 8.30 and 8.4 is the same as 8.40. The number halfway between 8.30 and 8.40 is 8.35.

Multiply by 1,000 because the population is given "in thousands."

$1000 \times 8.35$

8.350.

*Answer:* 8,350

**EXAMPLE 3** Make a line graph to show the information given in the table.

| Mr. Harris' Systolic Blood Pressure | |
|---|---|
| July 4 | 140 |
| July 5 | 138 |
| July 6 | 138 |
| July 7 | 135 |
| July 8 | 139 |
| July 9 | 140 |
| July 10 | 137 |
| July 11 | 135 |
| July 12 | 132 |
| July 13 | 132 |

Choose a scale that will fit the given numbers. The scale need not begin at zero. Mark off the scale on graph paper. Use labels.

Place a point on the graph for each piece of given information.

Draw line segments between consecutive points.

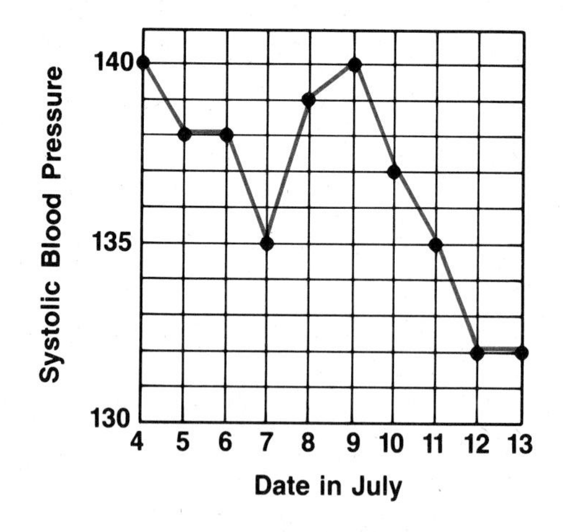

## CLASS EXERCISES

1. This graph shows the average price of one share of stock of the Bionic Dog Food Co. over a period of 8 months.

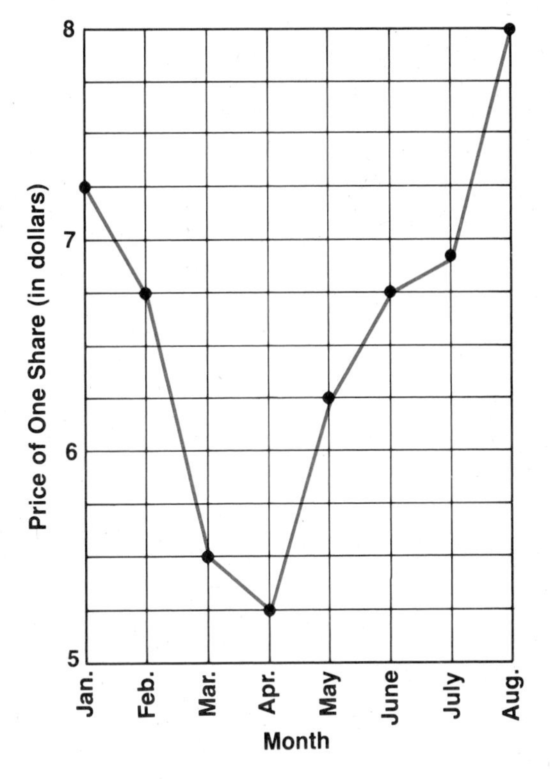


   a. When did the stock reach its lowest average price during this period?
   b. What was the lowest average price of the stock?
   c. When was the average price of the stock the highest?
   d. What was the highest average price of the stock?
   e. What was the average price of one share of stock in February?
   f. Between which two months did the average price fall the most?
   g. During which month was the average price of the stock the same as it was in June?
   h. What is the best estimate of the average price of the stock in July?
      (a) $6.50 (c) $7.10
      (b) $6.75 (d) $6.90
   i. In which month was the average price of the stock $6.25?
   j. How much did the average price of the stock drop from January to February?

2. This line graph shows the attendance figures at Beefburg High School football games for a period of 7 weeks.

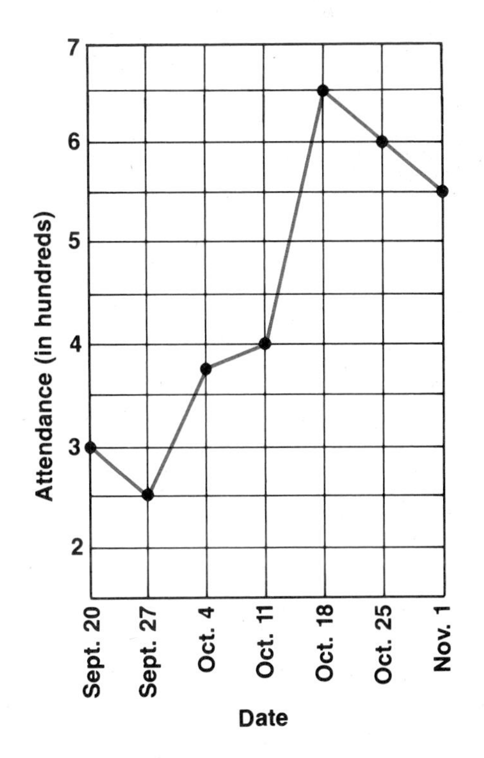


   a. Which game had the lowest attendance?
   b. What was the lowest attendance?
   c. What was the highest attendance?
   d. Which game had 550 in attendance?
   e. How many fewer people attended the September 27 game than had attended the September 20 game?
   f. Estimate the attendance at the October 4 game.
   g. How many games had fewer than 500 in attendance?
   h. What was the total attendance for the 4-week period beginning October 11?
   i. By how much did the attendance for the 2-week period beginning October 11 exceed the attendance for the 2-week period beginning September 20?

3. Make a line graph to show the information in each table.

a.

| Ms. William's Weight | |
|---|---|
| August 1 | 137 |
| September 1 | 135 |
| October 1 | 138 |
| November 1 | 135 |
| December 1 | 135 |
| January 1 | 133 |

b.

| Per Share Value of Capitol Stock | |
|---|---|
| January 3 | $10.25 |
| January 10 | 10.25 |
| January 17 | 10.25 |
| January 24 | 10.40 |
| January 31 | 10.40 |
| February 7 | 10.30 |
| February 14 | 10.30 |
| February 21 | 10.35 |
| February 28 | 10.40 |

## HOMEWORK EXERCISES

1. This line graph shows how the price of gasoline changed over a six-month period.

   a. What was the difference in price of a gallon of gasoline from November to December?
   (a) 20¢ (b) $2.00 (c) $10 (d) 10¢

   b. In which month did the price of gasoline begin to rise?

   c. Between which months was the price of gasoline falling?

   d. For which months did the price of gasoline remain the same?

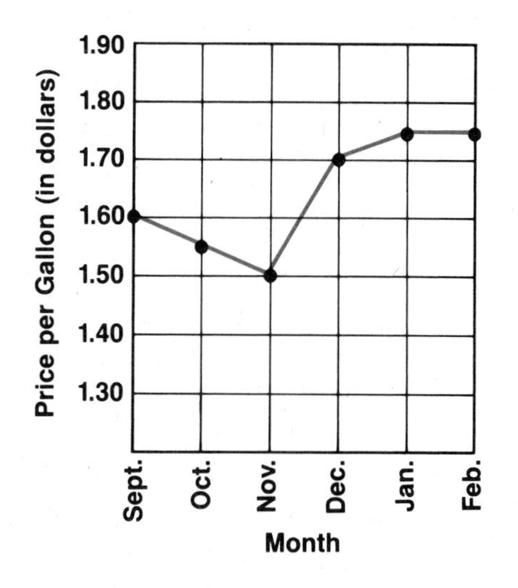


2. This line graph shows the number of births at Umberland Hospital over a five-year period.

   a. Between which two years did the number of births increase the most?
   (a) 1960–1961 (c) 1962–1963
   (b) 1961–1962 (d) 1963–1964

   b. During which interval did the number of births double?
   (a) 1960–1962 (c) 1961–1963
   (b) 1960–1963 (d) 1961–1964

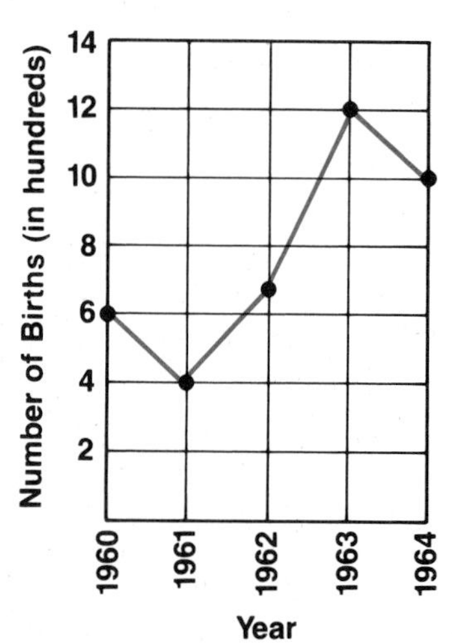

3. This line graph shows the number of tickets sold each night at the Bijou Theatre during one week.

   a. How many tickets were sold on Wednesday?

   b. How many more tickets were sold on Saturday than on Wednesday?
   (a) 45 (b) 450 (c) 30 (d) 300

   c. What was the total number of tickets sold that week?

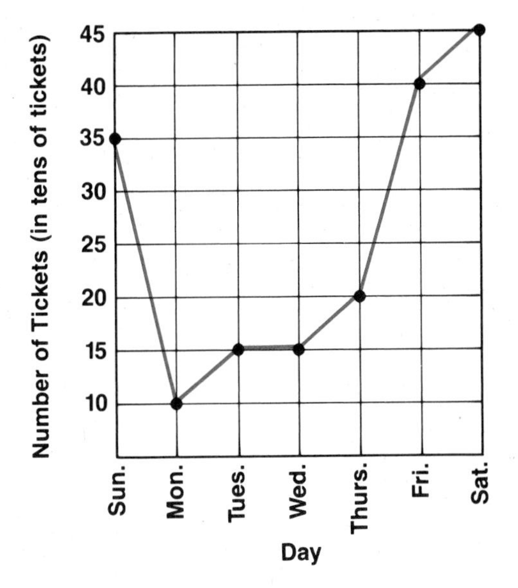


4. This line graph shows Manny's scores on weekly reading tests.

   a. What was Manny's score in week 5?

   b. What was Manny's lowest score?

   c. What was his highest score?

   d. What was the difference between Manny's highest score and his lowest score?

   e. Estimate Manny's score in week 4.

   f. What was the increase in Manny's score from week 6 to week 7?

   g. What was the greatest decrease in Manny's reading score from one week to the next?

   h. In which week did Manny score 30 points higher than he did in week 1?

   i. For how many weeks did Manny score above 70?

   j. In which week did Manny's score drop by about 2 points?

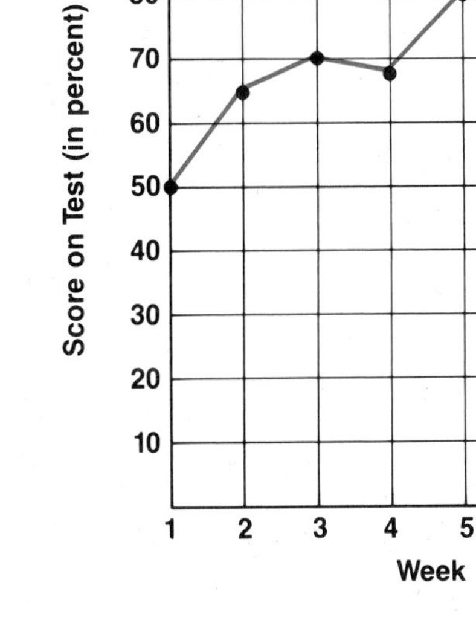


5. Make a line graph to show the given information.

| | Mon. | Tues. | Wed. | Thurs. | Fri. | Sat. | Sun. |
|---|---|---|---|---|---|---|---|
| **Number of Miles Cindy Bicycled** | 5 | 3 | 7 | 1 | 0 | 5 | 4 |

## SPIRAL REVIEW EXERCISES

1. This bar graph shows the number of miles traveled by six commuters.

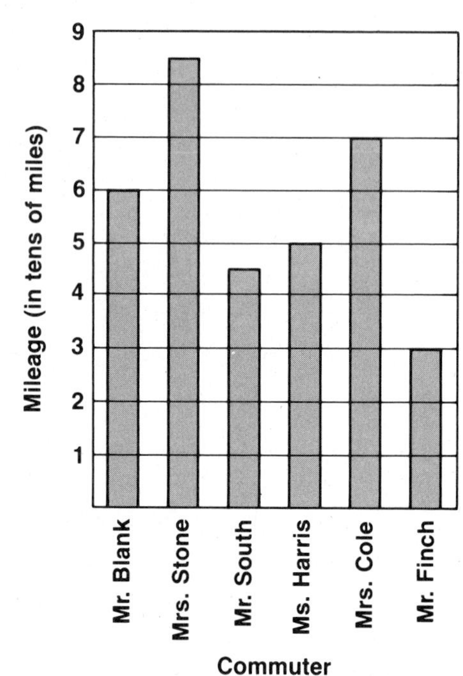


Estimate the number of miles traveled by Mr. South.

(a) $4\frac{1}{2}$ (b) 5 (c) 40 (d) 45

2. The cost of a steak dinner at five different restaurants is as follows: \$11, \$12, \$9, \$18, and \$16. What is the mean cost?

3. The heights of the five tallest girls in Eastern School are 5 ft. 6 in., 5 ft. 8 in., 5 ft. 9 in., 5 ft. 11 in., and 6 ft. What is the median height of these girls?

4. What is the probability of picking an ace at random from a standard deck of 52 cards?

5. Find 28% of 200.

6. If 9 cans of soda cost \$2.79, what is the cost of 5 cans?

7. What is $\frac{7}{8}$ of 72?

8. Subtract $3\frac{3}{4}$ from $5\frac{7}{8}$.

9. .35 is equivalent to

(a) $\frac{35}{1000}$ (b) $\frac{7}{20}$ (c) $\frac{1}{4}$ (d) $\frac{2}{5}$

10. The percent of the figure that is shaded is

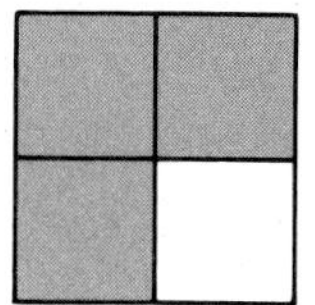

(a) 25% (c) $66\frac{2}{3}$%

(b) 50% (d) 75%

11. Which numeral represents one million, fifty-two thousand?

(a) 1,520,000 (c) 1,052,000

(b) 105,200 (d) 1,005,200

12. Find the sale price of a \$24 dress that is being sold at a 10% discount.

13. Use a signed number to show each situation.

**a.** a \$50 deposit **b.** 10° below zero

**c.** a 20° decrease in temperature

**d.** a loss of \$10

14. Mr. Syms buys a television set by making a \$50 down payment and paying 20 installments of \$18 each. Find the cost of the television set.

15. To the quotient of −20 and 4, add −6.

## CHALLENGE QUESTION

This graph shows the number of cars sold by Happy Harry's Car Lot during an eight-week period.

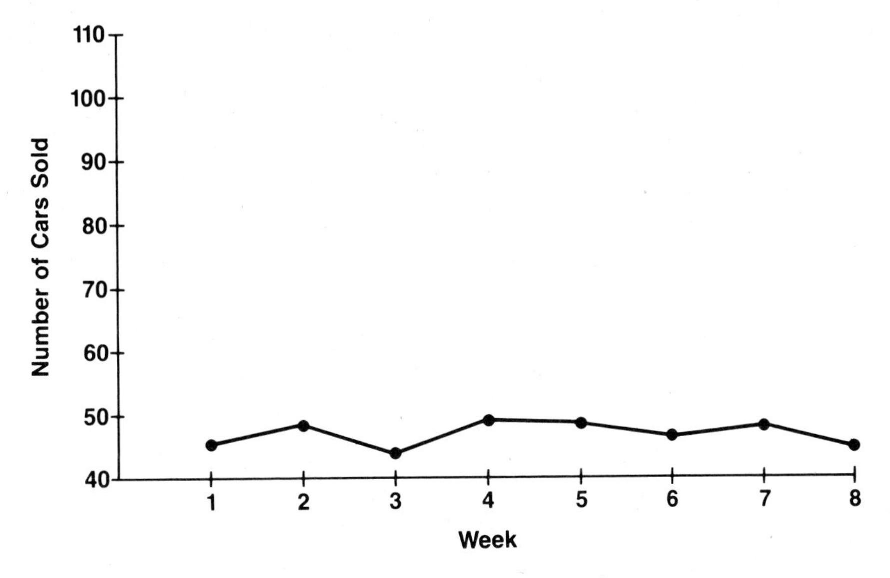


Explain what you would do to improve this graph. Be specific.

# UNIT 16–3 Circle Graphs

## THE MAIN IDEA

1. A ***circle graph***, also called a ***pie graph***, shows the way in which a whole quantity is divided into parts called ***sectors***.
2. The parts are compared in two ways:
   a. by comparing the fractions, percents, or other numerical data, or
   b. by comparing the sizes of the sectors.

**EXAMPLE 1** The circle graph represents the kinds of cheese made in a dairy plant.

**a.** What kind of cheese is produced in the smallest quantity?

Look for the smallest sector. The smallest sector, with $\frac{1}{12}$ of the total, represents Colby cheese.

*Answer:* Colby cheese is produced in the smallest quantity.

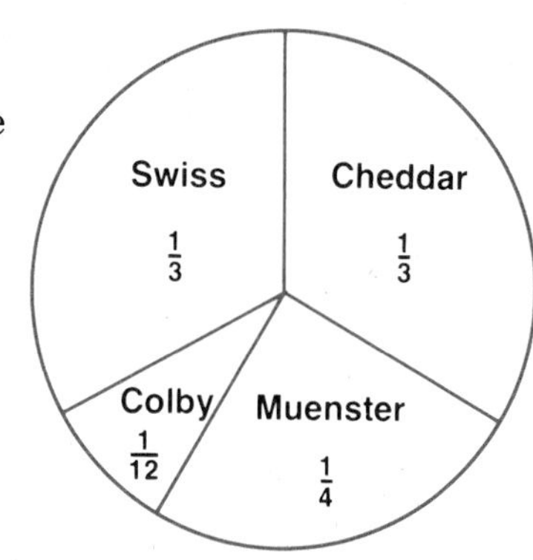


**b.** What kind of cheese is made in the same quantity as cheddar?

The graph shows that cheddar cheese is $\frac{1}{3}$ of the total production. Swiss cheese is also $\frac{1}{3}$ of the total.

*Answer:* Swiss cheese is produced in the same quantity as cheddar cheese.

**c.** Of a total of 168 pounds of cheese, how many pounds are Muenster cheese?

Of the total production, $\frac{1}{4}$ is Muenster cheese. Find $\frac{1}{4}$ of 168.

$$\frac{1}{\cancel{4}_1} \times \frac{\cancel{168}^{42}}{1} = 42$$

*Answer:* There are 42 pounds of Muenster cheese.

**EXAMPLE 2** This circle graph represents the Martino family budget of $1,000 for a month.

**a.** What is the largest part of the Martino budget?

Look for the largest sector. This is also the sector with the largest percent.

*Answer:* The largest part of the budget is for rent.

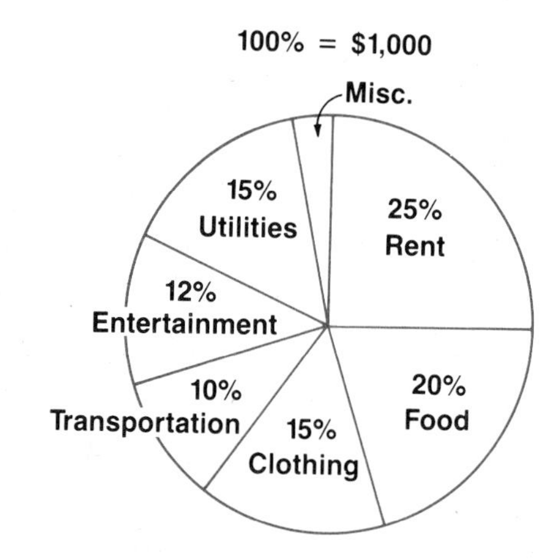

**b.** How much money do the Martinos spend on transportation each month?

The percent for transportation is 10%.

Write 10% as a decimal. 10% = .10

Multiply the whole amount of money in the budget, $1,000, by .10.

$$\begin{array}{r} \$1000 \\ \times .10 \\ \hline 00\ 00 \\ 100\ 0 \\ \hline \$100.00 \end{array}$$

*Answer:* The Martinos spend $100 per month on transportation.

**c.** What percent of the budget is set aside for miscellaneous?

The sum of all the sectors must be 100%.

Add the known percents.

$$\begin{array}{r} 25\% \\ 20\% \\ 15\% \\ 10\% \\ 12\% \\ 15\% \\ \hline 97\% \end{array}$$

Subtract the sum from 100%.

$$\begin{array}{r} 100\% \\ -97\% \\ \hline 3\% \end{array}$$

*Answer:* 3% of the budget is for miscellaneous.

## CLASS EXERCISES

**1.** The circle graph shows the categories of the 15,000 books in a public library.

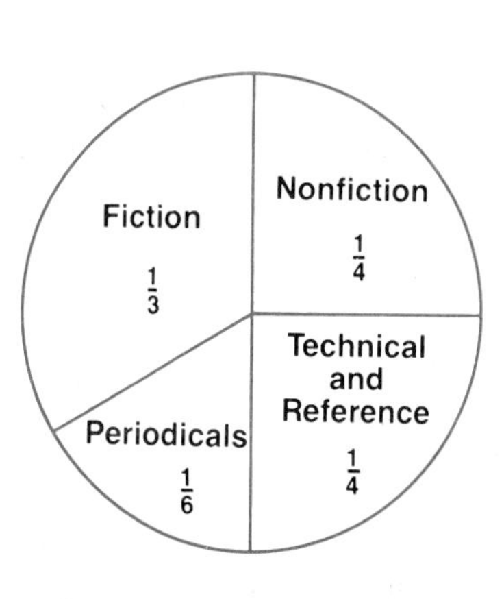


**a.** Which is the largest category of books?

**b.** How many nonfiction books does the library have?

**c.** Which category has fewer books than nonfiction books?

**d.** How many periodicals are there?

**e.** How many more nonfiction books are there than periodicals?

**2.** This circle graph shows the results of polling 250 people to find out their favorite sport.

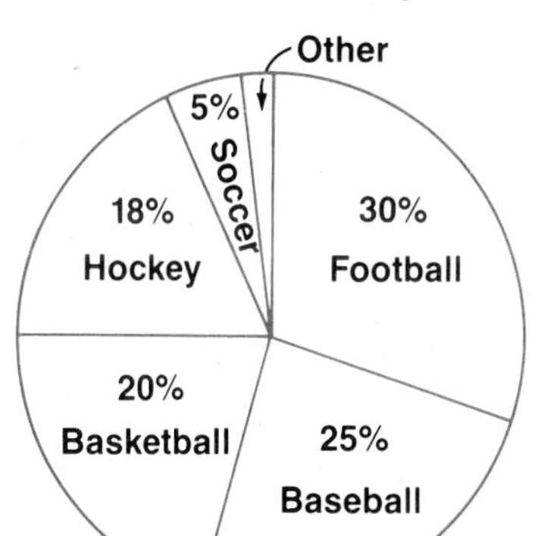


**a.** What percent of the people polled had a favorite sport other than football, baseball, basketball, hockey, or soccer?

**b.** How many of the people had a favorite sport other than those mentioned above?

**c.** How does the number of people whose favorite sport is soccer compare with the number of people whose favorite sport is basketball?

3. This circle graph shows how the voters in Apex City voted in the last election.

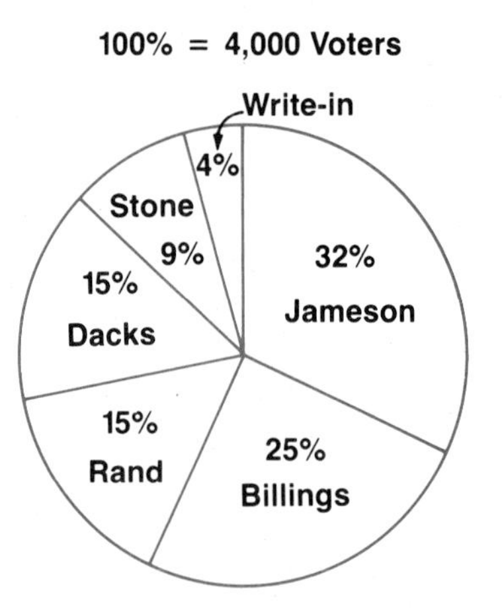


   a. Which candidate received the greatest number of votes?

   b. How many people voted for Stone?
   (a) 36 (b) 360 (c) 900 (d) 9

   c. How many more people voted for Billings than voted for Rand?

   d. Suppose the rule in this election was that to be the winner a candidate had to receive a *majority* of the votes. Would there have been a winner? Explain.

## HOMEWORK EXERCISES

1. This circle graph shows the level of school completed by 1,000 people questioned in a poll.

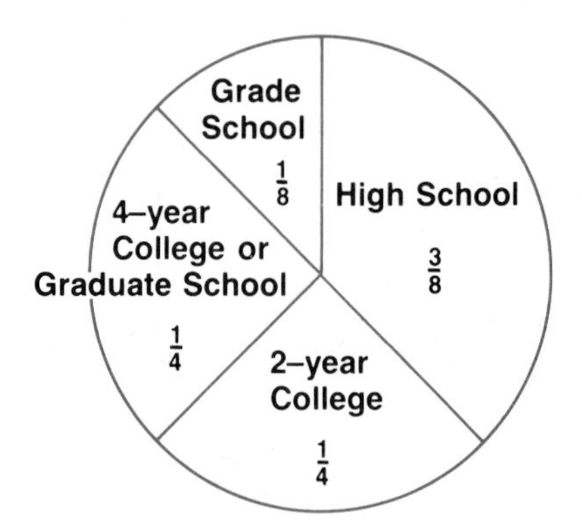


   a. What fraction of the people completed only grade school?

   b. How many people completed only 2 years of college?
   (a) 2.5 (b) 25 (c) 250 (d) 2,500

   c. What fraction of the people completed *at least* 2 years of college?

2. This circle graph shows how the monthly sales income for a pet store was obtained.

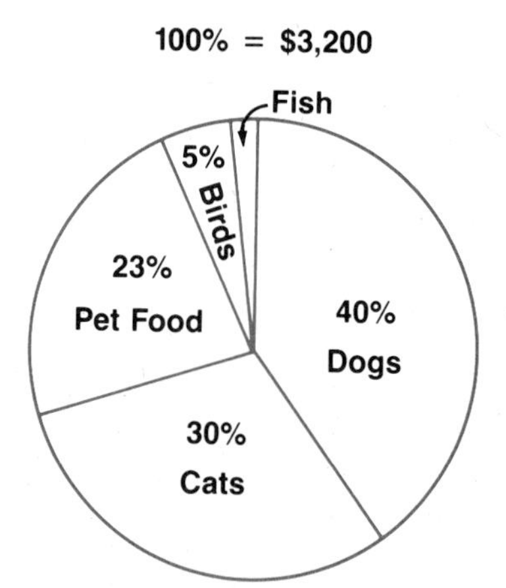


   a. What percent of the monthly income was from the sale of birds?

   b. What was the income, in dollars, from the sale of cats and pet food?

   c. What part of the monthly income was from the sale of fish?
   (a) none (b) 1% (c) 2% (d) 5%

3. This circle graph represents the results of a survey of 250 students at Riding High School who were asked to name their favorite food.

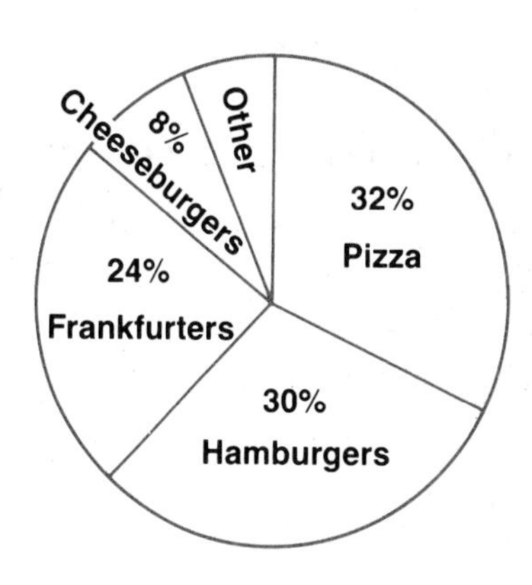


a. How many students listed pizza as their favorite?

b. What was the second most popular lunch food?

c. What percent of the students surveyed had a favorite lunch food other than pizza, hamburgers, frankfurters, or cheeseburgers?

d. Of those listed, which was the least popular lunch food?

e. Which lunch food had $\frac{1}{4}$ the number of votes as pizza?

f. How many students voted for hamburgers? for frankfurters? for cheeseburgers?

g. Which food had 3 times the number of votes as cheeseburgers?

h. How many more students voted for hamburgers than for frankfurters?

i. What percent of the students surveyed preferred *either* pizza or hamburgers?

## SPIRAL REVIEW EXERCISES

1. This line graph shows the temperature in Mr. Florio's greenhouse for 5 hourly readings.

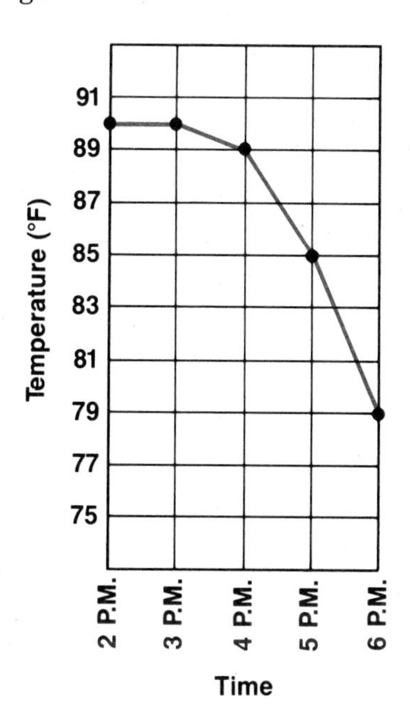


By how many degrees did the temperature drop from 4 P.M. to 5 P.M.?
(a) 2 (b) 4 (c) 40 (d) 20

2. This bar graph shows the number of times five students were absent from school in one year.

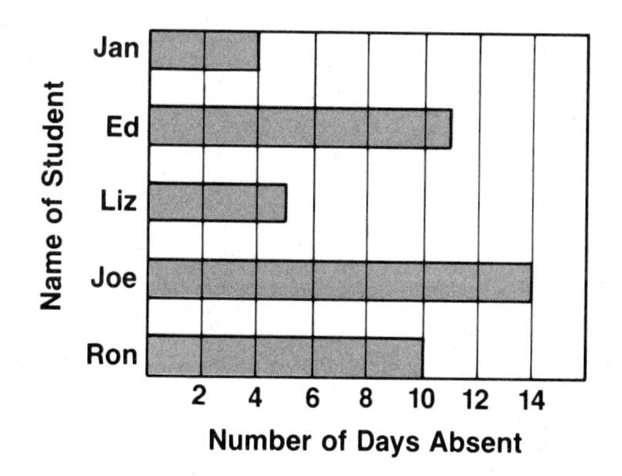


How many more days than Liz was Joe absent?
(a) 20 (b) 9 (c) 11 (d) 8

3. Find the mode of these temperatures.
56°, 59°, 59°, 60°, 63°, 64°, 70°

4. Find the mean of these weekly salaries.
$150, $185, $225, $280

5. A box contains 5 red marbles, 7 green marbles, and 8 yellow marbles. What is the probability of randomly picking a red marble?

6. What percent of 40 is 8?

7. If the sales tax is 7%, how much tax would you pay on a $56 purchase?

8. A $60 jacket is on sale for $30. The percent of discount is
   (a) 25%  (c) 50%
   (b) $33\frac{1}{3}$%  (d) 75%

9. Cynthia bought a fur coat by making a down payment of $500 and 10 monthly payments of $300 each. How much did the coat cost her?

10. Mr. Sachs withdraws $125 from his savings account that had a balance of $2,515. What is the new balance?

11. Add: $2.7 + 15.06 + 7.93 + .8$

12. Multiply: $5\frac{1}{3} \times 3\frac{3}{8}$

13. What is the greatest common factor of 36 and 45?

## CHALLENGE QUESTION

These circle graphs show the family budgets for the Greene family and the Dawson family.

Greene Family

Misc.
5%
10% Carfare
20% Clothing
40% Rent
25% Food

Dawson Family

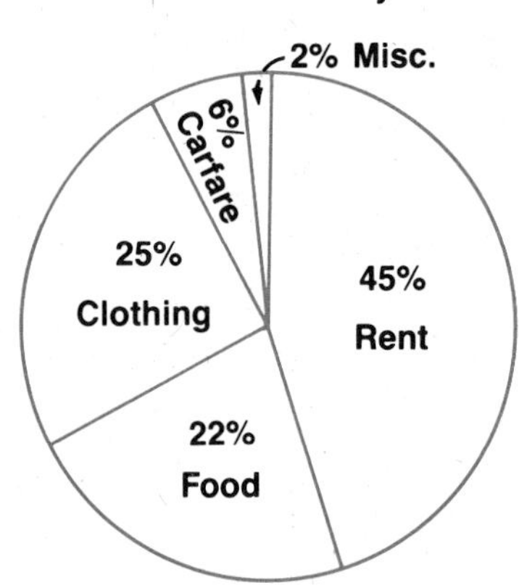


If the weekly income of the Greene family is $650 and the weekly income of the Dawson family is $580, which family has budgeted more money each week for rent?

# UNIT 16–4 Pictographs

## THE MAIN IDEA

1. In a ***pictograph,*** numbers are represented by symbols.
2. There are two types of pictographs:
   a. pictographs in which each small symbol stands for a given amount, and numbers are shown as rows of little symbols, and
   b. pictographs in which different amounts are shown with proportionately sized symbols.

**EXAMPLE 1** This pictograph shows the number of books circulated by the local library over a period of six months.

**Each** = **50 Books**

**May**

**June**

**July**

**August**

**September**

**October**

**a.** How many books were circulated in June?

There are 10 symbols for books shown for June. Each symbol stands for 50 books.

Multiply the number of symbols by the value of each symbol, 50 books.

$$\begin{array}{r} 50 \\ \times 10 \\ \hline 500 \end{array}$$

*Answer:* 500 books in June

**b.** How many books were circulated in August?

There are $3\frac{1}{2}$ symbols shown for August.

Multiply $3\frac{1}{2}$ by 50.

$$3\frac{1}{2} \times 50 = \frac{7}{2} \times 50$$

$$= \frac{7}{\not{2}_{1}} \times \frac{\not{50}^{25}}{1} = \frac{175}{1} = 175$$

*Answer:* 175 books in August

**c.** How many more books were circulated in October than in September?

To find the number of books for October, multiply $6\frac{1}{2}$ by 50.

$$6\frac{1}{2} \times 50$$

$$= \frac{13}{2} \times \frac{50}{1}$$

$$= \frac{13}{\cancel{2}_{1}} \times \frac{\cancel{50}^{25}}{1}$$

$$= \frac{13}{1} \times \frac{25}{1}$$

$$= \frac{325}{1} = 325 \leftarrow \text{books in October}$$

To find the number of books in September, multiply 5 by 50.

$$\begin{array}{r} 50 \\ \times 5 \\ \hline 250 \end{array} \leftarrow \text{books in September}$$

Subtract to find the difference.

$$\begin{array}{r} 325 \\ -250 \\ \hline 75 \end{array}$$

*Answer:* 75 more books were circulated in October than in September.

**d.** During which month were 350 books in circulation?

Since each symbol stands for 50 books, divide 350 by 50 to find the number of symbols needed.

The month that shows 7 symbols is May.

$$\begin{array}{r} 7 \\ 50\overline{)350} \\ 350 \\ \hline 0 \end{array}$$

*Answer:* During May, 350 books were in circulation.

**EXAMPLE 2** This pictograph shows the number of homes in Sunnyside that local planners hope to heat by solar power each year from 1995 to 2000.

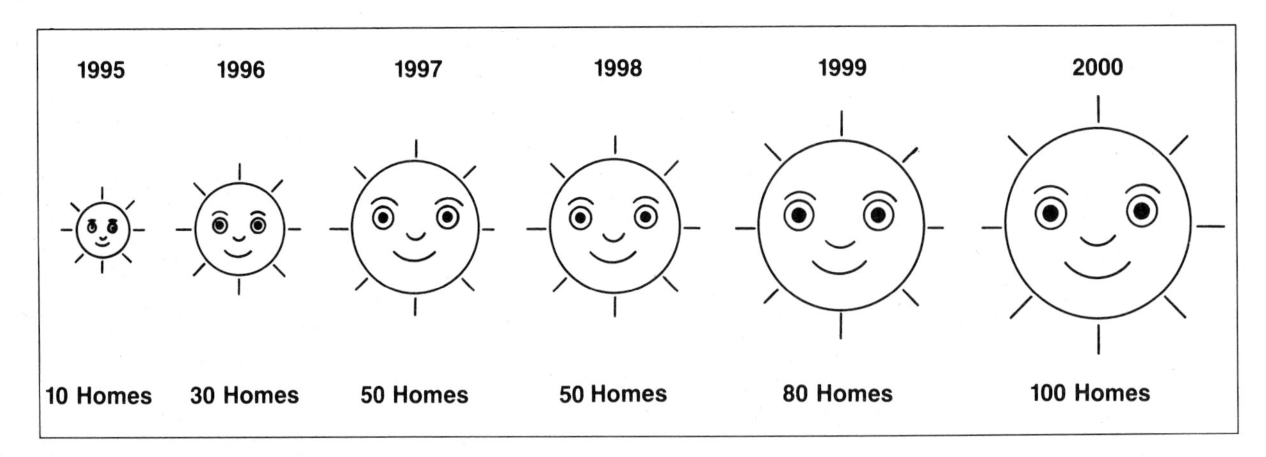


**a.** Will the number of homes heated by solar power increase or decrease from 1995 to 2000?

Since the size of the sun picture grows larger, this shows an increase.

*Answer:* Increase

**b.** In which year will there be no change from the previous year in the number of homes with solar heat?

The pictograph shows 50 homes in 1998, as there are in 1997.

*Answer:* 1998

**c.** In which year will the number of solar homes be 8 times the number in 1995?

The pictograph shows 10 homes in 1995.
8 times 10 is 80. The year that will have 80 solar homes is 1999.

*Answer:* 1999

EXAMPLE 3 Make a pictograph to show the given information.

| Number of Trees Planted in New Township | |
|---|---|
| March | 400 |
| April | 500 |
| May | 700 |
| June | 250 |
| July | 150 |
| August | 50 |
| September | 450 |

Decide upon a symbol related to the subject of the graph.

Determine how many whole symbols and parts of a symbol are needed to picture the given information.

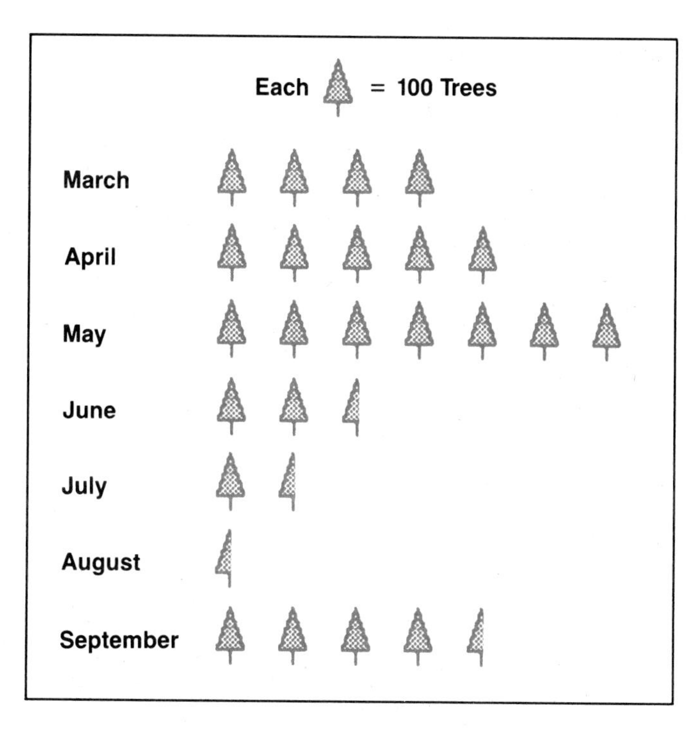

## CLASS EXERCISES

1. This pictograph shows the number of ice cream cones sold by the Dipit Ice Cream Parlor over a period of six days.

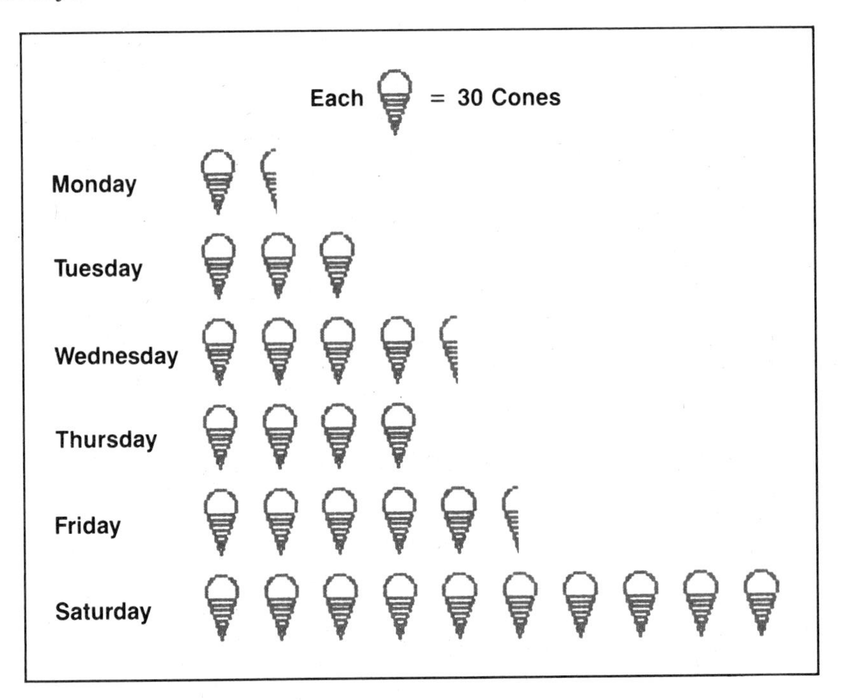


   a. How many more cones were sold on Friday than on Thursday?

   b. On which day were twice as many cones sold as on Monday?

   c. On which day were 30 fewer cones sold than on Thursday?

   d. What was the total number of cones sold during the six days?

2. This pictograph represents the amount of rainfall in a state over a four-month period.

   a. What was the smallest amount of rainfall in one month?

   b. Which month had the greatest amount of rainfall for the four-month period?

   c. What was the total amount of rainfall for the four months?

   d. Which month had 4 times the amount of rainfall that there was in August?

   e. How much less rainfall was there in June than in July?

   f. In which month did the rainfall decrease from the previous month?

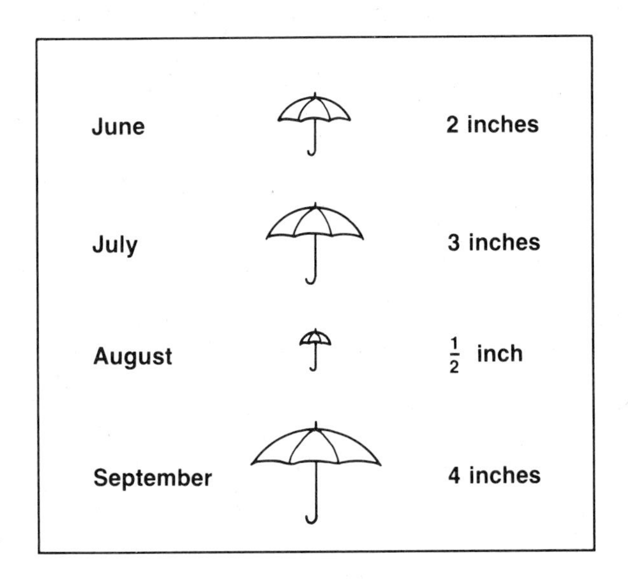

3. Make a pictograph to show the given information.

a.

| Number of Cats Rescued by the Animal Shelter | |
|---|---|
| January | 300 |
| February | 500 |
| March | 200 |
| April | 250 |
| May | 100 |
| June | 50 |

b.

| Number of Hours of Exercise per Week | |
|---|---|
| Mary | 14 |
| Gina | 28 |
| Melanie | 8 |
| Alexia | 20 |
| Marion | 12 |
| Hope | 18 |

## HOMEWORK EXERCISES

1. This pictograph shows the number of traffic accidents in Laurelton for a period of five months.

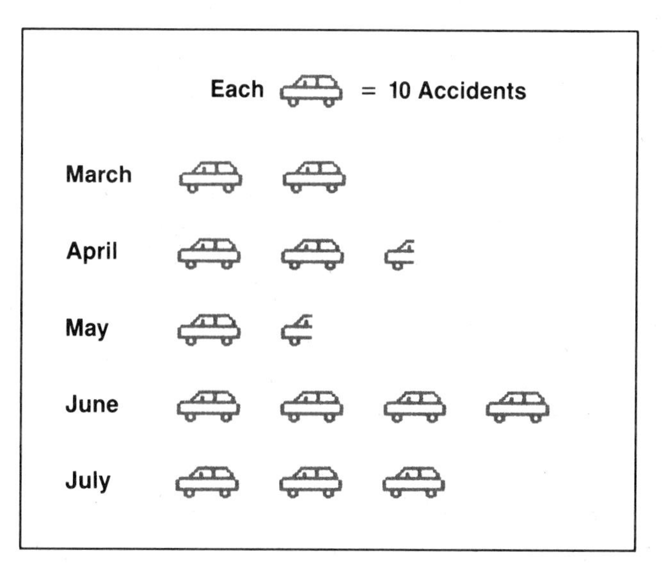


   a. How many accidents were there in April?
   (a) 2 (b) 3 (c) 20 (d) 25

   b. How many more accidents were there in June than in July?

   c. For which two months combined was the total number of accidents about equal to the number for June?

2. This pictograph represents the number of pizzas sold by Pizza Palace during the first five months of business.

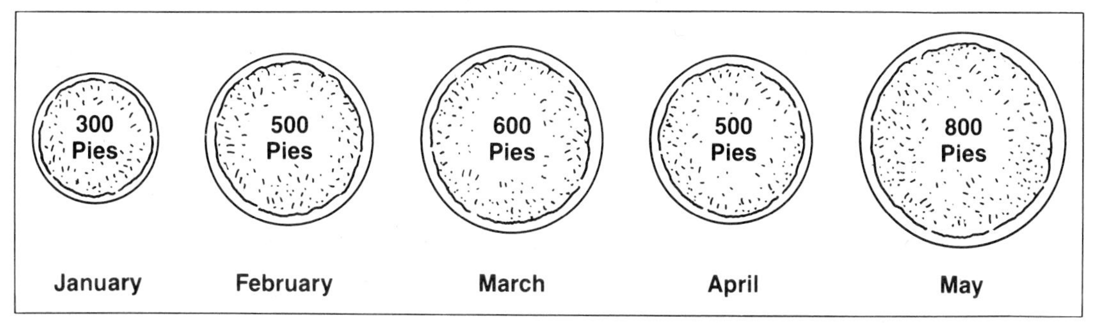


   a. How many more pies were sold in May than in January?
   (a) 800 (b) 500 (c) 300 (d) 1,100

   b. What was the total number of pies sold?

   c. In which month was the number of pies sold equal to the sum of those sold in January and February?

3. This pictograph shows the number of hours of television watched per week by 5 children.

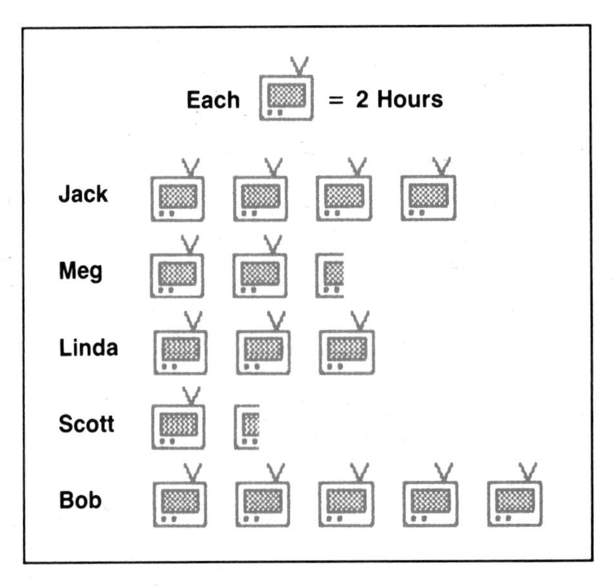


a. Who watches the most television per week?

b. Who watches the least television per week?

c. How many hours a week does Meg watch television?

d. Who watches twice as much television as Scott?

e. Who watches television 2 hours less than Jack each week?

f. How many children watch more television than Linda?

g. How many more hours a week does Bob watch television than Linda?

h. Who watches more television each week than Scott and Linda combined?

4. Make a pictograph to show the given information.

| Bushels of Apples Shipped by Citrus Coop | |
|---|---|
| June | 3,000 |
| July | 4,000 |
| August | 1,500 |
| September | 7,500 |
| October | 8,000 |
| November | 500 |

## SPIRAL REVIEW EXERCISES

1. This bar graph shows the average number of minutes per day that 4 students spend on the phone.

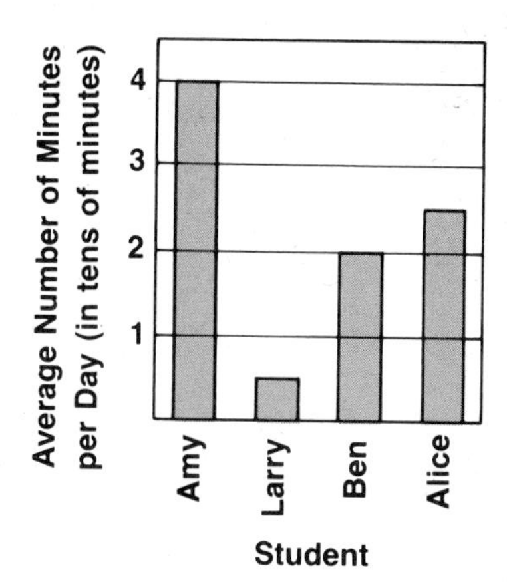


What is the average number of minutes that Alice spends on the phone each day?

(a) 3 (b) 30 (c) 25 (d) $2\frac{1}{2}$

2. From the line graph below, estimate the number of sit-ups that Charlie did on Wednesday.

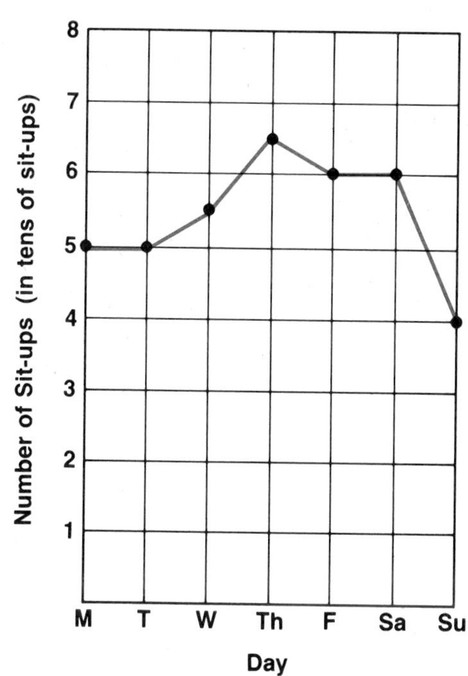


3. What is the probability that a die will show a number less than 3 when it is tossed?

4. This circle graph shows the kinds of vehicles in a high school parking lot.

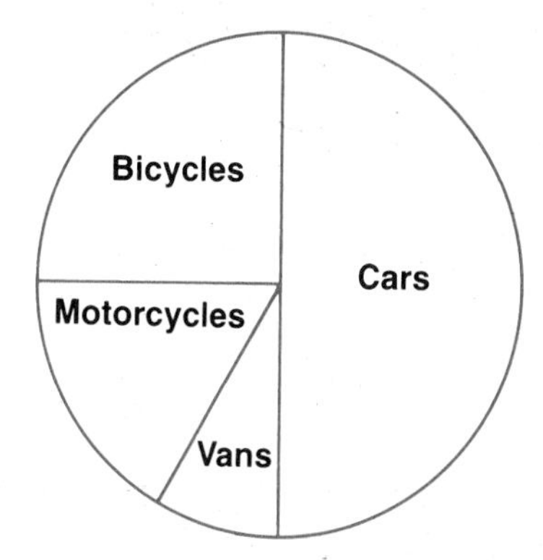


What kind of vehicle is about 25% of the total?

5. The highest temperature recorded for each of the first 5 days of July was 81°, 83°, 84°, 85°, and 88°. What was the median high temperature for these days?

6. The cost of a pound of chopped meat at 3 different butcher shops is $2.20, $2.60, and $3.00. What is the average cost per pound?

7. Mr. Sanchez invested $5,000 at 11% interest. How much interest will he earn in one year?

8. The price of a stock increased from $20 a share to $30 a share. What was the percent of increase?

9. Subtract: $12\frac{3}{8} - 9\frac{11}{16}$

10. Divide: $\frac{3}{5} \div 3$

11. Find $\frac{7}{8}$ of 720.

12. Find the mode of this set of numbers.
9, 9, 11, 14, 15, 15, 15, 17, 19

13. Joseph was absent from school 18 days and present 162 days. The ratio of the number of days present to the total number of days is

(a) 1:9 (c) 9:10
(b) 9:1 (d) 10:9

## CHALLENGE QUESTION

Mark wants to make a pictograph to show the given information about the expected number of families in his neighborhood with 2 or more jobs in the family.

| Year | Number of Families with 2 or More Jobs |
|---|---|
| 1990 | 23 |
| 1991 | 38 |
| 1992 | 47 |
| 1993 | 52 |
| 1994 | 42 |
| 1995 | 50 |

He decides to use the symbol to stand for one family. What advice would you give Mark before he begins to make the graph?

# Introduction to Algebra

## MODULE 17



## MODULE 18

# UNIT 17–1 The Meaning of Variable

## THE MAIN IDEA

1. In mathematics, a ***sentence*** is a *complete statement* about two quantities.

   $13 + 7 = 4 \times 5$ is a sentence.

2. For a statement to be a sentence, it must use one of these symbols:

   | | | | | | |
   |---|---|---|---|---|---|
   | $=$ | is equal to | $<$ | is less than | $\leq$ | is less than or equal to |
   | $\neq$ | is not equal to | $>$ | is greater than | $\geq$ | is greater than or equal to |

   Think of these symbols as the *verbs* of the sentences.

3. When a sentence contains a blank space for a number, the sentence is called an ***open sentence***.

   $3 \times \underline{?} = 45$ is an open sentence.

4. The blank space in an open sentence is called a ***variable***. Variables are shown as follows:

   a. as a blank space with a question mark:

   $\underline{?} + 10 = 21$

   b. as a symbol, such as an empty square, circle, or triangle:

   $\triangle + 10 = 21$

   c. as a letter, which stands for an empty space:

   $x + 10 = 21$

5. In algebra, variables are shown by using letters.

6. The variable in an open sentence usually stands for the words "a number" or "an unknown number."

**EXAMPLE 1** Which statements are complete sentences?

| | *Statement* | *"Verb" Symbol* | *Complete Sentence?* |
|---|---|---|---|
| **a.** | $7 \times 3 = 16 + 5$ | $=$ | yes |
| **b.** | $6 + 3 - 5 < 2 \times 3$ | $<$ | yes |
| **c.** | $6 + 5 \times 2$ | none | no |
| **d.** | $45 \div 9 \leq 10$ | $\leq$ | yes |

**EXAMPLE 2** Which of the given sentences are open sentences?

| | *Sentence* | *Variable* | *Open Sentence?* |
|---|---|---|---|
| **a.** | $5 + \underline{\ ?\ } = 12$ | $\underline{\ ?\ }$ | yes |
| **b.** | $9 \div 3 = 2 + 1$ | none | no |
| **c.** | $20 \times \square > 100$ | $\square$ | yes |
| **d.** | $15 \times a = 60$ | $a$ | yes |
| **e.** | $12 + 7 - 19 < 25 - 12 - 3$ | none | no |

**EXAMPLE 3** Which sentence means "a number plus 3 equals 18"?

(a) $15 + 3 = 18$ (b) $15 + 3 = 6 \times 3$ (c) $x + 3 = 18$ (d) $15 = 18 - 3$

Sentence (c) uses the variable $x$ to stand for "a number."

*Answer:* (c)

## CLASS EXERCISES

**1.** Which statements are complete sentences?

**a.** $100 + 20 - 40$ **b.** $5 \times 3 > 4 \times 3$ **c.** $144 \div 12 = 12$ **d.** $100 - 2 \geq 50 + 40$

**e.** $5 \times 60$ **f.** $5 \times 60 = 3 \times 100$ **g.** $2 \times 2 \times 2 = 8$ **h.** $40 \div 8 \neq 4$ **i.** $40 \div 8 + 4$

**2.** Which of these sentences are open sentences?

**a.** $6 \times m = 24$ **b.** $\triangle - 12 = 36$ **c.** $144 \div \underline{\ ?\ } = 6$ **d.** $120 \div 6 = 20$

**e.** $5 + \bigcirc < 20$ **f.** $x - 10 \geq 40$ **g.** $4 \times 3 \leq 6 \times 8$ **h.** $25 + 15 - 5 > \underline{\ ?\ }$

**i.** $3 \times 3 \times 3 < 30$ **j.** $4 \times 3 + 2 \neq 4 + 3 \times 2$

**3.** Which expression means "a number minus 2"?
(a) $x - 2$ (b) $2 - x$ (c) $8 - 2$ (d) $2 - 8$

**4.** Which expression means "4 times an unknown number"?
(a) $4 \times 12$ (b) $x + 4$ (c) $7 + 7 + 7 + 7$ (d) $4 \times n$

**5.** Which sentence means "a number plus 12 is less than 20"?
(a) $8 + 12 < 20$ (b) $x + 12 > 20$ (c) $a + 12 < 20$ (d) $n + 12 \leq 20$

**6.** Which sentence means "2 times a number is equal to the number plus 8"?
(a) $2 \times a = 4 + 8$ (b) $2 \times 8 = 8 + 8$ (c) $2 \times 5 = 2 + 8$ (d) $2 \times b = b + 8$

**7.** Which sentence means "an unknown number minus 5 is greater than or equal to 5"?
(a) $m - 5 \leq 5$ (b) $y - 5 \geq 5$ (c) $10 - 5 \geq 5$ (d) $10 - 5 \leq 5$

## HOMEWORK EXERCISES

**1.** Tell whether each statement is a complete sentence.

**a.** $15 \times \frac{2}{3} = 14 - 4$ **b.** $96 \div 12$ **c.** $34 \times 3 > 14 \times 5$ **d.** $5 + 9 \neq 8 \times 2$

**e.** $32 \times 7$ **f.** $50 \div 2 + 17 \times 5$ **g.** $30 \div 3 = 15 - 5$ **h.** $32 - 7 \leq 40$

2. Tell whether each sentence is an open sentence.

a. $9 + 7 = 16$   b. $x - 2 = 6$   c. $12 \times 3 < 40$   d. $\square + 4 = 15$

e. $19 + 17 = \triangle$   f. $60 \div 5 \neq 42$   g. $25 \div 5 + 8 \times 3 < 100$   h. $\underline{?} = 50$

i. $3 \times y = 36$   j. $1{,}001 > 50 \times 20$   k. $\square \div 6 = 8$   l. $5 \times a > 40$

3. Which expression means "a number increased by 5"?
(a) $8 + 5$ (b) $y - 5$ (c) $y + 5$ (d) $8 - 5$

4. Which expression means "an unknown number multiplied by 11"?
(a) $11 \times 12$ (b) $11 + 11 + 11 + 11$ (c) $x + 11$ (d) $11 \times a$

5. Which expression means "9 subtracted from a number"?
(a) $9 - n$ (b) $n - 9$ (c) $9 \times n$ (d) $n \div 9$

6. Which sentence means "5 times a number is equal to 15"?
(a) $5 \times n = 15$ (b) $5 + n = 15$ (c) $15 \times n = 5$ (d) $n - 5 = 15$

7. Which sentence means "a number decreased by 7 is greater than 10"?
(a) $x + 7 \geq 10$ (b) $x + 7 < 10$ (c) $x - 7 > 10$ (d) $x + 7 > 10$

8. Which sentence means "20 is greater than an unknown number plus 6"?
(a) $20 + n > 6$ (b) $20 \times n < 6$ (c) $20 > n - 6$ (d) $20 > n + 6$

9. Which sentence means "a number divided by 3 is equal to 5 less than the number"?
(a) $3 \times n = 5 - n$ (b) $3 \times n = n - 5$ (c) $n \div 3 = n - 5$ (d) $3 \div n = 5 - n$

## SPIRAL REVIEW EXERCISES

1. The bar graph shows the number of graduates at Apex High School each year for a period of 5 years, in hundreds of students.

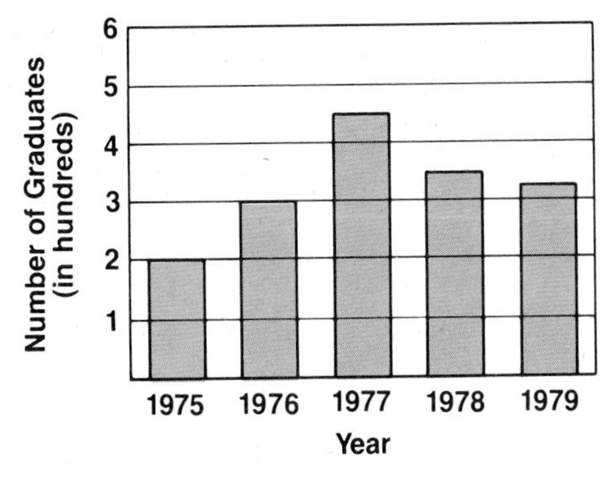


The number of graduates in 1979 was closest to
(a) 3 (b) 4 (c) 350 (d) 325

2. Find the number of centimeters in 25 meters.

3. Find the mean of 8.9, 4.6, 12.1, and 9.8.

4. A box contains 3 red marbles and 7 green marbles. The probability of picking a red marble is

(a) 3 (b) 7 (c) $\frac{3}{7}$ (d) $\frac{3}{10}$

5. Add: $-10 + (-4)$

6. What is the sum of $\frac{2}{3}$ and $\frac{3}{5}$?

(a) $\frac{2}{5}$ (b) $1\frac{4}{15}$ (c) $\frac{5}{8}$ (d) $\frac{6}{15}$

7. .14 has the same value as

(a) 14% (b) 1.4% (c) $\frac{14}{10}$ (d) $\frac{14}{1000}$

### CHALLENGE QUESTION

Write an open sentence for each given sentence.

a. The sum of a number and 7 is less than 10.

b. Two less than a number is equal to 12.

c. Twice a number is greater than 4.

d. A number increased by 3 is not equal to 5.

# UNIT 17–2 Writing Algebraic Expressions

## THE MAIN IDEA

1. You *translate* word phrases into algebraic expressions by writing symbols for the words.
2. You have been using the symbols $+$, $-$, $\times$, and $\div$ for the operations of arithmetic, and you know certain key words that are associated with these operations.
3. Since it could be confusing to write $2 \times x$, there are other ways to write "2 times the number $x$":

   $2 \cdot x \qquad 2(x) \qquad (2)(x) \qquad 2x$

4. Parentheses ( ) are used to group numbers.

**EXAMPLE 1** Translate each word phrase into an algebraic expression.

| | *Word Phrase* | *Answer* |
|---|---|---|
| **a.** | the sum of $x$ and 2 ("the sum of" → $+$) | $x + 2$ |
| **b.** | the difference of $n$ and 4 ("the difference of" → $-$) | $n - 4$ |
| **c.** | the product of $y$ and 3 ("the product of" → $\times$) | $3y$, $3 \cdot y$, $3(y)$, or $(3)(y)$ |
| **d.** | the quotient of $b$ and 7 ("the quotient of" → $\div$) | $\frac{b}{7}$ or $b \div 7$ |
| **e.** | a number increased by 7 | $x + 7$ |
| **f.** | 4 more than a number | $y + 4$ |
| **g.** | a number decreased by 2 | $a - 2$ |
| **h.** | 5 less than a number | $b - 5$ |
| **i.** | the difference of 4 and $n$ | $4 - n$ |

Note: The expression "a number" can be represented by any variable.

**EXAMPLE 2** Use parentheses to group numbers.

*Answer*

**a.** 6 times the sum of $x$ and 4 — $6(x + 4)$

$6 \quad \cdot \quad (x + 4)$

**b.** twice the difference of $y$ and 3 — $2(y - 3)$

$2 \cdot \quad (y - 3)$

**c.** the product of 7 and $m$, divided by 3 — $(7m) \div 3$ or $\dfrac{7m}{3}$

$(7m) \quad \div \quad 3$

## CLASS EXERCISES

**1.** Use the variable $n$ to translate each expression into an algebraic expression.

**a.** a number increased by 11 **b.** the product of 6 and a number

**c.** a number divided by 10 **d.** 7 less than a number

**e.** 7 decreased by a number **f.** the sum of a number and 15

**g.** the difference between a number and 8 **h.** $\frac{3}{4}$ of a number

**i.** 20% of a number **j.** 24 divided by a number

**2.** Write each expression algebraically.

**a.** 2 times the sum of a number and 5

**b.** 6 more than 4 times a number

**c.** the product of 5 and a number, decreased by 11

**d.** the difference between a number and 5, divided by 3

**e.** the sum of a number and 6, multiplied by 7 more than the number

**3.** Which expression means "a number increased by 10"?
(a) $n + 10$ (b) $10n$ (c) $(10)(n)$ (d) $10 \cdot n$

**4.** Which expression means "a certain number less than 8"?
(a) $x \div 8$ (b) $\dfrac{8}{x}$ (c) $n - 8$ (d) $8 - a$

**5.** Which expression means "7 more than twice a number"?
(a) $7 \cdot (2y)$ (b) $2y + 7$ (c) $n + (2)(7)$ (d) $7y + 2$

**6.** Which expression means "the sum of 5 times a number and 9"?
(a) $x(9 + 5)$ (b) $5b + 9$ (c) $(9)(5a)$ (d) $y + (a)(5)$

**7.** Which expression means "the sum of a number and 3, multiplied by the difference between the number and 2"?
(a) $(a - 3)(a + 2)$ (b) $(a + 3)(a - 2)$ (c) $(a - 2) + (a + 3)$ (d) $3a - 2$

8. Write an algebraic expression for each word phrase.

   a. Mary's age decreased by 3
   b. twice Mr. Jones' salary
   c. the sum of the price of a book and $3
   d. 4 inches more than Mike's height
   e. $\frac{1}{5}$ of the number of students in the class

## HOMEWORK EXERCISES

1. Use the variable $n$ to translate each expression into an algebraic expression.

   a. a number divided by 7
   b. a number diminished by 30
   c. a number increased by 75
   d. the product of 15 and a number
   e. 5 times a number
   f. 85 subtracted from a number
   g. 35 divided by a number
   h. the sum of a number and 37
   i. $\frac{5}{8}$ of a number
   j. the difference between 20 and a number

2. Write each expression algebraically.

   a. 8 increased by 4 times a number
   b. 10 subtracted from 9 times a number
   c. 3 times the sum of a number and 4
   d. 7 less than 5 times a number
   e. 15 more than a number, divided by 2
   f. the difference of 15 and a number, multiplied by 3
   g. 50 divided by the sum of a number and 2
   h. the difference of a number and 7, multiplied by 5 more than the number
   i. 15 more than twice a number, multiplied by 3 less than the number

3. Which expression means "a number decreased by 7"?
   (a) $n + 7$ (b) $n - 7$ (c) $7n$ (d) $\frac{n}{7}$

4. Which expression means "5 less than twice a number"?
   (a) $5 - 2n$ (b) $5 + 2n$ (c) $2n - 5$ (d) $2(n - 5)$

5. Which expression means "the product of 3 and 7 less than a number"?
   (a) $n - 21$ (b) $3(7 - n)$ (c) $3(7 + n)$ (d) $3(n - 7)$

6. Which expression means "10 more than a number, divided by 5"?
   (a) $\frac{(n + 10)}{5}$ (b) $\frac{(n - 10)}{5}$ (c) $n + 10 \div 5$ (d) $n - 10 \div 5$

7. Which expression means "a number increased by 7, multiplied by 2 more than the number"?
   (a) $(a + 7)(2a)$ (b) $(a - 7)(2a)$ (c) $(a + 7)(a + 2)$ (d) $(a - 7)(a + 2)$

8. Write an algebraic expression for each word phrase.

   a. June's wages increased by $30
   b. Joseph's weight decreased by 5 pounds
   c. 3 times the distance that Carlos walked
   d. $\frac{1}{2}$ the price of a movie ticket
   e. The sum of the cost of a jacket and a $20 shirt

9. Maria earned $d$ dollars on Monday and \$28 on Tuesday. Write algebraically the total amount of money Maria earned.

10. James has $x$ dollars and spends \$50 for a radio. Write algebraically the amount of money that James has left.

11. Mr. Jones divides $w$ dollars evenly among his 5 children. Write algebraically the amount of money given to each child.

## SPIRAL REVIEW EXERCISES

1. Which statement is a complete sentence?
(a) $3(5 + 4)$ (c) $5 + 4 = 9$
(b) $5(3 + 4)$ (d) $5 + 4 \times 9$

2. Which expression means "3 less than a number"?
(a) $3 - x$ (c) $3 \div x$
(b) $x - 3$ (d) $x \div 3$

3. James bought 2 quarts of milk. How many fluid ounces did he buy?

4. What is the mode of $1\frac{1}{8}$, $2\frac{1}{4}$, $1\frac{1}{2}$, $1\frac{1}{8}$, $2\frac{5}{16}$, $1\frac{1}{4}$, $1\frac{1}{8}$, and $2\frac{1}{4}$?

5. The remainder when 225 is divided by 17 is
(a) 0 (b) 1 (c) 13 (d) 4

6. $\frac{24}{15}$ is equivalent to
(a) $1\frac{7}{15}$ (b) $1\frac{2}{3}$ (c) $1\frac{3}{5}$ (d) $1\frac{11}{15}$

7. $4 - (-5)$ is equal to
(a) $-1$ (b) 9 (c) 1 (d) $-9$

8. Andy spent an average of 140 minutes a day on homework. The circle graph shows the average number of minutes spent on each subject.

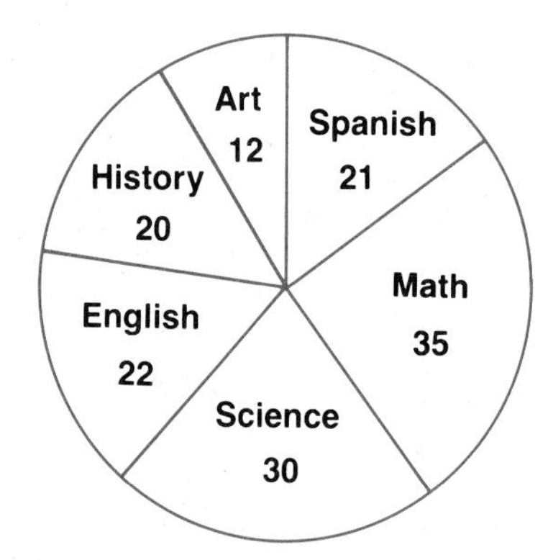


What percent of Andy's homework time was spent on Math?
(a) 15% (b) 20% (c) 25% (d) 35%

9. If the sales tax is 6%, how much tax is paid on a \$35 purchase?

10. If 7 boxes of cereal cost \$9.10, what is the cost of 3 boxes of cereal?

## CHALLENGE QUESTION

Explain why you cannot tell which is the greater quantity, $x + 5$ or $n + 20$.

## UNIT 17–3 Evaluating Algebraic Expressions

### THE MAIN IDEA

1. To ***evaluate*** (find the value of) an algebraic expression, replace the variables by their numerical values. Then perform the operations that are shown.
2. If there is more than one operation, the operations follow the order of operations that you learned.
   a. Do the operations inside parentheses first.
   b. Do the multiplications and divisions next.
   c. Do the additions and subtractions last.
   d. If there are several additions and subtractions in a row, do these from left to right.

**EXAMPLE 1** Find the value of $3x$ when $x = 12$.

**THINKING ABOUT THE PROBLEM**

$3x$ means "3 times a number." If the number is 12, the expression $3x$ means "3 times 12." You can write this expression as $3 \times 12$ or $(3)(12)$ or $3(12)$.

Avoid using the dot between numerals $(3 \cdot 12)$, because it could be mistaken for a decimal point. Do not write numerals next to each other without any sign, as we do with variables, because that would make a product look like a different number (312).

$3(12) = 36$ *Ans.*

**EXAMPLE 2** Evaluate $2 + 3a$ when $a = 5$.

| | |
|---|---|
| Replace the variable $a$ by the numeral 5. | $2 + 3a$<br>$= 2 + 3(5)$ |
| Perform the multiplication first. | $= 2 + 15$ |
| Perform the addition. | $= 17$ *Ans.* |

**EXAMPLE 3** Find the value of $5(a + b)$ when $a = 2$ and $b = 7$.

| | |
|---|---|
| Replace the variables by the numerals. | $5(a + b)$<br>$= 5(2 + 7)$ |
| Perform the operation inside the parentheses first. | $= 5(9)$ |
| Perform the multiplication. | $= 45$ *Ans.* |

**EXAMPLE 4** Evaluate $2x - 5y + 3z$ when $x = 10$, $y = 3$, and $z = 1$.

| | |
|---|---|
| Replace the variables by the numerals. | $2x - 5y + 3z$<br>$= 2(10) - 5(3) + 3(1)$ |
| Do the multiplications first. | $= 20 - 15 + 3$ |
| Do the addition and subtraction from left to right. | $= 5 + 3$<br>$= 8$ *Ans.* |

## CLASS EXERCISES

**1.** Evaluate each expression, using the given values of the variables.

**a.** $3 + w$ when $w = 11$ **b.** $a - 12$ when $a = 27$ **c.** $6x$ when $x = 25$

**d.** $\frac{b}{3}$ when $b = 144$ **e.** $23 - y + 2$ when $y = 19$ **f.** $\frac{6z}{18}$ when $z = 12$

**2.** Evaluate each expression, using the values $x = 13$ and $y = 7$.

**a.** $x + y$ **b.** $x - y$ **c.** $xy$ **d.** $\frac{x}{y}$ **e.** $2x + y$ **f.** $x + 2y$

**g.** $2x + 2y$ **h.** $4x - 3y$ **i.** $x + y - x - y$

**3.** Use the values $a = 5$ and $b = 3$ to evaluate each expression.

**a.** $3(a + b)$ **b.** $(a + b) - 7$ **c.** $4(2a + b)$ **d.** $6(a - b) + 5$

**e.** $(a + b)(a - b)$ **f.** $(2a + b)(2a - b)$

**4.** Find the value of $2\ell + 2w$, using the given values of the variables.

**a.** $\ell = 2, w = 3$ **b.** $\ell = 3, w = 2$ **c.** $\ell = 2\frac{1}{2}, w = 3\frac{1}{2}$ **d.** $\ell = 4, w = 1$

**5.** If $m = 12$ and $p = 1$, the value of $3m - 2p$ is
(a) 33 (b) 34 (c) 36 (d) 38

## HOMEWORK EXERCISES

**1.** Evaluate each expression, using the given values of the variables.

**a.** $x + 19$ when $x = 20$ **b.** $y - 26$ when $y = 54$ **c.** $31 + z$ when $z = 29$

**d.** $20x$ when $x = 5$ **e.** $5y$ when $y = 6$ **f.** $\frac{w}{7}$ when $w = 28$

**g.** $35 - a$ when $a = 20$ **h.** $11c$ when $c = 9$

**2.** Find the value of $85 - w$ for each of the given values of $w$.

**a.** 5 **b.** 0 **c.** 50 **d.** 85 **e.** 12.9 **f.** $80\frac{5}{8}$

**3.** Find the value of $30x$ for each of the given values of $x$.

**a.** 3 **b.** 9 **c.** 0 **d.** $\frac{1}{2}$ **e.** 5.8 **f.** $\frac{2}{3}$

**4.** Find the value of $50 + 3y$ for each of the given values of $y$.

**a.** 2 **b.** 7 **c.** 16 **d.** 0 **e.** $5\frac{1}{3}$ **f.** 10.7

**5.** Evaluate each expression, using the values $x = 20$ and $y = 30$.

**a.** $x + y$ **b.** $y - x$ **c.** $\frac{x}{y}$ **d.** $3x + y$ **e.** $3(x + y)$

**f.** $4x - 2y$ **g.** $\frac{60}{x} + \frac{30}{y}$ **h.** $9(y - x)$

**6.** Use the values $a = 12$ and $b = 4$ to evaluate each expression.

**a.** $3a - b$ **b.** $3(a - b)$ **c.** $(a - b)(a + b)$ **d.** $3b - a$

**e.** $5b + 2a$ **f.** $\frac{1}{3}a + b$ **g.** $10(a + b)$ **h.** $10(a - 3b)$

**7.** Find the value of $4y - 3x$, using the given values of the variables.

**a.** $y = 2, x = 1$ **b.** $y = 5, x = 6$ **c.** $y = 0, x = 0$ **d.** $y = \frac{1}{2}, x = \frac{1}{3}$

**8.** If $d = 10$ and $t = 5$, then the value of $2d + 4t$ is
(a) 120 (b) 40 (c) 28 (d) 255

**9.** If $k = 20$ and $m = 15$, then the value of $3k - 4m$ is
(a) 0 (b) 15 (c) 5 (d) 35

## SPIRAL REVIEW EXERCISES

**1.** Which expression says "5 decreased by twice a number"?
(a) $5 + 2n$ (c) $5 - 2n$
(b) $2n - 5$ (d) $2 - 5n$

**2.** Which sentence says "twice a number is greater than or equal to 12"?
(a) $2x > 12$ (c) $2x \leq 12$
(b) $2x \geq 12$ (d) $2x < 12$

**3.** The number of liters in 4,500 mL is
(a) 450 (b) 45 (c) 4.5 (d) .45

**4.** $-7 + (-4)$ is equal to
(a) $-11$ (b) $+11$ (c) $-3$ (d) $+3$

**5.** Find the median of 35, 41, 52, 58, 63, 65, and 68.

**6.** If the rate of sales tax is 8%, how much sales tax is charged on a $28 purchase?

**7.** Ms. Lorenzo deposited $432.85 into her checking account. If her balance before the deposit was $950.75, what is her new balance?

**8.** Multiply: 2.9 and .7

**9.** This line graph shows Mandy's temperature each hour for 5 hours.

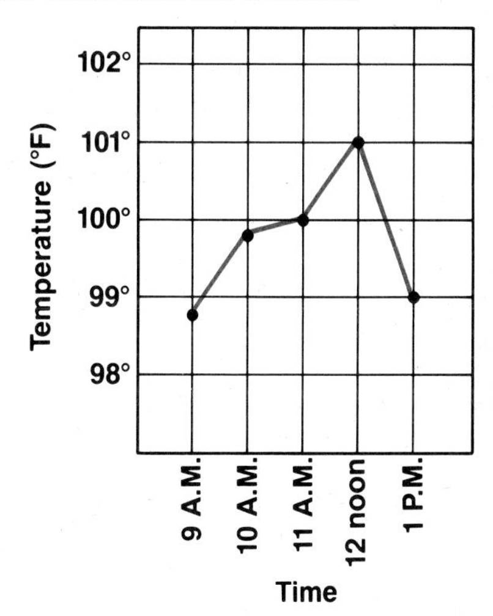


How much did Mandy's temperature drop from 12 noon to 1 P.M.?

**10.** On a mathematics test, Sandy answered 20 of 25 questions correctly. What percent of the questions did Sandy get correct?

## CHALLENGE QUESTIONS

**1.** If $a = 5$, $b = 2$, and $c = 0$, which of the following algebraic expressions cannot be evaluated?

(a) $b - c + a$ (b) $\frac{a + b}{c}$ (c) $ac + ab$ (d) $c(a + b)$

**2.** If $x = 2$ and $y = -1$, the value of $3x - y$ is
(a) 5 (b) 6 (c) 7 (d) 8

# UNIT 17–4 Exponents; Scientific Notation

## EXPONENTS

### THE MAIN IDEA

1. An *exponent* shows how many times a *base* is used as a factor.

   exponent ↘

   $6^3$ means $6 \times 6 \times 6$

   base ↗

   and is read "6 raised to the third power," "6 to the third power," or "6 cubed."

2. To evaluate an expression that contains an exponent:
   a. Do the work with the exponent first.
   b. Then follow the usual order of operations.

**EXAMPLE 1**

**a.** The number 5 is to be used 4 times as a factor. Write this product in two ways.

*Answer:* $5 \times 5 \times 5 \times 5$ or $5^4$

**b.** The number 4 is to be used 5 times as a factor. Write this product in two ways.

*Answer:* $4 \times 4 \times 4 \times 4 \times 4$ or $4^5$

**EXAMPLE 2**

**a.** Write the expression $a \cdot a \cdot a \cdot b \cdot b$, using exponents.

*Answer:* $a^3 \cdot b^2$ or $a^3b^2$

**b.** Write the expression $x^4y^2z$ without exponents.

*Answer:* $x \cdot x \cdot x \cdot x \cdot y \cdot y \cdot z$

**EXAMPLE 3** Write each expression in words.

| | *Expression* | *Answer* |
|---|---|---|
| **a.** | $5^2$ | five raised to the second power<br>or five to the second power<br>or five squared |
| **b.** | $a^3$ | *a* raised to the third power<br>or *a* to the third power<br>or *a* cubed |
| **c.** | $4^5$ | four raised to the fifth power<br>or four to the fifth power |

**EXAMPLE 4** Evaluate each expression.

| | *Expression* | *Arithmetic* |
|---|---|---|
| **a.** | $3^4$ | $3 \times 3 \times 3 \times 3$ <br> $= 81$ *Ans.* |
| **b.** | $5^2 \cdot 2^3$ | $5 \times 5 \times 2 \times 2 \times 2$ <br> $= 25 \times 8$ <br> $= 200$ *Ans.* |
| **c.** | $5(3)^3$ | $5 \times 3 \times 3 \times 3$ <br> $= 5 \times 27$ <br> $= 135$ *Ans.* |
| **d.** | $(5 \cdot 3)^2$ | $(15)^2$ <br> $= 15 \times 15$ <br> $= 225$ *Ans.* |
| **e.** | $5 + 3^2$ | $5 + 3 \times 3$ <br> $= 5 + 9$ <br> $= 14$ *Ans.* |

**EXAMPLE 5** Evaluate each expression when $x = 3$ and $y = 4$.

| | *Expression* | *Arithmetic* |
|---|---|---|
| **a.** | $x(y)^2$ | $3(4)^2$ <br> $= 3(4 \times 4)$ <br> $= 3(16)$ <br> $= 48$ *Ans.* |
| **b.** | $(xy)^2$ | $(3 \times 4)^2$ <br> $= (12)^2$ <br> $= 12 \times 12$ <br> $= 144$ *Ans.* |
| **c.** | $x^2y^2$ | $3^2 \times 4^2$ <br> $= 3 \times 3 \times 4 \times 4$ <br> $= 9 \times 16$ <br> $= 144$ *Ans.* |
| **d.** | $x^2y$ | $3^2 \times 4$ <br> $= 3 \times 3 \times 4$ <br> $= 9 \times 4$ <br> $= 36$ *Ans.* |

**EXAMPLE 6** What is the value of $7 \times 10^3 + 5 \times 10^2 + 3 \times 10^1 + 4$?
(a) 7,534 (b) 753.4 (c) 75.34 (d) 75,340

| | |
|---|---|
| Do the work with exponents first. | $7 \times 10^3 + 5 \times 10^2 + 3 \times 10^1 + 4$ <br> $7 \times 1{,}000 + 5 \times 100 + 3 \times 10 + 4$ |
| Do the multiplication next. | $7{,}000 + 500 + 30 + 4$ |
| Do the addition. | 7,534 |

*Answer:* (a)

## CLASS EXERCISES

**1.** Write each product in two ways, one of which uses exponents.

**a.** 6 used as a factor 3 times **b.** $x$ multiplied by $x$

**c.** $a$ used as a multiplier 7 times **d.** $w$ times $w$ times $w$

**2.** Write each expression, using exponents.

**a.** $x \cdot x \cdot x$ **b.** $m \cdot m \cdot m \cdot m \cdot m$ **c.** $2 \cdot a \cdot a$ **d.** $10 \cdot y \cdot y \cdot y \cdot y$ **e.** $(p)(p)(q)(q)(q)$

**f.** $a \cdot a \cdot a \cdot b \cdot b \cdot b$ **g.** $3(r)(r)(r)(r)(s)(s)$ **h.** $v \cdot v \cdot v \cdot w \cdot w \cdot x \cdot x \cdot x \cdot x \cdot x$

**3.** Write each expression without exponents.

**a.** $a^3b^2$ **b.** $a^2b^3$ **c.** $x^3y^3z$ **d.** $xy^3z^3$

**4.** Write each expression in words.

**a.** $5^2$ **b.** $6^3$ **c.** $8^4$ **d.** $4^8$ **e.** $x^5$ **f.** $a^3$ **g.** $5 \cdot x^4$ **h.** $m^2 \cdot n^3$ **i.** $a^4 \cdot b^2 \cdot c^6$

**5.** Evaluate each expression.

**a.** $3^2$ **b.** $2^3$ **c.** $4^3$ **d.** $3^4$ **e.** $10^2$ **f.** $10^3$ **g.** $2^6$ **h.** $\left(\frac{1}{2}\right)^2$ **i.** $(.1)^3$

**6.** Evaluate each expression.

**a.** $2 \times 3^2$ **b.** $2^2 \times 3^2$ **c.** $3 \times 2^3$ **d.** $3^3 \times 2^3$ **e.** $3^2 \times 2^3$ **f.** $(3 \times 2)^3$

**g.** $2^2 + 3^2$ **h.** $3^2 + 4^2$ **i.** $10^2 - 7^2$

**7.** Evaluate each expression when $x = 2$.

**a.** $x^2$ **b.** $x^3$ **c.** $2 \cdot x^2$ **d.** $x^2 \cdot x^3$ **e.** $x^5$ **f.** $x^6$ **g.** $5x^2$ **h.** $5^2 \cdot x^2$ **i.** $(5x)^2$

**8.** What is the value of $2 \times 10^3 + 8 \times 10^2 + 5 \times 10^1 + 1$?
(a) 28.51 (b) 285.1 (c) 2,851 (d) 28,510

**9.** Evaluate:

**a.** $3 \times 10^3 + 4 \times 10^2 + 7$ **b.** $9 \times 10^4 + 7 \times 10^2 + 5 \times 10^1 + 8$

## SCIENTIFIC NOTATION

### THE MAIN IDEA

1. A number in the form $2.5 \times 10^3$ is in ***scientific notation***. The first factor is a number between 1 and 10, and the second factor is a power of 10.
2. To change a number *from* scientific notation, work with the exponent first, then multiply the two factors.
3. To change a number *to* scientific notation:
   a. Place a decimal point so that the first factor is a number between 1 and 10.
   b. Count how many places the decimal point must be moved to change the new first factor into the original number. The number of places is the number used as the exponent of 10.
   c. Show the two factors as a product.

**EXAMPLE 7** For which number is $3.7 \times 10^5$ the scientific notation?
(a) 3.700000 (b) 3,700,000 (c) 370,000 (d) $37^5$

Work with the exponent first. $3.7 \times 10^5$

$3.7 \times 100{,}000$

Multiply. Count the number of zeros in the power of 10, and move the decimal point that many places to the right.

3. 7 0 0 0 0 .
→→→→→

$= 370{,}000$

*Answer:* (c)

**EXAMPLE 8** Write the number 56,000 in scientific notation.

| | |
|---|---|
| Place a decimal point so that the first factor is a number between 1 and 10. | 5.6000 ↑ 5.6 is between 1 and 10. |
| Count how many places the decimal point must be moved to the right to change 5.6 to 56,000. | 5. 6 0 0 0 . →→→→ Move 4 places. |
| Use 4 as the exponent of 10. | $10^4$ ← exponent |
| Show the product. | $5.6 \times 10^4$ *Ans.* |

## CLASS EXERCISES

1. Write each number without using scientific notation.
   **a.** $1.8 \times 10^2$ **b.** $2.45 \times 10^4$ **c.** $8.3 \times 10^6$

2. Write each number in scientific notation.
   **a.** 5,700 **b.** 620,000 **c.** 1,250

3. For which number is $2.5 \times 10^3$ the scientific notation?
   (a) 2,500 (b) $25^3$ (c) 2.5000 (d) 250

4. Which is the scientific notation for 175,000?
   (a) $175 \times 10^3$ (b) $17.5 \times 10^4$ (c) $1.75 \times 10^5$ (d) $1.75 \times 10^3$

## HOMEWORK EXERCISES

1. Write each product in two ways, one of which uses exponents.
   **a.** 10 used as a factor 4 times **b.** *a* times *a* times *a*
   **c.** *m* used as a multiplier 2 times **d.** 12 multiplied by 12

2. Write each expression, using exponents.
   **a.** $w \cdot w$ **b.** $x \cdot x \cdot x \cdot x$ **c.** $a \cdot a \cdot b \cdot b$ **d.** $5 \cdot y \cdot y \cdot y$ **e.** $4(a)(b)(b)(b)$
   **f.** $x \cdot x \cdot y \cdot y \cdot y \cdot z \cdot z$ **g.** $12m \cdot m \cdot m \cdot r \cdot r$ **h.** $7(a)(a)(a)(a)(b)(b)(b)(c)(c)$

3. Write each expression without exponents.
   **a.** $x^4y^2$ **b.** $x^2y^4$ **c.** $a^2b^3c$ **d.** $ab^2c^3$

4. Write each expression in words.
   **a.** $8^2$ **b.** $5^3$ **c.** $9^5$ **d.** $5^9$ **e.** $w^6$ **f.** $y^3$ **g.** $3x^3$ **h.** $x^2y^4$ **i.** $x^3y^4z^5$

5. Evaluate each expression.
   **a.** $6^2$ **b.** $5^3$ **c.** $3^5$ **d.** $10^4$ **e.** $10^5$ **f.** $2^5$ **g.** $11^2$ **h.** $\left(\frac{1}{3}\right)^3$ **i.** $(.1)^4$

6. Evaluate each expression.
   **a.** $4 \times 2^2$ **b.** $5 \times 3^2$ **c.** $4^2 \times 5^2$ **d.** $(4 \times 5)^2$ **e.** $6 \times 2^3$
   **f.** $6^2 \times 2^3$ **g.** $6^2 + 8^2$ **h.** $5^2 + 12^2$ **i.** $20^2 - 12^2$

**7.** Evaluate each expression when $a = 3$.

**a.** $a^2$ **b.** $a^3$ **c.** $2a^2$ **d.** $a^2 \cdot a^3$ **e.** $a^5$ **f.** $6a^2$ **g.** $4^2 + a^2$ **h.** $(4 + a)^2$

**8.** What is the value of $9 \times 10^3 + 6$?
(a) 906 (b) 960 (c) 9,006 (d) 9,060

**9.** Find the value of each expression.

**a.** $4 \times 10^3 + 7 \times 10^2 + 1 \times 10^1 + 5$ **b.** $8 \times 10^2 + 7$

**c.** $5 \times 10^3 + 3 \times 10^1 + 2$ **d.** $7 \times 10^3 + 1 \times 10^2 + 4$

**10.** Write each number without using scientific notation.

**a.** $7.3 \times 10^2$ **b.** $3.1 \times 10^3$ **c.** $5.42 \times 10^5$

**11.** Write each number in scientific notation.

**a.** 98,000 **b.** 870,000 **c.** 4,720

**12.** For which number is $1.2 \times 10^4$ the scientific notation?

(a) 1.20000 (b) 1,200 (c) $12^4$ (d) 12,000

**13.** Which is the scientific notation for 28,400?

(a) $.284 \times 10^5$ (b) $2.84 \times 10^4$ (c) $28.4 \times 10^3$ (d) $284 \times 10^2$

## SPIRAL REVIEW EXERCISES

**1.** $-8 - (-5)$ is equal to
(a) $-13$ (b) $+13$ (c) $-3$ (d) $+3$

**2.** Write an algebraic expression that means "a person's age decreased by 12 years."

**3.** If the probability that it will rain is .4, what is the probability that it will not rain?

**4.** In a group of 20 athletes, 8 were professionals and 12 were amateurs. The ratio of professional athletes to the total number of athletes was

(a) $\frac{12}{20}$ (b) $\frac{2}{5}$ (c) $\frac{2}{3}$ (d) $\frac{4}{5}$

**5.** The price of a shirt was reduced from \$20 to \$12. The percent of decrease is

(a) 40% (b) $66\frac{2}{3}$% (c) 8% (d) 20%

**6.** Find $\frac{4}{5}$ of $\frac{2}{7}$. **7.** Divide: $1.04 \div .02$

**8.** This circle graph shows the kinds of vehicles passing through a toll booth in one day.

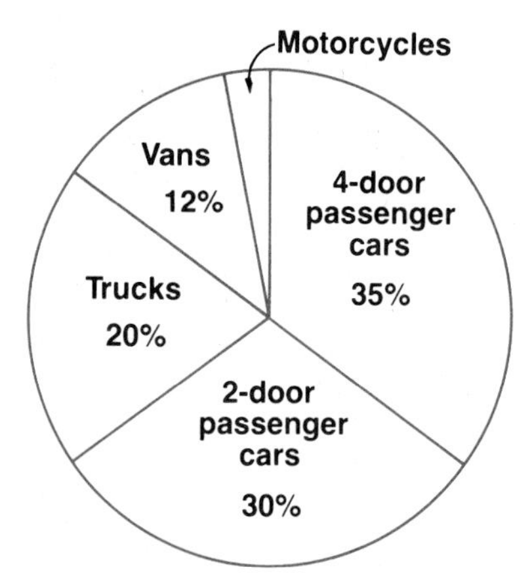


What percent of the vehicles passing through the toll booth were motorcycles?

### CHALLENGE QUESTION

When is the square of a number less than the number?

## UNIT 17–5 Squares and Square Roots

### THE MAIN IDEA

1. When a number is raised to the second power:
   a. The number is squared.
      $5^2$ is 5 squared.
   b. The result is called the *square* of the original number.
      25 is the square of 5.
   c. The original number is called the *square root* of the result.
      5 is the square root of 25.
2. The words "square root" can be written by using the symbol $\sqrt{\phantom{x}}$. For example, the sentence "7 is the square root of 49" can be written: $7 = \sqrt{49}$
3. Some whole numbers have square roots that are also whole numbers. These are called *perfect squares*. 36 is a perfect square because the square root of 36 is the whole number 6.
4. You should become familiar with the perfect squares through $12^2$:

| | | | |
|---|---|---|---|
| $1^2 = 1$ | $1 = \sqrt{1}$ | $7^2 = 49$ | $7 = \sqrt{49}$ |
| $2^2 = 4$ | $2 = \sqrt{4}$ | $8^2 = 64$ | $8 = \sqrt{64}$ |
| $3^2 = 9$ | $3 = \sqrt{9}$ | $9^2 = 81$ | $9 = \sqrt{81}$ |
| $4^2 = 16$ | $4 = \sqrt{16}$ | $10^2 = 100$ | $10 = \sqrt{100}$ |
| $5^2 = 25$ | $5 = \sqrt{25}$ | $11^2 = 121$ | $11 = \sqrt{121}$ |
| $6^2 = 36$ | $6 = \sqrt{36}$ | $12^2 = 144$ | $12 = \sqrt{144}$ |

**EXAMPLE 1** For the sentence $8^2 = 64$, write equivalent sentences using the word "square," the words "square root," and the symbol $\sqrt{\phantom{x}}$.

*Answer:* 64 is the square of 8, 8 is the square root of 64, $8 = \sqrt{64}$

**EXAMPLE 2** Find the square of 11.

The square of 11 means $11^2$. $11^2 = 11 \times 11 = 121$ *Ans.*

**EXAMPLE 3** What number is $\sqrt{81}$?

$\sqrt{81}$ means "the square root of 81." We want a number which, when it is squared, will have 81 as the result. This number is 9, because $9^2 = 9 \times 9 = 81$.

*Answer:* $\sqrt{81} = 9$

**EXAMPLE 4** Which number is a perfect square?
(a) 90 (b) 144 (c) 50 (d) 2

**THINKING ABOUT THE PROBLEM**

Take each choice and test it to see if it is a perfect square.

90 is not a perfect square because its square root is not a whole number. Since $9^2 = 81$ and $10^2 = 100$, $\sqrt{90}$ is between 9 and 10. $\sqrt{50}$ is between 7 and 8. $\sqrt{2}$ is between 1 and 2.

$\sqrt{144}$ is exactly 12, because $12^2 = 12 \times 12 = 144$. Thus, 144 is a perfect square.

*Answer:* (b)

## CLASS EXERCISES

1. For each sentence, write equivalent sentences, using the word "square," the words "square root," and the symbol $\sqrt{\phantom{x}}$.

   **a.** $4^2 = 16$ **b.** $10^2 = 100$ **c.** $121 = 11^2$ **d.** $1^2 = 1$ **e.** $169 = 13^2$ **f.** $20^2 = 400$

2. Find the value of each square.

   **a.** $6^2$ **b.** $4^2$ **c.** $14^2$ **d.** $32^2$

3. Find the value of each square root.

   **a.** $\sqrt{81}$ **b.** $\sqrt{100}$ **c.** $\sqrt{144}$ **d.** $\sqrt{1600}$ **e.** $\sqrt{484}$ **f.** $\sqrt{625}$

4. Tell whether or not each number is a perfect square.

   **a.** 100 **b.** 1,000 **c.** 40 **d.** 49 **e.** 36 **f.** 10 **g.** 121 **h.** 5 **i.** 0

## HOMEWORK EXERCISES

1. For each sentence, write equivalent sentences, using the word "square," the words "square root," and the symbol $\sqrt{\phantom{x}}$.

   **a.** $7^2 = 49$ **b.** $100 = 10^2$ **c.** $2^2 = 4$ **d.** $196 = 14^2$ **e.** $9^2 = 81$
   **f.** $50^2 = 2{,}500$ **g.** $10{,}000 = 100^2$ **h.** $6{,}889 = 83^2$ **i.** $0^2 = 0$

2. Find the value of each square.

   **a.** $1^2$ **b.** $20^2$ **c.** $12^2$ **d.** $11^2$ **e.** $13^2$ **f.** $22^2$

3. Find the value of each square root.

   **a.** $\sqrt{121}$ **b.** $\sqrt{49}$ **c.** $\sqrt{16}$ **d.** $\sqrt{9}$ **e.** $\sqrt{900}$ **f.** $\sqrt{225}$
   **g.** $\sqrt{3{,}600}$ **h.** $\sqrt{324}$ **i.** $\sqrt{0}$

4. Tell whether or not each number is a perfect square.

   **a.** 1 **b.** 2 **c.** 3 **d.** 12 **e.** 24 **f.** 49 **g.** 121 **h.** 640 **i.** 6,400

## SPIRAL REVIEW EXERCISES

1. Find the value of each numerical expression.

   a. $4 \times 10^3$  b. $3.6 \times 10^2$

   c. $8.5 \times 10^5$  d. $9.52 \times 10^6$

2. The value of $7 \times 10^2 + 3 \times 10 + 5$ is
   (a) 7,035 (c) 375
   (b) 735 (d) 7,350

3. Find the value of $50 - x^2$ when $x$ is 7.

4. Find the value of $3y^4$ when $y$ is 2.

5. The opposite of 5 is
   (a) $\frac{1}{3}$ (b) $-\frac{1}{5}$ (c) $-5$ (d) 0

6. A dozen crates of eggs weighed 240 pounds. Find the weight of one crate.

7. Mr. Solomon invests $3,000 at 8% interest per year. How much interest will he earn in the first year?

8. Divide: $40 \div \frac{1}{2}$

9. At a pet show, there were 5 dogs for every 3 cats. If there were 80 dogs, how many cats were at the pet show?

10. Find the value of $16 + 4 \times 8 \div 2$.

11. The mode for the set of numbers 5, 5, 10, 15, 35 is
    (a) 5 (b) 10 (c) 12 (d) 14

12. The value of 20 kilometers is
    (a) 200 m (c) 20,000 m
    (b) 2,000 m (d) 200,000 m

## CHALLENGE QUESTION

Explain why $\sqrt{-16}$ is not $-4$.

## UNIT 18–1 Formulas

### THE MAIN IDEA

1. A *formula* is a mathematical sentence that shows how variables are related to each other.
2. The variable that stands alone in a formula is called the *subject* of the formula. To find the value of the subject of a formula, replace the other variables by their values, and perform the operations.
3. You can write a formula if you are given either a word sentence or a table of values that shows how the variables are related.

**EXAMPLE 1** What is the subject of each formula?

The subject of a formula is the variable that stands alone.

| *Formula* | *Subject* |
|---|---|
| $P = 2\ell + 2w$ | $P$ |
| $\ell \times w = A$ | $A$ |

**EXAMPLE 2** Find the value of the subject of the formula $P = 2\ell + 2w$ when $\ell = 5$ and $w = 6$.

The subject of the formula is $P$. Replace the variables $\ell$ and $w$ by their values.

$$P = 2\ell + 2w$$
$$P = 2(5) + 2(6)$$

Perform the operations in the correct order: multiplication first, followed by addition.

$$P = 10 + 12$$
$$P = 22 \quad \textit{Ans.}$$

**EXAMPLE 3** The formula $C = 50 + 30p$ gives the cost $C$, in cents, of developing a roll of film that has $p$ photographs. Find the cost of developing a roll of film that has 24 photographs.

$$C = 50 + 30p$$

Substitute 24 for $p$. $\quad C = 50 + 30(24)$

Perform the operations. $\quad C = 50 + 720$

$$C = 770 \text{ cents}$$

*Answer:* The cost of developing the film is \$7.70.

**EXAMPLE 4** Write a formula that says: "The distance is equal to the product of the rate and the time."

Choose variables to stand for the quantities. Let $D$ represent Distance, $R$ represent Rate, $T$ represent Time.

Use "=" to stand for "is equal to." $D =$

Write "the product of the rate and the time," using one of the ways of showing multiplication. $R \times T$ or $R(T)$ or $RT$

*Answer:* $D = RT$

**EXAMPLE 5** Write a formula that states the relationship between $x$ and $y$.

| $x$ | 3 | 4 | 5 | 6 |
|---|---|---|---|---|
| $y$ | 21 | 28 | 35 | 42 |

**THINKING ABOUT THE PROBLEM**

The table shows the values of $x$ for different values of $y$. Look for a pattern. Ask yourself questions such as: Is the same number always added on to $x$? Is $y$ always multiplied by the same number?

Note that in this table, $x$ is always multiplied by 7 to produce the value of $y$. Thus, $y$ is always equal to 7 times $x$. For example: $28 = 7(4)$.

$$y = 7x$$

Also, $y$ is always divided by 7 to produce the value of $x$. Thus, $x$ is always equal to $y$ divided by 7. For example: $5 = \frac{35}{7}$

$$x = \frac{y}{7}$$

*Answer:* $y = 7x$ or $x = \frac{y}{7}$

**EXAMPLE 6** Mrs. Burns buys some meat selling for \$3 a pound for a total cost of $x$ dollars. Which formula can be used to find the number of pounds $(n)$ of meat she bought?

(a) $n = 3x$ (b) $n = 3 + x$ (c) $n = \frac{3}{x}$ (d) $n = \frac{x}{3}$

The number of pounds $(n)$ of meat is found by dividing the total cost $(x)$ by the price per pound (3). Thus, $n$ equals $x$ divided by 3.

*Answer:* (d)

## CLASS EXERCISES

**1.** What is the subject of each formula?

**a.** $c = 2.54 \times I$ **b.** $\frac{I}{12} = F$ **c.** $P = a + b + c$ **d.** $4s = P$ **e.** $\frac{1}{2}bh = A$

**f.** $I = PRT$ **g.** $\ell wh = V$ **h.** $A = \pi r^2$ **i.** $2\ell + 2w = P$

**2.** Find the value of the subject of each formula, using the given values of the variables.

**a.** $D = rt$ — $r = 45, t = 3$

**b.** $I = prt$ — $p = 2{,}000, r = .05, t = 2$

**c.** $2\ell + 2w = p$ — $\ell = 7, w = 11$

**d.** $A = s^2$ — $s = 12$

**e.** $A = \frac{1}{2}bh$ — $b = 8, h = 5$

**f.** $P = a + b + c$ — $a = 2.3, b = 5, c = 4.3$

**g.** $S = 6e^2$ — $e = 2$

**h.** $C = P + .05P$ — $P = 50$

**3.** The formula $S = 200 + 10t$ gives Mr. Smith's salary $S$, in dollars, for a week in which he works $t$ hours overtime. Find $S$ for each of the given values of $t$.

**a.** 0 **b.** 5 **c.** 10 **d.** 18

**4.** Write a formula for each sentence.

**a.** The time of a train trip is equal to 400 miles divided by the train's rate.

**b.** The number of quarts is equal to 4 times the number of gallons.

**c.** A father's age is 22 years more than his son's age.

**d.** The total price of a bag of apples is equal to .49 times the number of pounds of apples.

**e.** The cost of a cab ride is equal to the number of miles driven multiplied by the rate per mile.

**5.** Write a formula that states the relationship between the variables in each table of values.

**a.**

| $x$ | 7 | 8 | 9 | 10 | 11 |
|---|---|---|---|---|---|
| $y$ | 11 | 12 | 13 | 14 | 15 |

**b.**

| $m$ | 20 | 22 | 24 | 26 | 28 |
|---|---|---|---|---|---|
| $n$ | 27 | 29 | 31 | 33 | 35 |

**c.**

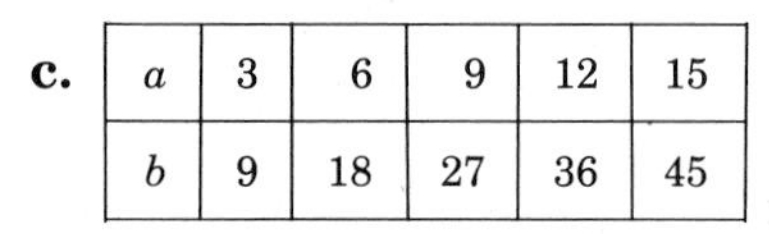

| $a$ | 3 | 6 | 9 | 12 | 15 |
|---|---|---|---|---|---|
| $b$ | 9 | 18 | 27 | 36 | 45 |

**d.**

| $q$ | 30 | 60 | 90 | 120 | 150 |
|---|---|---|---|---|---|
| $r$ | 5 | 10 | 15 | 20 | 25 |

**e.**

| $k$ | 1 | 2 | 3 | 4 | 5 |
|---|---|---|---|---|---|
| $l$ | 1 | 4 | 9 | 16 | 25 |

**f.**

| $t$ | 0 | 1 | 2 | 3 | 4 |
|---|---|---|---|---|---|
| $v$ | 1 | 4 | 7 | 10 | 13 |

**6.** Write a formula for each relationship.

**a.** The relationship between the total cost of a box of pastries and the number of pastries if the cost of each pastry is \$.80.

**b.** The relationship between a train's speed and its traveling time if the train makes the trip between two cities that are 300 miles apart.

**c.** The relationship between liters and milliliters if there are 1,000 milliliters in every liter.

**d.** The relationship between weeks and days if the number of days divided by 7 equals the number of weeks.

## *HOMEWORK EXERCISES*

**1.** Name the subject of each formula.

**a.** $A = \ell w$ **b.** $P = 2\ell + 2w$ **c.** $S = 4\ell w + 2wd$ **d.** $C = 2\pi r$

**e.** $12 \times I = F$ **f.** $P - D = C$ **g.** $S = 6e^2$ **h.** $R = \frac{d}{t}$

**2.** Find the value of the subject of each formula, using the given values of the variables.

**a.** $V = \ell wh$ $\quad \ell = 10, w = 8, h = 6$

**b.** $A = \frac{1}{2}bh$ $\quad b = 8, h = 6$

**c.** $P = 2(\ell + w)$ $\quad \ell = 15, w = 5$

**d.** $V = e^3$ $\quad e = 5$

**e.** $C = 2\pi r$ $\quad \pi = \frac{22}{7}, r = 14$

**f.** $P = 3s$ $\quad s = 20$

**g.** $F = \frac{9}{5}C + 32$ $\quad C = 100$

**h.** $C = \frac{5}{9}(F - 32)$ $\quad F = 32$

**i.** $A = \frac{1}{2}h(b + c)$ $\quad h = 4, b = 6, c = 10$

**3.** The formula $C = 70 + 30(m - 3)$ gives the cost, $C$, of a long-distance telephone call of $m$ minutes. Find $C$ for each of the given values of $m$.

**a.** 3 **b.** 5 **c.** 10

**4.** The formula $C = 20s + 10t$ gives the cost $C$, in dollars, of $s$ shirts and $t$ ties. Find $C$ for the given values of $s$ and $t$.

**a.** $s = 5, t = 2$ **b.** $s = 10, t = 6$ **c.** $s = 0, t = 9$

**5.** Write a formula for each sentence.

**a.** The number of centimeters is equal to 100 times the number of meters.

**b.** The number of days is equal to 7 times the number of weeks.

**c.** The salary is equal to 10.25 times the number of hours worked.

**d.** The number of feet is equal to the number of inches divided by 12.

**e.** The perimeter of a rectangle is equal to twice the sum of the measures of its length and width.

**6.** Write a formula that states the relationship between the variables in each table of values.

**a.**

| $x$ | 4 | 5 | 6 | 7 | 8 |
|---|---|---|---|---|---|
| $y$ | 9 | 10 | 11 | 12 | 13 |

**b.**

| $m$ | 22 | 20 | 16 | 12 | 8 |
|---|---|---|---|---|---|
| $n$ | 18 | 16 | 12 | 8 | 4 |

**c.**

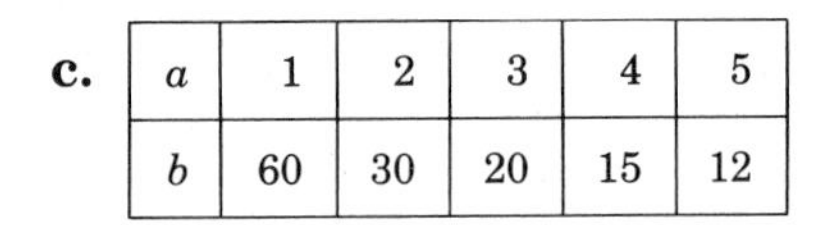

| $a$ | 1 | 2 | 3 | 4 | 5 |
|---|---|---|---|---|---|
| $b$ | 60 | 30 | 20 | 15 | 12 |

**d.**

| $r$ | 18 | 27 | 36 | 45 | 54 |
|---|---|---|---|---|---|
| $s$ | 6 | 9 | 12 | 15 | 18 |

**e.**

| $c$ | 7 | 9 | 11 | 13 | 15 |
|---|---|---|---|---|---|
| $d$ | 7 | 9 | 11 | 13 | 15 |

**f.**

| $h$ | 10 | 20 | 30 | 40 | 50 |
|---|---|---|---|---|---|
| $b$ | 6 | 11 | 16 | 21 | 26 |

**7.** Write a formula for each relationship.

**a.** The relationship between the total cost of a box of candies and the number of candies if each candy costs $.42.

**b.** The relationship between an airplane's speed and its traveling time if the plane makes the trip between two cities that are 2,000 miles apart.

**c.** The relationship between grams and kilograms if there are 1,000 grams in every kilogram.

**d.** The relationship between years and months if the number of years multiplied by 12 equals the number of months.

## SPIRAL REVIEW EXERCISES

**1.** Which number is not a perfect square?
(a) 10 (b) 100 (c) 25 (d) 49

**2.** Find the value of each square root.

**a.** $\sqrt{1}$ **b.** $\sqrt{9}$ **c.** $\sqrt{81}$ **d.** $\sqrt{169}$

**3.** Evaluate: $5 \times 10^3 + 6 \times 10 + 9$

**4.** What is the value of $4x^3$ when $x = 2$?

**5.** What is the opposite of $-3$?

(a) $+3$ (b) $\frac{1}{3}$ (c) $-\frac{1}{3}$ (d) 0

**6.** What is the probability that the spinner shown will stop on a multiple of 3?

**7.** Ribbon costs $.19 per foot. If David bought 4 yards of ribbon, how much did he spend?

**8.** Subtract $11\frac{5}{8}$ from $18\frac{1}{4}$.

## CHALLENGE QUESTION

For each table of values, write a formula that states the relationship between the values.

**a.**

| $x$ | 1 | 2 | 3 | 4 | 5 |
|---|---|---|---|---|---|
| $y$ | 1 | 8 | 27 | 64 | 125 |

**b.**

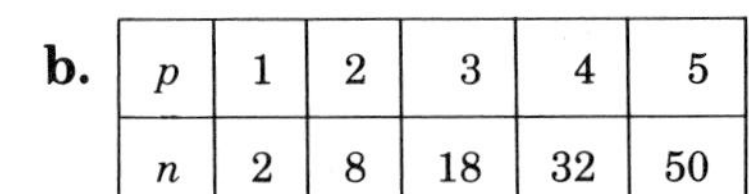

| $p$ | 1 | 2 | 3 | 4 | 5 |
|---|---|---|---|---|---|
| $n$ | 2 | 8 | 18 | 32 | 50 |

# UNIT 18–2 Solving Equations by Adding

## THE MAIN IDEA

1. An ***equation*** is a mathematical sentence stating that *two quantities are equal.*
2. When an equation contains a variable, *a value of the variable that makes the equation true* is called a ***solution*** of the equation. Finding the solution of an equation is called ***solving*** the equation.
3. To solve an equation like

   $$x + 15 = 7, \quad 17 + q = 20, \quad \text{or} \quad y - 9 = 21$$

   add the opposite of the number that accompanies the variable to both sides of the equation. This keeps the equation "in balance."
4. To check a solution, substitute the number for the variable in the original equation and see if a true sentence results.

**EXAMPLE 1** Tell whether 3 is a solution of the equation $7x + 12 = 33$.

$$7x + 12 = 33$$

In the original equation, substitute 3 for $x$. $\quad 7(3) + 12 \stackrel{?}{=} 33$

Evaluate, following the order of operations. $\quad 21 + 12 \stackrel{?}{=} 33$

Since a true sentence results, 3 is a solution. $\quad 33 = 33 \checkmark$

*Answer:* 3 is a solution of $7x + 12 = 33$.

**EXAMPLE 2** Solve for $x$: $x + 15 = 7$

$$x + 15 = 7$$

Add the opposite of 15 to both sides of the equation. (Remember: $+15 + (-15) = 0$)

$$x + 15 + (-15) = 7 + (-15)$$
$$x = -8$$

*Check:* In the original equation, substitute $-8$ for $x$ and evaluate.

$$x + 15 = 7$$
$$-8 + 15 \stackrel{?}{=} 7$$
$$7 = 7 \checkmark$$

*Answer:* The solution is $-8$.

**EXAMPLE 3** Solve for $q$: $17 + q = 20$

$$17 + q = 20$$

Add the opposite of 17 to both sides of the equation.

$$(-17) + 17 + q = 20 + (-17)$$
$$q = 3$$

*Check:* In the original equation, substitute 3 for $q$ and evaluate.

$$17 + q = 20$$
$$17 + 3 \stackrel{?}{=} 20$$
$$20 = 20 \checkmark$$

*Answer:* The solution is 3.

**EXAMPLE 4** Solve for $y$: $y - 9 = 21$

| | |
|---|---|
| | $y - 9 = 21$ |
| Rewrite the subtraction. | $y + (-9) = 21$ |
| Add the opposite of $-9$ to both sides. | $y + (-9) + (9) = 21 + (9)$ |
| | $y = 30$ |
| *Check:* In the original equation, substitute 30 for $y$ and evaluate. | $y - 9 = 21$ |
| | $30 - 9 \stackrel{?}{=} 21$ |
| | $21 = 21$ ✓ |

*Answer:* The solution is 30.

## CLASS EXERCISES

1. Tell whether the given value of the variable is a solution of the given equation.

**a.** $x + 7 = 28;\ x = 21$ **b.** $x - 13 = 13;\ x = 26$ **c.** $5 + a = 21;\ a = 26$

**d.** $3n = 120;\ n = 40$ **e.** $2x + 11 = 27;\ x = 15$ **f.** $3x - 10 = 17;\ x = 9$

2. Solve each equation and check.

**a.** $x + 14 = 23$ **b.** $y - 20 = 60$ **c.** $35 + x = 43$ **d.** $a + 1 = -1$

**e.** $z - 18 = -3$ **f.** $y + 19 = 19$ **g.** $m - 87 = 101$ **h.** $73 + b = 98$

**i.** $x - 25 = -25$ **j.** $y + 19 = -19$ **k.** $31 + z = 13$ **l.** $n - 42 = -39$

## HOMEWORK EXERCISES

1. Tell whether the given value of the variable is a solution of the given equation.

**a.** $x + 17 = 24;\ x = 7$ **b.** $a - 9 = 27;\ a = 36$ **c.** $8 + y = 21;\ y = 17$

**d.** $3m = 33;\ m = 11$ **e.** $z + 15 = 15;\ z = 0$ **f.** $n - 14 = 1;\ n = 13$

**g.** $35 + x = 25;\ x = -10$ **h.** $x - 1 = 1;\ x = 1$

2. Solve each equation and check.

**a.** $x + 13 = 25$ **b.** $x - 19 = 11$ **c.** $y - 25 = 35$ **d.** $1 + x = 1$

**e.** $y - 13 = 13$ **f.** $14 + y = -14$ **g.** $x + 67 = 93$ **h.** $n - 47 = 100$

**i.** $y - 39 = -38$ **j.** $44 + x = 22$ **k.** $x - 15 = -32$ **l.** $y + 26 = -51$

**m.** $y - 26 = 51$ **n.** $100 + a = 5$ **o.** $b - 150 = -2$

## SPIRAL REVIEW EXERCISES

1. Using the formula $A = 2B - 3C$, find the value of $A$ when $B = 10$ and $C = 5$.

2. Which statement is correct?
   (a) 16 is the square root of 4.
   (b) 25 is the square of 5.
   (c) $\frac{1}{4}$ is the square root of $\frac{1}{2}$.
   (d) 100 is the square root of 1,000.

3. Which number is a perfect square?
   (a) 5 (b) 55 (c) 9 (d) 99

4. 36 is what percent of 50?

5. The value of $5 \times 10^3 + 2 \times 10 + 9$ is
   (a) 5,290 (c) 5,029
   (b) 5,209 (d) 52,009

6. The mean of 12, 12, 20, 30, and 56 is
   (a) 12 (b) 20 (c) 24 (d) 26

7. The product of $-18$ and 3 is
   (a) $-15$ (b) $-21$ (c) $-54$ (d) $-6$

8. As shown in the bar graph, the number of cars sold in March is closest to
   (a) 2 (b) 150 (c) 175 (d) 200

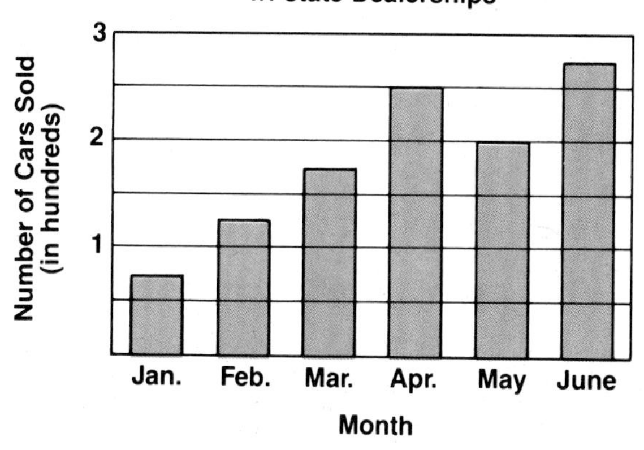


9. The number of meters in 3 kilometers is
   (a) 3 (b) 30 (c) 300 (d) 3,000

**CHALLENGE QUESTION**

Write and solve an equation that says "4 more than a number is 12."

# UNIT 18–3 Solving Equations by Multiplying

## THE MAIN IDEA

To solve an equation like

$$3x = 15 \quad \text{or} \quad \frac{1}{4}y = 7$$

multiply both sides of the equation by the reciprocal of the number that multiplies the variable.

**EXAMPLE 1** Solve for $x$: $9x = 72$

$$9x = 72$$

Multiply both sides of the equation by the reciprocal of 9.

(Remember: $9 \times \frac{1}{9} = 1$)

$$\frac{1}{9}(9x) = \frac{1}{9}(72)$$

$$x = 8$$

*Check:* In the original equation, substitute 8 for $x$ and evaluate.

$$9x = 72$$
$$9(8) \stackrel{?}{=} 72$$
$$72 = 72 \checkmark$$

*Answer:* The solution is 8.

**EXAMPLE 2** Solve for $x$: $\frac{x}{7} = -5$

$$\frac{x}{7} = -5$$

Rewrite the division.

$$\frac{1}{7} \cdot x = -5$$

Multiply both sides of the equation by the reciprocal of $\frac{1}{7}$.

$$7\left(\frac{1}{7} \cdot x\right) = 7(-5)$$
$$x = -35$$

*Check:* In the original equation, substitute $-35$ for $x$ and evaluate.

$$\frac{x}{7} = -5$$
$$\frac{-35}{7} \stackrel{?}{=} -5$$
$$-5 = -5 \checkmark$$

*Answer:* The solution is $-35$.

**EXAMPLE 3** Solve for $y$: $6y = 3$

$$6y = 3$$

Multiply both sides of the equation by the reciprocal of 6.

$$\frac{1}{6}(6y) = \frac{1}{6}(3)$$
$$y = \frac{3}{6}$$
$$y = \frac{1}{2}$$

*Check:* In the original equation, substitute $\frac{1}{2}$ for $y$ and evaluate.

$$6y = 3$$
$$6\left(\frac{1}{2}\right) \stackrel{?}{=} 3$$
$$3 = 3 \checkmark$$

*Answer:* The solution is $\frac{1}{2}$.

## CLASS EXERCISES

1. Solve each equation and check.

   a. $8x = 48$ b. $7a = 49$ c. $-11x = 132$ d. $12b = 84$ e. $9x = -36$

   f. $10y = -70$ g. $-45a = 15$ h. $14x = 7$ i. $42x = 0$

2. Solve each equation and check.

   a. $\frac{x}{8} = 4$ b. $\frac{y}{3} = 96$ c. $\frac{a}{5} = 1$ d. $\frac{x}{9} = -4$ e. $\frac{b}{-7} = -6$ f. $\frac{y}{2} = -1$

3. Solve each equation and check.

   a. $x + 4 = 12$ b. $4x = 12$ c. $\frac{1}{4}x = 12$ d. $x - 10 = -10$

   e. $12 + x = 4$ f. $\frac{x}{12} = 4$

## HOMEWORK EXERCISES

1. Solve each equation and check.

   a. $5x = 85$ b. $3x = -48$ c. $-7x = 98$ d. $11x = -99$ e. $4x = -24$

   f. $6x = 54$ g. $6x = -72$ h. $10x = 5$ i. $15x = -5$

2. Solve each equation and check.

   a. $\frac{x}{4} = 80$ b. $\frac{y}{6} = 48$ c. $\frac{a}{5} = 105$ d. $\frac{x}{7} = 14$ e. $\frac{n}{2} = 33$ f. $\frac{x}{9} = 1$

   g. $\frac{x}{8} = -1$ h. $\frac{y}{8} = -96$ i. $\frac{m}{12} = -144$

3. Solve each equation and check.

   a. $5x = 10$ b. $5 + x = 10$ c. $x - 5 = 10$ d. $\frac{1}{5}x = 10$ e. $10x = 5$

## SPIRAL REVIEW EXERCISES

1. Tell whether the given value of the variable is a solution of the given equation.

   a. $x + 13 = -13;\ x = -1$

   b. $a - 17 = 27;\ a = 44$

   c. $3m = -18;\ m = -6$

   d. $\frac{y}{8} = -4;\ y = -2$

2. Solve each equation and check.

   a. $x + 34 = 52$ b. $y - 11 = 74$

   c. $a - 13 = -12$ d. $y + 25 = -34$

3. Evaluate $5x + 2y$ when $x = 4$ and $y = 6$.

4. If it costs \$1.85 for the first 15 words of a telegram and each additional word costs \$.10, how much does it cost to send a telegram of 22 words?

5. Jason deposits $34.80 into his savings account that has a balance of $192.75. What is the new balance?

6. If a pair of shoes that sells for $25 is on sale for 30% off, what is the sale price?

7. If $m = 3$ and $p = 2$, find the value of $m^2 + p^3$.

8. Find $\frac{3}{4}$ of 440.

9. Subtract 81.9 from 90.82.

10. Find the mode of the numbers 2.8, 2.9, 3.1, 3.3, 3.7, 3.7, and 3.7.

11. Evaluate:

    a. $5 + 8 \times 3$

    b. $2 \times 7 + 9 \times 4$

## CHALLENGE QUESTION

Write and solve an equation that says:

"3 times a number is 4.5"

## UNIT 18–4 Solving Equations by Two Operations

### THE MAIN IDEA

To solve an equation like

$$5x + 8 = 38, \quad 2x - 3 = -13, \quad \text{or} \quad \frac{y}{4} + 6 = 5$$

1. First, work on the additions by adding an opposite to both sides of the equation.
2. Then, work on the multiplication by multiplying both sides of the equation by the reciprocal of the number that multiplies the variable.

**EXAMPLE 1** Solve for $x$: $5x + 8 = 38$

$$5x + 8 = 38$$

Add the opposite of 8 to both sides of the equation.

$$5x + 8 + (-8) = 38 + (-8)$$
$$5x = 30$$

Multiply both sides of the equation by the reciprocal of 5.

$$\frac{1}{5}(5x) = \frac{1}{5}(30)$$
$$x = 6$$

*Check:* In the original equation, substitute 6 for $x$ and evaluate.

$$5x + 8 = 38$$
$$5(6) + 8 \stackrel{?}{=} 38$$
$$30 + 8 \stackrel{?}{=} 38$$
$$38 = 38 \ \checkmark$$

*Answer:* The solution is 6.

**EXAMPLE 2** Solve for $x$: $2x - 3 = -13$

$$2x - 3 = -13$$

Rewrite the subtraction.

$$2x + (-3) = -13$$

Add 3 to both sides.

$$2x + (-3) + (3) = -13 + (3)$$
$$2x = -10$$

Multiply both sides by $\frac{1}{2}$.

$$\frac{1}{2}(2x) = \frac{1}{2}(-10)$$
$$x = -5$$

*Check:* In the original equation, substitute $-5$ for $x$ and evaluate.

$$2x - 3 = -13$$
$$2(-5) - 3 \stackrel{?}{=} -13$$
$$-10 - 3 \stackrel{?}{=} -13$$
$$-13 = -13 \ \checkmark$$

*Answer:* The solution is $-5$.

**EXAMPLE 3** Solve for $y$: $\frac{y}{4} + 6 = 5$

$$\frac{y}{4} + 6 = 5$$

Add $-6$ to both sides.

$$\frac{y}{4} + 6 + (-6) = 5 + (-6)$$

$$\frac{y}{4} = -1$$

Rewrite the division.

$$\frac{1}{4}y = -1$$

Multiply both sides by 4.

$$4\left(\frac{1}{4}y\right) = 4(-1)$$

$$y = -4$$

*Check:* In the original equation, substitute $-4$ for $y$ and evaluate.

$$\frac{y}{4} + 6 = 5$$

$$\frac{-4}{4} + 6 \stackrel{?}{=} 5$$

$$-1 + 6 \stackrel{?}{=} 5$$

$$5 = 5 \checkmark$$

*Answer:* The solution is $-4$.

## CLASS EXERCISES

**1.** Solve each equation and check.

**a.** $3x + 4 = 10$ **b.** $6 + 2x = 12$ **c.** $5y + 7 = 27$ **d.** $\frac{a}{4} + 2 = 34$

**e.** $9 + \frac{1}{3}z = 3$ **f.** $5x + 15 = 0$ **g.** $5x + 6 = 26$ **h.** $3 + 8x = 51$

**i.** $\frac{y}{7} + 7 = -21$ **j.** $57 + 3q = 39$ **k.** $\frac{1}{2}x + 21 = 3$ **l.** $11m + 9 = -79$

**2.** Solve each equation and check.

**a.** $2x - 6 = 10$ **b.** $3x - 5 = -11$ **c.** $8y - 12 = 68$ **d.** $5a - 1 = 14$

**e.** $7m - 9 = -2$ **f.** $10x - 14 = 36$

## HOMEWORK EXERCISES

**1.** Solve each equation and check.

**a.** $2x + 7 = 15$ **b.** $6y + 8 = 38$ **c.** $1 + 9z = 82$ **d.** $8a + 5 = 53$

**e.** $3b + 14 = 29$ **f.** $4q + 8 = 12$ **g.** $5x + 5 = 45$ **h.** $2m + 22 = 30$

**i.** $7 + 12a = 31$ **j.** $8y + 17 = 97$ **k.** $26 + 6x = 14$ **l.** $5y + 53 = 108$

**m.** $4z + 20 = 0$ **n.** $15 + 12x = -9$ **o.** $3x + 2 = 2$

2. Solve each equation and check.

a. $3x - 4 = 11$ b. $2y - 8 = 18$ c. $5a - 15 = 5$ d. $6m - 6 = 0$

e. $9x - 3 = -21$ f. $10y - 10 = 100$ g. $11a - 44 = -22$ h. $-3 + 14t = -3$

3. Solve each equation and check.

a. $3x + 1 = 10$ b. $\frac{x}{3} + 5 = 14$ c. $8m - 3 = 21$ d. $\frac{y}{3} - 12 = -2$

e. $11 + 4r = 15$ f. $5x + 12 = -13$ g. $\frac{a}{2} + 17 = 12$ h. $\frac{1}{4}x - 1 = 0$

i. $2 + 2a = 16$ j. $\frac{1}{3}x + 9 = 9$ k. $\frac{n}{5} - 1 = -2$ l. $\frac{y}{10} - 8 = -16$

4. Solve each equation and check.

a. $\frac{x}{4} + 2 = 5$ b. $\frac{1}{5}y - 3 = 1$ c. $\frac{a}{3} + 12 = 9$ d. $\frac{1}{7}x + 5 = 5$ e. $\frac{n}{4} - 4 = 1$

f. $\frac{1}{3}x - 1 = -1$ g. $\frac{y}{10} - 5 = -45$ h. $\frac{1}{12}x + 1 = -3$

## SPIRAL REVIEW EXERCISES

1. Solve each equation and check.

a. $8x = 96$ b. $-6x = 66$

c. $5x = -35$ d. $21a = 7$

2. Solve each equation and check.

a. $y + 53 = 91$ b. $a - 51 = 50$

c. $x - 17 = 34$ d. $18 + x = 22$

3. Solve each equation and check.

a. $\frac{a}{9} = 7$ b. $\frac{x}{5} = -6$

c. $\frac{-m}{8} = 64$ d. $\frac{y}{4} = -144$

4. Tell whether the given value of the variable is a solution of the given equation.

a. $x + 47 = 53$; $x = -6$

b. $3m = -12$; $m = -4$

c. $x - 15 = 15$; $x = 0$

d. $3x - 7 = 17$; $x = 8$

5. The sum of $-9$ and $-3$ is
(a) 27 (b) 3 (c) $-6$ (d) $-12$

6. The expression $a^3$ means
(a) $a \cdot a \cdot a$ (c) $a + 3$
(b) $3 \cdot a$ (d) $\frac{3}{a}$

7. If 4 pounds of potatoes cost $1.38, what is the cost of 9 pounds?

8. If 9 out of 25 students are absent, what percent of the students are present?

9. The price of a shirt was decreased from $15 to $10. What was the percent of decrease?

10. The relationship between $A$ and $B$ in the table is expressed by the formula
(a) $B = A + 4$ (c) $A = B$
(b) $A + B = 10$ (d) $B = A - 4$

| A | 3 | 4 | 5 | 6 | 7 |
|---|---|---|---|---|---|
| B | 7 | 6 | 5 | 4 | 3 |

## CHALLENGE QUESTION

Write and solve an equation that says:
"3 times a number decreased by 5 is the same as 2 times the number increased by 8"

# UNIT 18–5 Graphing Number Sentences on a Number Line

## THE MAIN IDEA

1. Recall that a number is represented by a point on a number line.
   a. All numbers greater than the number are to its right.
   b. All numbers less than the number are to its left.

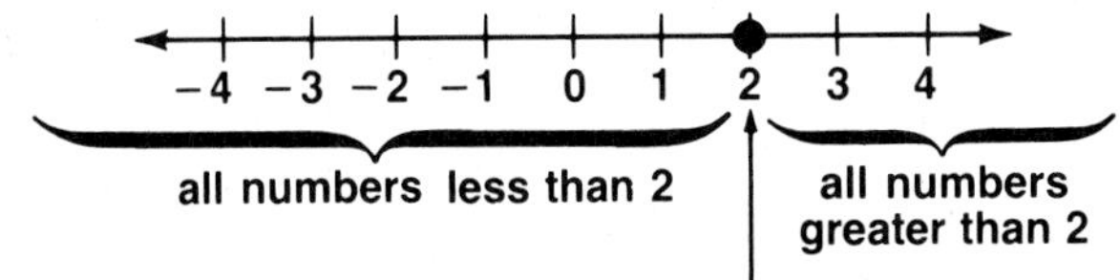


2. The number sentences that you will graph:

| *Specify* | *Number Sentence* | *Graph* |
|---|---|---|
| a. a given number. | $x = 2$ | −4 −3 −2 −1 0 1 2 3 4 |
| b. all numbers greater than a given number. | $x > 2$ | −4 −3 −2 −1 0 1 2 3 4<br>The *open circle* at 2 shows that the value 2 *is not included* in the graph. |
| c. all numbers greater than or equal to a given number. | $x \geq 2$ | −4 −3 −2 −1 0 1 2 3 4<br>The *closed circle* at 2 shows that the value 2 *is included* in the graph. |
| d. all numbers less than a given number. | $x < 2$ | −4 −3 −2 −1 0 1 2 3 4 |
| e. all numbers less than or equal to a given number. | $x \leq 2$ | −4 −3 −2 −1 0 1 2 3 4 |
| f. all numbers between two given numbers. The sentence tells whether or not to include the end values. | $-1 \leq x \leq 2$ | −4 −3 −2 −1 0 1 2 3 4 |
| | $-1 < x < 2$ | −4 −3 −2 −1 0 1 2 3 4 |
| | $-1 \leq x < 2$ | −4 −3 −2 −1 0 1 2 3 4 |
| | $-1 < x \leq 2$ | −4 −3 −2 −1 0 1 2 3 4 |

## CLASS EXERCISES

1. Graph each of the given open sentences on a number line.

   a. $x < -5$ b. $x \geq -2$ c. $x > 7$ d. $x \leq -3$ e. $-4 \leq x < 1$

   f. $-10 < x \leq -5$ g. $x > -1$ h. $x \leq -5$ i. $-1 \leq x < 1$ j. $x < -3$

   k. $x > -5$ l. $-2 \leq x \leq 0$ m. $x \geq -2$ n. $-1 < x < 2$ o. $-7 < x \leq -5$

2. What is the inequality that is represented by the graph?
   (a) $x > -3$ (b) $x < -3$ (c) $x \geq -3$ (d) $x \leq -3$

3. Which open sentence is represented by the graph?
   (a) $x \leq 4$ (c) $-3 \leq x < 4$
   (b) $x > -3$ (d) $-3 < x \leq 4$

4. The graph that represents the open sentence $-6 < x \leq -3$ is

(a) 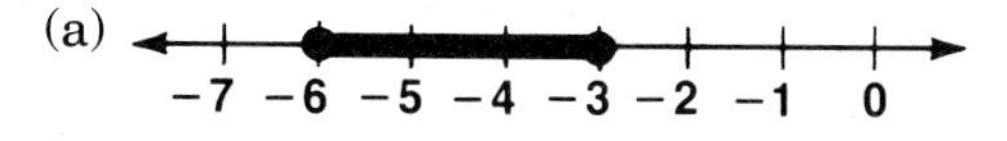


(c) 

(b) 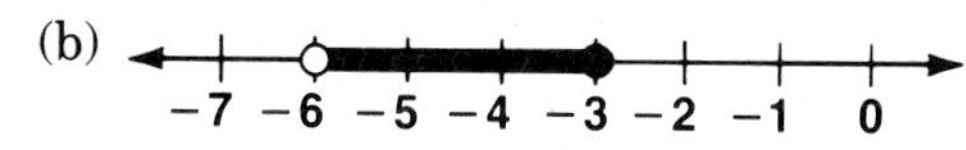


(d) 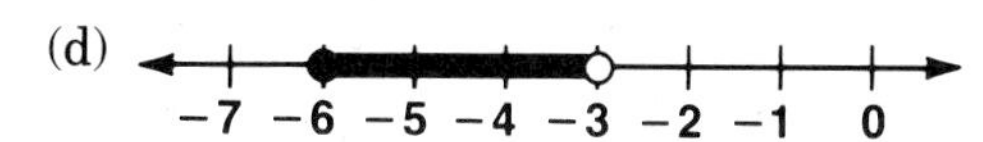


## HOMEWORK EXERCISES

1. Graph each of the given open sentences on a number line.

   a. $x > -3$ b. $x < 2$ c. $x \geq -4$ d. $x < -5$ e. $x > -7$

   f. $x \leq 0$ g. $x \geq -6$ h. $x \geq 3$ i. $x \leq -7$ j. $x > -6$

   k. $0 < x < 5$ l. $-3 \leq x < 1$ m. $-2 \leq x \leq 3$ n. $-5 < x \leq -1$ o. $-6 \leq x \leq 0$

2. What is the inequality that is represented by the graph?
   (a) $x > -2$ (b) $x < -2$ (c) $x < 2$ (d) $x \leq -2$

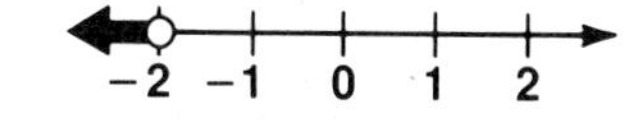


3. What is the inequality that is represented by the graph?
   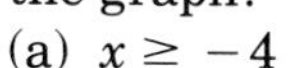
   (a) $x \geq -4$ (c) $x > -4$
   (b) $x \leq -4$ (d) $x < 4$

4. Which open sentence is represented by the graph?
   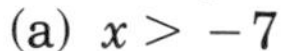

   (a) $x > -7$ (c) $-7 < x < 1$
   (b) $x < 1$ (d) $-7 \leq x < 1$

5. Which open sentence is represented by the graph?

   (a) $-5 < x < 0$ (c) $x \geq -5$
   (b) $-5 \leq x \leq 0$ (d) $x \leq 0$

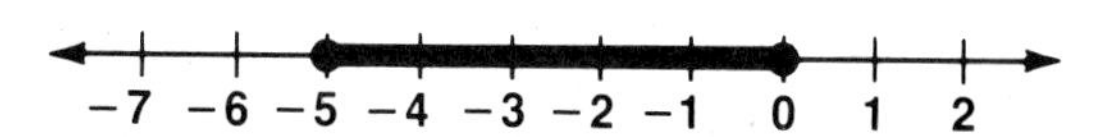

**6.** The graph that represents the open sentence $x < 3$ is

(a)

(b)

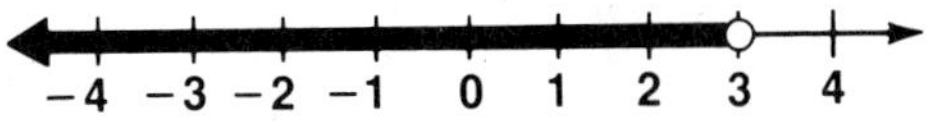

(c)

(d)

**7.** The graph that represents the open sentence $-2 < x \leq 3$ is

(a)
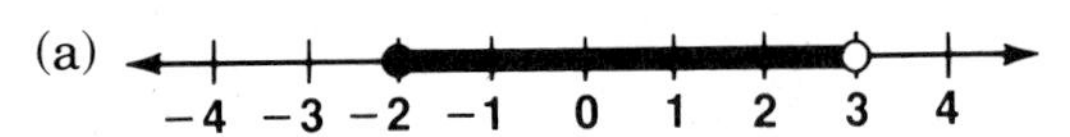

(b)

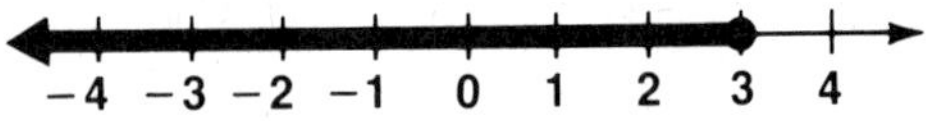

(c)

(d)

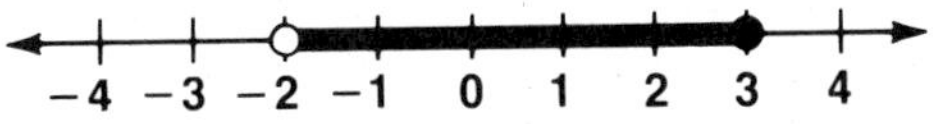

## SPIRAL REVIEW EXERCISES

**1.** Replace ? with < or > to make a true comparison.

**a.** $0 \; ? \; -6$ **b.** $-26 \; ? \; -10$

**c.** $-112 \; ? \; 3$ **d.** $-40 \; ? \; -100$

**e.** $-8.4 \; ? \; -9$ **f.** $-\frac{1}{4} \; ? \; -\frac{1}{3}$

**2.** Name the opposite of each signed number.

**a.** $-\frac{3}{4}$ **b.** $+9$ **c.** $-11.5$

**d.** $0$ **e.** $-2\frac{1}{2}$

**3.** Mr. Graham invests \$500 at 9% interest. How much interest does he earn after one year?

**4.** Solve the equation $5x - 12 = 93$ and check.

**5.** Which number is a solution of the equation $2x + 5 = -1$?

(a) $-3$ (b) $3$ (c) $1\frac{1}{2}$ (d) $-1\frac{1}{2}$

**6.** Add: $\frac{2}{3} + \frac{3}{5}$

**7.** Multiply: $\frac{2}{3} \times \frac{3}{5}$

**8.** What percent of 40 is 8?

**9.** If in a class there are 20 boys and 15 girls, the ratio of the number of boys to the total number of students in the class is

(a) $\frac{3}{4}$ (b) $\frac{4}{3}$ (c) $\frac{3}{7}$ (d) $\frac{4}{7}$

**10.** The product of .8 and .7 is
(a) 1.5 (b) .1 (c) .56 (d) 5.6

## CHALLENGE QUESTION

On a number line, show all the values of $x$ that make the statement true.

**a.** $x \leq -2$ or $x > 3$ **b.** $x \leq -2$ and $x > 3$

# THEME 10

# Introduction to Geometry

## MODULE 19



## MODULE 20

MODULE 19

# UNIT 19–1 Angles

## THE MEANING OF AN ANGLE

### THE MAIN IDEA

1. An *angle* (symbol ∠) is the figure formed by two *rays* that have the same endpoint called the *vertex* (plural *vertices*).

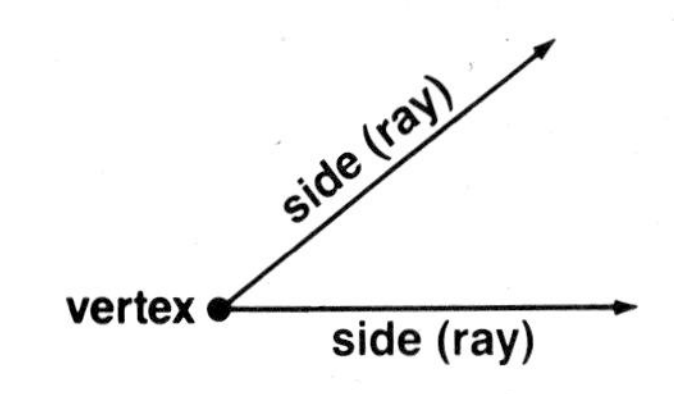


2. The ways of naming an angle are:

with a capital letter at its vertex

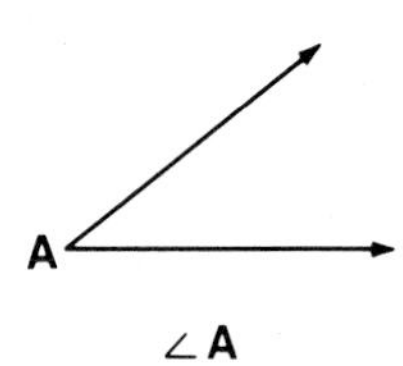


∠A

with a number inside the angle

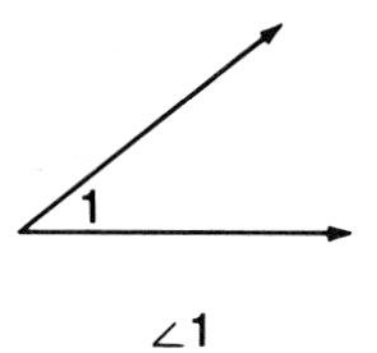


∠1

with three capital letters, of which the middle letter is the vertex

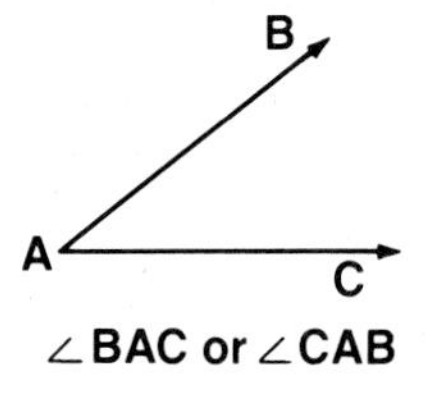


∠BAC or ∠CAB

**EXAMPLE 1** Name each angle by using a single letter, and by using three letters.

| | *Angle* | *Single-Letter Name* | *Three-Letter Name* |
|---|---|---|---|
| a. | 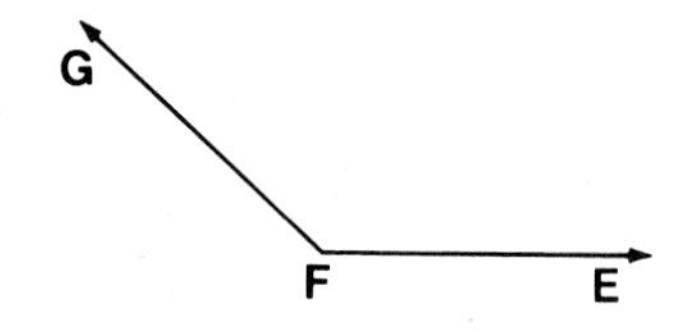 | ∠*F* | ∠*GFE* or ∠*EFG* (The vertex must be the middle letter.) |
| b. | 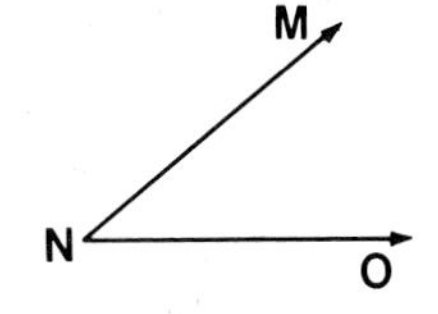 | ∠*N* | ∠*MNO* or ∠*ONM* |

**EXAMPLE 2** Use three letters to name $\angle 1$, $\angle 2$, and $\angle 3$.

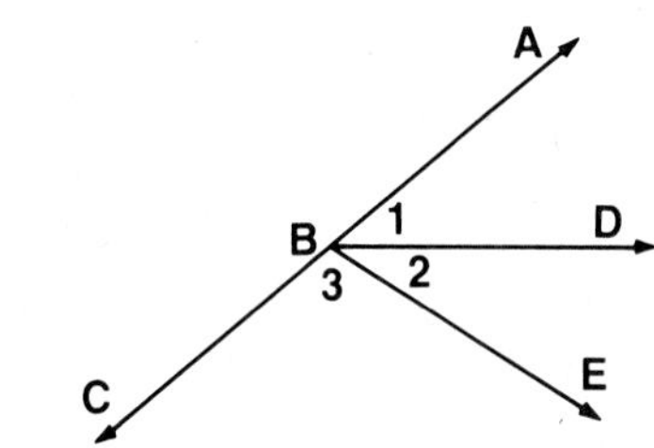


| *Angle* | *Name* |
|---|---|
| $\angle 1$ | $\angle ABD$ or $\angle DBA$ |
| $\angle 2$ | $\angle DBE$ or $\angle EBD$ |
| $\angle 3$ | $\angle EBC$ or $\angle CBE$ |

## CLASS EXERCISES

**1.** Name each angle by using a single letter and by using three letters.

a. T P A

b.
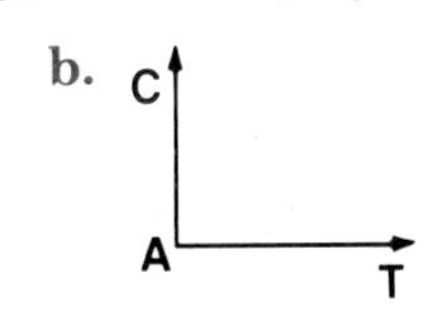


c.
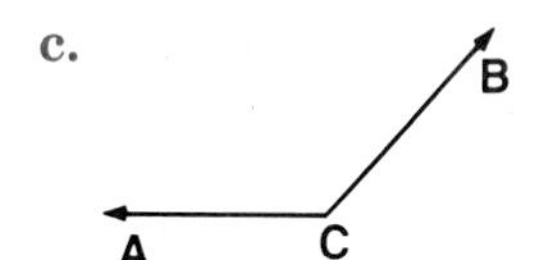


d.
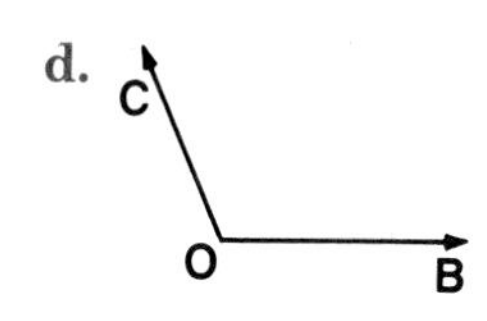


e.
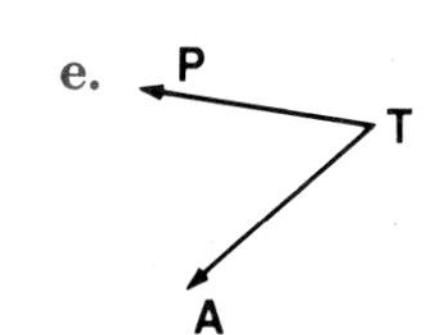


**2.** Use three letters to name the numbered angles.

a. X Y 1 Z

b.
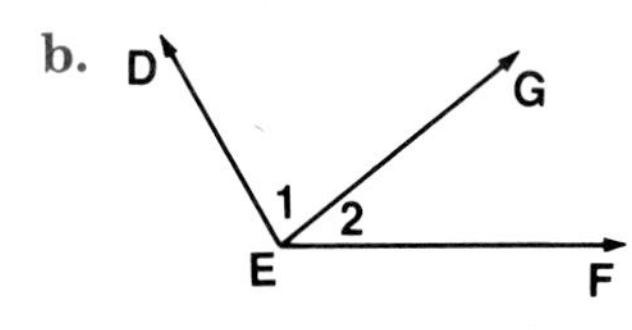


c.
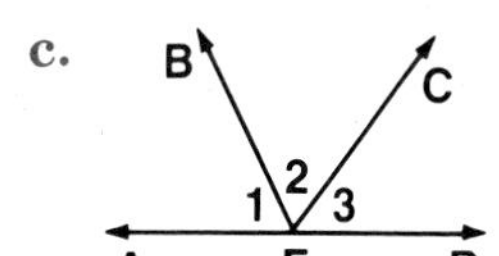


d.
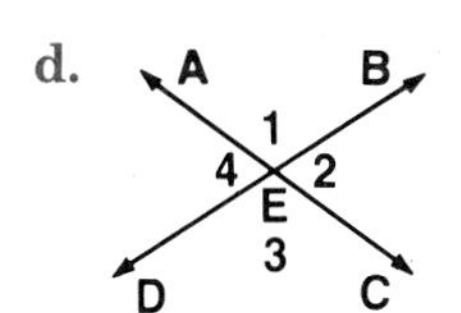

## MEASURING ANGLES

### THE MAIN IDEA

1. The *degree* is the unit of measure for angles.
   There are 360 degrees, written 360°, in a *complete rotation*.

   **Complete Rotation**

   Measures exactly 360°.

2. Angles are classified according to their measure.

   **Right Angle**

   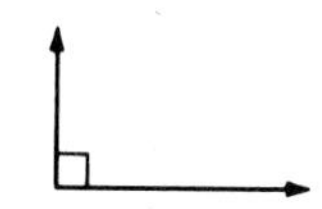

   Measures exactly 90°.
   (Looks like a square corner.)

   **Straight Angle**

   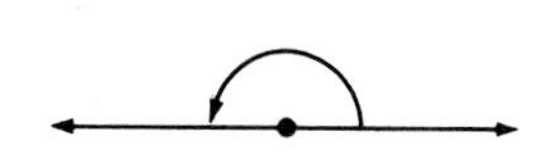

   Measures exactly 180°.
   (Both rays lie on the same line.)

   **Acute Angle**

   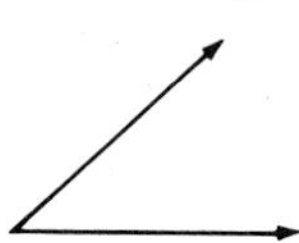

   Measures more than 0° and less than 90°. (Looks less open than a right angle.)

   **Obtuse Angle**

   Measures more than 90° and less than 180°. (Looks more open than a right angle.)

3. A *protractor* is an instrument that is used to measure angles. It has two scales, each starting at 0° and ending at 180°, to make it convenient to measure an angle regardless of its position.

   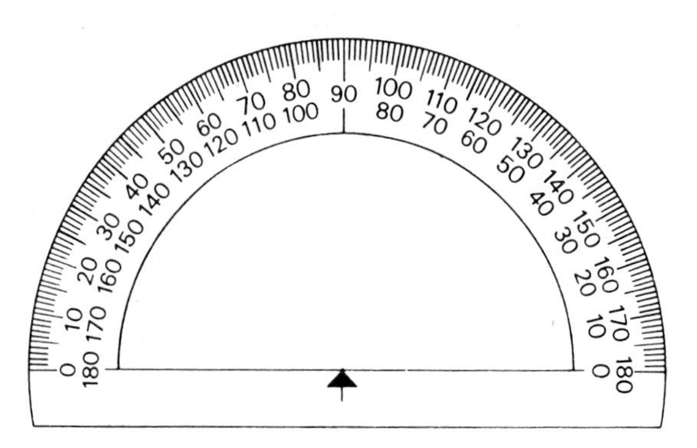
   

4. The measure of an angle is determined only by the number of degrees in the angle, not by the lengths of the sides.

**EXAMPLE 3** Tell whether each of the given angles is a right angle, an acute angle, an obtuse angle, or a straight angle.

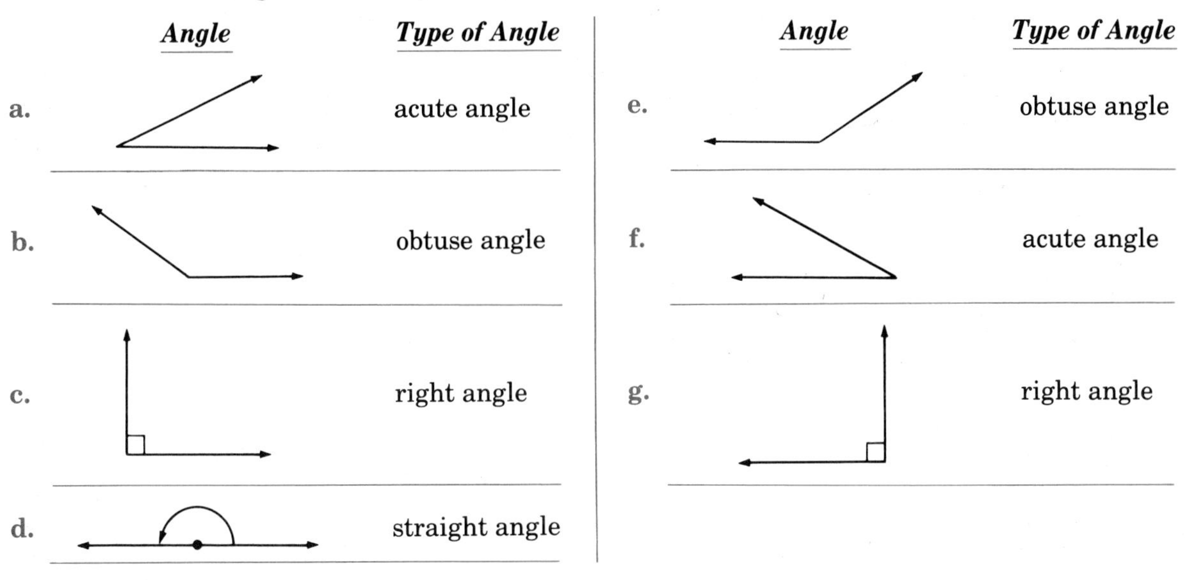

| | Angle | Type of Angle |
|---|---|---|
| a. | | acute angle |
| b. | | obtuse angle |
| c. | | right angle |
| d. | | straight angle |
| e. | | obtuse angle |
| f. | | acute angle |
| g. | | right angle |

**EXAMPLE 4** Tell from the measure of each angle whether the angle is a right angle, an acute angle, an obtuse angle, or a straight angle.

| | *Measure of the Angle* | *Type of Angle* |
|---|---|---|
| **a.** | 50° | acute angle |
| **b.** | 98° | obtuse angle |
| **c.** | 180° | straight angle |
| **d.** | 90° | right angle |

**EXAMPLE 5** Find the measure of $\angle ABC$.

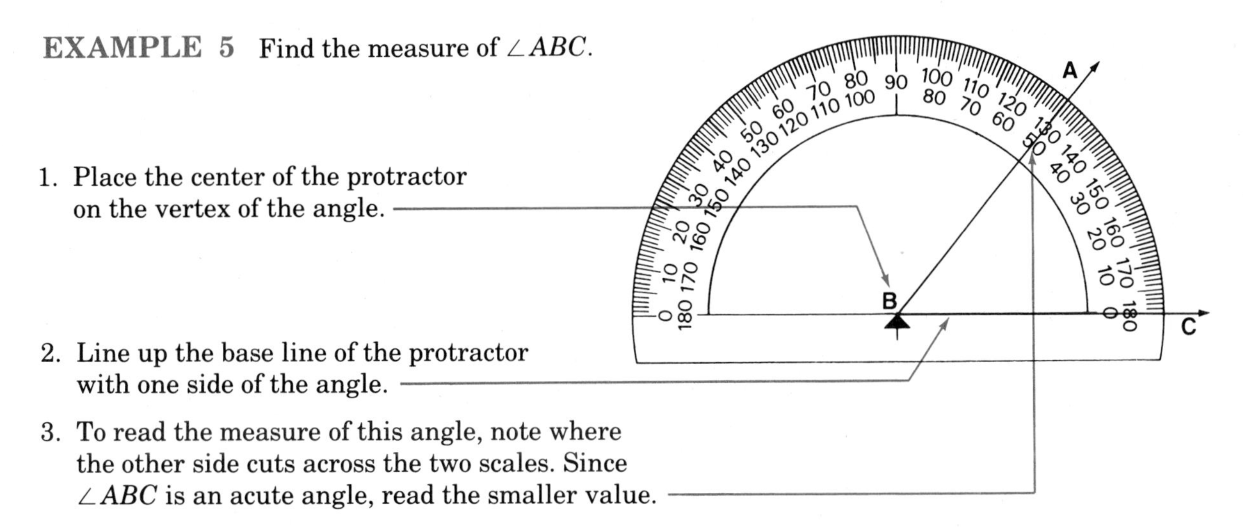


1. Place the center of the protractor on the vertex of the angle.
2. Line up the base line of the protractor with one side of the angle.
3. To read the measure of this angle, note where the other side cuts across the two scales. Since $\angle ABC$ is an acute angle, read the smaller value.

*Answer:* The measure of $\angle ABC$ is 50°.

**EXAMPLE 6** Find the measure of $\angle LMN$.

Since $\angle LMN$ is an obtuse angle, read the larger value.

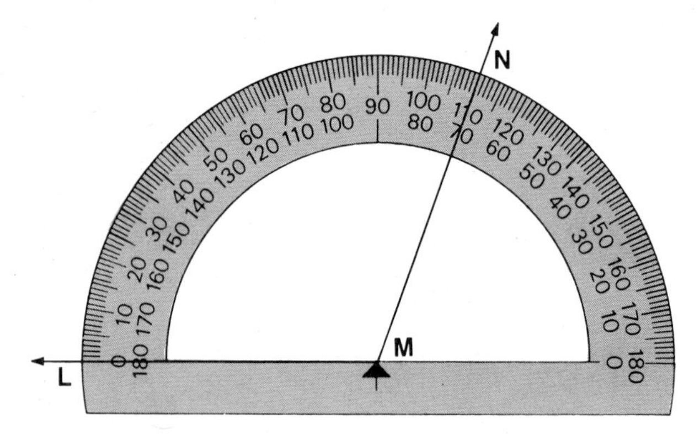


*Answer:* The measure of $\angle LMN$ is 110°.

**EXAMPLE 7** Find the number of degrees in each given angle.

| | *Given Angle* | *Answer* |
|---|---|---|
| **a.** | $\frac{3}{10}$ of a right angle | $\frac{3}{10} \times 90° = 27°$ |
| **b.** | $\frac{2}{9}$ of a straight angle | $\frac{2}{9} \times 180° = 40°$ |

## CLASS EXERCISES

**1.** Tell whether each angle is a right angle, an acute angle, an obtuse angle, or a straight angle.

a. 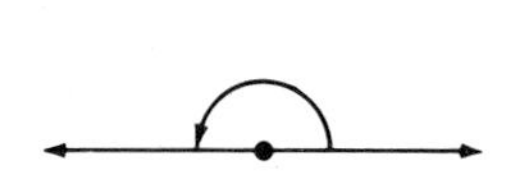

b. 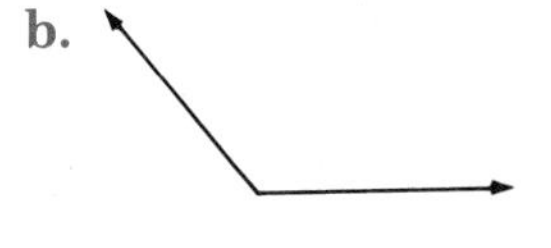

c. 

d. 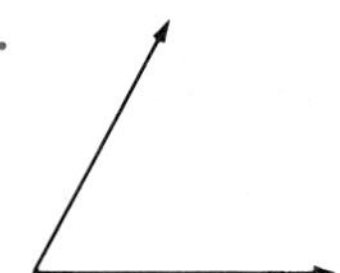

e.

**2.** Tell from the measure of each angle whether the angle is a right angle, an acute angle, an obtuse angle, or a straight angle.

**a.** 25° **b.** 89° **c.** 146° **d.** 180° **e.** 79° **f.** 46° **g.** 90° **h.** 101°

**3.** Trace each angle on your paper and use a protractor to find the measure of each angle. If necessary, extend the sides of the angle.

a. 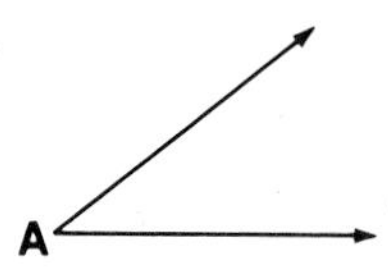


b. 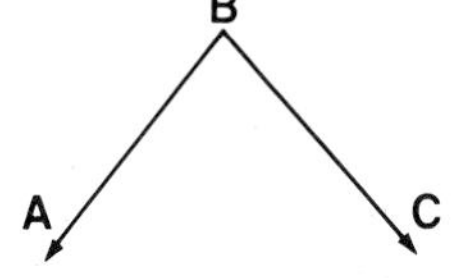


c. 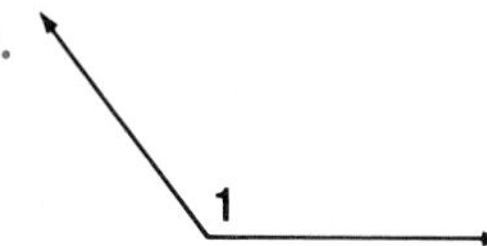


d. 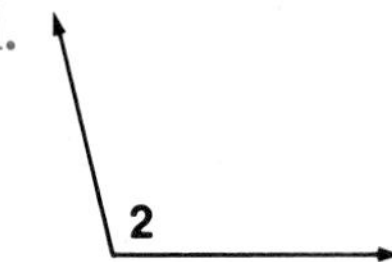


e. 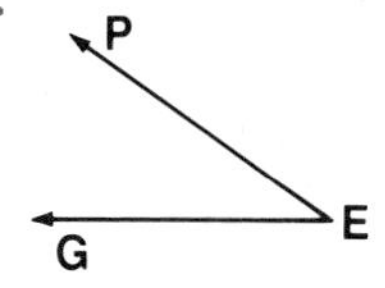


f. 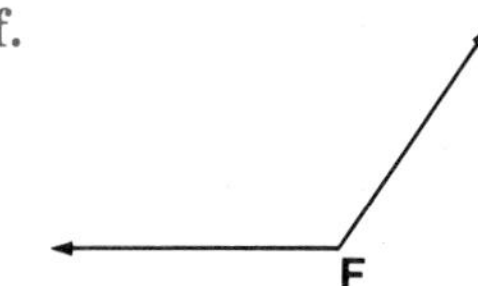

4. Tell whether each of the angles in Exercise 3 is acute, right, or obtuse.

5. Find the number of degrees in each given angle.

**a.** $\frac{1}{3}$ of a right angle  **b.** $\frac{5}{6}$ of a straight angle  **c.** $\frac{7}{12}$ of a complete rotation

## HOMEWORK EXERCISES

1. Name each angle by using a single letter and by using three letters.

a. 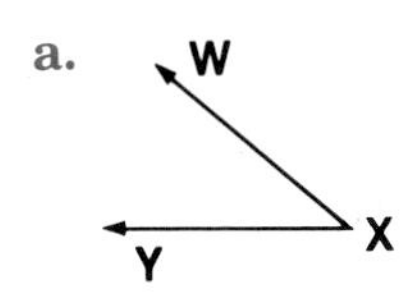


b. 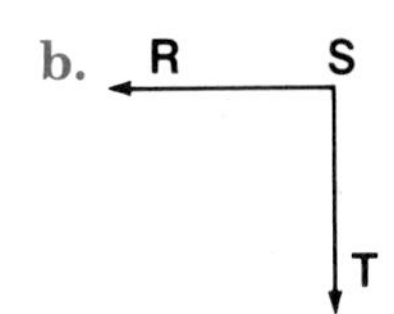


c. 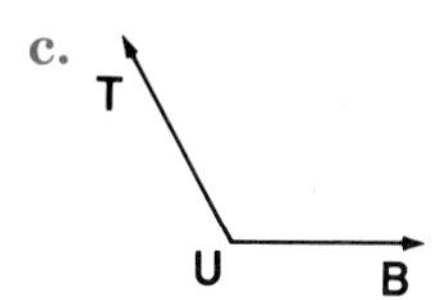


d. 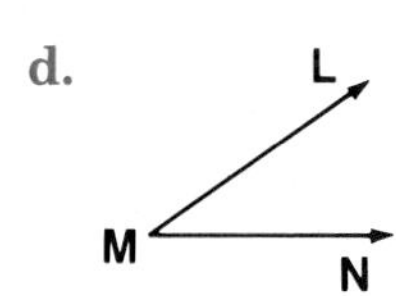


e. 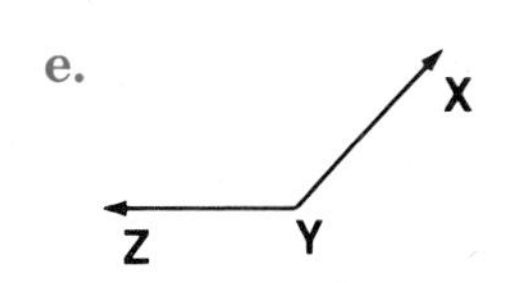


2. Use three letters to name the numbered angles.

a. 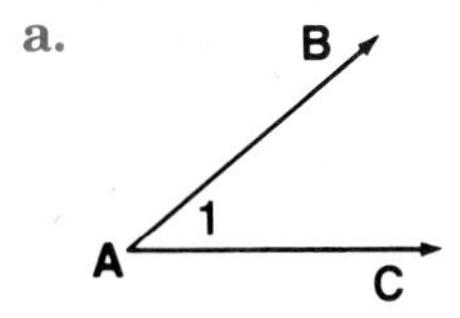


b. 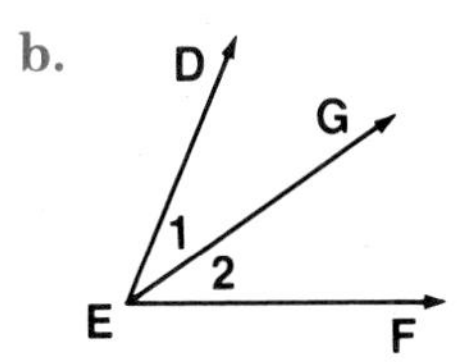


c. 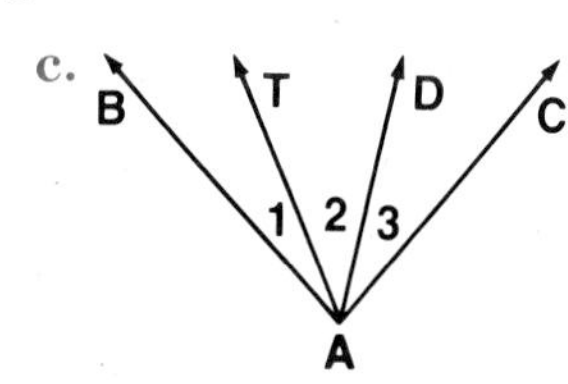


d. 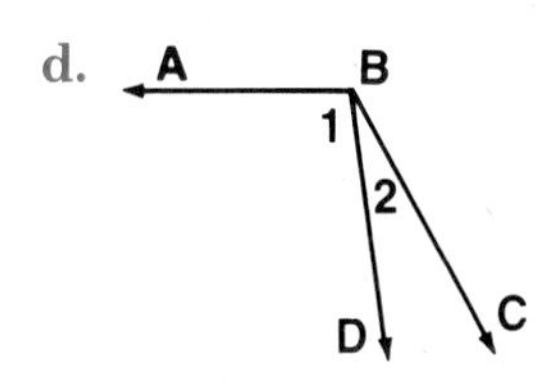


3. Tell whether each angle is a right angle, an acute angle, an obtuse angle, or a straight angle.

a. 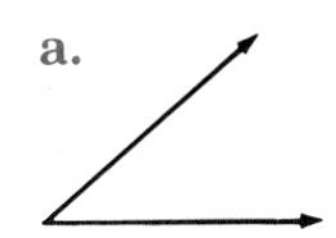

b. 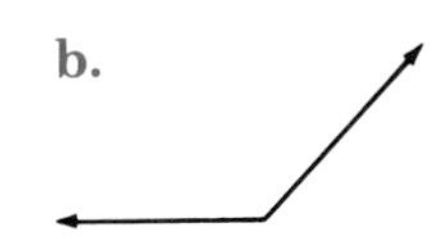

c. 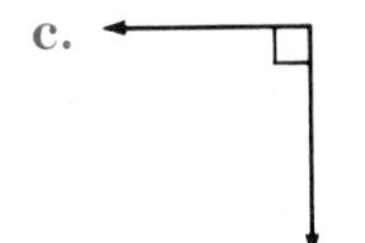

d. 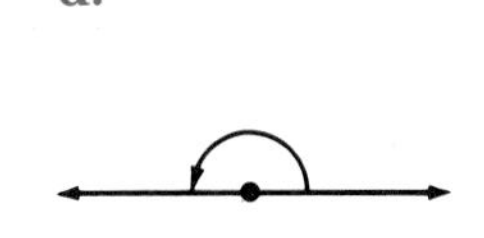

e. 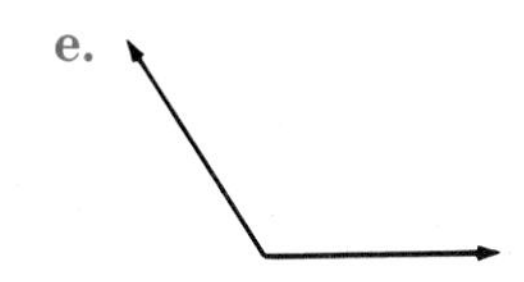

4. Tell from the measure of each angle whether the angle is a right angle, an acute angle, an obtuse angle, or a straight angle.

**a.** 90°  **b.** 37°  **c.** 105°  **d.** 175°  **e.** 180°  **f.** 49°  **g.** 68°  **h.** 178°

5. Trace each angle on your paper and use a protractor to find the measure of each angle. If necessary, extend the sides of the angle.

a. 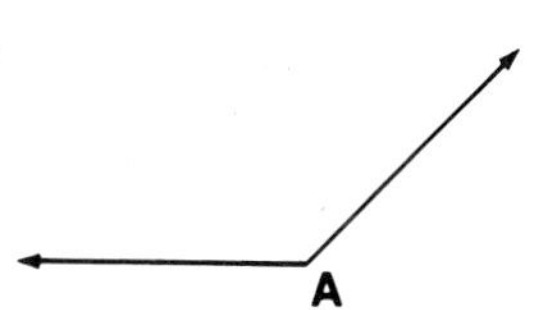


b. 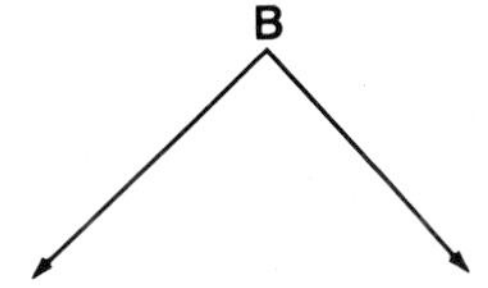


c. 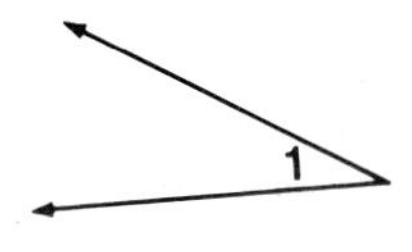


d. 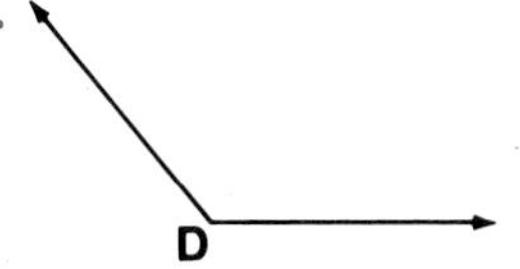


e. 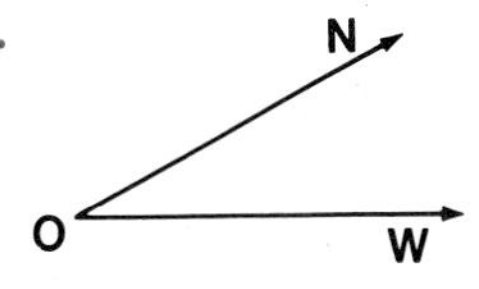


f. 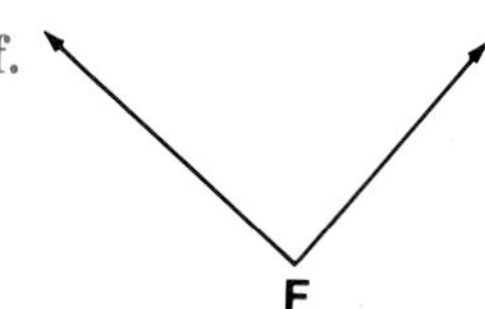

6. Tell whether each of the angles in Exercise 5 is acute, right, or obtuse.

7. The number of degrees contained in $\frac{2}{3}$ of a right angle is

(a) 30° (b) 60° (c) 45° (d) 120°

8. The number of degrees contained in $\frac{1}{4}$ of a straight angle is

(a) $22\frac{1}{2}°$ (b) 45° (c) 30° (d) 40°

9. The ratio of the number of degrees contained in a right angle to the number of degrees contained in a straight angle is
(a) 1 : 3 (b) 2 : 3 (c) 2 : 1 (d) 1 : 2

## SPIRAL REVIEW EXERCISES

1. A length of 3 meters is equivalent to
(a) 30 cm (c) 300 cm
(b) 30 mm (d) 3,000 cm

2. Graph $x < -3$ on a number line.

3. If $n$ is used to represent a number, then "5 less than $n$" is

(a) $5n$ (b) $n - 5$ (c) $5 - n$ (d) $\frac{n}{5}$

4. The value of $2x + 5$ when $x = 3$ is
(a) 28 (b) 16 (c) 8 (d) 11

5. If $I = prt$, find the value of $I$ when $p = 1{,}000$, $r = .05$, and $t = 2$.

6. Solve: $4x = 36$

7. Evaluate: $\frac{1}{10} + \frac{2}{5} \times \frac{1}{2}$

### CHALLENGE QUESTION

From the diagram, name six angles that have vertex $P$.

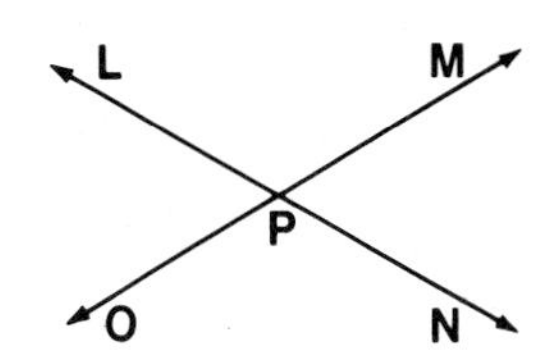

# UNIT 19–2 Triangles

## CLASSIFYING TRIANGLES

### THE MAIN IDEA

1. A ***triangle*** (symbol △) is a flat, closed figure that has three sides and three angles.
2. A triangle is usually named by three capital letters, one for each vertex. There is no special order in which to read the vertices.
3. Triangles can be classified according to the types of angles they contain.

**Acute Triangle**

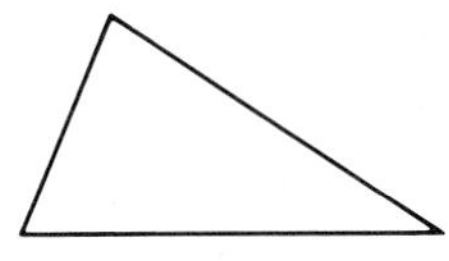

Contains three acute angles.

**Right Triangle**

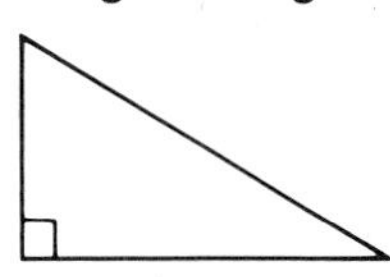

Contains one right angle and two acute angles.

**Obtuse Triangle**

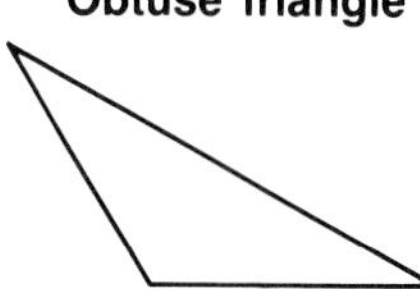

Contains one obtuse angle and two acute angles.

4. Triangles can also be classified according to the number of sides that are equal in length.

**Scalene Triangle**

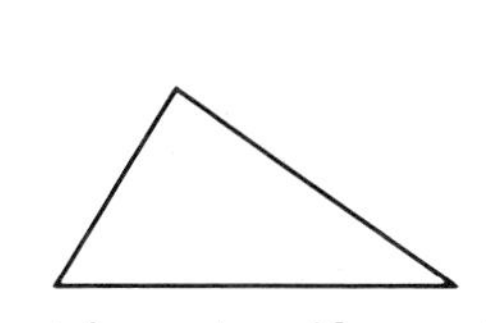

Contains no two sides equal in length.

**Isosceles Triangle**

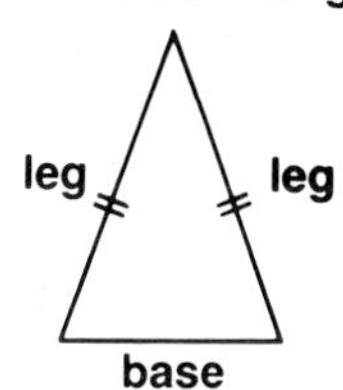


Contains two sides equal in length.

**Equilateral Triangle**

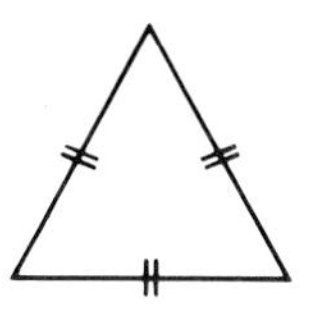

Contains three sides equal in length.

**EXAMPLE 1** From the diagram, name the triangle, its sides, and its angles.

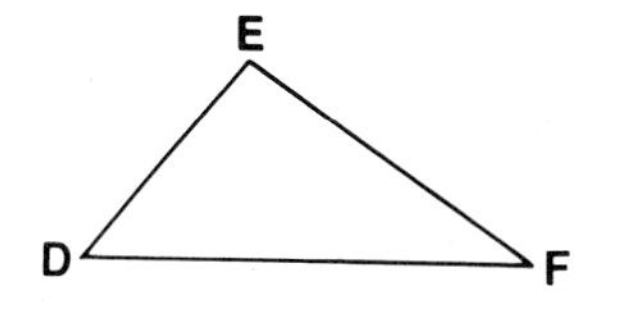


Since the vertices are $D$, $E$, and $F$, the triangle can be called by any arrangement of the three letters.

*Answer:* The triangle is named $\triangle DEF$ or $\triangle DFE$ or $\triangle EDF$ or $\triangle EFD$ or $\triangle FED$ or $\triangle FDE$.

The sides of a triangle are ***line segments***, which are named by two letters, often written with a bar over them.

*Answer:* The sides of the triangle are $\overline{DE}$ (or $\overline{ED}$), $\overline{EF}$ (or $\overline{FE}$), and $\overline{FD}$ (or $\overline{DF}$).

Each of the three angles of the triangle is named in either of the two usual ways.

*Answer:* The angles of the triangle are $\angle D$ (or $\angle EDF$ or $\angle FDE$), $\angle E$ (or $\angle DEF$ or $\angle FED$), and $\angle F$ (or $\angle DFE$ or $\angle EFD$).

**EXAMPLE 2** Classify each triangle according to its angles.

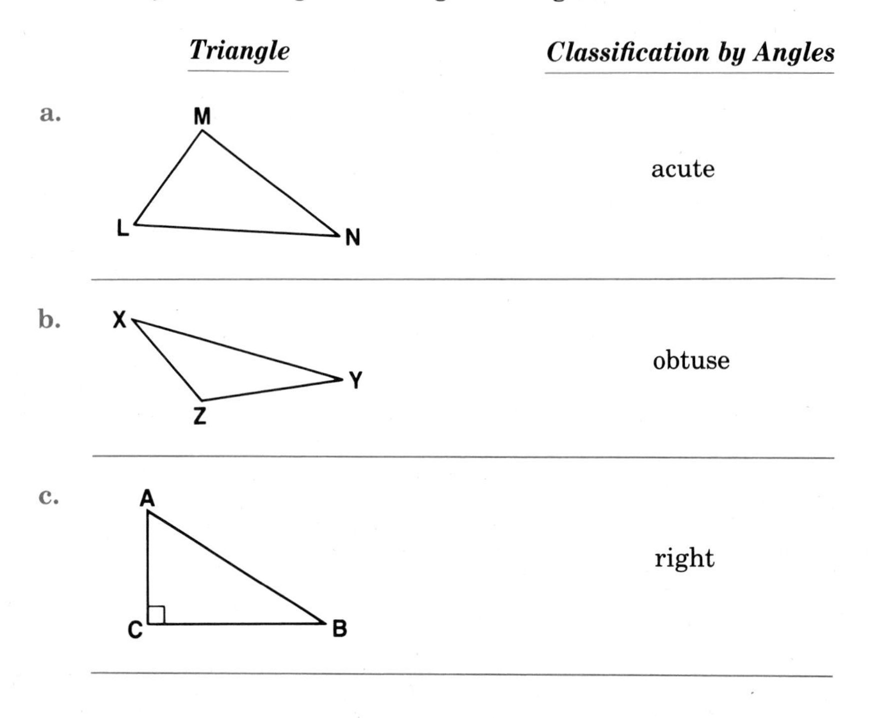


**EXAMPLE 3** Classify each triangle according to its sides.

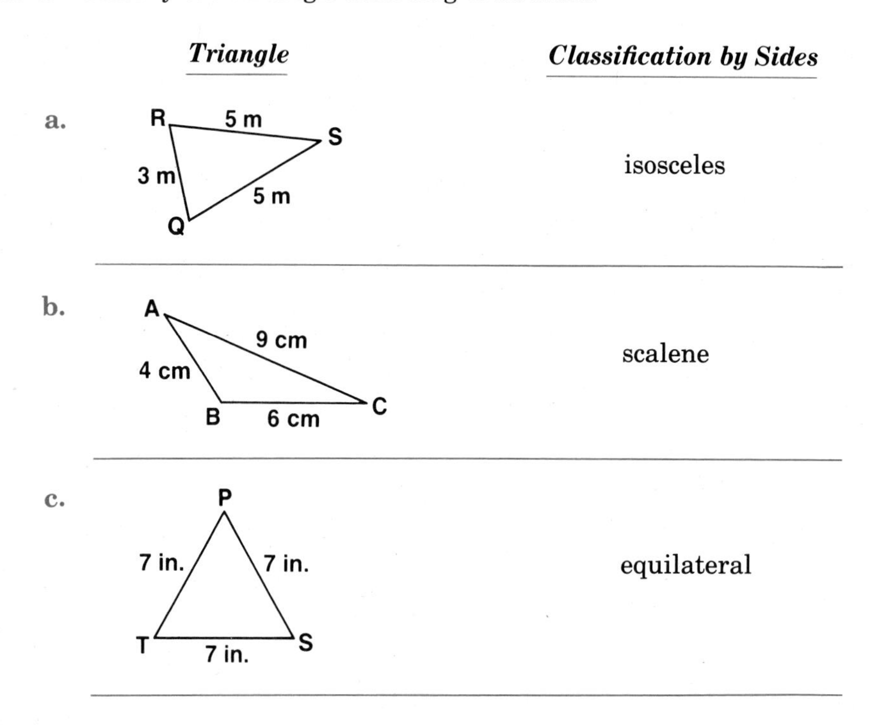

## CLASS EXERCISES

1. Name each triangle, its sides, and its angles.

a. 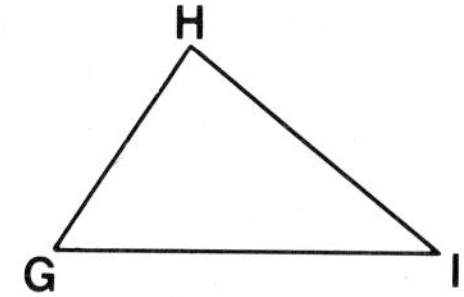


b. 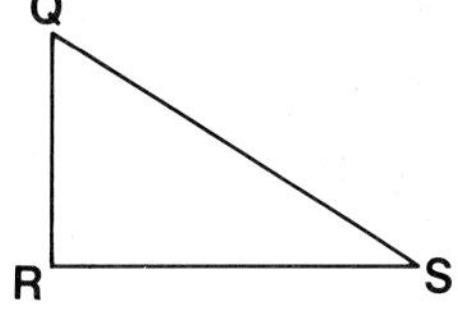


c. 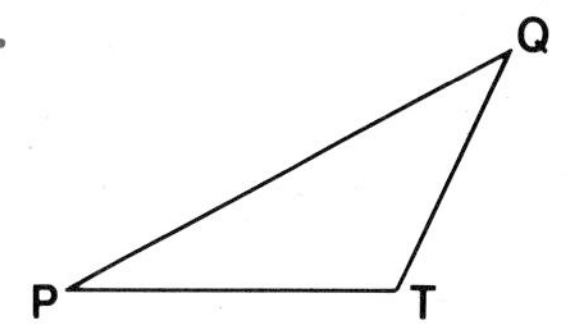


2. Classify each triangle according to its angles.

a. 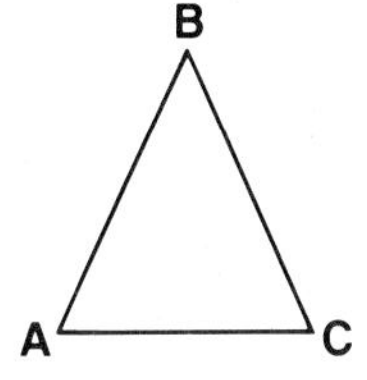


b. 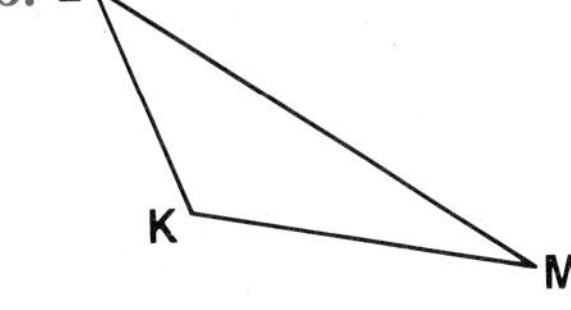


c. 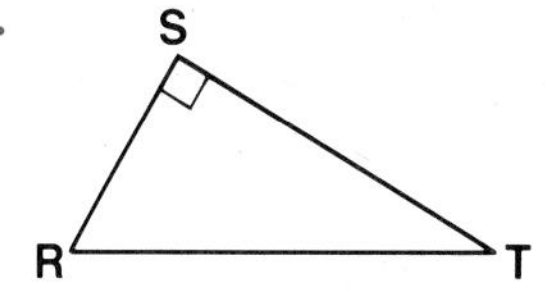


3. Classify each triangle according to its sides.

a. 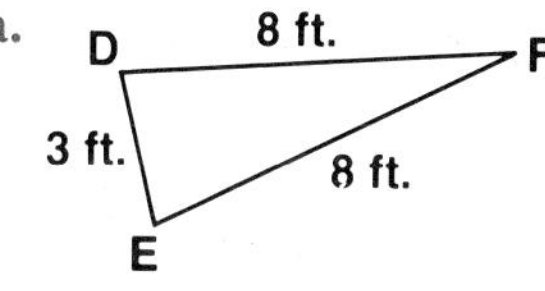


b. 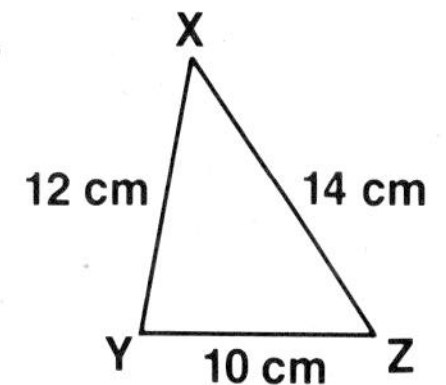


c. 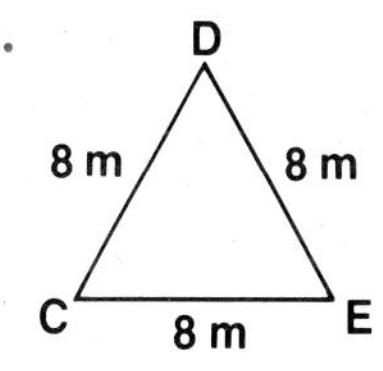


4. Classify each triangle according to its angles if the measures of the angles are:

**a.** 40°, 50°, 90° **b.** 125°, 25°, 30° **c.** 70°, 80°, 30°

5. Classify each triangle according to its sides if the measures of the sides are:

**a.** 7 in., 7 in., 4 in. **b.** 5 cm, 12 cm, 13 cm **c.** 9 cm, 9 cm, 9 cm

## THE SUM OF THE MEASURES OF THE ANGLES OF A TRIANGLE

### THE MAIN IDEA

The sum of the measures of the angles of a triangle is 180°.

**EXAMPLE 4** Two angles of a triangle measure 60° and 38°. Find the measure of the third angle.

Find the sum of the measures of the two given angles.

$$\begin{array}{r} 60^\circ \\ +38^\circ \\ \hline 98^\circ \end{array}$$

Subtract the sum from 180°.

$$\begin{array}{r} 180^\circ \\ -98^\circ \\ \hline 82^\circ \end{array}$$ *Ans.*

**EXAMPLE 5** Is it possible to draw a triangle that has angles whose measures are 42°, 67°, and 81°?

## THINKING ABOUT THE PROBLEM

In every triangle, the sum of the measures of the angles is 180°. For three measures to be the measures of the angles of a triangle, their sum must be 180°.

Find the sum of the measures. $42° + 67° + 81° = 190°$

*Answer:* Since the sum of the measures does not equal 180°, it is not possible to draw a triangle that has the given angles.

**EXAMPLE 6** The maximum number of right angles that a triangle can have is
(a) 0 (b) 1 (c) 2 (d) 3

## THINKING ABOUT THE PROBLEM

Choose the answer by eliminating, or ruling out, the wrong choices. Since you know that a triangle can have a right angle, choice (a) is wrong. Choice (b) may be the answer. If a triangle had two right angles, as in choice (c), the sum of the measures of its angles would be greater than 180°. This is not possible. For the same reason, eliminate choice (d).

*Answer:* (b)

## CLASS EXERCISES

1. Find the measure of ∠1 in each triangle.

a.

b. 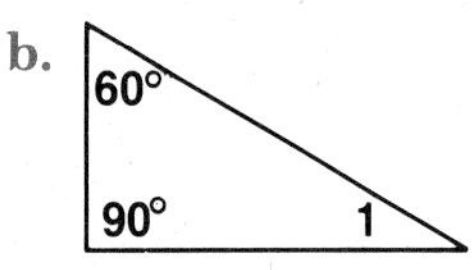


c. 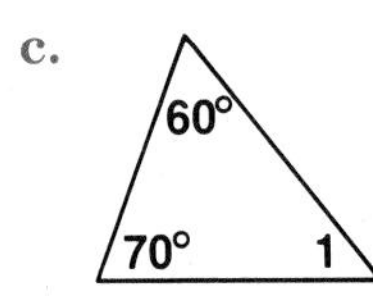


d.

e. 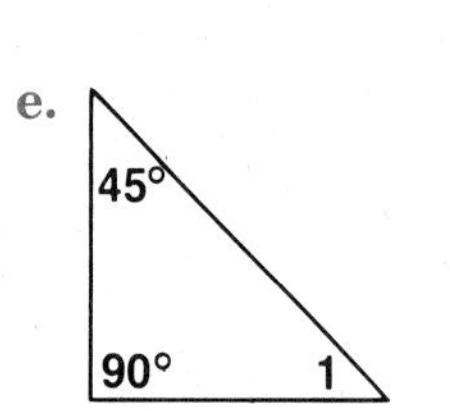


f. 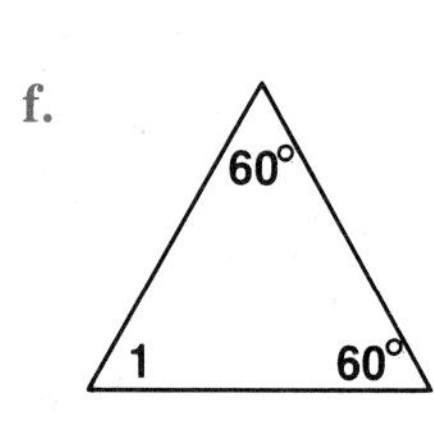


2. Find the measure of the third angle of a triangle if the measures of the other two angles are:

**a.** 75° and 20° **b.** 15° and 105° **c.** 18° and 72° **d.** 47° and 90°

**e.** 56° and 52° **f.** 165° and 5°

3. Tell whether or not you can draw a triangle that has three angles that measure:

**a.** 48°, 100°, and 32° **b.** 65°, 65°, and 50° **c.** 110°, 30°, and 95° **d.** 60°, 50°, and 40°

**e.** 60°, 60°, and 60° **f.** 90°, 90°, and 1°

4. The maximum number of obtuse angles that a triangle can have is
(a) 0 (b) 1 (c) 2 (d) 3

## HOMEWORK EXERCISES

**1.** Name each triangle, its sides, and its angles.

a. C A B

b. 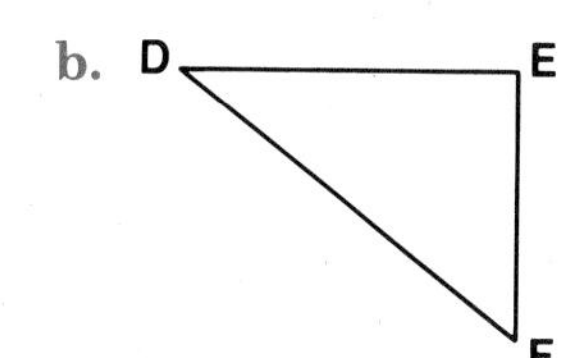


c. 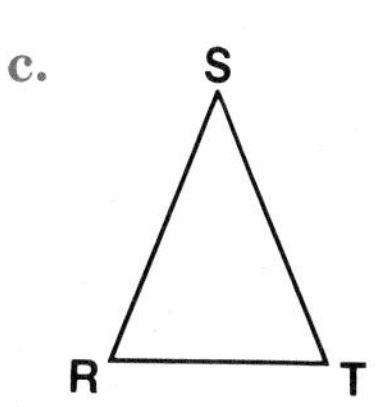


**2.** Classify each triangle according to its angles.

a. 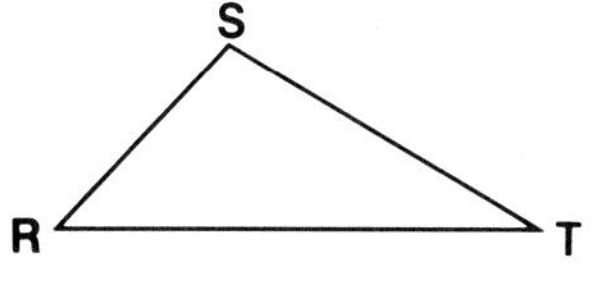


b. 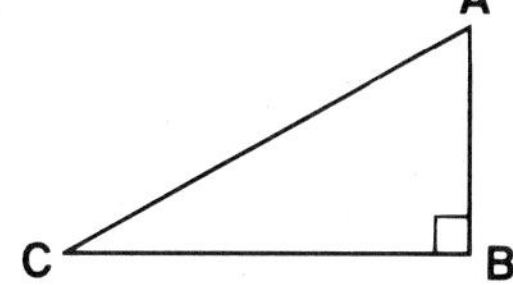


c. 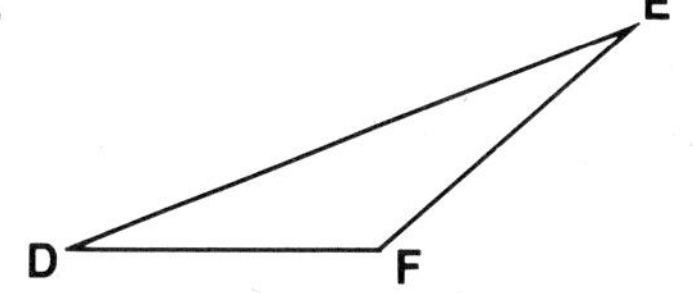


**3.** Classify each triangle according to its sides.

a. 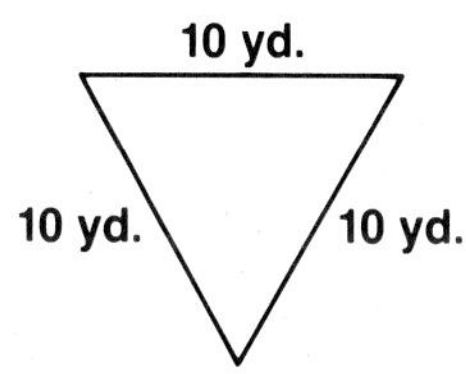


b. 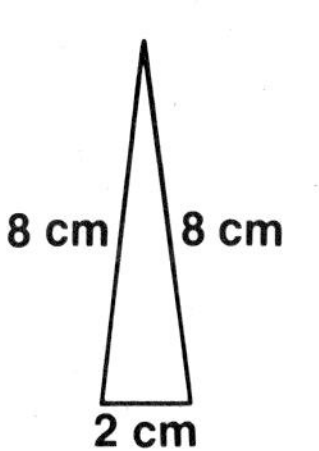


c. 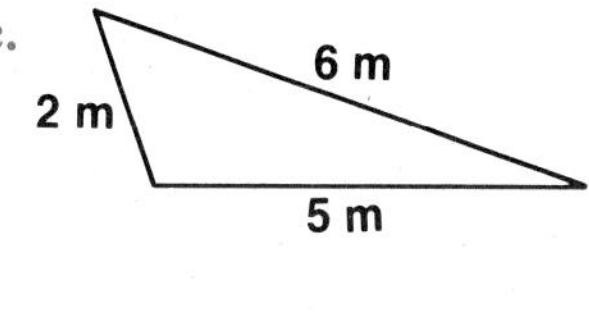


**4.** Classify each triangle according to its angles if the measures of the angles are:

**a.** 20°, 75°, 85° **b.** 30°, 50°, 100° **c.** 30°, 60°, 90°

**5.** Classify each triangle according to its sides if the measures of the sides are:

**a.** 3 in., 4 in., 5 in. **b.** 3 m, 3 m, 5 m **c.** 15 ft., 15 ft., 15 ft.

**6.** Find the measure of $\angle A$ in each triangle.

a. 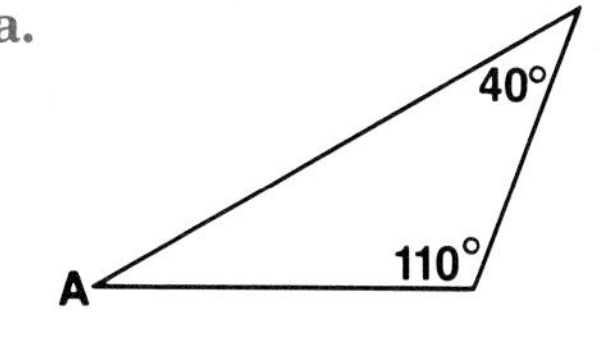


b. 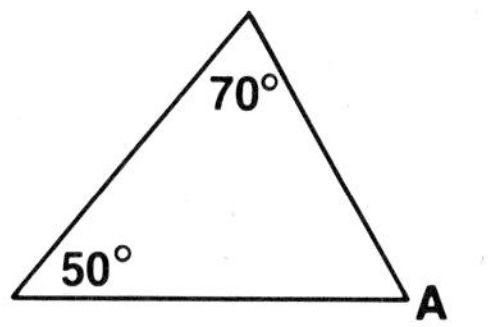


c. 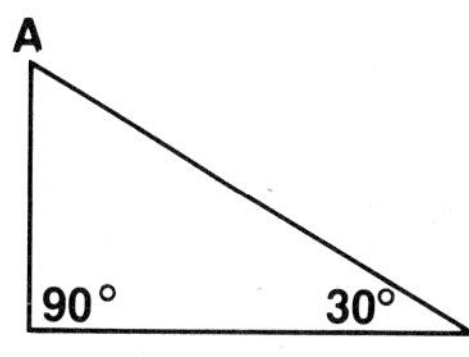


d. 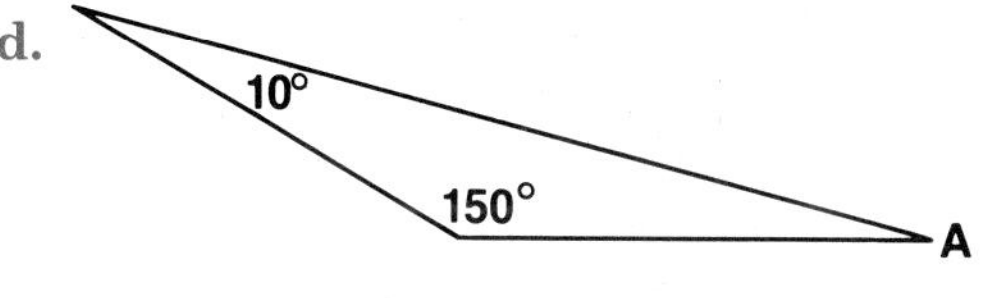


e. 

f. 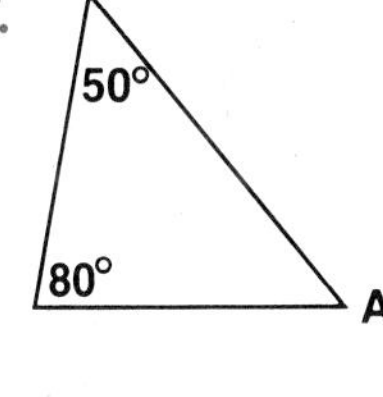

**7.** Find the measure of the third angle of a triangle if the measures of the other two angles are:

**a.** 95° and 45° **b.** 80° and 60° **c.** 37° and 42° **d.** 58° and 109°

**e.** 49° and 120° **f.** 67° and 75°

**8.** Tell whether or not you can draw a triangle that has three angles that measure:

**a.** 28°, 42°, and 100° **b.** 50°, 50°, and 60° **c.** 70°, 60°, and 50°

**d.** 110°, 35°, and 35° **e.** 105°, 105°, and 30° **f.** 45°, 45°, and 90°

**9.** The maximum number of acute angles that a triangle can have is
(a) 0 (b) 1 (c) 2 (d) 3

**10.** The measure of each angle of an equiangular triangle (three angles of equal measure) is
(a) 30° (b) 45° (c) 60° (d) 90°

## SPIRAL REVIEW EXERCISES

**1.** The number of degrees contained in $\frac{4}{5}$ of a right angle is
(a) 18° (b) 36° (c) 72° (d) 144°

**2.** The kind of angle that has the smallest measure is
(a) right (c) obtuse
(b) acute (d) straight

**3.** Graph $x \geq 0$ on a number line.

**4.** The measure of $\angle ABC$ is
(a) 65° (b) 115° (c) 75° (d) 125°

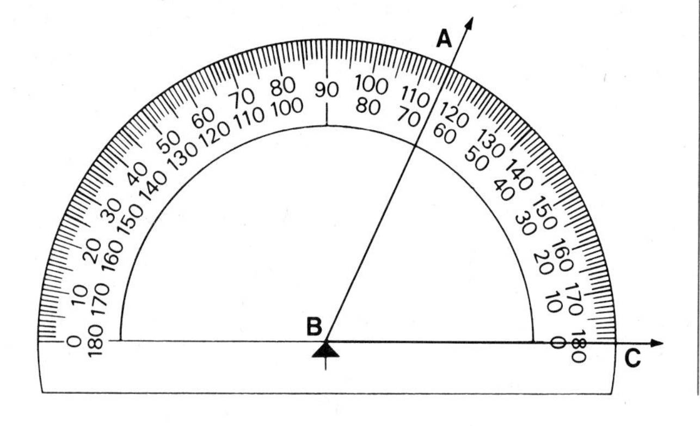


**5.** A quart is equivalent to
(a) $\frac{1}{4}$ gallon (c) 16 ounces
(b) 4 pints (d) 64 ounces

**6.** 2,500 mL is equivalent to
(a) 25 L (c) 2.5 kL
(b) 2.5 L (d) .25 kL

**7.** If $a$ represents a number, write an expression to represent "three more than the number."

**8.** How much time elapses between 11:52 P.M. and 6:12 A.M.?

**9.** If $A = \frac{1}{2}bh$, find the value of $A$ when $b = 10$ and $h = 7$.

**10.** Solve: $3w - 4 = 7$

**11.** Multiply: $.9 \times .02$

**12.** How much greater is $3\frac{5}{8}$ than $1\frac{1}{2}$?

### CHALLENGE QUESTION

If the measure of one acute angle of a right triangle is twice the measure of the other acute angle, find the measure of each angle.

# UNIT 19–3 Similar Figures

## THE MAIN IDEA

1. *Similar figures* are figures that have the same shape.
2. A pair of sides in one figure compares in the same way as the corresponding pair of sides in the other figure. The ratios are equal.
3. To find an unknown length in a pair of similar figures, write and solve a proportion.

**EXAMPLE 1** Find the width of the smaller figure in the pair of similar figures shown.

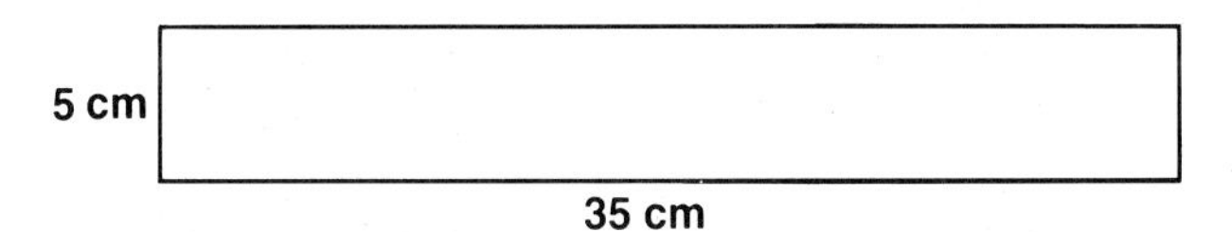


Represent the width of the smaller figure by $n$ and write a proportion.

$$\frac{5}{35} = \frac{n}{7}$$

Solve the proportion.

$$35n = 35$$

$$\frac{1}{35}(35n) = \frac{1}{35}(35)$$

$$n = 1$$

*Answer:* 1 cm

**EXAMPLE 2** From the choices below, select the triangle that is similar to the one at the right. (The choices are not drawn to scale.)

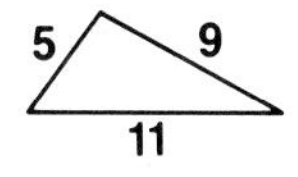


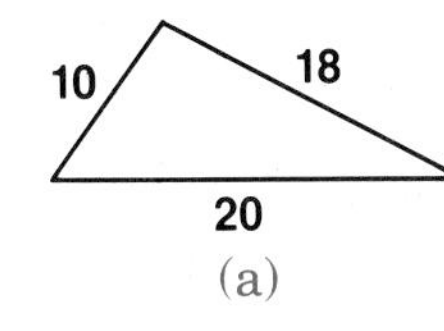

(a)

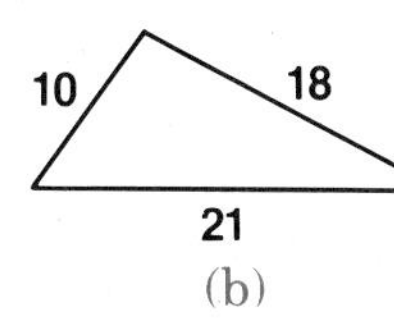

(b)

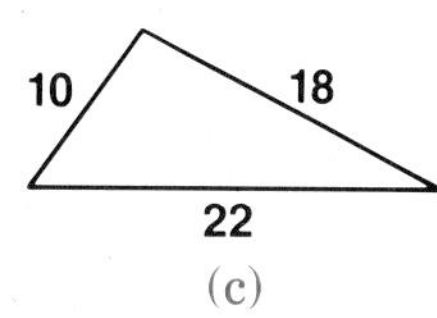

(c)

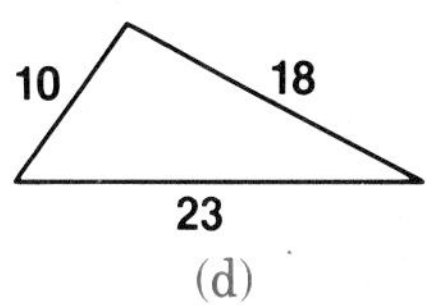

(d)

### THINKING ABOUT THE PROBLEM

From the choices, you see that the measures of two sides of the triangle are repeated: 10 and 18 appear in all the choices. Thus, the correct answer depends on the measure of the third side.

From the original triangle, form a ratio that involves the third side: $\frac{5}{11}$

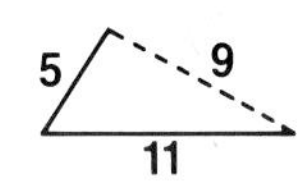


For each choice, consider the corresponding ratio to see which is equal to $\frac{5}{11}$:

$$\frac{10}{20} \quad \frac{10}{21} \quad \frac{10}{22} \quad \frac{10}{23}$$

*Answer:* (c)

## CLASS EXERCISES

**1.** Find the value of $n$ in each pair of similar figures.

a.

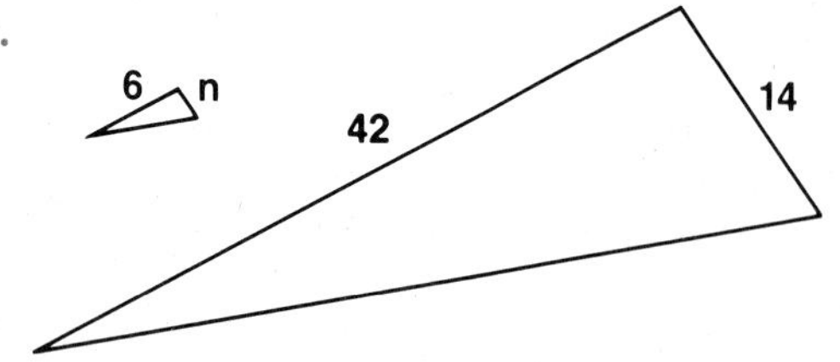


b.

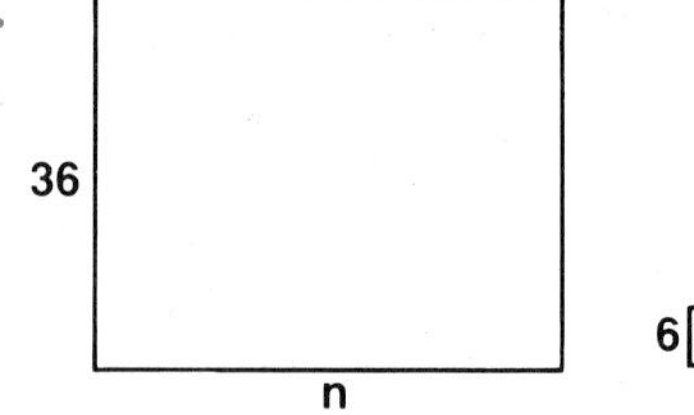


c.

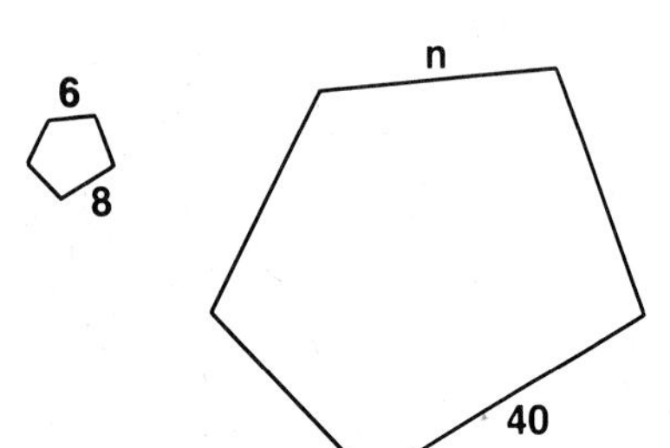


d.

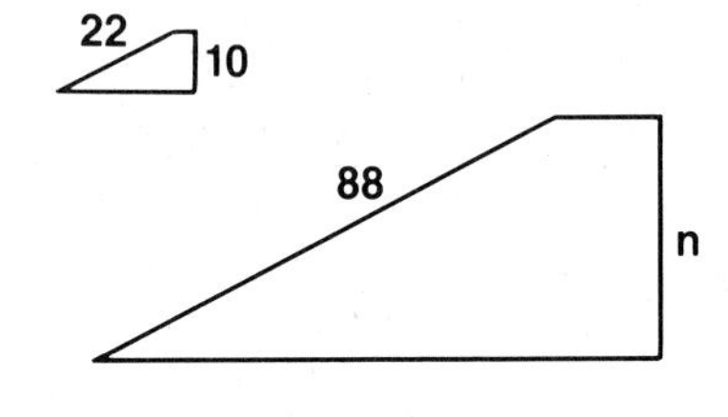


**2.** Find the value of $n$ in each pair of similar figures.

a.

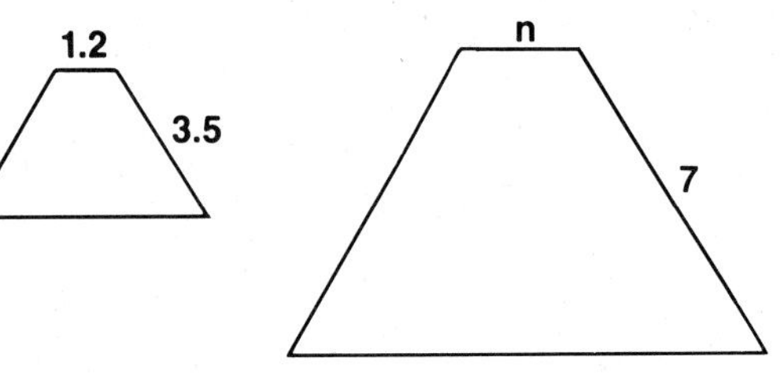


b.

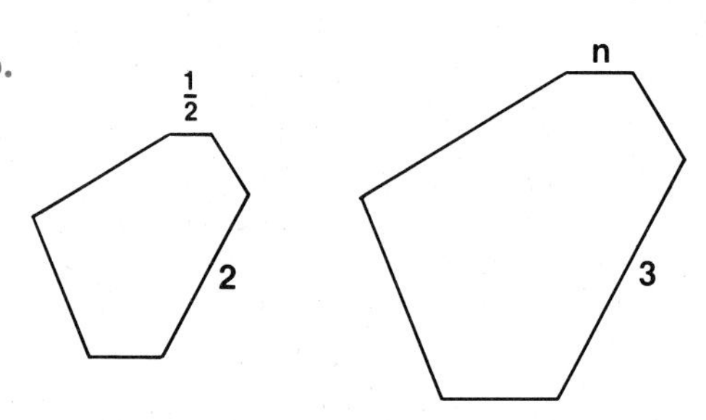


c.

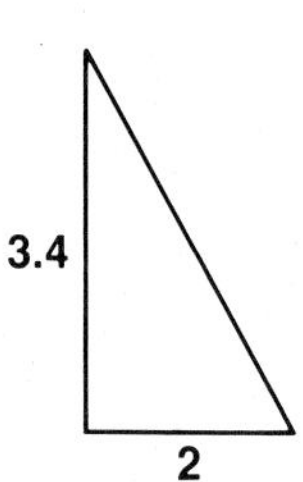

d.

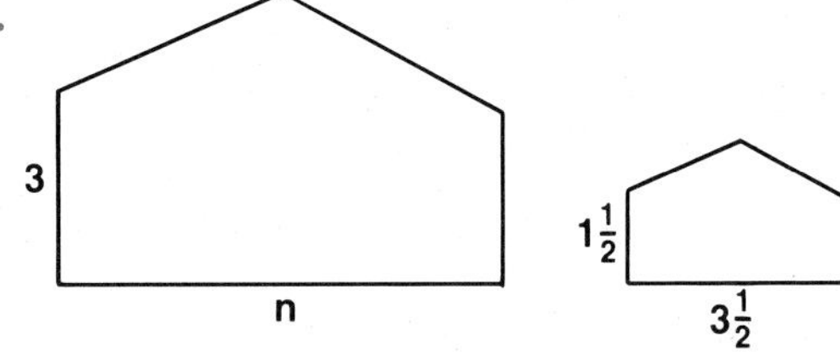


**3.** From the choices below, select the triangle that is similar to the one at the right. (The choices are not drawn to scale.)

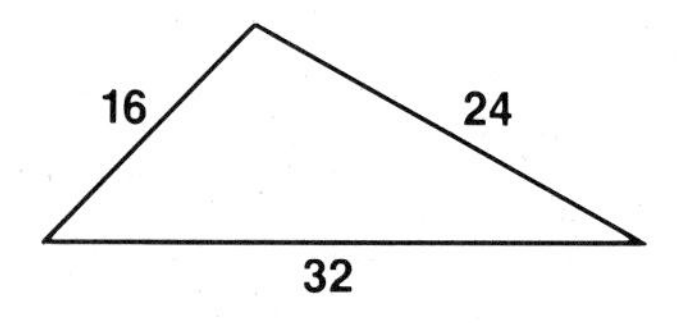


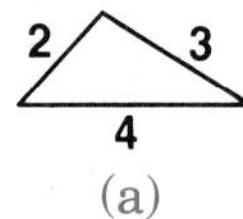


(a)

2, 3, 3
(b)

2, 3, 2
(c)

1, 3, 3
(d)

**4.** From the choices below, select the figure that is similar to the one at the right. (The choices are not drawn to scale.)

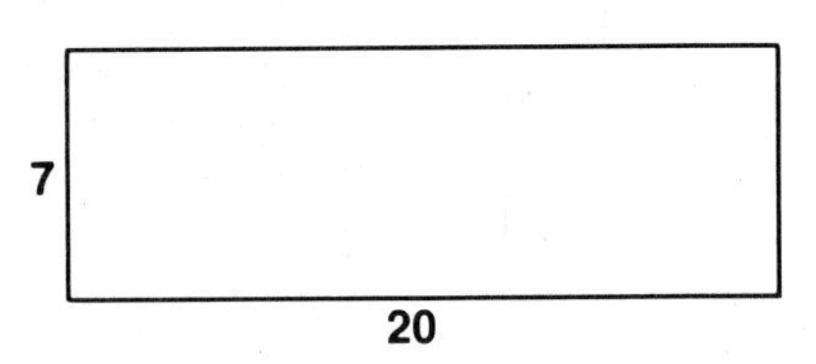


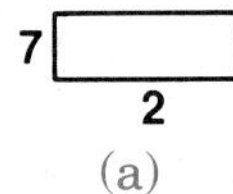


(a)

.7, 2
(b)

.07, 2
(c)

.007, 2
(d)

## HOMEWORK EXERCISES

**1.** Find the value of $n$ in each pair of similar figures.

a.

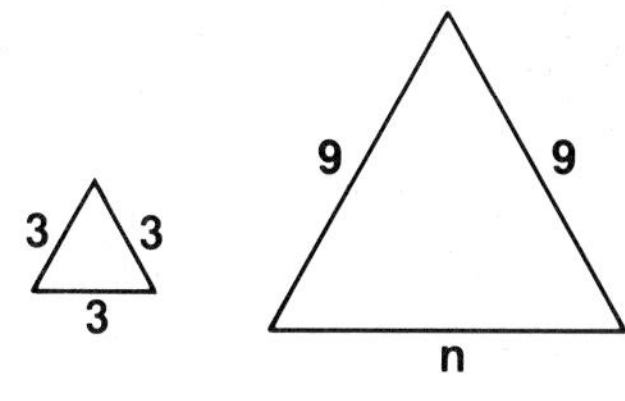


b.

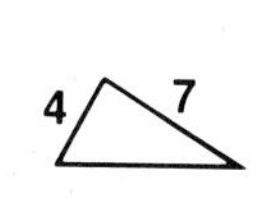


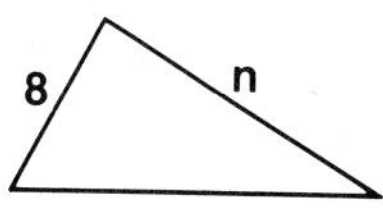


c.

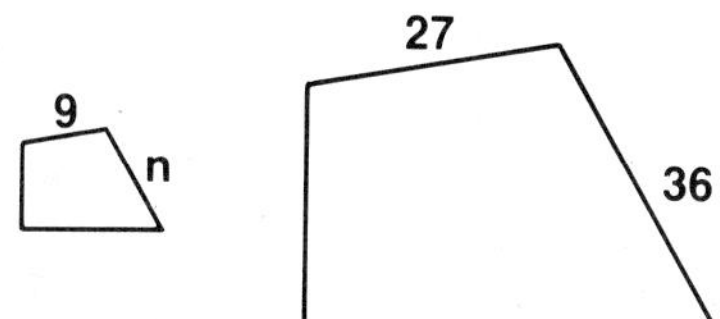


d.

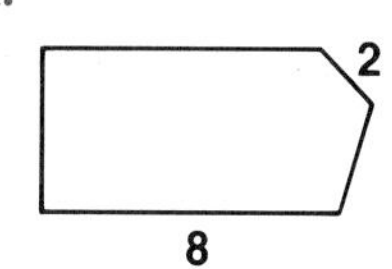


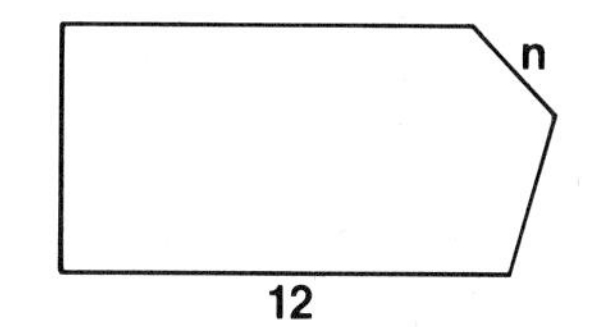


**2.** Find the value of $n$ in each pair of similar figures.

a.

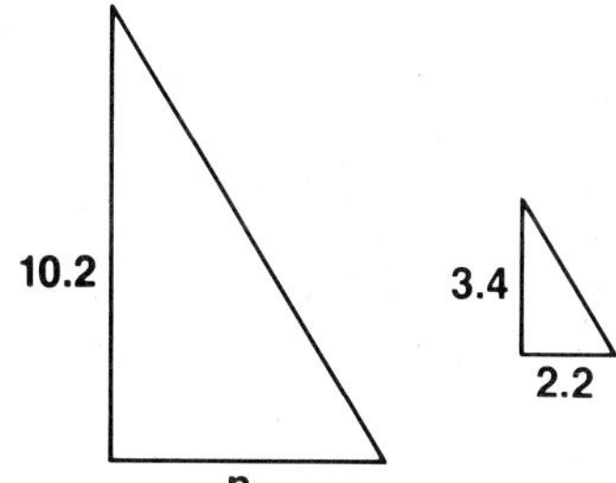


b.

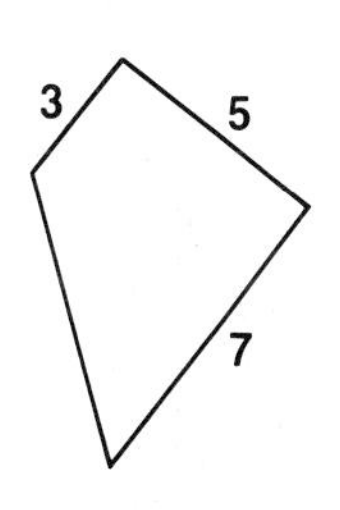


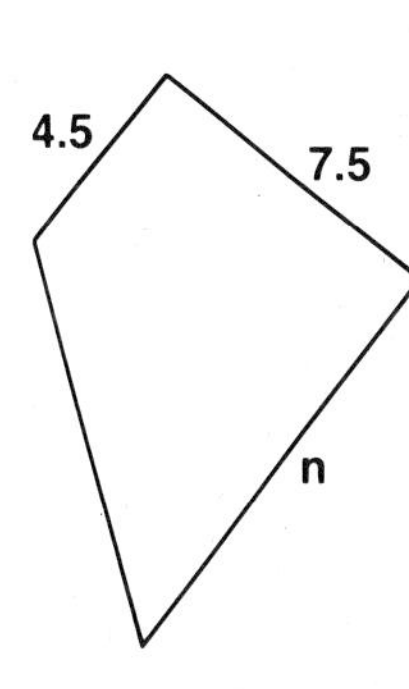


c.

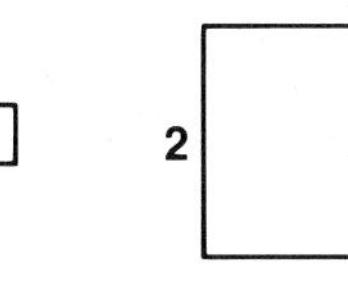


d.

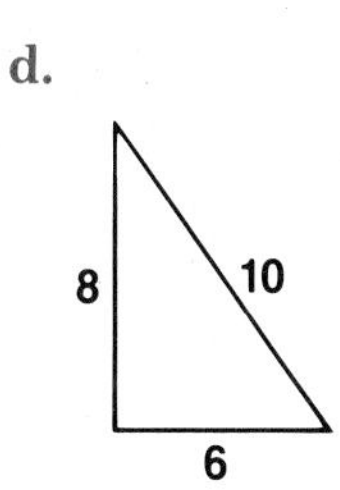


n

7.2

**3.** From the choices below, select the triangle that is similar to the one at the right. (The choices are not drawn to scale.)

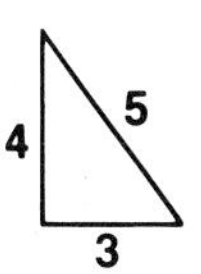


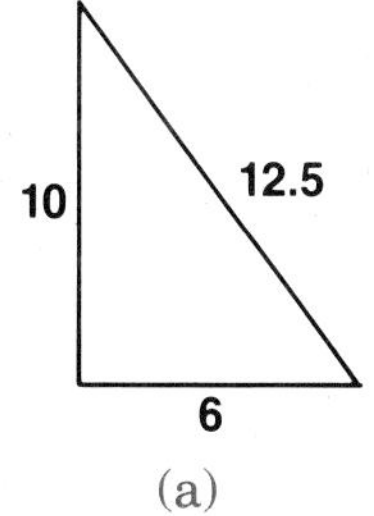

(a)

10 12.5 6.5
(b)

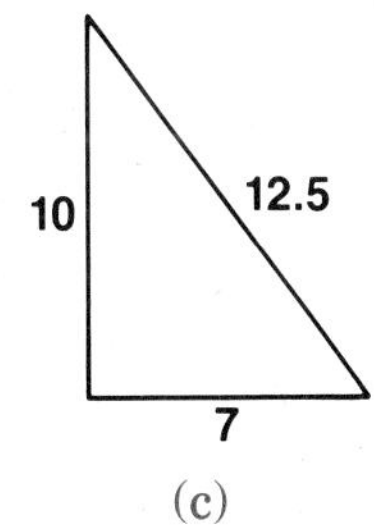

(c)

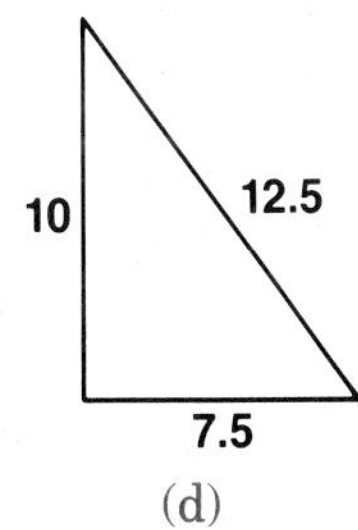

(d)

## SPIRAL REVIEW EXERCISES

1. The measure of angle $C$ is
   (a) 80° (c) 100°
   (b) 90° (d) 110°

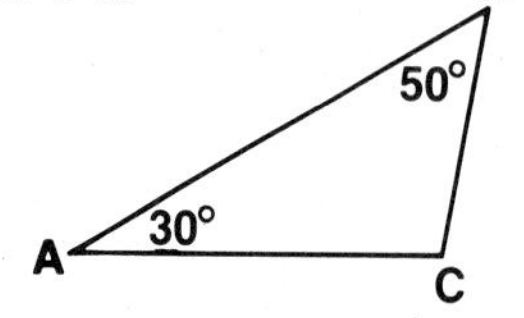


2. The angle whose measure is closest to 30° is

(a)
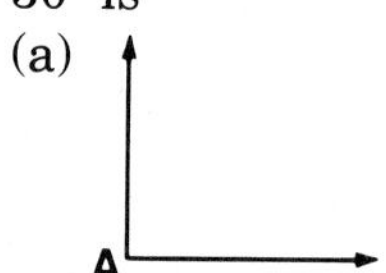


(c)
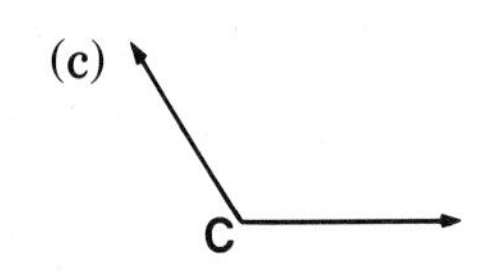


(b)
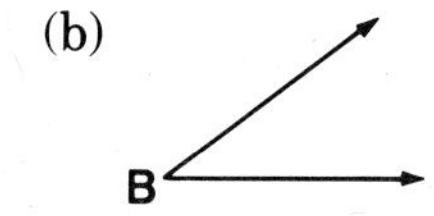


(d)

3. Which of the following is the graph of the inequality $-4 < x \leq 0$?

(a)
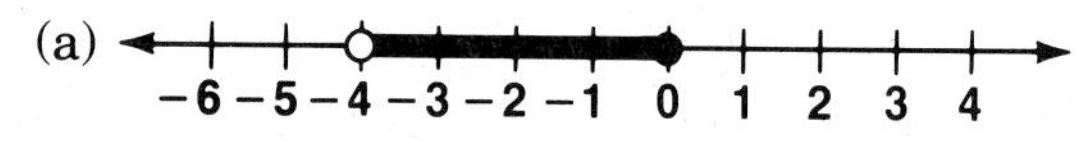


(b)
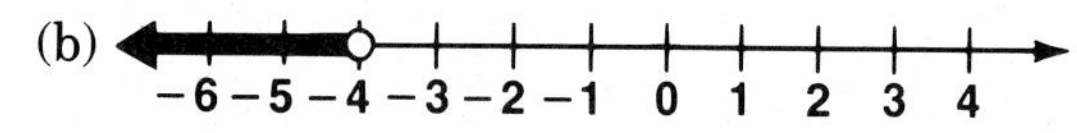


(c)
−6 −5 −4 −3 −2 −1 0 1 2 3 4

(d)
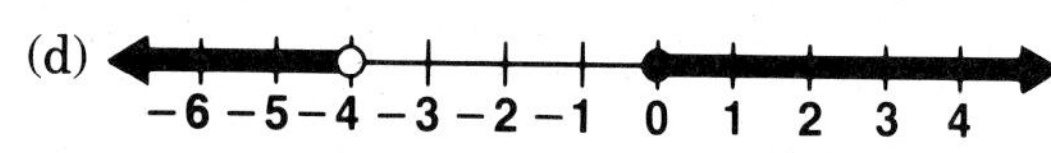


4. Solve the equation: $5x - 11 = -11$

5. Using the formula $A = \frac{1}{2}bh$, find $A$ when $b = 4.2$ and $h = 20$.

6. In this circle graph, what percent of Mr. Jonah's paycheck is spent on entertainment?

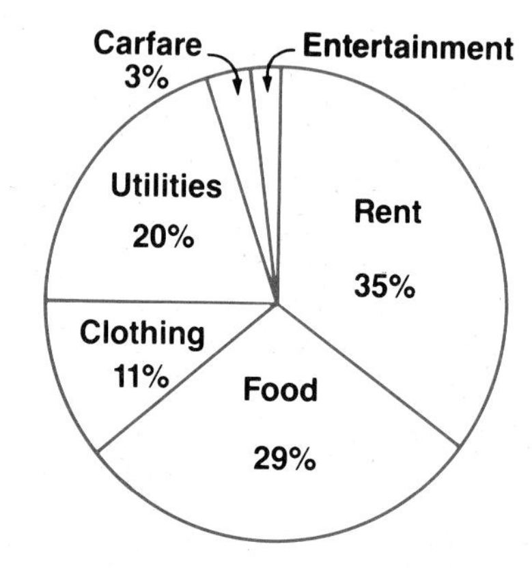


7. Evaluate $xy^2$ when $x = 2$ and $y = 5$.

8. Divide 8,596 by 28.

9. Add: $\frac{1}{2} + \frac{1}{3}$

10. If the temperature increased from $-2$°F to $+7$°F, by how many degrees did it increase?

### CHALLENGE QUESTION

Given:

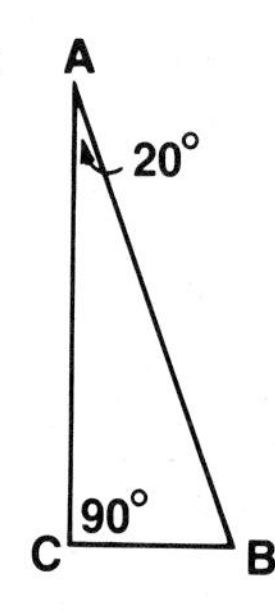


Which triangle is similar to triangle $ABC$?

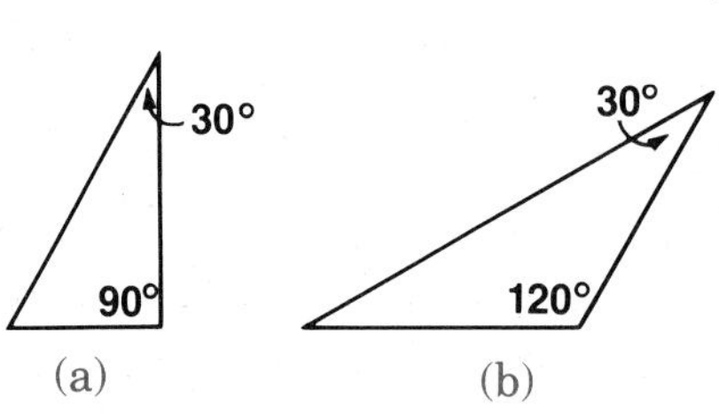


(a) (b)

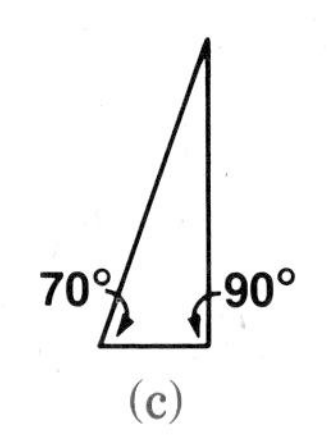


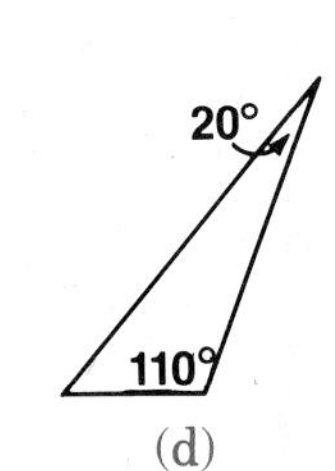


(c) (d)

# UNIT 19–4 The Pythagorean Relationship

## THE MAIN IDEA

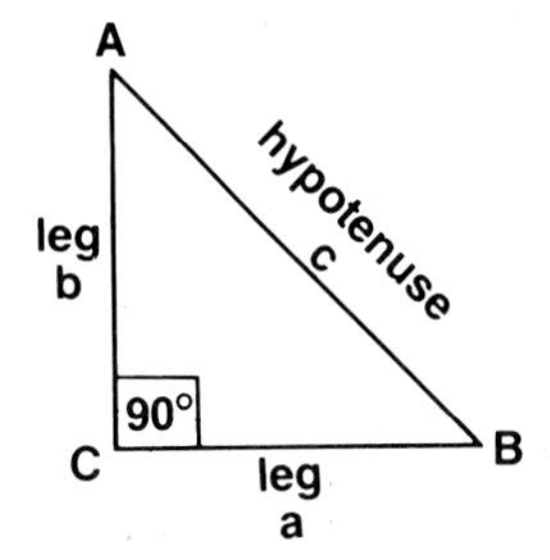


1. The *hypotenuse* of a right triangle is the side opposite the right angle. The hypotenuse is longer than either of the two *legs* (sides) that form the right angle.

2. The Greek mathematician Pythagoras showed the following relationship for the measures of the sides of a right triangle:

$$(leg)^2 + (leg)^2 = (hypotenuse)^2$$

3. Using the letters shown in the diagram above, the Pythagorean relationship is:

$$a^2 + b^2 = c^2$$

4. A triangle is a right triangle if the measures of its three sides satisfy the Pythagorean relationship.

**EXAMPLE 1** Find the length of the hypotenuse of a right triangle whose other two sides measure 6 mm and 8 mm.

### THINKING ABOUT THE PROBLEM

Making a diagram and labeling it with the given information is helpful. The hypotenuse must be $c$, but either leg can be $a$ or $b$.

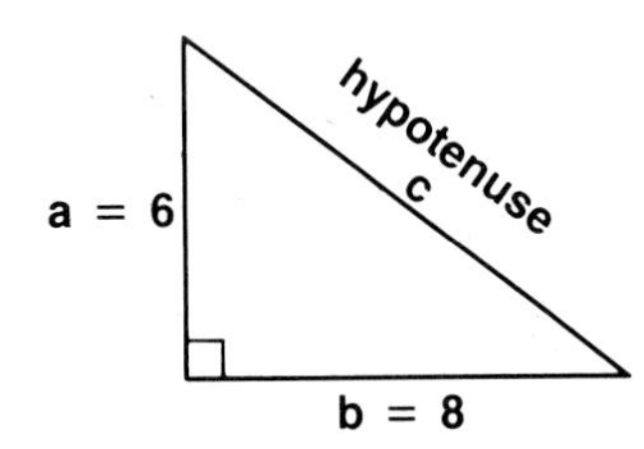


Use the Pythagorean relationship. $a^2 + b^2 = c^2$

Substitute 6 for $a$ and 8 for $b$. $6^2 + 8^2 = c^2$

Evaluate, following the order of operations.

$$6 \times 6 + 8 \times 8 = c^2$$
$$36 + 64 = c^2$$
$$100 = c^2$$

Since 100 is $c$ squared, $c$ is the square root of 100: $\sqrt{100} = 10$

*Answer:* The length of the hypotenuse is 10 mm.

**EXAMPLE 2** Find the value of $b$ in the right triangle shown.

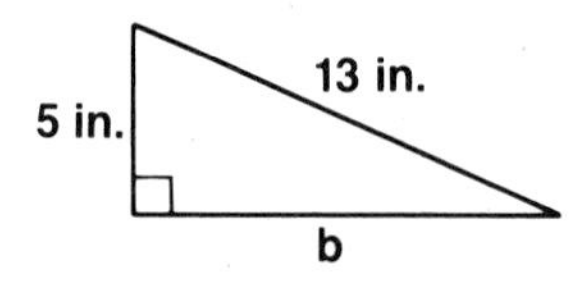


Use the Pythagorean relationship. $a^2 + b^2 = c^2$

Since 13 is the hypotenuse, substitute 13 for $c$ and 5 for $a$. $5^2 + b^2 = 13^2$

Evaluate the squares. $25 + b^2 = 169$

Subtract 25 from each side of the equation. $b^2 = 144$

Find the square root of 144. $b = 12$

*Answer:* 12 in.

**EXAMPLE 3** Tell whether a triangle that has sides that measure 6 inches, 5 inches, and 9 inches is a right triangle.

See if the measures satisfy the Pythagorean relationship.

$$a^2 + b^2 = c^2$$

Since 9 is the length of the longest side, substitute 9 for $c$. $6^2 + 5^2 \stackrel{?}{=} 9^2$

Evaluate on each side of the equals sign. $36 + 25 \stackrel{?}{=} 81$

$$61 \neq 81$$

*Answer:* Since the triangle does not satisfy the Pythagorean relationship, the triangle is not a right triangle.

**EXAMPLE 4** In the drawing, a ladder that is placed 9 feet from the base of a wall touches the wall 12 feet above the ground. What is the length of the ladder?

Use the Pythagorean relationship. $a^2 + b^2 = c^2$

Substitute 9 for $a$ and 12 for $b$. $9^2 + 12^2 = c^2$

Evaluate. $81 + 144 = c^2$

$$225 = c^2$$

Find the square root of 225. $15 = c$

*Answer:* The length of the ladder is 15 feet.

## CLASS EXERCISES

1. Find the length of the hypotenuse of a right triangle whose two legs measure:

   **a.** 3 in. and 4 in. **b.** 5 cm and 12 cm **c.** 8 yd. and 15 yd. **d.** 12 m and 16 m

2. Find each missing length.

a. 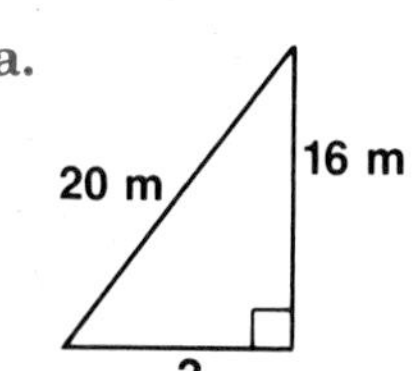


b. 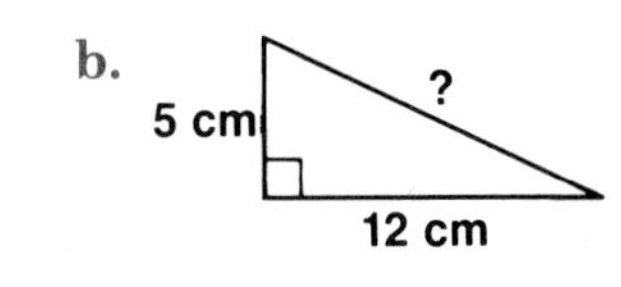


c. 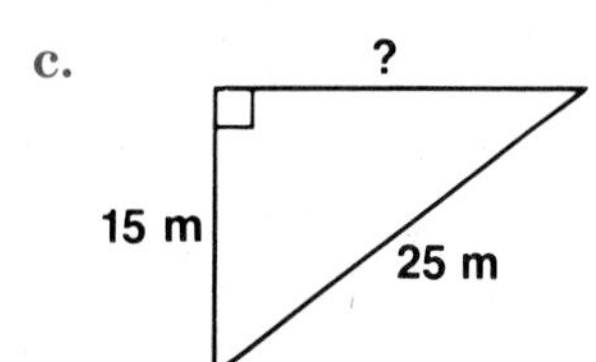


d. 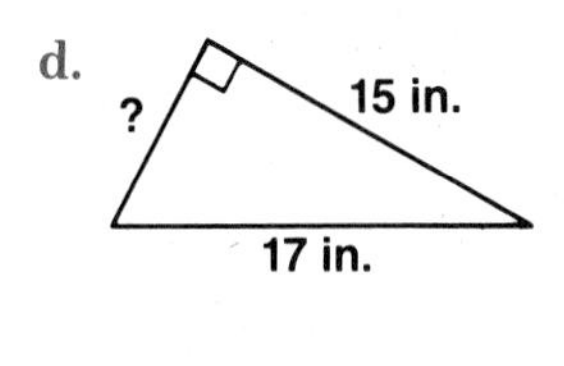


3. Tell whether each set of measurements can be the lengths of the sides of a right triangle.

   **a.** 5 in., 12 in., 13 in. **b.** 3 ft., 4 ft., 6 ft. **c.** 9 cm, 12 cm, 15 cm **d.** 8 m, 15 m, 20 m

4. In the drawing, a ladder 25 feet long leans against a wall. If the foot of the ladder is 15 feet from the wall, how far up the wall does the ladder reach?

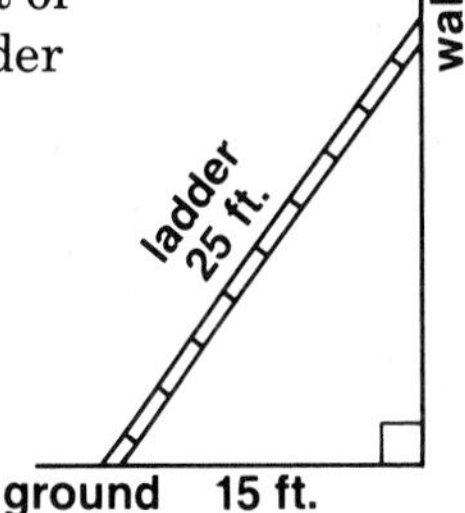


## HOMEWORK EXERCISES

1. Find the length of the hypotenuse of a right triangle whose two legs measure:

   **a.** 4 cm and 3 cm **b.** 9 in. and 12 in. **c.** 16 m and 30 m **d.** 10 yd. and 24 yd.

2. Find each missing length.

a. 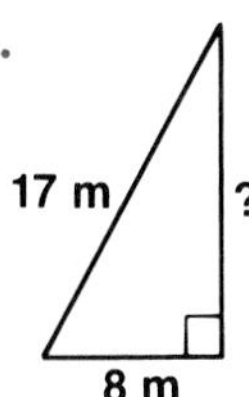


b. 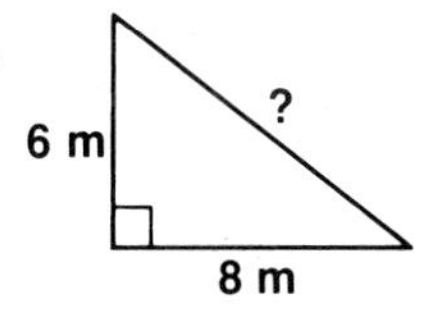


c. 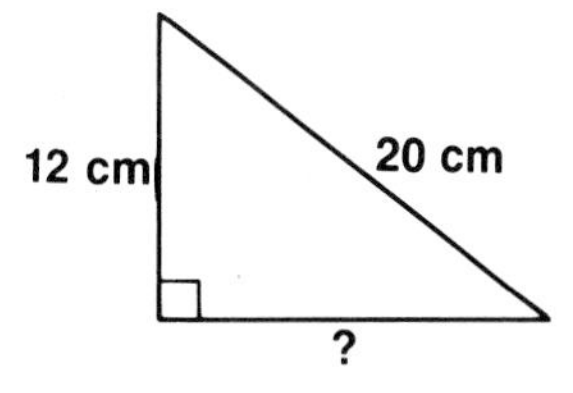


d. 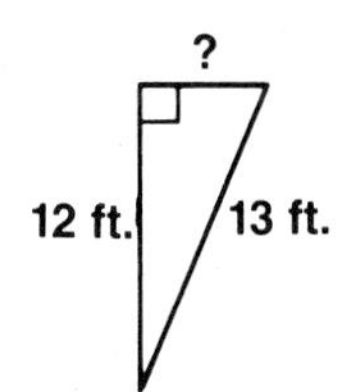


3. Tell whether each set of measurements can be the lengths of the sides of a right triangle.

   **a.** 2 cm, 3 cm, 4 cm **b.** 6 m, 8 m, 10 m **c.** 30 ft., 40 ft., 50 ft. **d.** 6 yd., 9 yd., 12 yd.

4. Find the length of the cable.

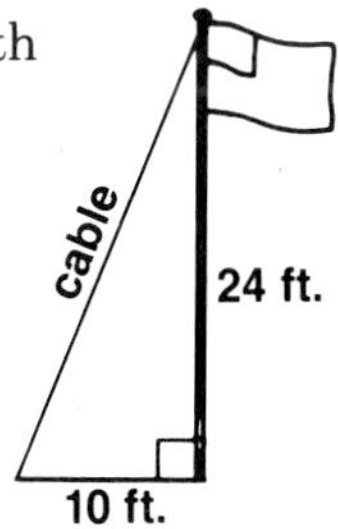


5. Find the distance between City A and City B.

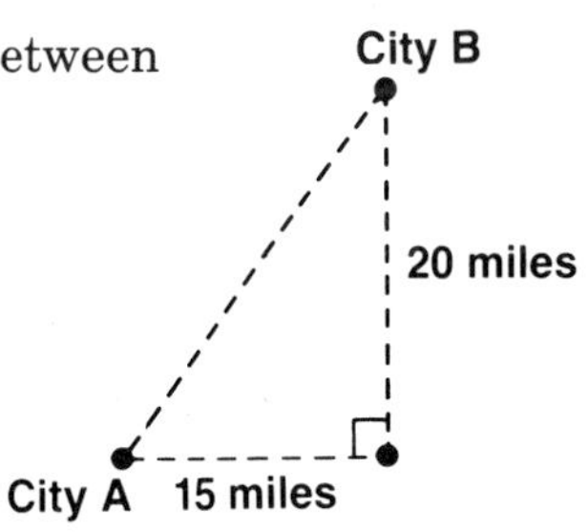

## SPIRAL REVIEW EXERCISES

1. If the measures of two angles of a triangle are 50° and 80°, find the measure of the third angle.

2. If the measure of $\angle A$ is 89°, then $\angle A$ is
   (a) acute (c) obtuse
   (b) right (d) straight

3. Graph $-2 \leq x < 5$ on a number line.

4. The probability that the spinner will stop on a space numbered with a prime is

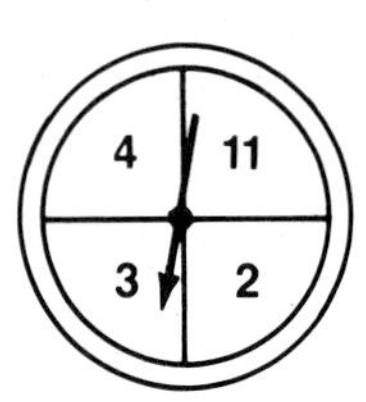


   (a) $\frac{1}{4}$ (c) $\frac{3}{4}$
   (b) $\frac{1}{2}$ (d) 1

5. If $A = bh$, the value of $A$ when $b = 10$ and $h = 8$ is
   (a) 18 (b) 108 (c) 80 (d) $\frac{5}{4}$

6. Solve: $5x + 12 = 27$

7. The value of $3x - 7$ when $x = 9$ is
   (a) 6 (b) 32 (c) 36 (d) 20

8. 35 meters is equivalent to
   (a) 35 cm (c) 350 cm
   (b) 3.5 cm (d) 3,500 cm

9. What number is 75% of 80?

10. Find 30% of 240.

11. Find the sum of $\frac{3}{4}$ and $\frac{7}{12}$.

## CHALLENGE QUESTION

In the diagram, $\angle ABD$ and $\angle DBC$ are right angles. Find the length of line segment $AC$.

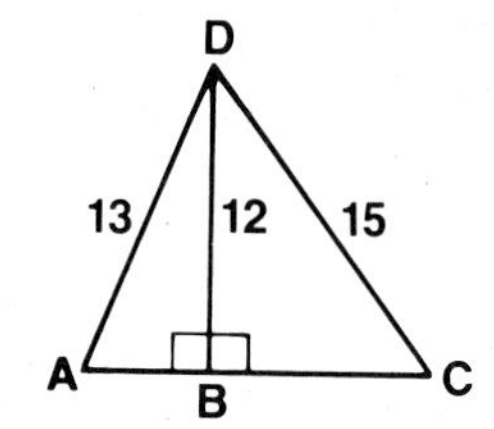

## UNIT 19–5 Using Coordinates to Graph Points

### THE MAIN IDEA

1. You can tell the location of points by using a horizontal number line and a vertical number line that cross at right angles to each other. For example, point $P$ corresponds to $-2$ on the horizontal number line and $+3$ on the vertical number line.

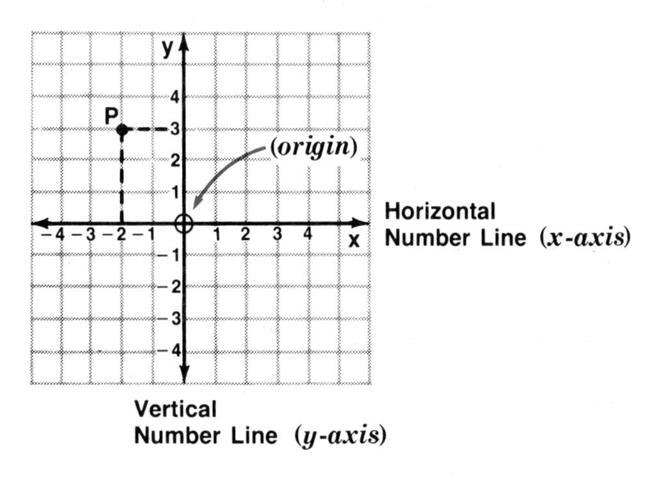


2. The numbers $-2$ and $+3$ are written as an ***ordered pair*** inside parentheses and separated by a comma. In an ordered pair, the first number gives the horizontal location and the second number gives the vertical location. The numbers in the ordered pair $(-2, 3)$ are called the ***coordinates*** of point $P$.
3. The horizontal number line is called the ***x-axis***. The vertical number line is called the ***y-axis***. The point where the number lines cross is called the ***origin***. The coordinates of the origin are $(0, 0)$.
4. To graph an ordered pair:
   a. Start at the origin, $(0, 0)$.
   b. Read the first number in the ordered pair and move to the left or to the right the number of units indicated.
   c. Read the second number in the ordered pair and move up or down the number of units indicated.
   d. Put a dot at the location reached and label the point with a capital letter.

**EXAMPLE 1** Use an ordered pair to name the location of point $A$.

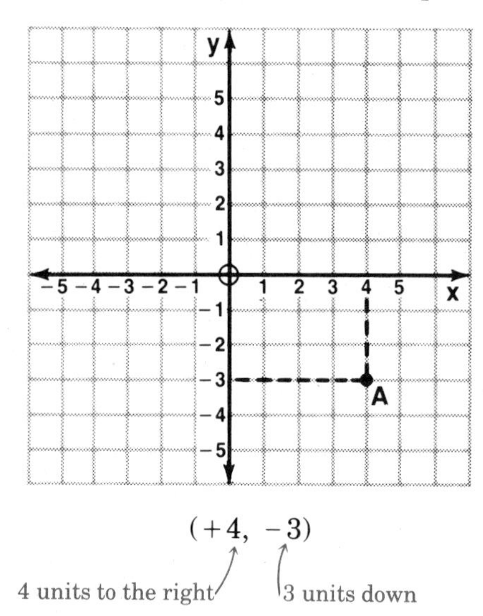


$(+4, \ -3)$

4 units to the right ↗ ↖ 3 units down

Notice that $(-3, 4)$ does *not* name the same location.

*Answer:* $A(4, -3)$

**EXAMPLE 2** Use an ordered pair to tell the coordinates of each of the given points.

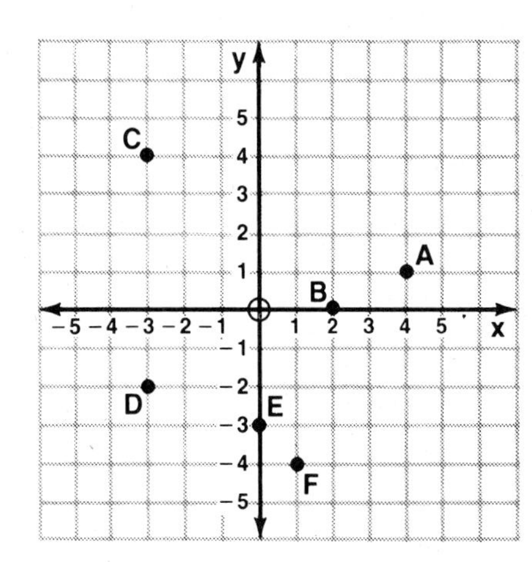


| | *Point* | *Coordinates* |
|---|---|---|
| **a.** | $A$ | $(4, 1)$ |
| **b.** | $B$ | $(2, 0)$ |
| **c.** | $C$ | $(-3, 4)$ |
| **d.** | $D$ | $(-3, -2)$ |
| **e.** | $E$ | $(0, -3)$ |
| **f.** | $F$ | $(1, -4)$ |

**EXAMPLE 3** Name the point that has the given coordinates.

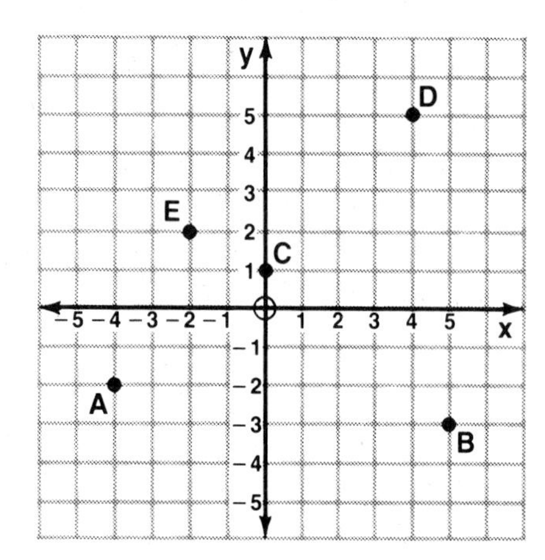


| | *Coordinates* | *Point* |
|---|---|---|
| **a.** | $(5, -3)$ | $B$ |
| **b.** | $(-2, 2)$ | $E$ |
| **c.** | $(4, 5)$ | $D$ |
| **d.** | $(0, 1)$ | $C$ |
| **e.** | $(-4, -2)$ | $A$ |

**EXAMPLE 4** Graph and label point $A$, which has the coordinates $(3, -2)$.

Start at $(0, 0)$. Read the first number, 3, and go 3 units to the right.

Read the second number, $-2$, and go 2 units down.

Label the point $A$.

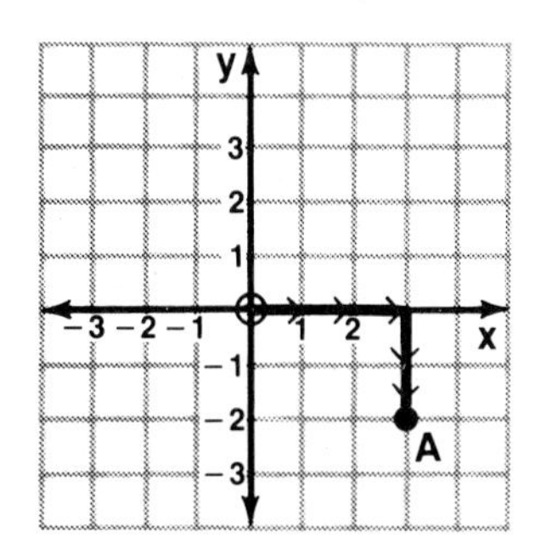


## CLASS EXERCISES

1. Use an ordered pair to tell the coordinates of each of the given points.

   a. $A$ b. $B$ c. $C$ d. $D$ e. $E$ f. $F$

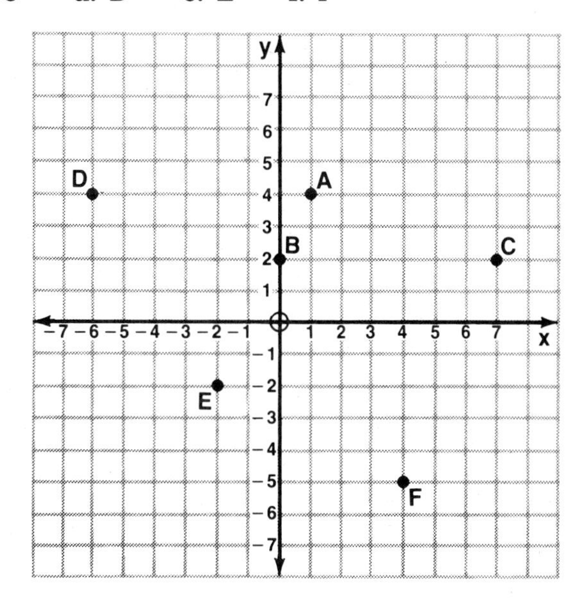


2. Tell the coordinates of each vertex of triangle $ABC$.

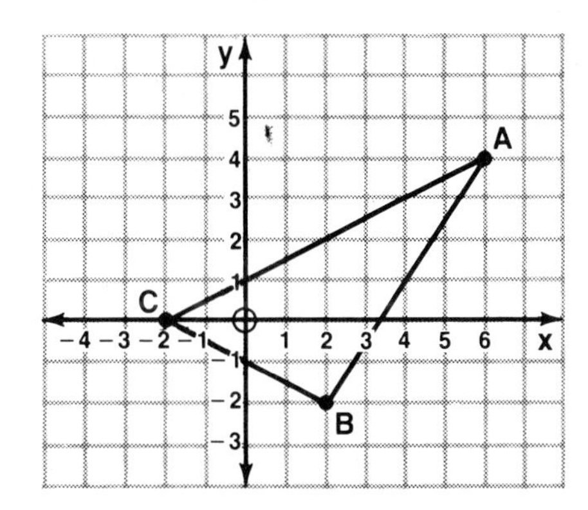

3. Name the point that has the given coordinates.

a. $(-1, 2)$ b. $(-3, 0)$ c. $(8, 2)$ d. $(-5, -2)$ e. $(2, 8)$ f. $(5, -4)$

g. $(0, -3)$ h. $(-2, -5)$

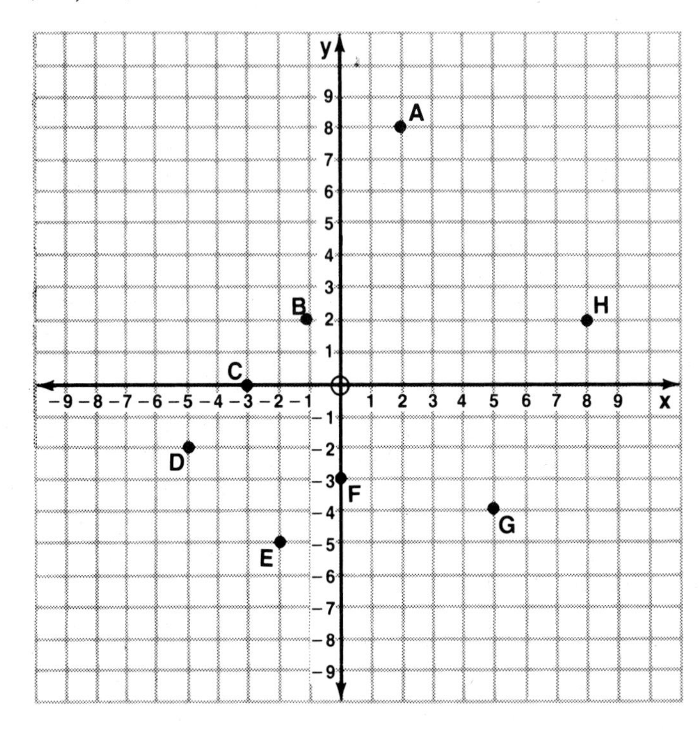


4. Graph and label the points having the given coordinates.

a. $A(2, 6)$ b. $B(-3, 0)$ c. $C(-4, 2)$ d. $D(-6, 6)$ e. $E(0, 2)$ f. $F(2, 9)$

g. $G(-7, -1)$ h. $H(-1, -7)$

## HOMEWORK EXERCISES

1. Use an ordered pair to tell the coordinates of each of the given points.

a. $A$ b. $B$ c. $C$ d. $D$

e. $E$ f. $F$ g. $G$

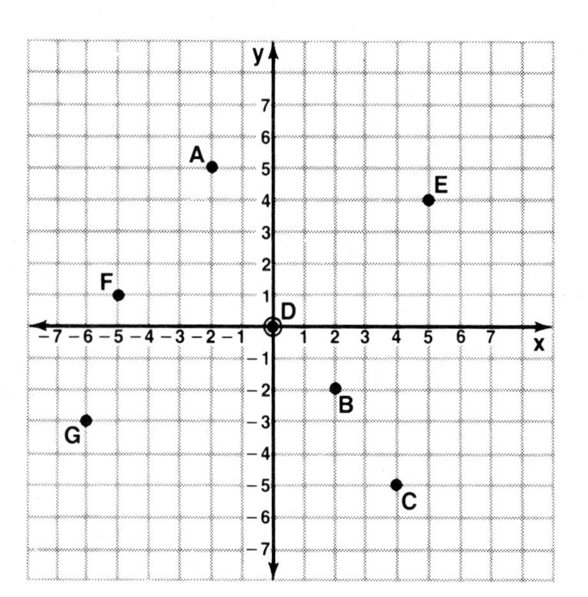

**2.** Tell the coordinates of each vertex of triangle *DEF*.

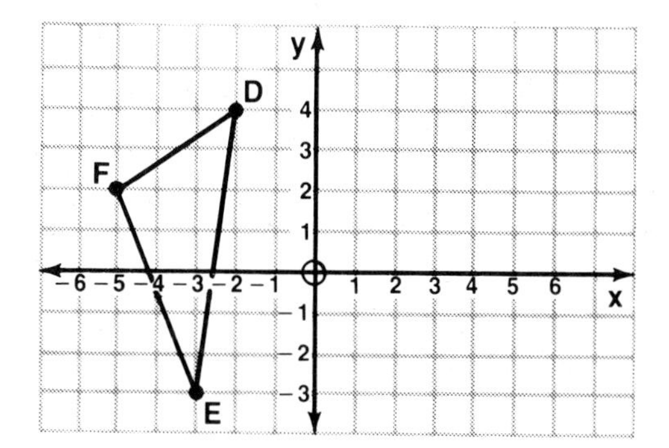


**3.** Tell the coordinates of each vertex of figure *ABCD*.

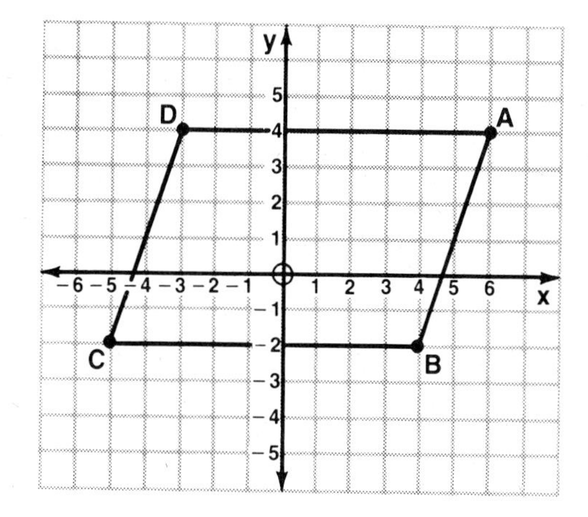


**4.** Name the point that has the given coordinates.

**a.** (6, 1) **b.** (1, 6) **c.** (3, −2)

**d.** (−4, 2) **e.** (−6, −3) **f.** (0, −4)

**g.** (0, 3) **h.** (−8, 0)

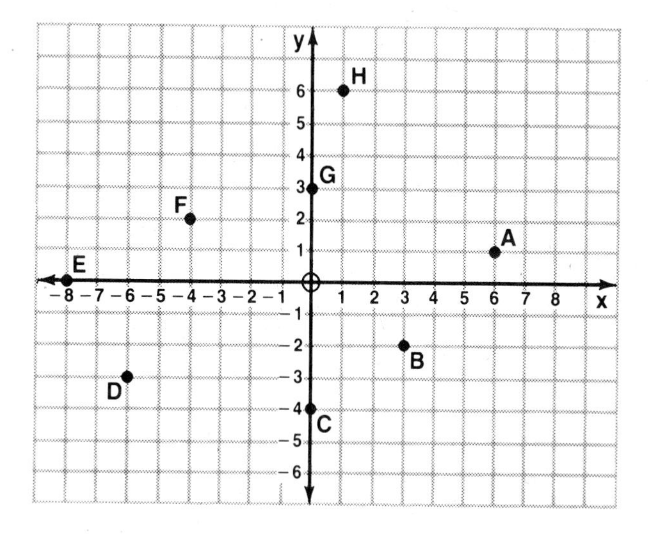


**5.** Graph and label each of the points having the given coordinates.

**a.** $A(7, 1)$ **b.** $B(0, 0)$ **c.** $C(-3, 5)$ **d.** $D(5, -3)$ **e.** $E(-2, -6)$ **f.** $F(2, 6)$

**g.** $G(-7, 2)$ **h.** $H(0, 6)$ **i.** $I(6, 0)$ **j.** $J(-8, -4)$ **k.** $K(-5, 0)$ **l.** $L(-8, 6)$

## SPIRAL REVIEW EXERCISES

**1.** Tell whether each statement is *true* or *false*.

**a.** $-9 < 0$ **b.** $0 > -100$

**c.** $-5 > -9$ **d.** $-29 < 2$

**e.** $-2.6 < -2.8$ **f.** $-\frac{7}{8} > -\frac{5}{8}$

**2.** Which open sentence is represented by the graph shown?

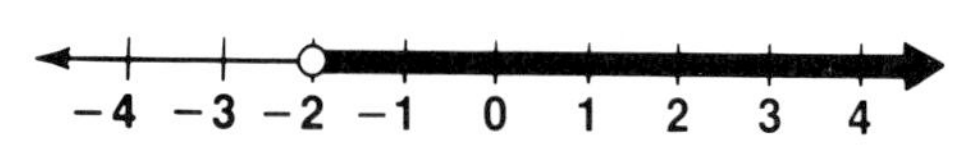


(a) $x < -2$ (c) $x \leq -2$
(b) $x > -2$ (d) $x \geq -2$

**3.** This pictograph shows the number of records sold at A & B Record Shop one week.

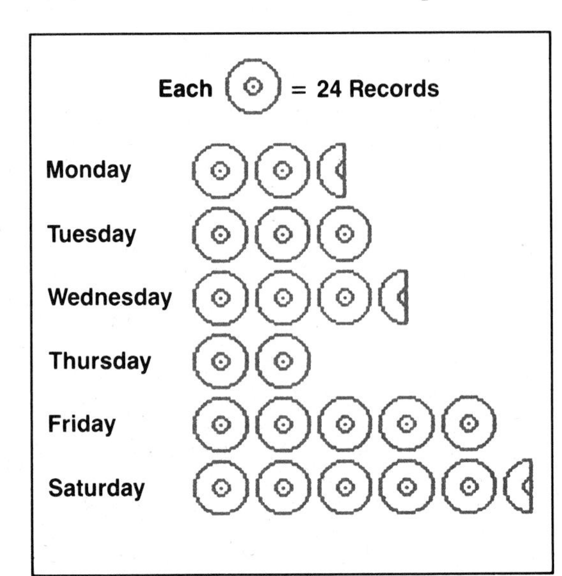


How many records were sold on Saturday?
(a) 5 (b) 6 (c) 132 (d) 55

**4.** The value of $7 \times 10^3 + 3 \times 10^2 + 5$ is
(a) 7,350 (c) 7,305
(b) 7,035 (d) 73,005

**5.** Divide: $\frac{8}{15} \div \frac{2}{5}$

**6.** Subtract: $\frac{3}{4} - \frac{7}{20}$

**7.** The distance between the cities of Winston and Clark is
(a) 125 miles
(b) 130 miles
(c) 150 miles
(d) 200 miles

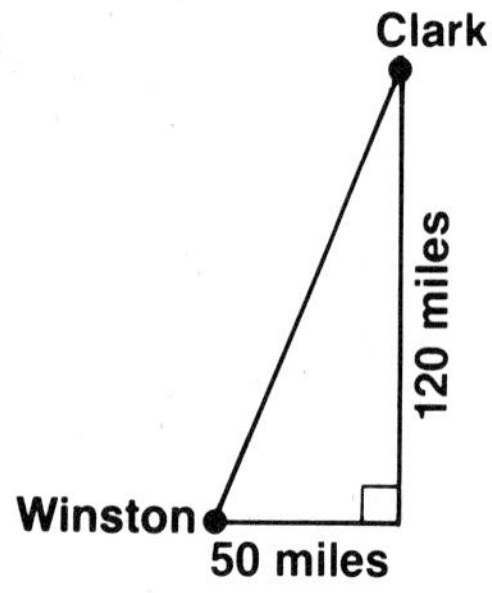


**8.** Change $7\frac{1}{2}\%$ to a decimal.

**9.** At 8 A.M. the temperature was $-5$°F, and at 12 noon the temperature was 20°F. What was the increase in temperature?

**10.** Mr. Jackson left a 15% tip for the waiter. How much money did he leave if his bill was $30?

**CHALLENGE QUESTION**

The coordinates of the vertices of a right triangle are $A(1, 2)$, $B(5, 2)$, and $C(1, 5)$. Find the length of hypotenuse $BC$.

# UNIT 20–1 Parallel Lines, Perpendicular Lines, and Polygons

## PARALLEL LINES AND PERPENDICULAR LINES

### THE MAIN IDEA

1. *Line segment* $AB$ (also written $\overline{AB}$) ends, but *line* $AB$ (also written $\overleftrightarrow{AB}$) does not end.

2. *Parallel lines* are different lines that go in exactly the same direction and, therefore, never meet. (The symbol $\parallel$ means "is parallel to.")

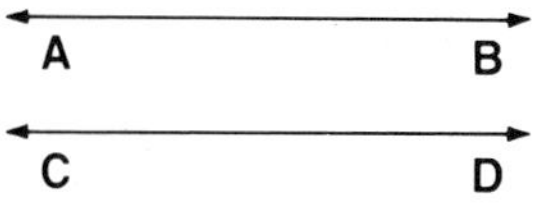


line $AB \parallel$ line $CD$

3. *Perpendicular lines* are lines that meet at right angles. (The symbol $\perp$ means "is perpendicular to.")

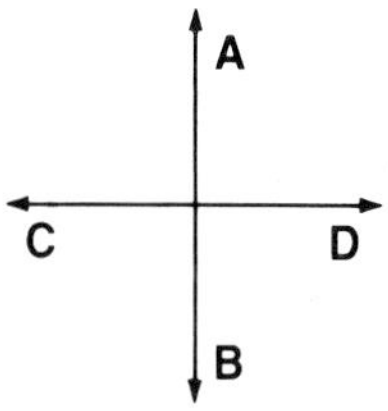


line $AB \perp$ line $CD$

**EXAMPLE 1** In each diagram, tell which pairs of lines (or line segments) are drawn parallel, and which pairs are drawn perpendicular.

| *Diagram* | *Answer* |
|---|---|
| a. 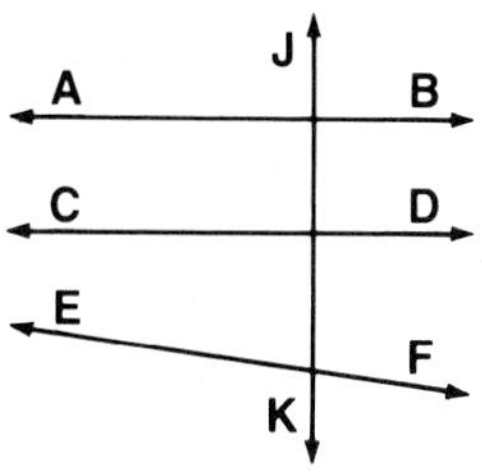 | Line $AB$ and line $CD$ are parallel.<br>Line $JK$ is perpendicular to both line $AB$ and line $CD$. |
| b. 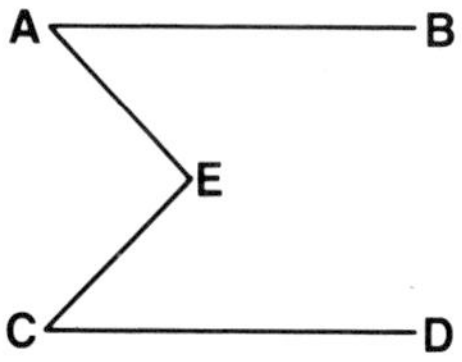 | Line segment $AB$ and line segment $CD$ are parallel.<br>Line segment $AE$ and line segment $CE$ are perpendicular. |

## CLASS EXERCISES

In each diagram, tell which pairs of lines (or line segments) are drawn parallel, and which pairs are drawn perpendicular.

1. 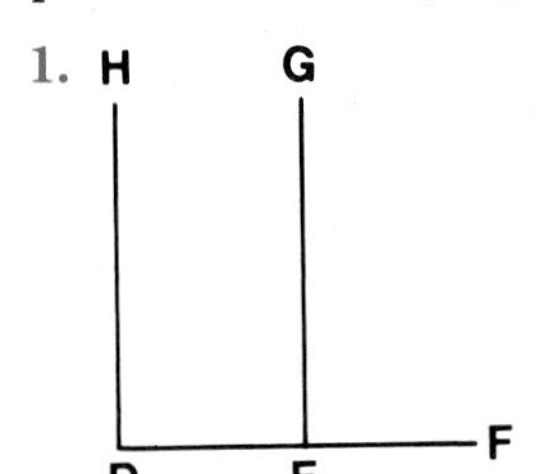


2. 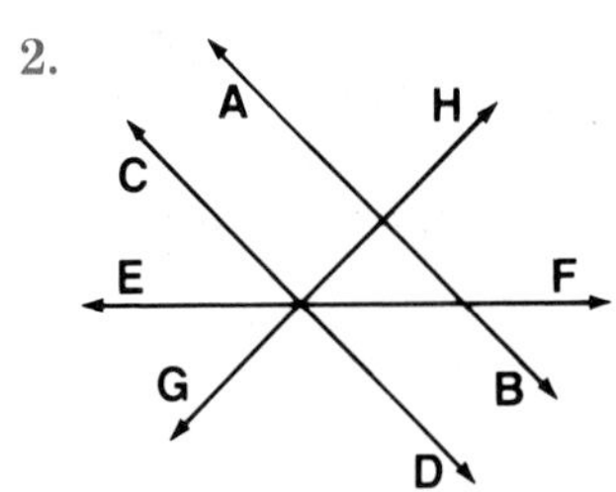


3. 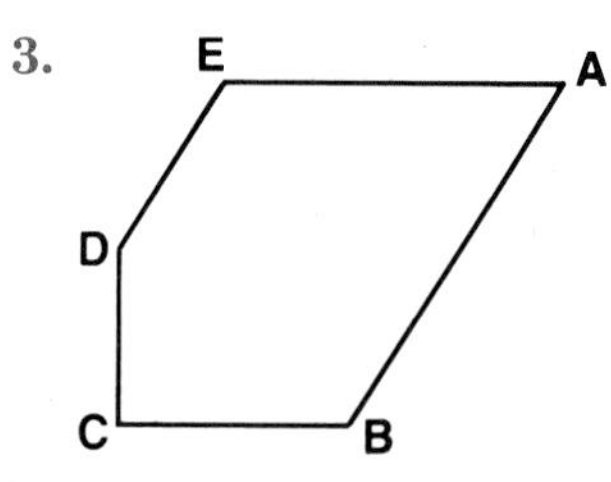


4. 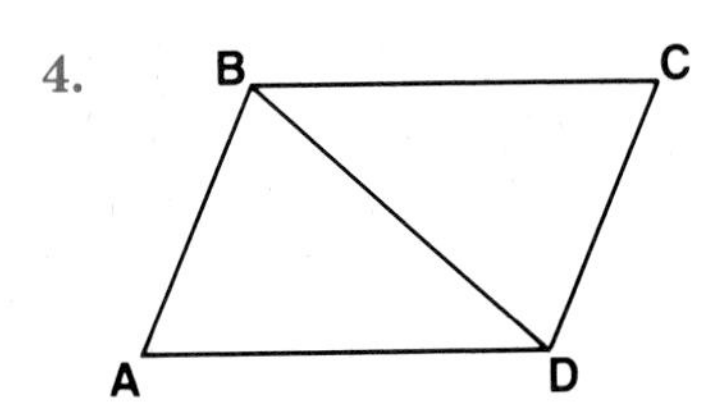


5. 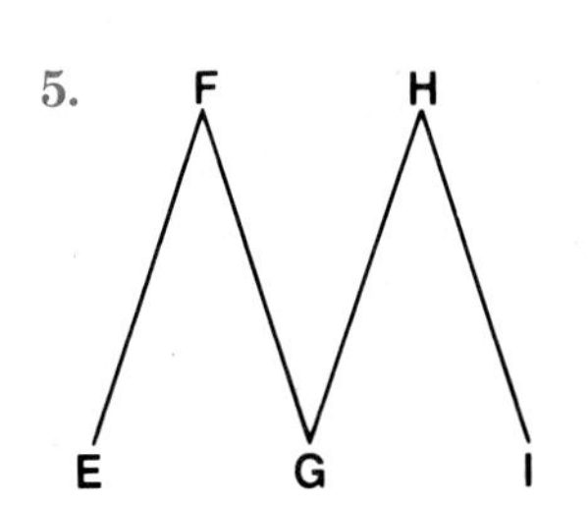


6. 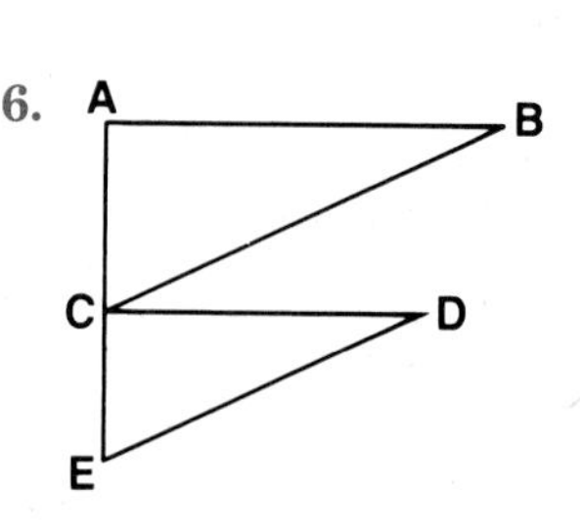


## POLYGONS

### THE MAIN IDEA

1. A ***polygon*** is a flat, closed figure whose sides are line segments. A polygon may be named according to the number of its sides. Some common names are:

| *Number of Sides* | *Name* |
|---|---|
| 3 | Triangle |
| 4 | Quadrilateral |
| 5 | Pentagon |
| 6 | Hexagon |
| 8 | Octagon |

2. Some ***quadrilaterals*** (4-sided polygons) have special properties and special names.

   a. A ***trapezoid*** is a quadrilateral that has only one pair of opposite sides parallel.

**Trapezoid ABCD**

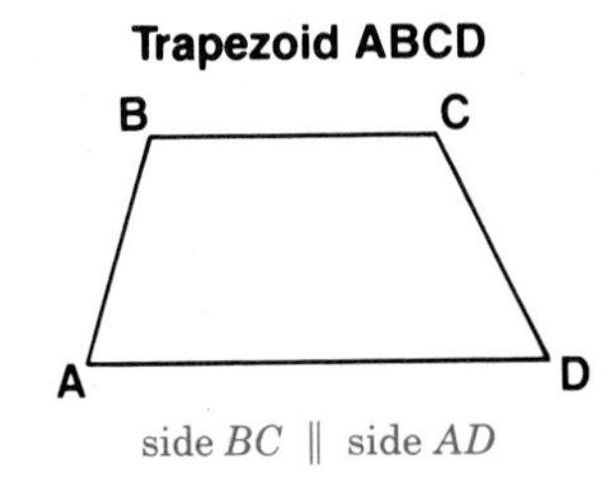


side $BC \parallel$ side $AD$

b. A ***parallelogram*** is a quadrilateral that has both pairs of opposite sides parallel. In a parallelogram, the opposite sides are also equal in measure.

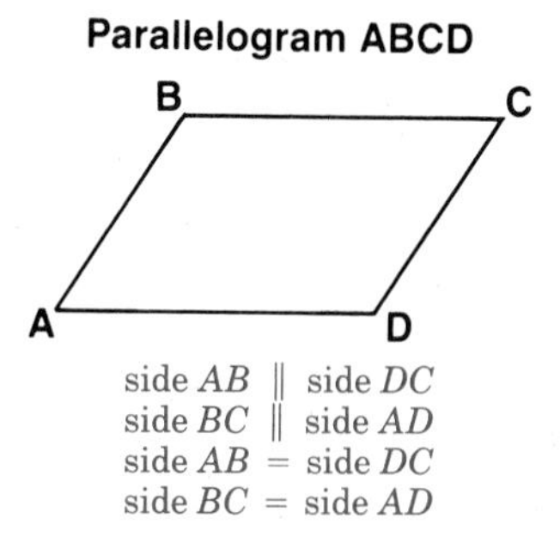


c. A ***rhombus*** is a parallelogram that has all four sides equal in measure.

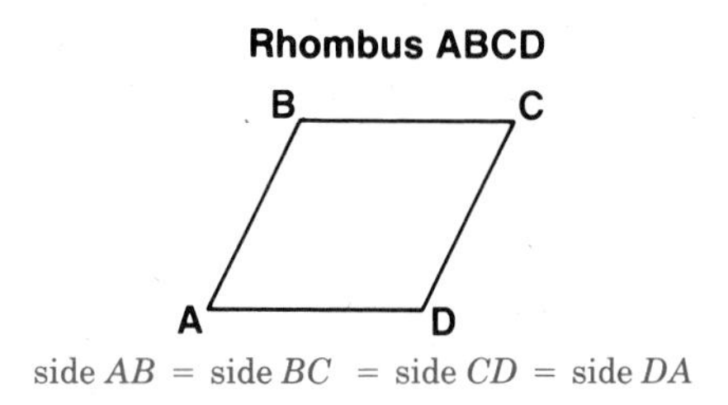


d. A ***rectangle*** is a parallelogram that has four right angles.

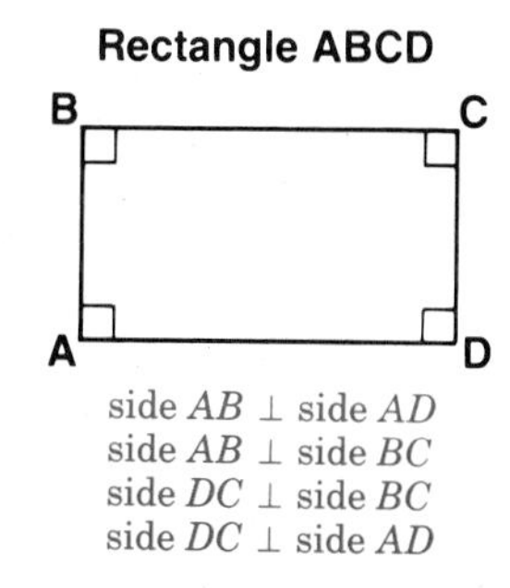


e. A ***square*** is a rectangle that has all four sides equal in measure.

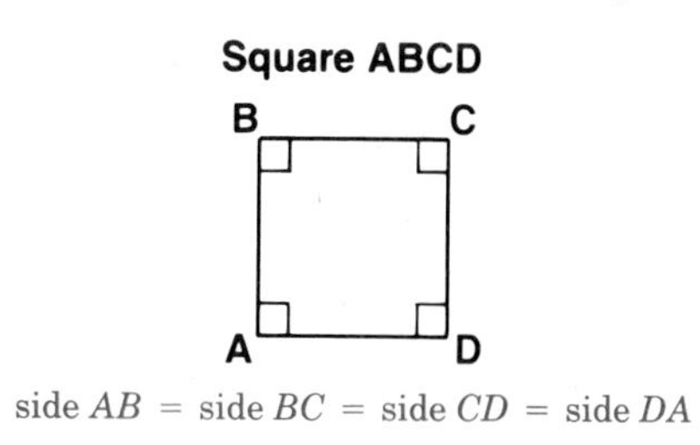

**EXAMPLE 2** In each quadrilateral, tell which pairs of sides are drawn parallel, and which are drawn perpendicular.

| *Quadrilateral* | *Answer* |
|---|---|
| a. XYZW | Side $XY$ and side $WZ$ are parallel. No sides are perpendicular. |
| b. RSTQ | Side $QR$ and side $ST$ are parallel. Side $QT$ and side $RS$ are parallel. No sides are perpendicular. |
| c. BCDA | No sides are parallel. Side $BC$ and side $CD$ are perpendicular. |
| d. EFGD | Side $DE$ and side $FG$ are parallel. Side $EF$ and side $DG$ are parallel. Side $DE$ and side $FG$ are each perpendicular to both side $EF$ and side $DG$. |

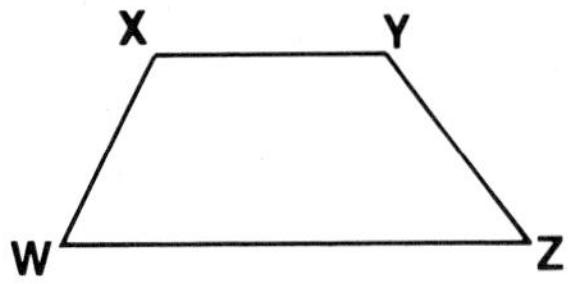

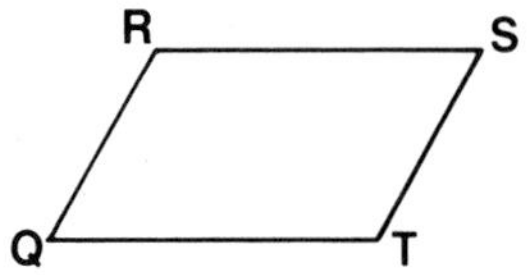

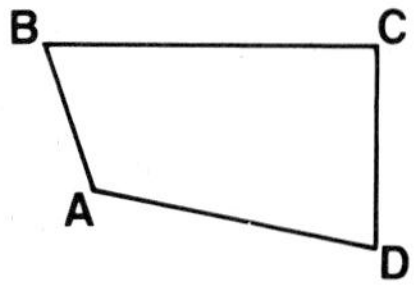

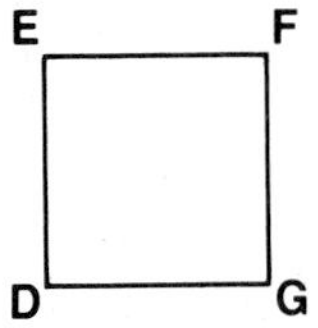

**EXAMPLE 3** For each quadrilateral, name the sides that are drawn parallel, name the sides that are drawn perpendicular, and find the missing lengths.

| *Quadrilateral* | *Answer* |
|---|---|
| a. Rectangle DEFG (EF = 10 in., FG = 6 in.) | Side $EF$ and side $DG$ are parallel. Side $DE$ and side $GF$ are parallel. Side $DE$ and side $GF$ are each perpendicular to both side $EF$ and side $DG$. Side $ED$ = 6 in.; side $DG$ = 10 in. |
| b. Rhombus ABCD (AB = 5 m) | Side $AD$ and side $BC$ are parallel. Side $AB$ and side $DC$ are parallel. No sides are perpendicular. Side $BC$ = 5 m, side $CD$ = 5 m, and side $AD$ = 5 m. |
| c. Square WXYZ (YZ = 12 cm) | Side $WX$ and side $ZY$ are parallel. Side $XY$ and side $WZ$ are parallel. Side $WX$ and side $ZY$ are each perpendicular to both side $XY$ and side $WZ$. Side $XY$ = 12 cm, side $WX$ = 12 cm, and side $WZ$ = 12 cm. |

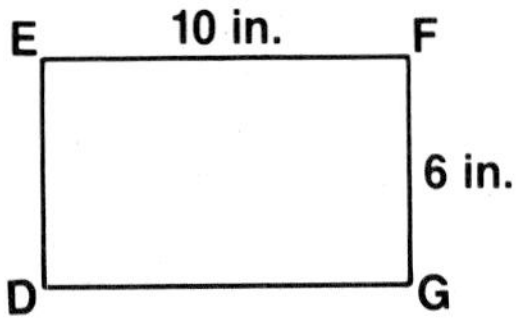

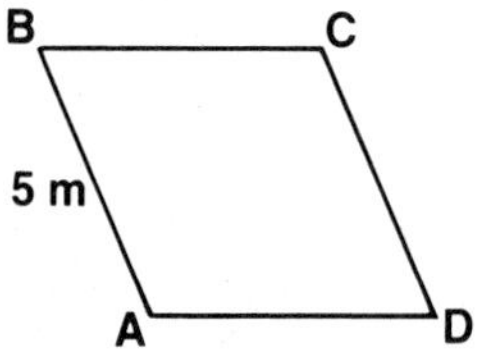

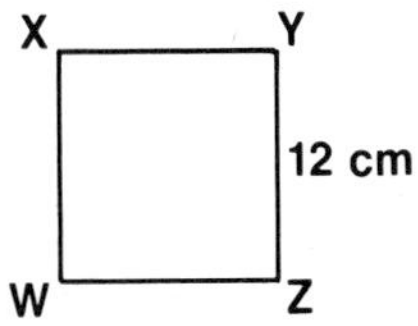

**EXAMPLE 4** In parallelogram *ABCD*, find the value of $x$.

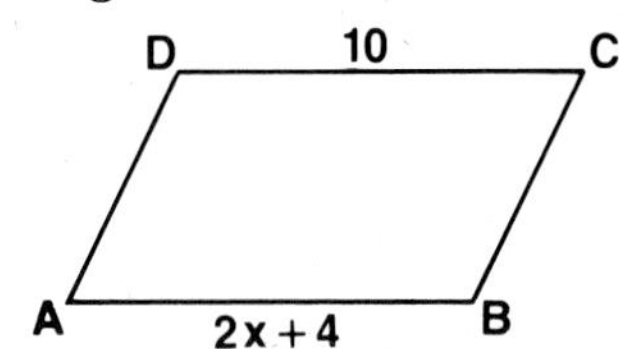


### THINKING ABOUT THE PROBLEM

Since the opposite sides of a parallelogram are equal in measure, you can write an equation involving $x$, and solve.

$$\begin{aligned} \text{side } AB &= \text{side } DC \\ 2x + 4 &= 10 \\ 2x + 4 + (-4) &= 10 + (-4) \\ 2x &= 6 \\ \frac{1}{2}(2x) &= (6)\frac{1}{2} \\ x &= 3 \quad \textit{Ans.} \end{aligned}$$

## CLASS EXERCISES

**1.** In each quadrilateral, tell which sides are drawn parallel, and which are drawn perpendicular.

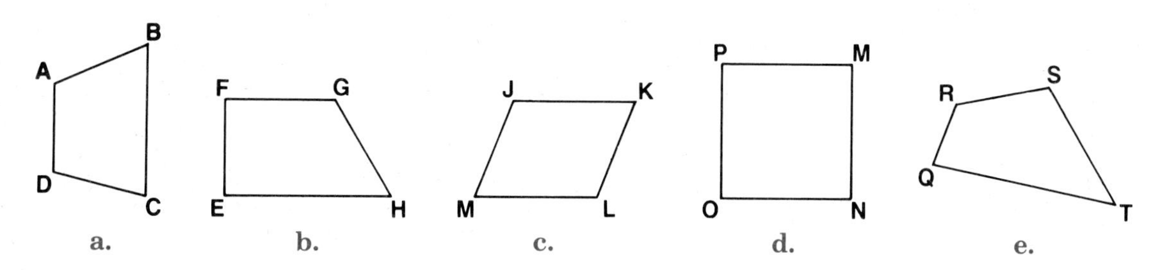


**2.** In each quadrilateral, name the parallel sides, name the perpendicular sides, and find the missing lengths.

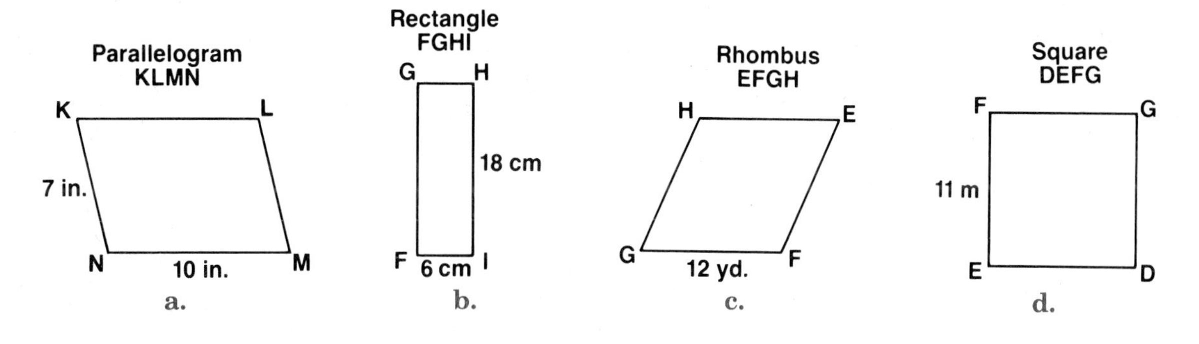

**3.** In each quadrilateral, find the value of $x$.

Parallelogram PQRS



a.

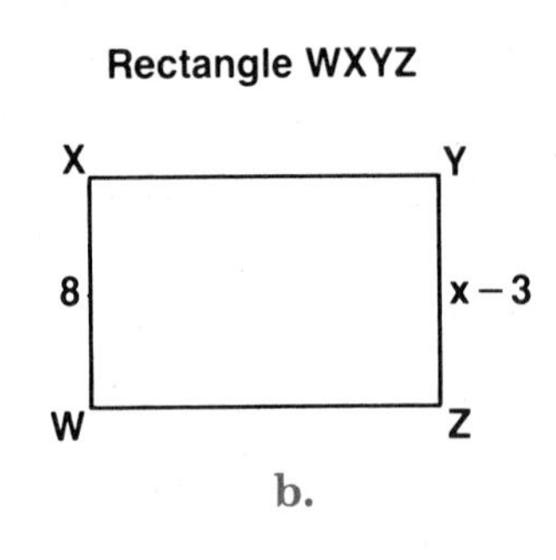


b.

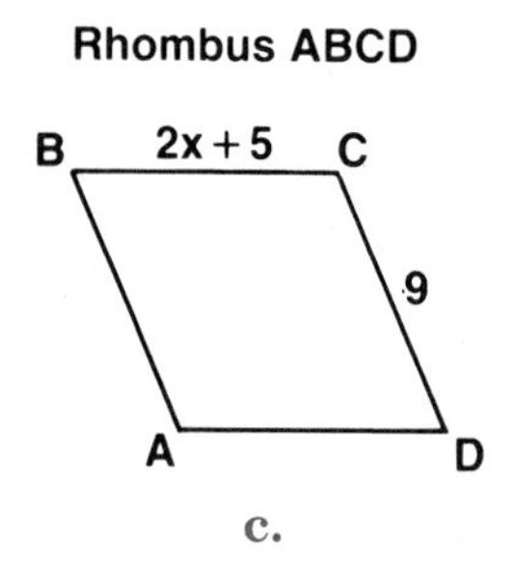


c.

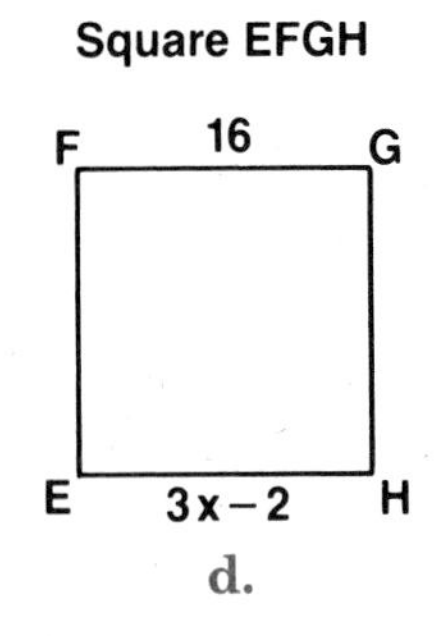


d.

## *HOMEWORK EXERCISES*

**1.** In each diagram, tell which pairs of lines (or line segments) are drawn parallel, and which pairs are drawn perpendicular.

a.

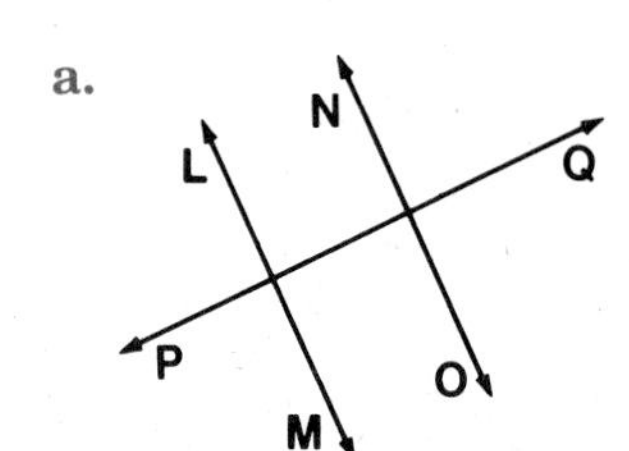


b.

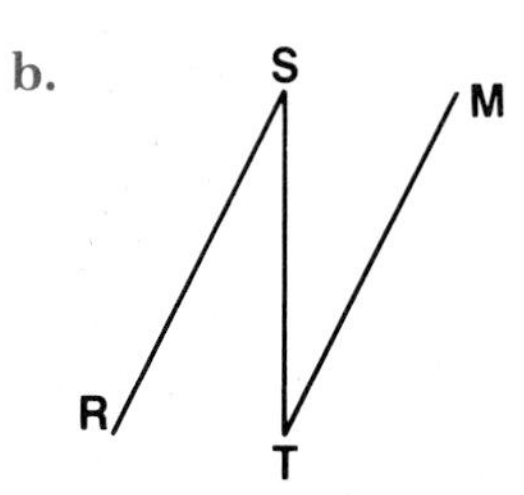


c.

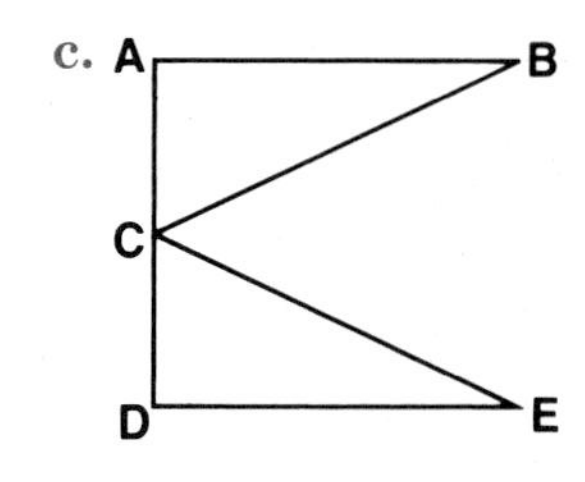


d.

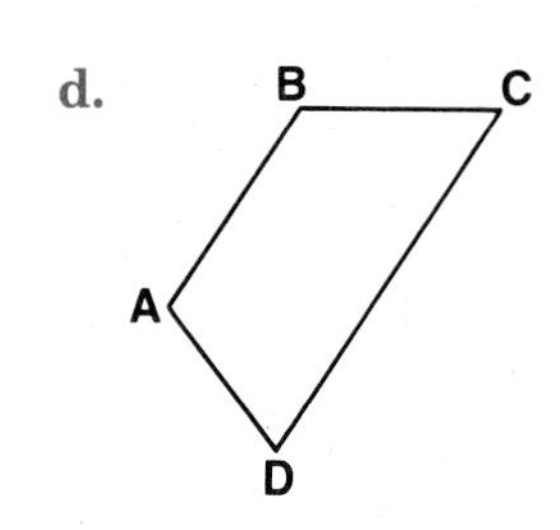


e.

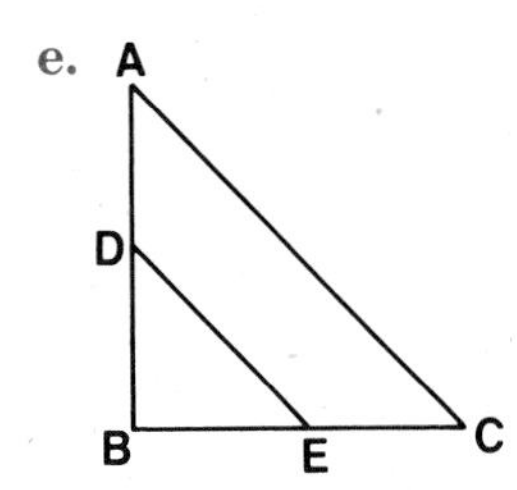


f.

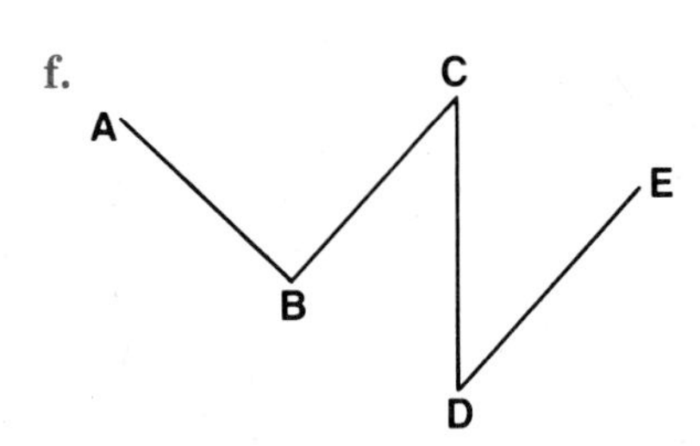


**2.** In each quadrilateral, tell which sides are drawn parallel, and which are drawn perpendicular.

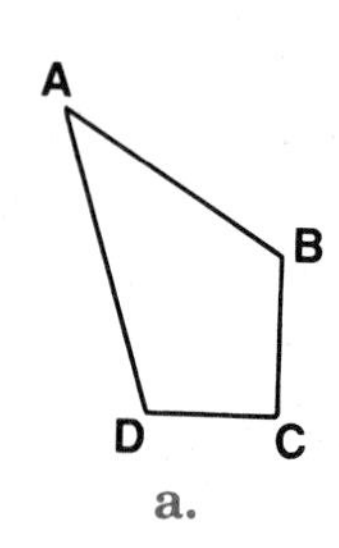


a.

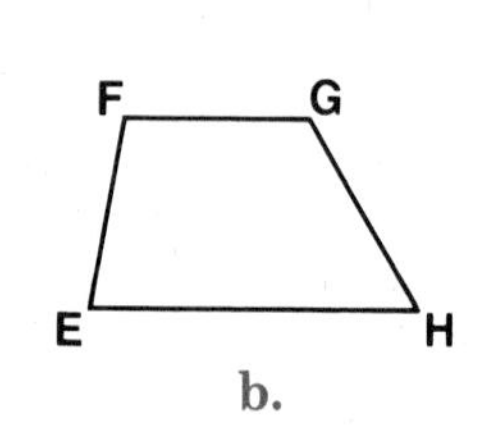


b.

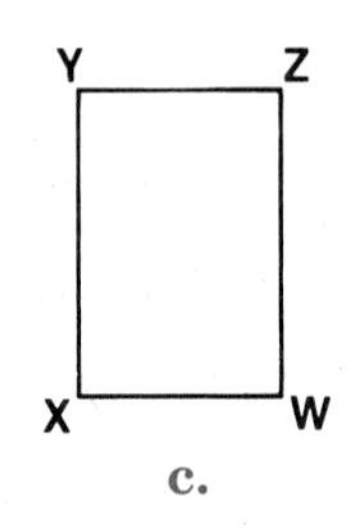


c.

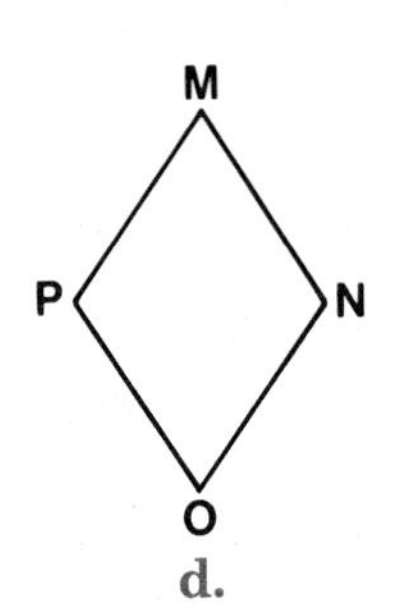


d.

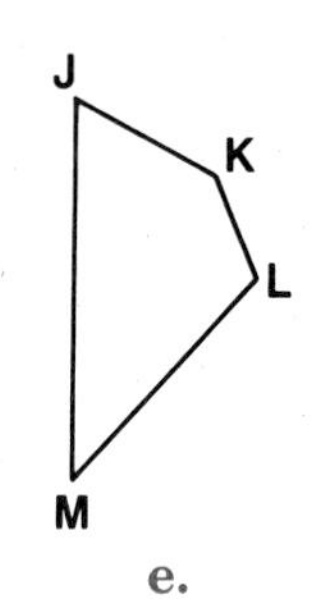


e.

3. In each quadrilateral, name the parallel sides, name the perpendicular sides, and find the missing lengths.

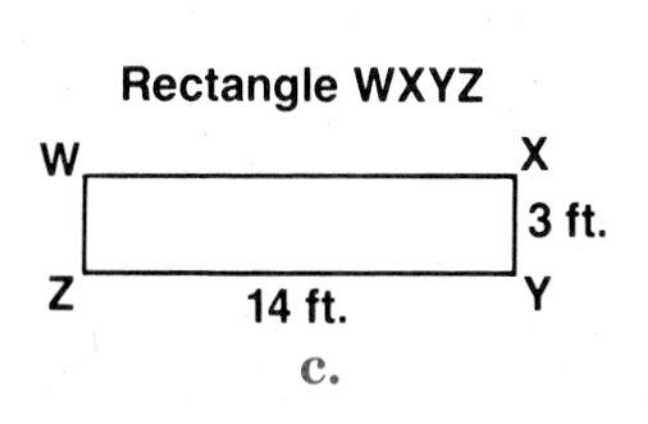


a.

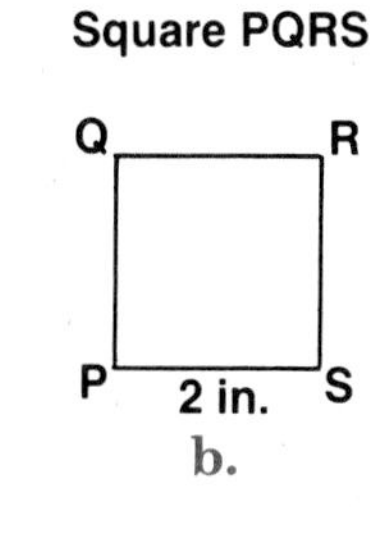


b.

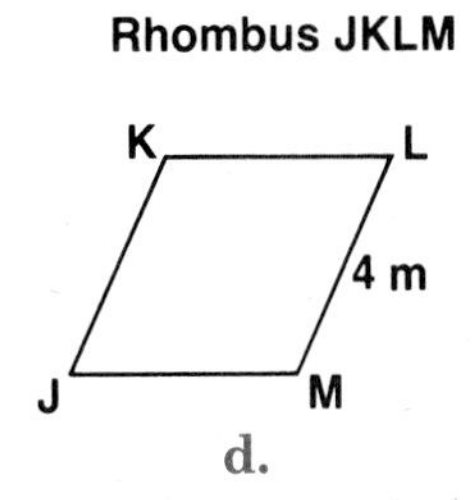


c.



d.

4. In each quadrilateral, find the value of $x$.

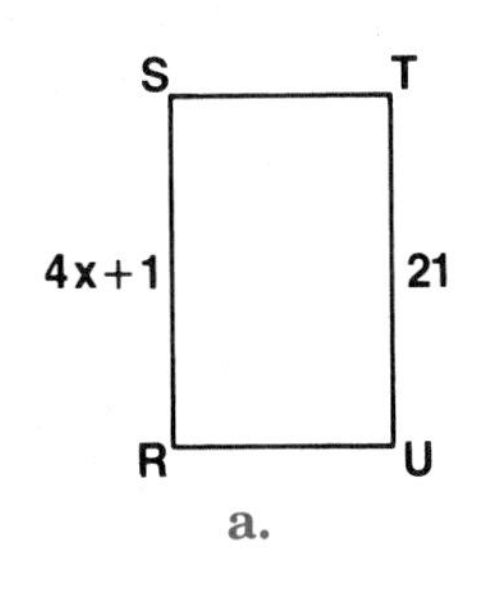


a.

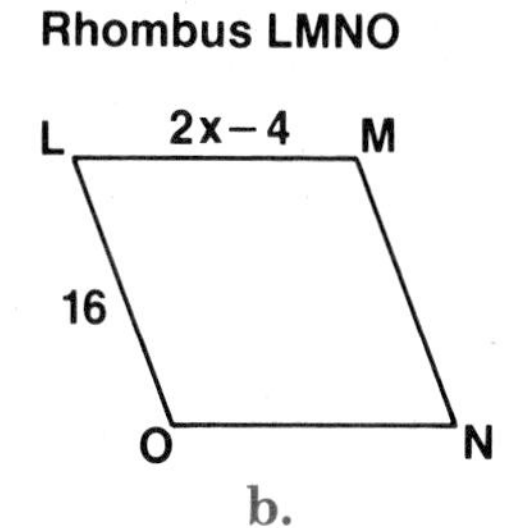


b.

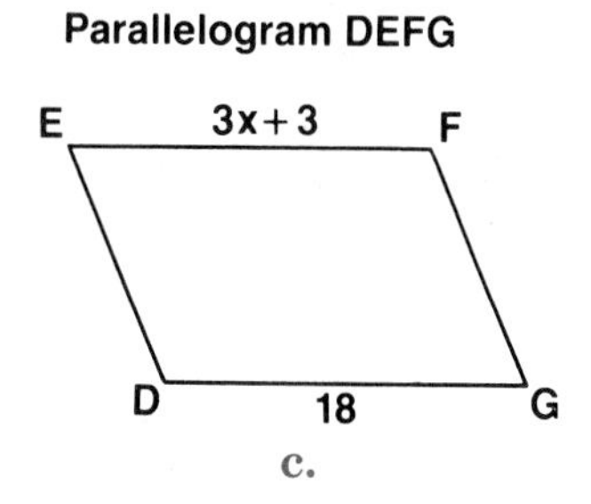


c.

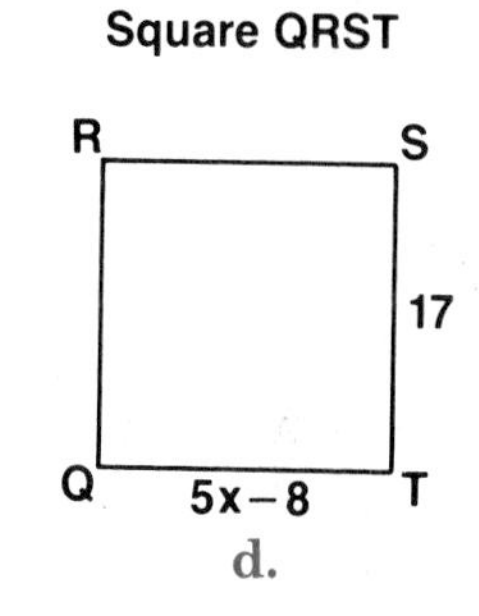


d.

## SPIRAL REVIEW EXERCISES

1. Use an ordered pair to tell the coordinates of each of the given points.

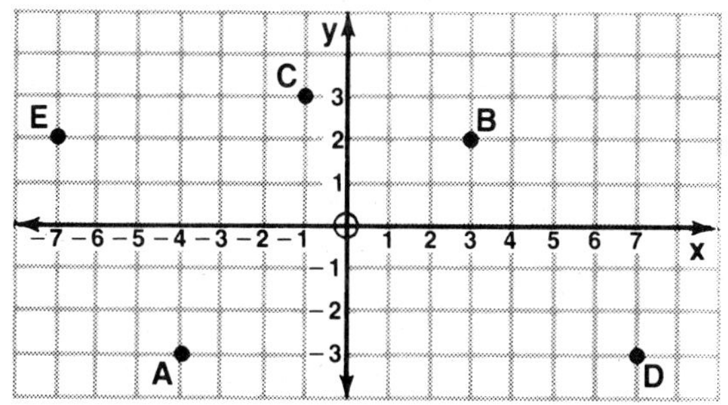


2. On the graph below, what are the coordinates of point A?
(a) (4, 1)
(b) (−1, 4)
(c) (−4, 1)
(d) (−4, −1)

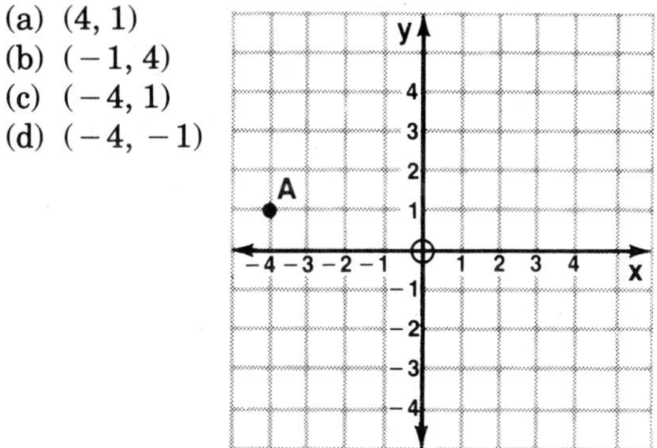


3. On the graph below, which point has coordinates (2, −3)?
(a) D
(b) E
(c) F
(d) G

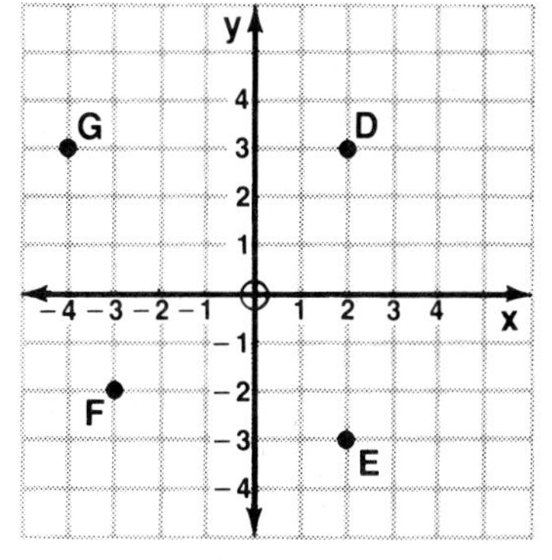


4. Find the measure of $\angle A$.

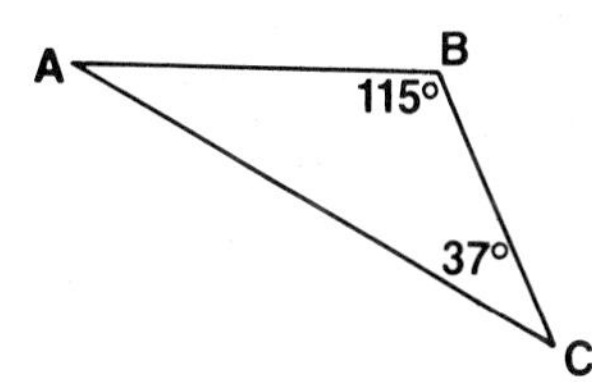


5. Tell whether a triangle whose sides measure 6 cm, 8 cm, and 12 cm is a right triangle.

**6.** 25 cm is equivalent to
(a) 25 m (c) .25 m
(b) 2.5 m (d) .025 m

**7.** If 3 candy bars sell for $.57, what is the cost of 7 candy bars?

**8.** If a coin is tossed 150 times, what is the number of times that you would expect it to come up "heads"?

**9.** 120% is equivalent to

(a) $\frac{1}{5}$ (b) $\frac{3}{5}$ (c) $\frac{4}{5}$ (d) $\frac{6}{5}$

**CHALLENGE QUESTION**

An ant started at point (0, 0) on a sheet of graph paper on which each space is 1 cm wide. It walked to point (0, 5), then to (3, 5), then to (3, 0), and then back to (0, 0). What was the total distance that the ant walked?

# UNIT 20–2 Perimeter

## THE MAIN IDEA

1. The *perimeter* of a polygon is the distance around the polygon.
2. To find the perimeter of a polygon, find the sum of the lengths of all its sides.

| *Figure* | *Find the Sum of* | *Formula* |
|---|---|---|
| Triangle (sides a, b, c) | 3 sides | $P = a + b + c$ |
| Rectangle (sides ℓ, w, w, ℓ) | 4 sides, two equal pairs | $P = 2\ell + 2w$ |
| Square (sides s, s, s, s) | 4 sides, all of which are equal in measure | $P = 4s$ |

**EXAMPLE 1** Find the perimeter of the polygon.

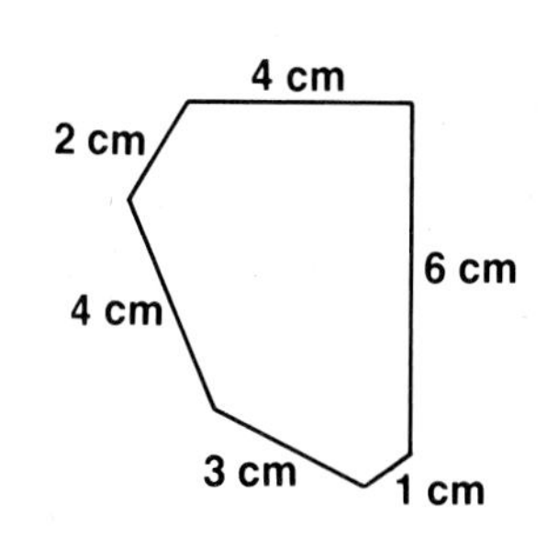


Find the sum of the lengths of all the sides.

$P = 2 + 4 + 6 + 1 + 3 + 4$
$P = 20$

*Answer:* The perimeter is 20 cm.

**EXAMPLE 2** Find the perimeter of each figure.

| *Figure* | *Diagram* | *By Formula* |
|---|---|---|
| **a.** a triangle whose sides measure 10 cm, 12 cm, and 20 cm | 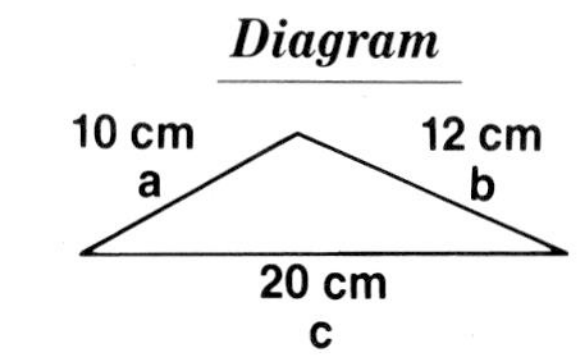  | $P = a + b + c$<br>$= 10 + 12 + 20$<br>$= 42$ cm *Ans.* |

| *Figure* | *Diagram* | *By Formula* |
|---|---|---|
| **b.** a rectangle whose length measures 9 in. and whose width measures 5 in. | 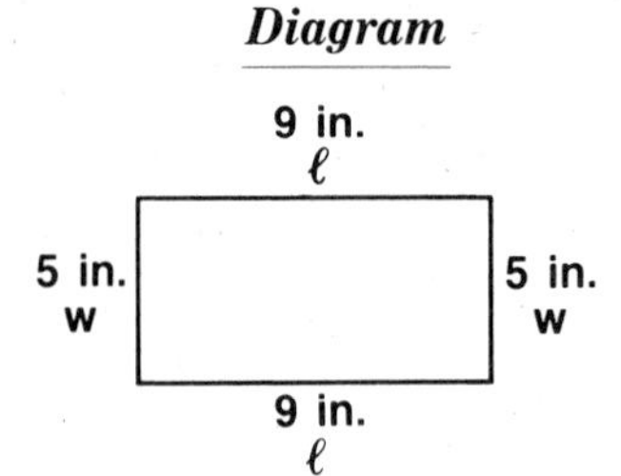  | $P = 2\ell + 2w$<br>$= 2(9) + 2(5)$<br>$= 18 + 10$<br>$= 28$ in. *Ans.* |
| **c.** a square, each of whose sides measures 22 m | 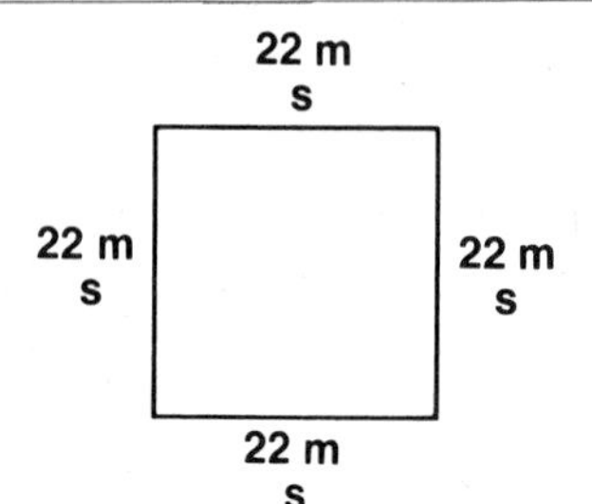  | $P = 4s$<br>$= 4(22)$<br>$= 88$ m *Ans.* |

**EXAMPLE 3** If the two legs of an isosceles triangle measure 20 cm each and the base measures 16 cm, what is the perimeter of the triangle?

### THINKING ABOUT THE PROBLEM

Draw a diagram and label it with the given information to help you see what has to be done.

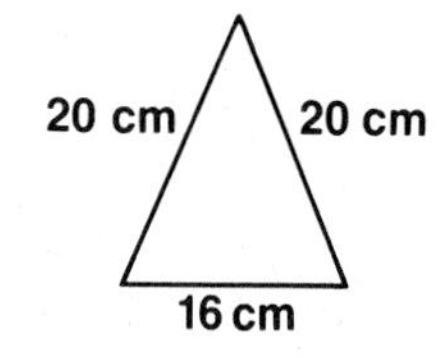


Add the lengths of the three sides. $P = 20 + 20 + 16$

$P = 56$

*Answer:* The perimeter is 56 cm.

**EXAMPLE 4** Find the measure of each side of a square whose perimeter is 100 cm.

### THINKING ABOUT THE PROBLEM

Work backwards from the perimeter to find the measure of one side. Since you multiply to find the perimeter, divide to find the measure of a side.

Divide 100 by 4. $4\overline{)100}$ = 25

*Answer:* The measure of each side of the square is 25 cm.

## CLASS EXERCISES

1. Find the perimeter of each of the following polygons.

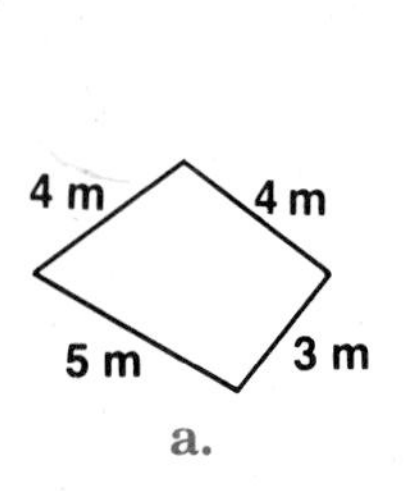


a.

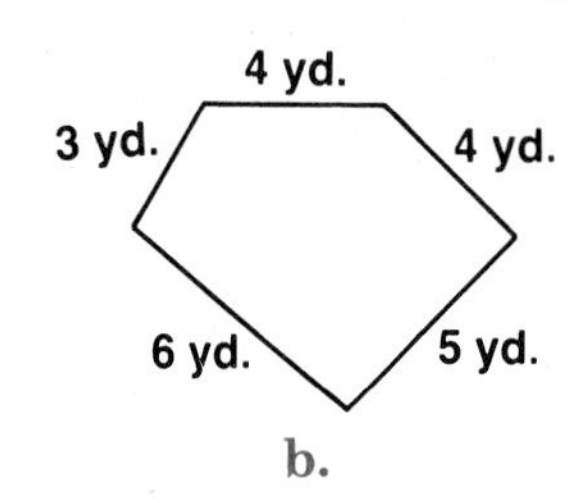


b.

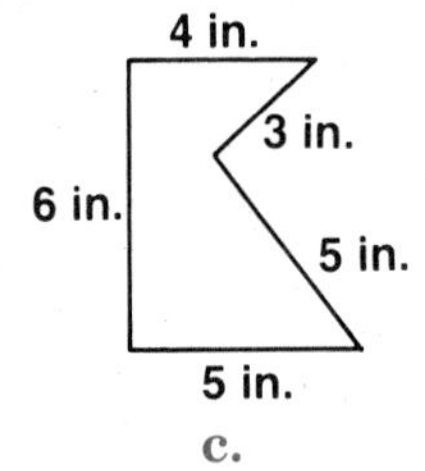


c.

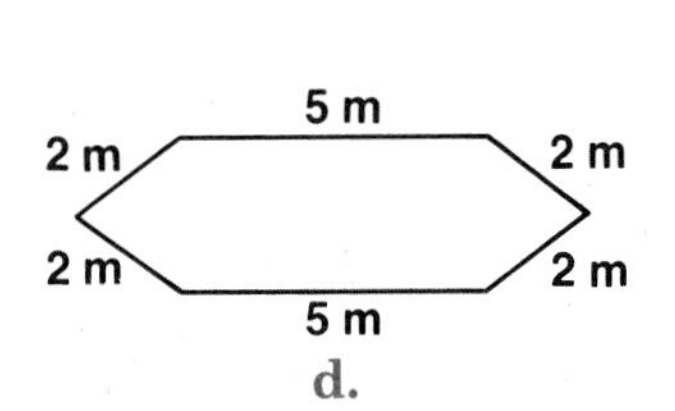


d.

2. Find the perimeter of a triangle whose sides measure:

   a. 5 in., 9 in., 7 in. b. 5 cm, 12 cm, 13 cm c. 3.1 m, 5.2 m, 4.5 m d. $1\frac{1}{2}$ ft., $3\frac{1}{3}$ ft., $2\frac{1}{6}$ ft.

3. Find the perimeter of a rectangle whose measurements are:

   a. $\ell = 5$ cm, $w = 2$ cm b. $\ell = 12$ m, $w = 7$ m

   c. $\ell = 10.5$ in., $w = 4.6$ in. d. $\ell = 5\frac{2}{3}$ yd., $w = 3\frac{1}{2}$ yd.

4. Find the perimeter of a square whose side measures:

   a. 9 in. b. 17 ft. c. 5.9 cm d. $8\frac{2}{3}$ m

5. What is the perimeter of an isosceles triangle whose base measures 10 ft. and each of whose legs measures 14.1 ft.?

6. Find the length of a side of a square whose perimeter is:

   a. 44 in. b. 60 cm c. 96 yd. d. 200 m

7. Find the length of a side of an equilateral triangle whose perimeter is:

   a. 36 cm b. 99 ft. c. 144 m d. 300 yd.

## *HOMEWORK EXERCISES*

1. Find the perimeter of each of the following polygons.

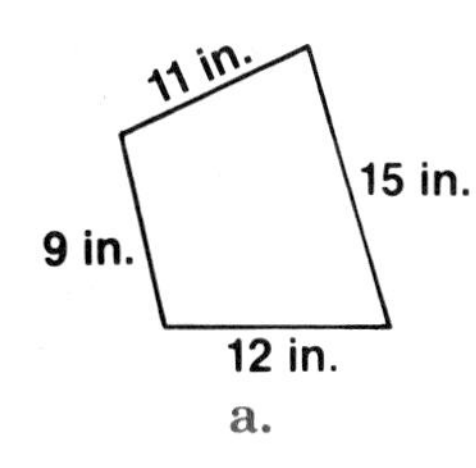

a.

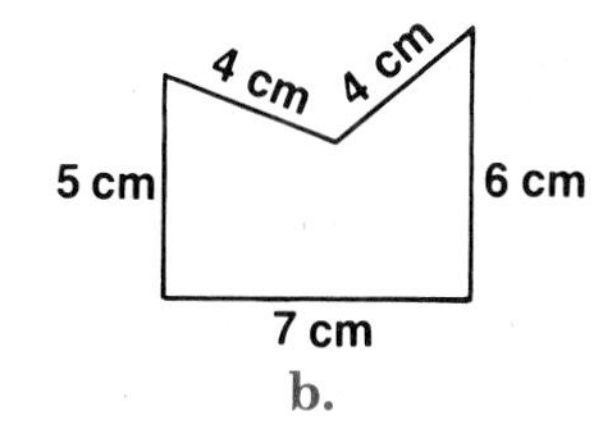

b.

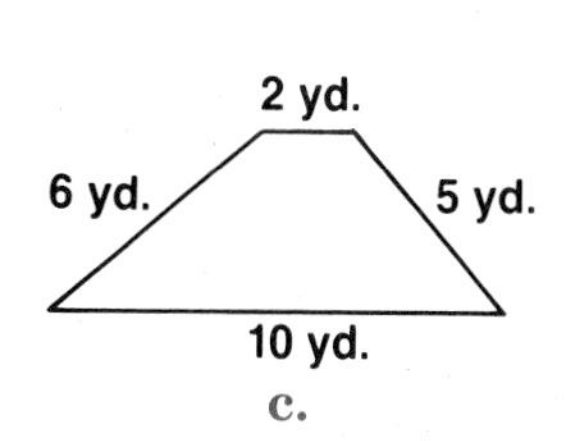

c.

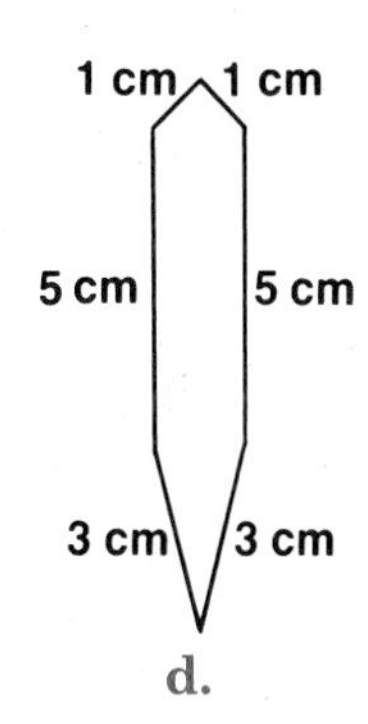

d.

2. Find the perimeter of a triangle whose sides measure:

   a. 15 cm, 20 cm, 25 cm b. 16 in., 30 in., 34 in.

   c. 6.3 m, 9.2 m, 14.9 m d. $3\frac{2}{3}$ yd., $4\frac{1}{8}$ yd., $6\frac{1}{2}$ yd.

3. Find the perimeter of a rectangle whose measurements are:

   a. $\ell = 12$ m, $w = 10$ m b. $\ell = 26$ cm, $w = 14$ cm

   c. $\ell = 9.6$ in., $w = 8.2$ in. d. $\ell = 2\frac{1}{8}$ ft., $w = 1\frac{3}{4}$ ft.

4. Find the perimeter of a square whose side measures:

   a. 11 cm b. 14 in. c. 9.9 ft. d. $7\frac{3}{8}$ yd.

5. Find the perimeter of an equilateral triangle whose side measures:

   a. 5 in. b. 15 in. c. $3\frac{1}{3}$ ft. d. 5.9 cm

**6.** What is the perimeter of an isosceles triangle whose legs each measure 7 inches and whose base measures $5\frac{1}{2}$ inches?

**7.** Find the length of a side of a square whose perimeter is:

**a.** 24 cm **b.** 48 yd. **c.** 14.4 m **d.** 4.48 ft.

**8.** Find the length of a side of an equilateral triangle whose perimeter is:

**a.** 12 in. **b.** 84 cm **c.** 150 ft. **d.** 180 m

## SPIRAL REVIEW EXERCISES

**1.** Which pair of line segments is drawn perpendicular?

(a)

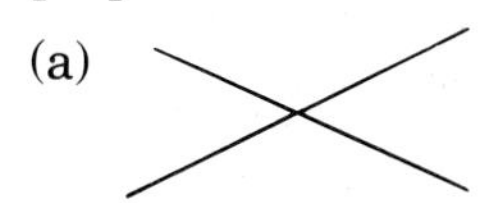

(c)

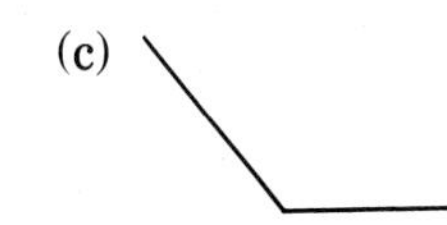

(b)

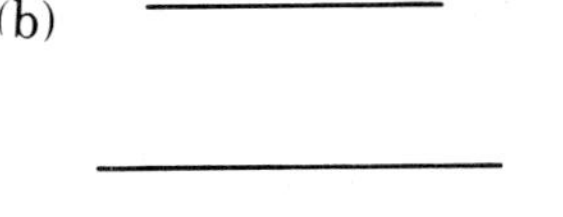

(d)

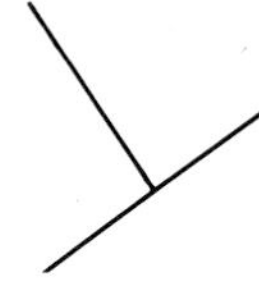

**2.** Find the measure of $\angle B$.

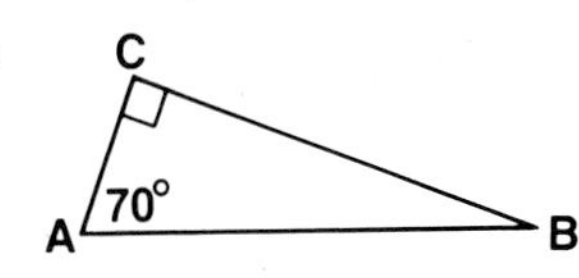


**3.** Find the missing length.

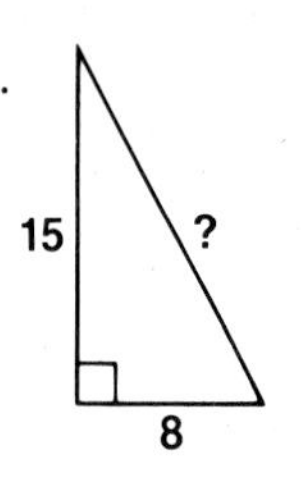


**4.** An angle that measures 98° is
(a) acute (c) obtuse
(b) right (d) straight

**5.** If 44 students out of 200 are girls, what percent are girls?

**6.** If the rates for a telephone call are:
$.75 for the first 5 minutes
$.10 for each additional minute
What is the cost of a 15-minute call?

**7.** A $98 jacket is on sale at 15% off. What is the sale price?

**8.** Jeff began working at 10:45 A.M. and finished at 1:15 P.M. How long did Jeff work?

**9.** 5 liters is equivalent to
(a) 5,000 mL (c) .5 kL
(b) 500 mL (d) .05 kL

**10.** 20 is 40% of
(a) 40 (b) 50 (c) 80 (d) 100

**11.** On the graph shown, which point has the coordinates (3, −4)?

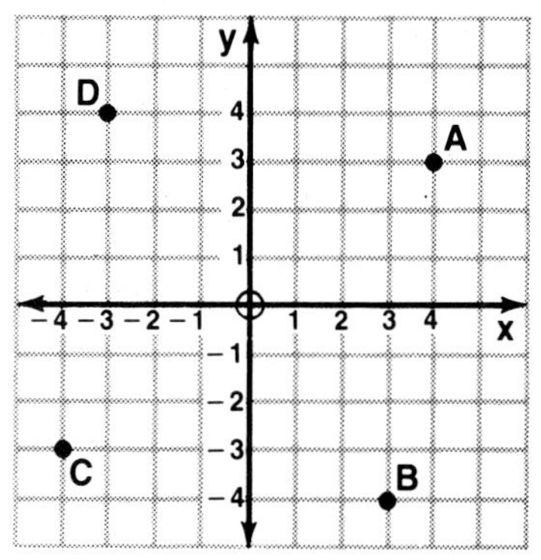

**CHALLENGE QUESTION**

Find the perimeter of a right triangle whose legs measure 15 cm and 20 cm.

# UNIT 20–3 Area: Rectangle and Square

## THE MAIN IDEA

1. *Area* means the measurement of the space contained within a flat, closed surface.

2. Area is measured in *square units*. A square centimeter (written 1 $cm^2$) is an example of a unit of area. It is the space contained in a square that measures 1 centimeter on each side.

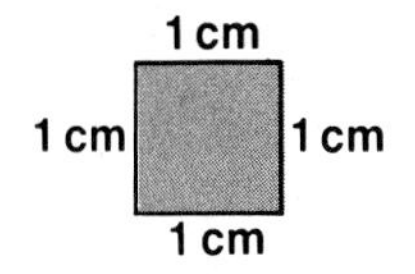


3. Other units of area include the square of any unit used to measure a line. For example: square meter ($m^2$) and square inch (sq. in. or in.$^2$).

4. To find the area of a rectangle, multiply the measure of its length (base) by the measure of its width (height or altitude).

$$A_{\text{rectangle}} = \ell w$$

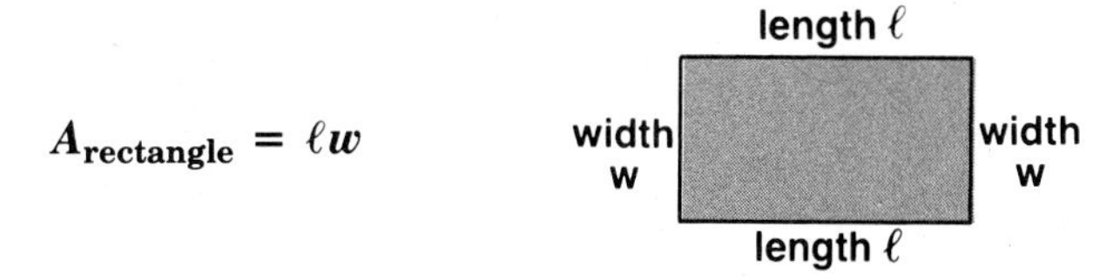


5. A square is a rectangle with four sides that are equal in measure. To find the area of a square, multiply the measure of its side by itself.

$$A_{\text{square}} = s^2$$

side s
side s
side s
side s

**EXAMPLE 1** Find the area of the rectangle.

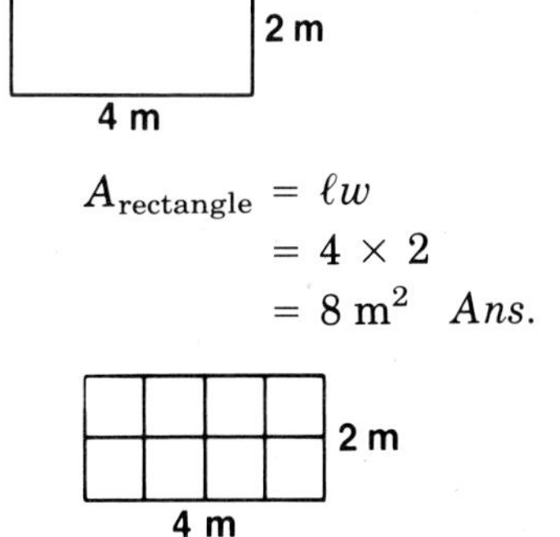


In the formula for the area of a rectangle, substitute 4 for $\ell$ and 2 for $w$.

$$\begin{aligned} A_{\text{rectangle}} &= \ell w \\ &= 4 \times 2 \\ &= 8 \text{ m}^2 \quad \textit{Ans.} \end{aligned}$$

If you count the boxes, you see that there are eight squares contained in the rectangle. Since each side of each square measures 1 meter, there are 8 square meters of area in this rectangle.

**EXAMPLE 2** How many square yards of carpet are needed to cover the floor of a rectangular room that is 6 yards wide and 8 yards long?

In the formula for the area of a rectangle, substitute 8 for $\ell$ and 6 for $w$.

$$\begin{aligned} A_{\text{rectangle}} &= \ell w \\ &= 8 \times 6 \\ &= 48 \text{ sq. yd.} \quad \textit{Ans.} \end{aligned}$$

**EXAMPLE 3** The area of a rectangle whose dimensions are 3 feet by 4 inches is
(a) 12 sq. in. (b) 144 sq. in. (c) 12 sq. ft. (d) 2 sq. ft.

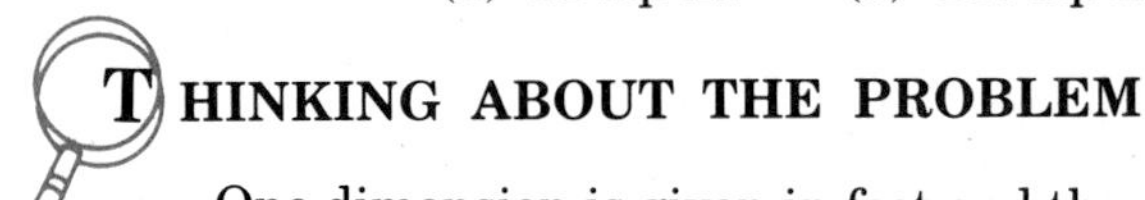

### THINKING ABOUT THE PROBLEM

One dimension is given in feet and the other is given in inches. To find the area, you must first write both dimensions in the same unit.

To change 3 feet to inches, multiply 3 ft. by 12. $3 \text{ ft.} \times 12 = 36 \text{ in.}$

In the formula for the area of a rectangle, substitute 36 in. for $\ell$ and 4 in. for $w$.

$$A_{rectangle} = \ell w$$
$$= 36 \text{ in.} \times 4 \text{ in.}$$
$$= 144 \text{ in.}^2$$

*Answer:* (b)

**EXAMPLE 4** The cost of carpet is $18.50 a square yard. How much will it cost to carpet a rectangular living room that is 3 yards wide and 5 yards long?

Find the number of square yards of carpet that are needed.

$$A_{rectangle} = \ell w$$
$$= 5 \times 3$$
$$= 15 \text{ sq. yd of carpet are needed}$$

To find the cost, multiply the number of square yards of carpet by the price per square yard. $\$18.50 \times 15 = \$277.50$ *Ans.*

**EXAMPLE 5** Find the area of a square each of whose sides measures 7 in.

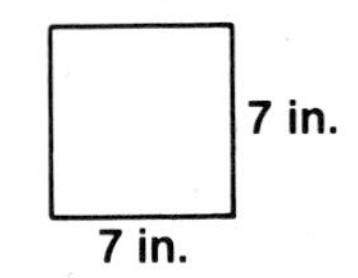


In the formula for the area of a square, substitute 7 for $s$.

$$A_{square} = s^2$$
$$= 7^2 = 7 \times 7 = 49 \text{ in.}^2 \quad \textit{Ans.}$$

**EXAMPLE 6** Find the measure of a side of a square whose area is 81 $cm^2$.

### THINKING ABOUT THE PROBLEM

Work backwards. Since you square the measure of a side to find the area, find the square root of the area to find the measure of a side.

In the formula for the area of a square, substitute 81 for $A$.

$$A_{square} = s^2$$
$$81 = s^2$$
$$\sqrt{81} = s$$
$$9 = s$$

*Answer:* Each side measures 9 cm.

## CLASS EXERCISES

1. Find the area of each rectangle.

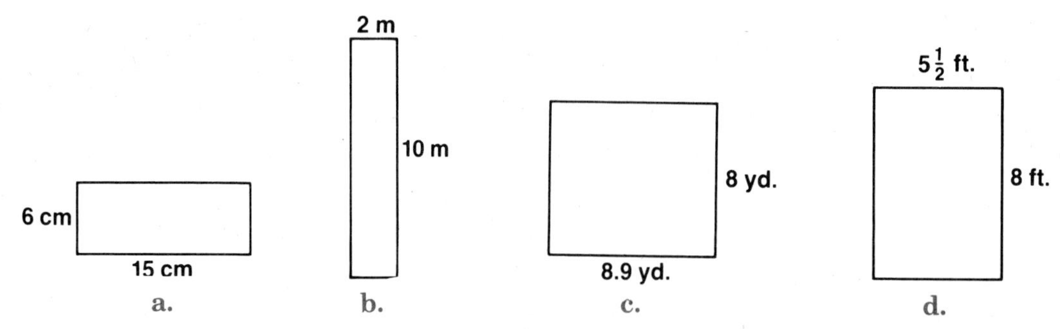

2. Find the area of a rectangle in which:

   a. $\ell = 9$ cm, $w = 4$ cm   b. $\ell = 15$ in., $w = 8$ in.   c. $\ell = 22.4$ m, $w = 8.7$ m

   d. $\ell = 15.9$ cm, $w = 4.2$ cm   e. $\ell = 4\frac{3}{8}$ ft., $w = 2$ ft.   f. $\ell = 10\frac{3}{4}$ in., $w = 5\frac{1}{2}$ in.

3. Find the number of square yards of linoleum needed to cover a kitchen floor that is 4 yards wide and 6 yards long.

4. How many square tiles, each with area of 1 sq. ft., are needed to cover a floor 9 ft. wide and 16 ft. long?

5. The area of a rectangle whose dimensions are 10 feet by 3 yards is
   (a) 30 square feet   (b) 30 square yards   (c) 9 square yards   (d) 90 square feet

6. The area of a rectangle whose dimensions are 25 cm by 1 m is
   (a) 2,500 $cm^2$   (b) 250 $cm^2$   (c) 25 $m^2$   (d) 250 $m^2$

7. The cost of carpet is $22 a square yard. How much will it cost to cover a rectangular floor that is 10 yards wide and 15 yards long?

8. The cost to clean a carpet is $1.15 a square foot. How much will it cost to clean a rectangular carpet 9 ft. by 12 ft.?

9. Find the area of a square each of whose sides measures:

   a. 5 in.   b. 10 cm   c. $\frac{3}{4}$ ft.   d. 9.7 m

10. Find the measure of a side of a square whose area is:

   a. 9 $cm^2$   b. 144 $ft.^2$   c. 36 $m^2$   d. 400 $cm^2$

## HOMEWORK EXERCISES

1. Find the area of each rectangle.

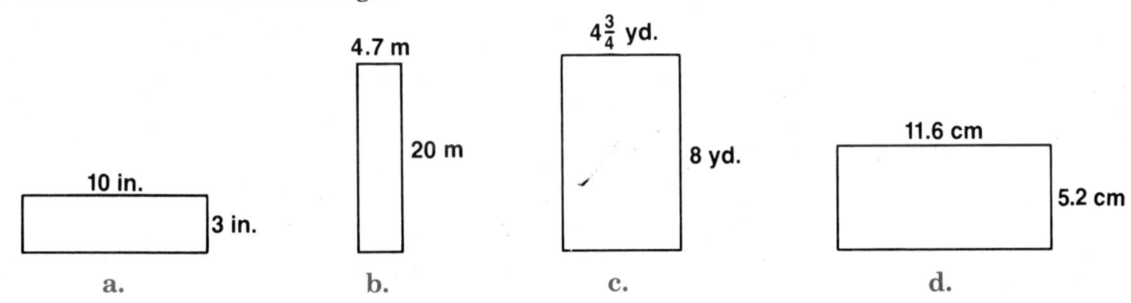


2. Find the area of a rectangle in which:

   a. $\ell = 5$ in., $w = 3$ in.   b. $\ell = 8$ cm, $w = 5$ cm   c. $\ell = 14$ m, $w = 10$ m

   d. $\ell = 6.7$ cm, $w = 3.4$ cm   e. $\ell = 5\frac{1}{2}$ in., $w = 2\frac{1}{4}$ in.   f. $\ell = 10$ ft., $w = 6\frac{3}{4}$ ft.

3. Find the number of square yards of carpet needed to cover a living room floor 3 yards wide and 7 yards long.

4. How many square tiles, each with area 1 sq. in., are needed to cover a countertop 14 inches wide and 30 inches long?

5. The area of a rectangle whose dimensions are 5 feet by 10 inches is
   (a) 50 square inches   (b) 600 square inches   (c) 50 square feet   (d) 60 square inches

6. The area of a rectangle whose dimensions are 2 m by 50 cm is
(a) 100 $cm^2$ (b) 1,000 $cm^2$ (c) 10,000 $cm^2$ (d) 10,000 $m^2$

7. The cost of carpet is $14.80 a square yard. How much will it cost to cover a rectangular floor that is 12 yards long and 8 yards wide?

8. A gardener charges $.30 per square foot to seed a lawn. How much will he charge to seed a rectangular lawn 20 feet by 12 feet?

9. Find the area of a square each of whose sides measures:

**a.** 6 cm **b.** 9 in. **c.** 11 m **d.** 15 yd.

**e.** 8.9 m **f.** 2.7 cm **g.** $\frac{1}{2}$ in. **h.** $3\frac{1}{4}$ ft.

10. Find the measure of a side of a square whose area is:

**a.** 16 $m^2$ **b.** 121 sq. in. **c.** 900 sq. ft. **d.** 64 $cm^2$

## SPIRAL REVIEW EXERCISES

1. Find the perimeter of a square each of whose sides measures 5.8 cm.

2. The measure of a side of an equilateral triangle that has a perimeter of 18 m is
(a) 9 m (b) 6 m (c) 54 m (d) 72 m

3. If the rate of sales tax is 7%, how much tax will be charged on a $20 purchase?

4. 17 meters is equal to
(a) 170 cm (c) 17,000 cm
(b) 1,700 cm (d) .17 cm

5. The value of $3x - 9$ when $x$ is 10 is
(a) 3 (b) 39 (c) 21 (d) 18

6. When written as a fraction, 55% is
(a) $\frac{11}{20}$ (b) $\frac{1}{2}$ (c) $\frac{2}{3}$ (d) $\frac{2}{5}$

7. $\frac{1}{2} + \frac{3}{8} - \frac{1}{8}$ is equal to
(a) $\frac{3}{8}$ (b) $\frac{1}{4}$ (c) $\frac{5}{8}$ (d) $\frac{3}{4}$

8. Tell the coordinates of each vertex of parallelogram *ABCD*.

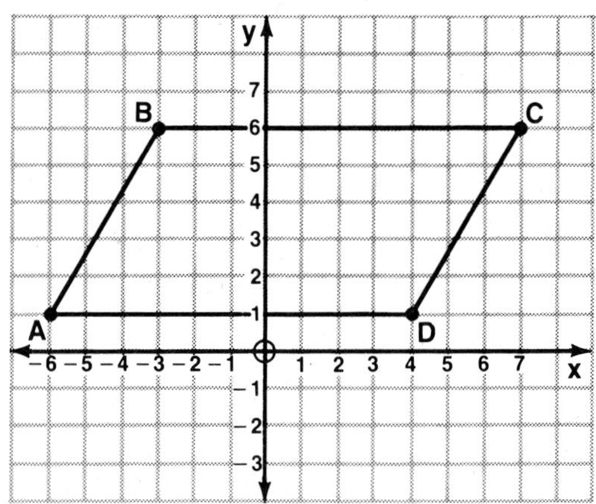


### CHALLENGE QUESTION

The sides of a square are doubled in length. What effect does this have on the area of the square?

# UNIT 20–4 Area: Parallelogram, Triangle, and Trapezoid

## THE AREA OF A PARALLELOGRAM

### THE MAIN IDEA

1. By moving $\triangle$I, a parallelogram can be changed into a rectangle.

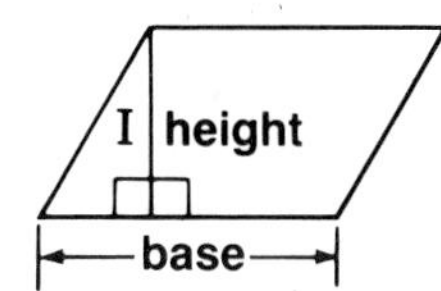


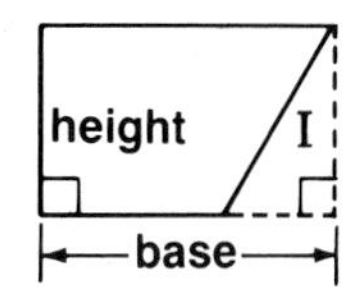


Thus, the area of a parallelogram is equal to the area of the rectangle that has the same dimensions.

2. $A_{\text{parallelogram}} = bh$

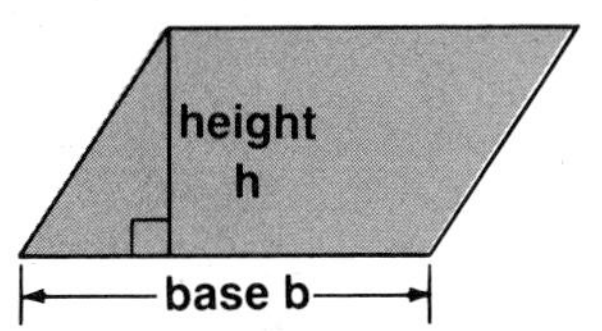


**EXAMPLE 1** Find the area of the parallelogram.

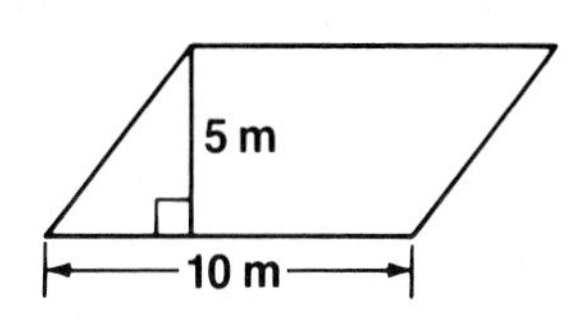


In the formula for the area of a parallelogram, substitute 10 for $b$ and 5 for $h$.

$$\begin{aligned} A_{\text{parallelogram}} &= bh \\ &= 10 \times 5 \\ &= 50 \text{ m}^2 \quad \textit{Ans.} \end{aligned}$$

## CLASS EXERCISES

1. Find the area of each parallelogram.

a. 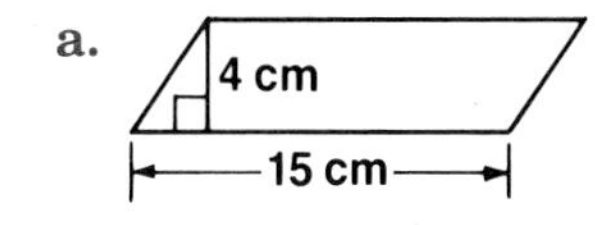


b. 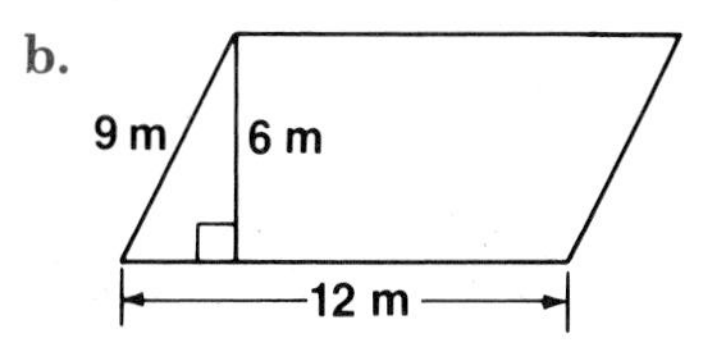


c. 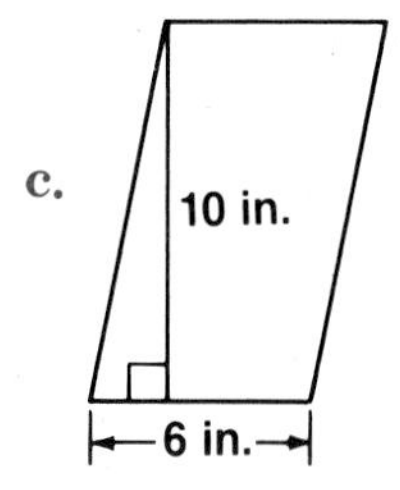


d. 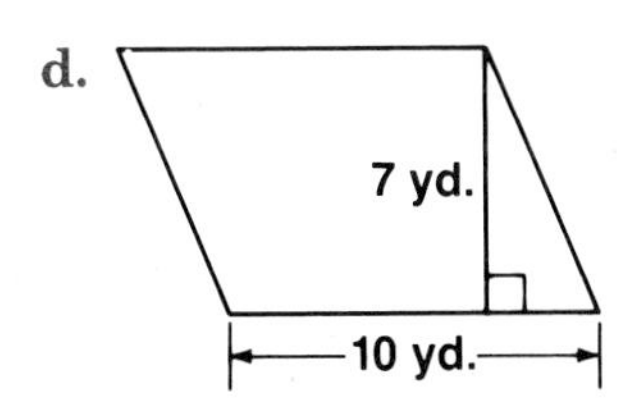


e. 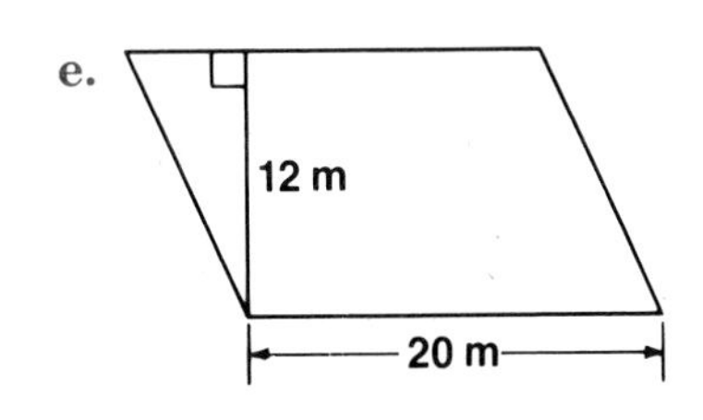


f. 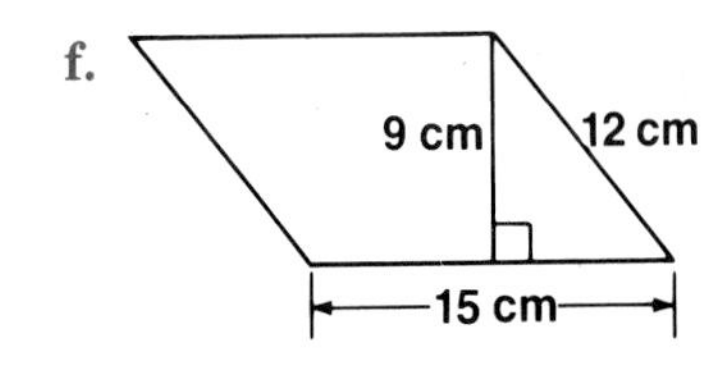

**2.** Find the area of a parallelogram in which:

**a.** $b = 9$ in., $h = 11$ in. **b.** $b = 11.3$ m, $h = 6.4$ m **c.** $b = 12$ ft., $h = 2\frac{1}{3}$ ft.

**d.** $b = 14$ cm, $h = 7$ cm **e.** $b = 20.7$ m, $h = 8.6$ m **f.** $b = 7\frac{5}{8}$ in., $h = 3\frac{1}{4}$ in.

## THE AREA OF A TRIANGLE

### THE MAIN IDEA

1. A ***diagonal*** of a parallelogram divides the parallelogram into two triangles that are equal in area. Thus, the area of one triangle is one-half the area of the parallelogram.

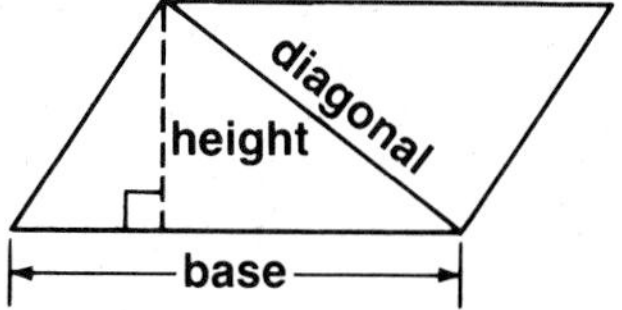

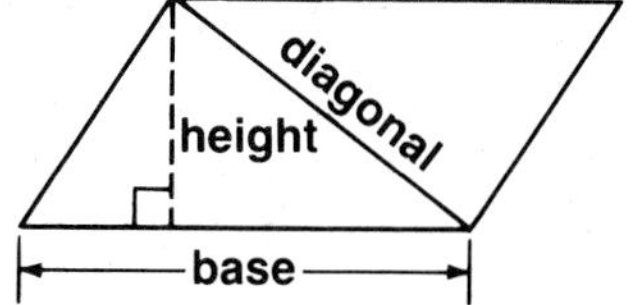

2. $A_{\text{triangle}} = \frac{1}{2}bh$

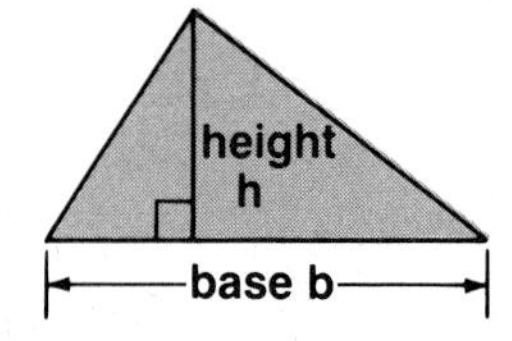


**EXAMPLE 2** Find the area of the triangle.

In the formula for the area of a triangle, substitute 12 for $b$ and 10 for $h$.

$$A_{\text{triangle}} = \frac{1}{2}bh$$
$$= \frac{1}{2} \times 12 \times 10$$
$$= \frac{1}{2} \times 120$$
$$= 60 \text{ cm}^2 \quad \textit{Ans.}$$

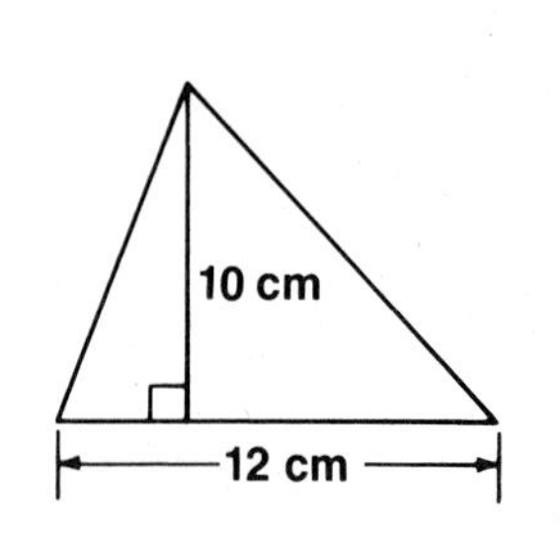


**EXAMPLE 3** Find the area of the triangle.

In an obtuse triangle, the height is measured outside the triangle.

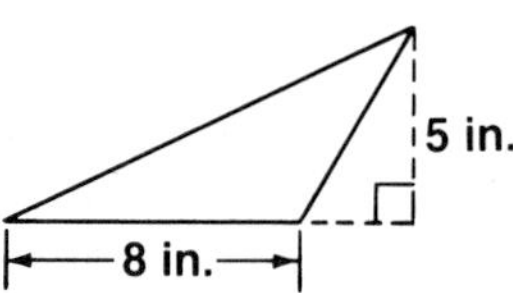


In the formula for the area of a triangle, substitute 8 for $b$ and 5 for $h$.

$$A_{\text{triangle}} = \frac{1}{2}bh$$
$$= \frac{1}{2} \times 8 \times 5$$
$$= \frac{1}{2} \times 40$$
$$= 20 \text{ in.}^2 \quad \textit{Ans.}$$

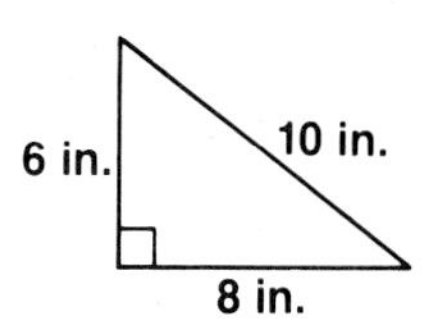


**EXAMPLE 4** Find the area of the right triangle.

In a right triangle, one leg is the base and the other leg is the height: $b = 8, h = 6$.

In the formula for the area of a triangle, substitute 8 for $b$ and 6 for $h$.

$$A_{\text{triangle}} = \frac{1}{2}bh$$
$$= \frac{1}{2} \times 8 \times 6$$
$$= \frac{1}{2} \times 48$$
$$= 24 \text{ in.}^2 \quad \textit{Ans.}$$

## CLASS EXERCISES

**1.** Find the area of each triangle.

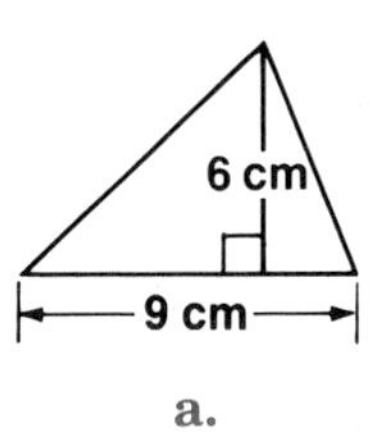


a.

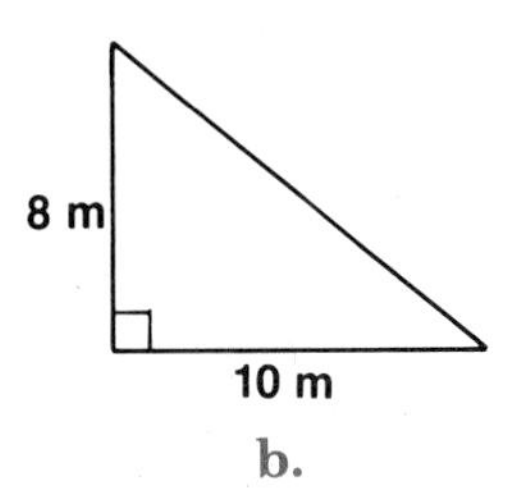


b.

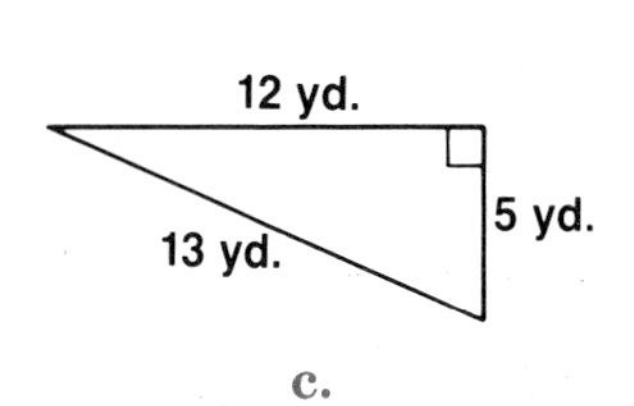


c.

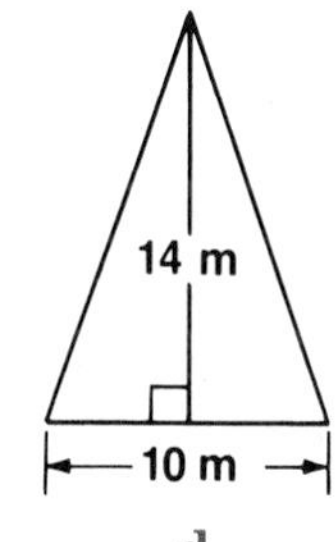


d.

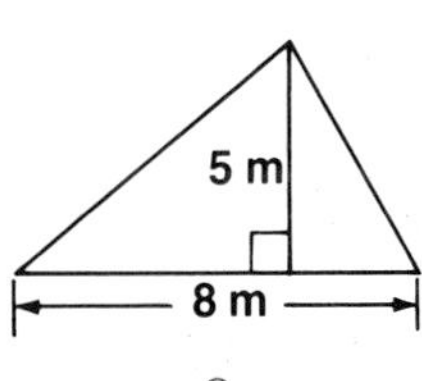


e.

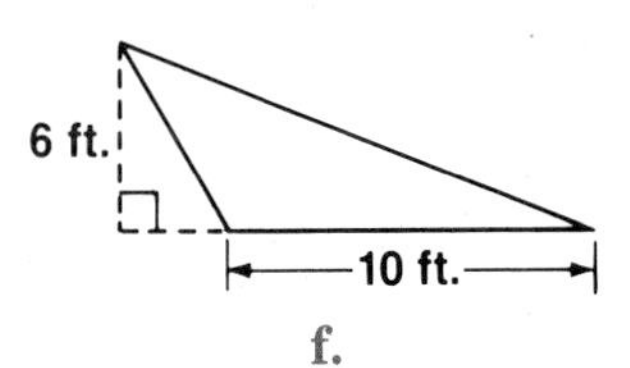


f.

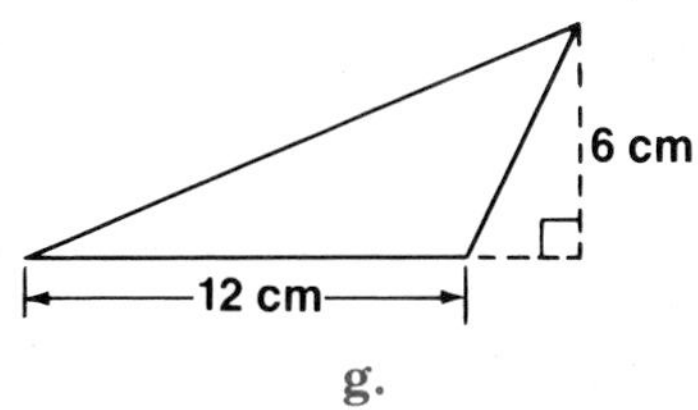


g.

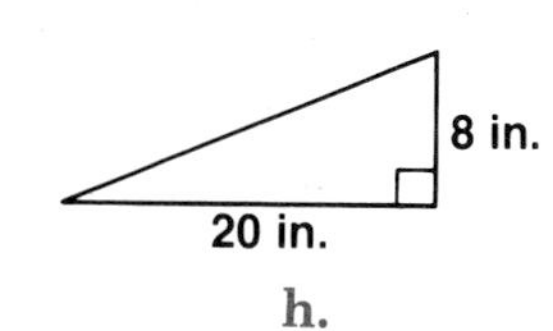


h.

**2.** Find the area of a triangle in which:

**a.** $b = 12$ cm, $h = 10$ cm **b.** $b = 5$ ft., $h = 4$ ft. **c.** $b = 2.5$ m, $h = 1.6$ m

**d.** $b = 5\frac{1}{2}$ ft., $h = 4$ ft. **e.** $b = 4.8$ m, $h = 2.2$ m **f.** $b = 8\frac{1}{4}$ in., $h = 5\frac{1}{2}$ in.

**3.** Find the area of each right triangle, using the given lengths of the legs.

**a.** 8 cm, 5 cm **b.** 8.4 m, 10 m **c.** 6.8 cm, 4.6 cm

## THE AREA OF A TRAPEZOID

### THE MAIN IDEA

1. A diagonal of a trapezoid divides the trapezoid into two triangles. Thus, the area of the trapezoid is the sum of the areas of the triangles.

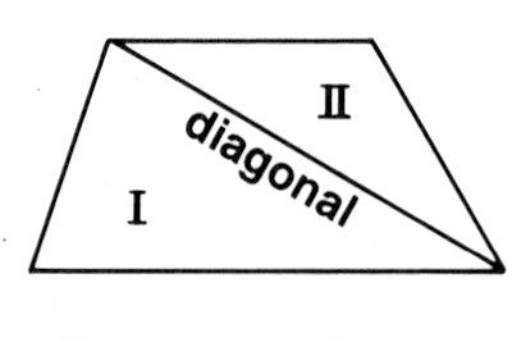


Each triangle contains one base of the trapezoid (the parallel sides). The height of each triangle is the same as the height of the trapezoid.

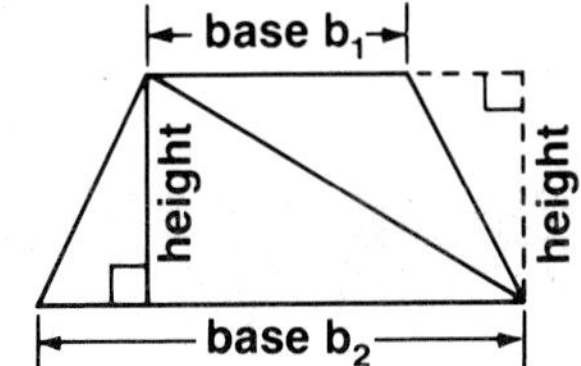


2. $A_{\text{trapezoid}} = \frac{1}{2}h(b_1 + b_2)$

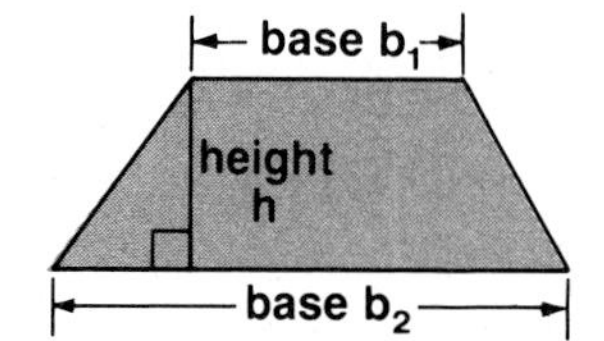


**EXAMPLE 5** Find the area of the given trapezoid.

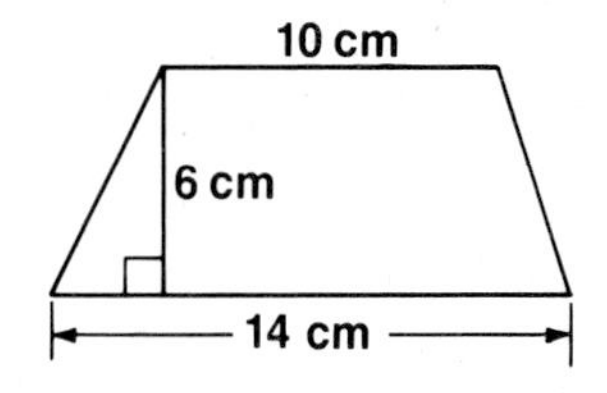


In the formula for the area of a trapezoid, substitute 10 for $b_1$, 14 for $b_2$, and 6 for $h$.

$$A_{\text{trapezoid}} = \frac{1}{2}h(b_1 + b_2)$$
$$= \frac{1}{2} \times 6 \times (10 + 14)$$

Following the order of operations, first add, then multiply.

$$= \frac{1}{2} \times 6 \times (24)$$
$$= 3(24)$$
$$= 72 \text{ cm}^2 \quad \textit{Ans.}$$

**EXAMPLE 6** Find the area of the trapezoid.

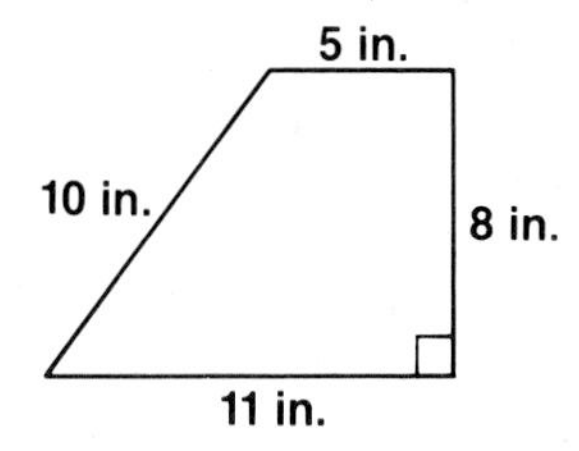


The right angle in the diagram tells that the 8-inch side is the height of the trapezoid.

In the formula for the area of a trapezoid, substitute 8 for $h$, 5 for $b_1$, and 11 for $b_2$. Evaluate, following the order of operations.

$$A_{\text{trapezoid}} = \frac{1}{2}h(b_1 + b_2)$$
$$= \frac{1}{2} \times 8 \times (5 + 11)$$
$$= \frac{1}{2} \times 8 \times (16)$$
$$= 4(16)$$
$$= 64 \text{ in.}^2 \quad \textit{Ans.}$$

## CLASS EXERCISES

1. Find the area of each trapezoid.

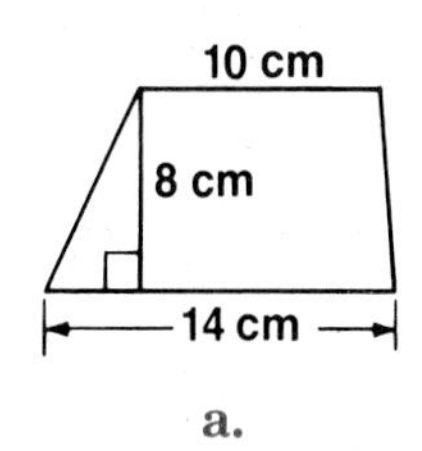


a.

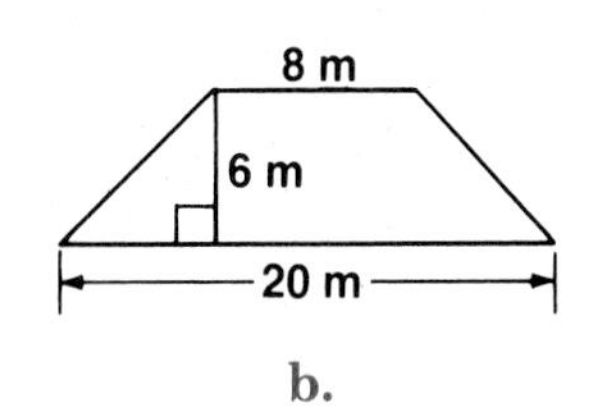


b.

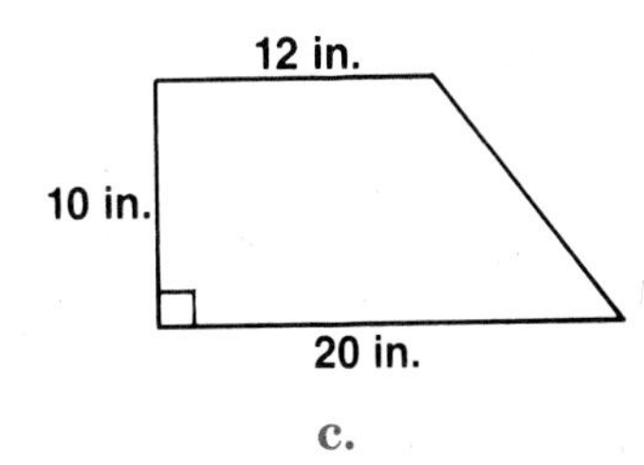


c.

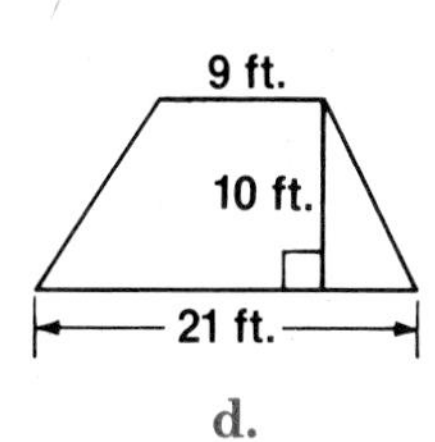


d.

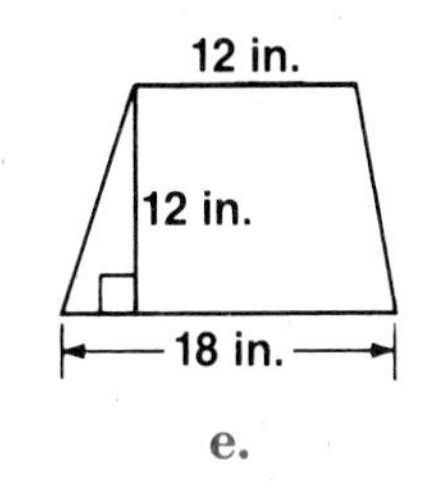


e.

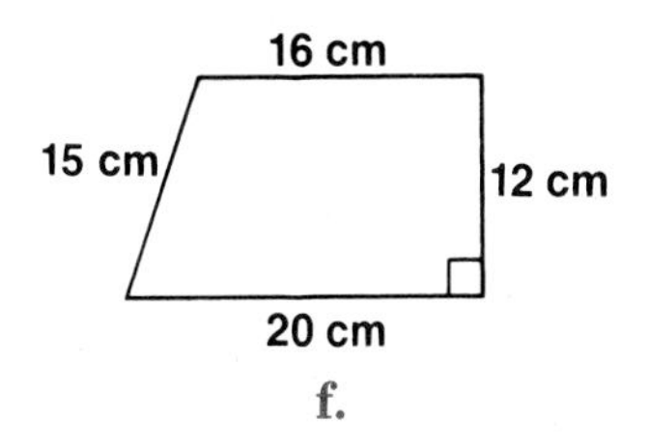


f.

2. Find the area of a trapezoid in which:

**a.** $h = 6$ m, $b_1 = 8$ m, $b_2 = 14$ m **b.** $h = 5$ in., $b_1 = 11$ in., $b_2 = 19$ in.

**c.** $h = 10$ ft., $b_1 = 9$ ft., $b_2 = 11$ ft. **d.** $h = 9$ m, $b_1 = 13$ m, $b_2 = 17$ m

**e.** $h = 10$ cm, $b_1 = 5.2$ cm, $b_2 = 12.8$ cm **f.** $h = 16$ in., $b_1 = 20$ in., $b_2 = 40$ in.

**g.** $h = 5.6$ m, $b_1 = 10.4$ m, $b_2 = 21.6$ m **h.** $h = 2\frac{1}{2}$ ft., $b_1 = 8$ ft., $b_2 = 12$ ft.

## HOMEWORK EXERCISES

1. Find the area of each parallelogram.

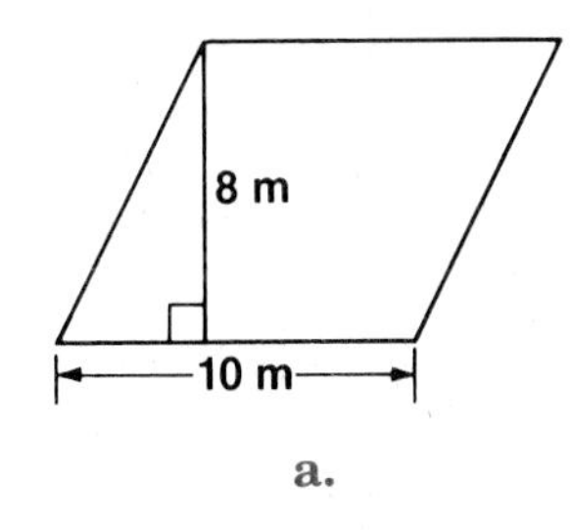


a.

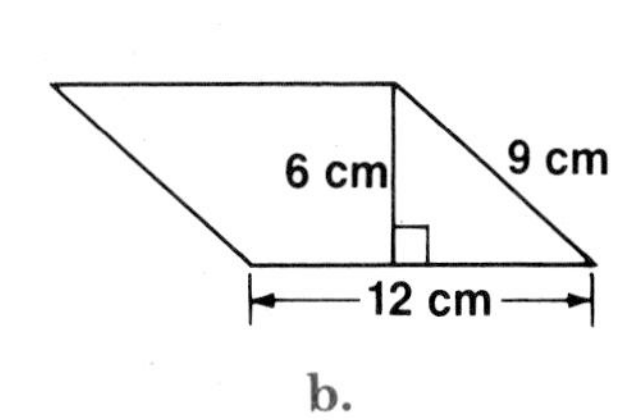


b.

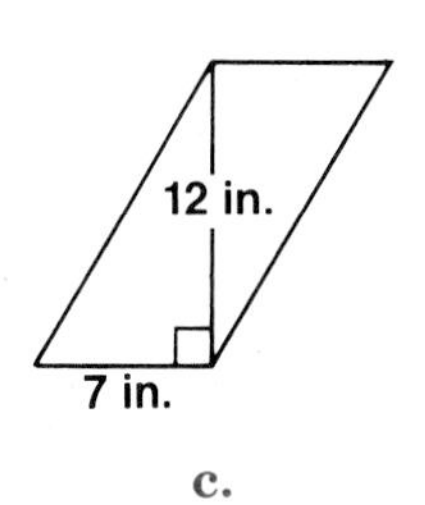


c.

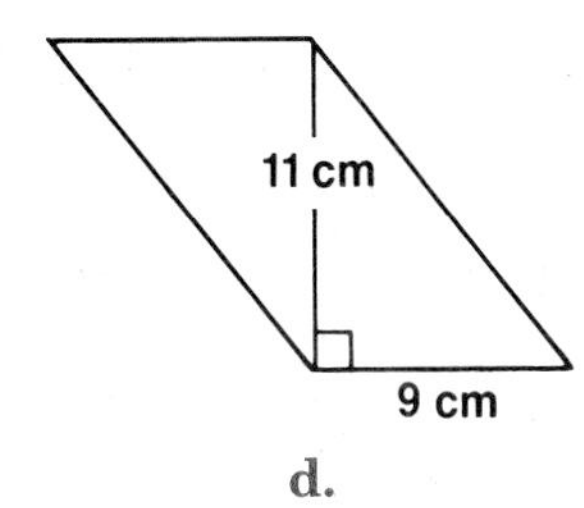


d.

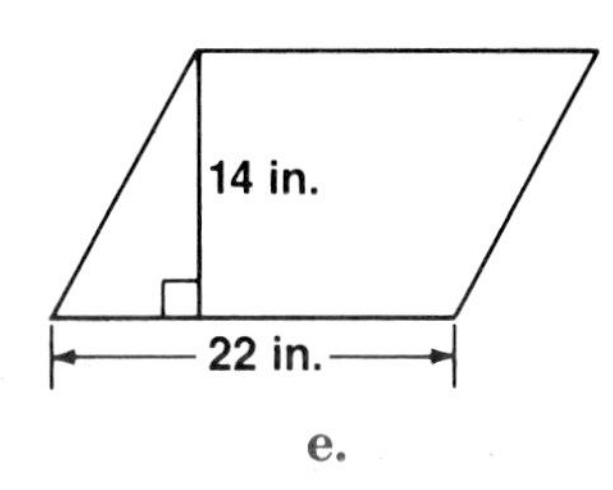


e.

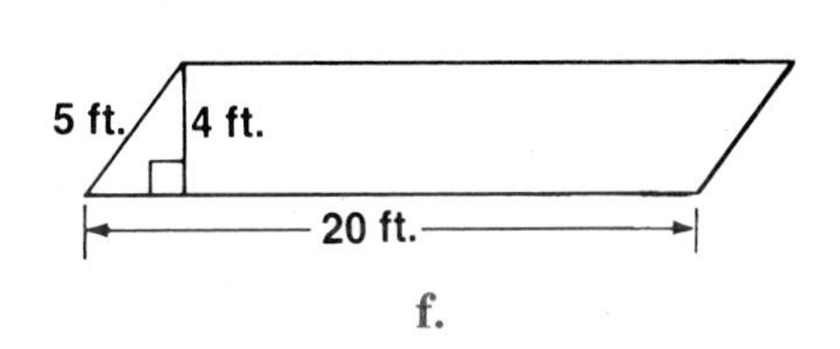


f.

2. Find the area of a parallelogram in which:

**a.** $b = 20$ in., $h = 11$ in. **b.** $b = 45$ cm, $h = 12.9$ cm **c.** $b = 4.8$ m, $h = 2.6$ m

**d.** $b = 24$ yd., $h = 10\frac{1}{2}$ yd. **e.** $b = 5\frac{1}{2}$ in., $h = 2\frac{3}{8}$ in. **f.** $b = 17.5$ cm, $h = 9.2$ cm

**3.** Find the area of each triangle.

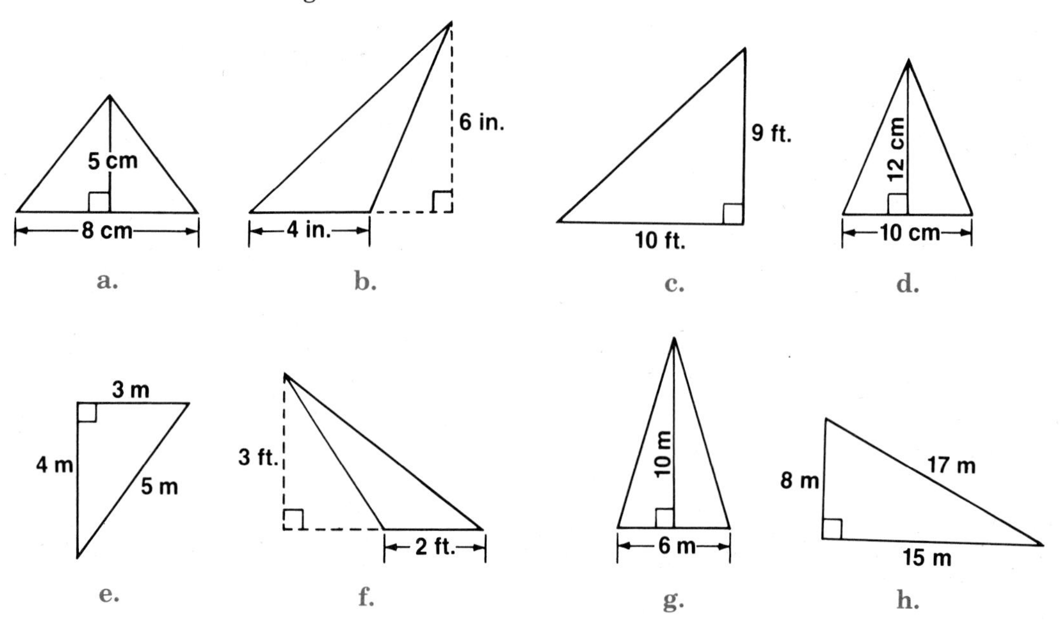


**4.** Find the area of a triangle in which:

**a.** $b = 8$ cm, $h = 6$ cm **b.** $b = 8.7$ cm, $h = 6.5$ cm **c.** $b = 14.2$ m, $h = 10.7$ m

**d.** $b = 18$ in., $h = 6\frac{1}{2}$ in. **e.** $b = 4\frac{1}{2}$ ft., $h = 2\frac{3}{4}$ ft. **f.** $b = 20$ in., $h = 10\frac{1}{4}$ in.

**5.** Find the area of each right triangle, using the given lengths of the legs.

**a.** 5 in., 6 in. **b.** 14 cm, 12 cm **c.** 9.2 m, 8.6 m **d.** 10 in., $4\frac{1}{2}$ in.

**e.** 15.8 cm, 12 cm **f.** $12\frac{1}{2}$ ft., 10 ft.

**6.** Find the area of each trapezoid.

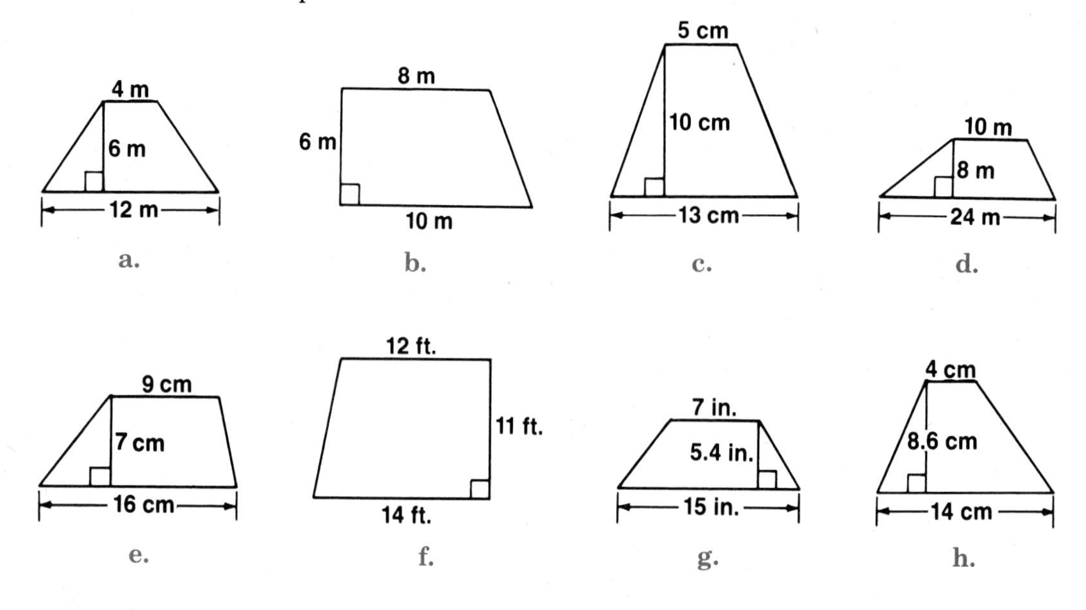

7. Find the area of a trapezoid in which:

a. $h = 4$ cm, $b_1 = 6$ cm, $b_2 = 12$ cm

b. $h = 6$ ft., $b_1 = 9$ ft., $b_2 = 13$ ft.

c. $h = 12$ cm, $b_1 = 18.6$ cm, $b_2 = 21.4$ cm

d. $h = 12$ in., $b_1 = 9\frac{1}{4}$ in., $b_2 = 20\frac{3}{4}$ in.

## SPIRAL REVIEW EXERCISES

1. Find the area of a rectangle whose dimensions are 25 cm by 10.6 cm.

2. Find the area of a square each of whose sides measures 23 inches.

3. The area of a square is 64 $cm^2$.

   a. What is the measure of a side of the square?

   b. Find the perimeter of the square.

4. Mr. Flint charges $3.50 per square foot to tile a floor. How much will he charge to tile a rectangular floor 15 feet by 11 feet?

5. The perimeter of this rectangle is
(a) 19 m (c) 42 m
(b) 38 m (d) 84 m

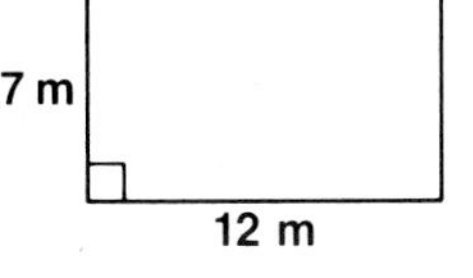


6. If 4 quarts of milk cost $2.88, what is the cost of 7 quarts of milk?

7. Find 80% of 6,000.

8. $\frac{15}{4}$ is equivalent to

   (a) $2\frac{3}{4}$ (b) $3\frac{3}{4}$ (c) $4\frac{1}{4}$ (d) $4\frac{1}{2}$

9. What is the length of the hypotenuse of a right triangle whose legs measure 6 cm and 8 cm?

10. Divide: $\frac{1}{6} \div 12$

11. Write $\frac{20}{42}$ in simplest form.

12. Find the mode of the following group of numbers: 1, 3, 6, 8, 9, 9, 9, 12

13. A box contains 25 balls, of which 7 are black. In a random drawing, what is the probability of picking a black ball?

14. Mr. Owens invests $2,000 at 11% interest. How much interest will he earn at the end of one year?

15. Find the mean of the following group of numbers: 84.2, 86.5, 92, 93.3

## CHALLENGE QUESTION

Find the area of the triangle whose vertices are $(-4, -3)$, $(-4, 5)$, and $(7, -3)$.

# UNIT 20–5 The Circle

## THE CIRCLE AND ITS LINE SEGMENTS

### THE MAIN IDEA

1. A *circle* is a flat, closed curve that has all of its points the same distance from an inside point called the *center*.

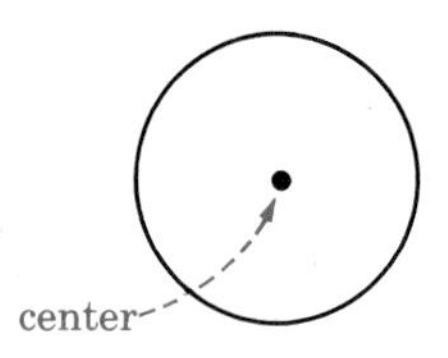


2. A *radius* of a circle is a line segment that has one endpoint at the center of the circle and the other endpoint on the circle. The plural of radius is *radii*.

   In a circle, the measures of all the radii are equal.

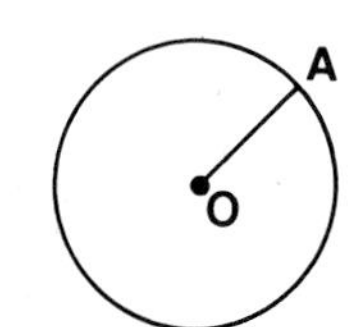


Line segment $OA$ is a radius.

3. A *chord* of a circle is a line segment that has both of its endpoints on the circle. A *diameter* of a circle is a chord that passes through the center of the circle.

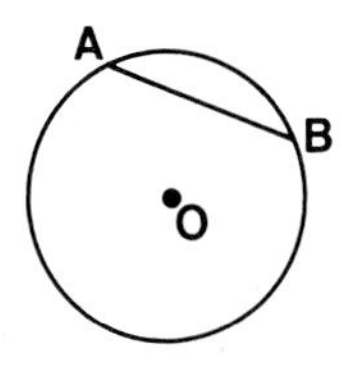


Line segment $AB$ is a chord.

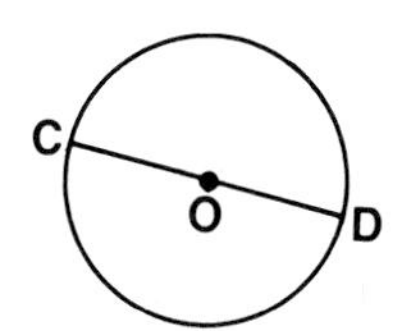


Chord $COD$ is a diameter.

4. A diameter of a circle is twice as long as a radius of that circle:

$$\text{diameter} = 2 \times \text{radius} \quad \text{or} \quad d = 2r$$

A radius of a circle is one-half as long as a diameter of that circle:

$$\text{radius} = \frac{1}{2} \times \text{diameter} \quad \text{or} \quad r = \frac{1}{2}d$$

**EXAMPLE 1** In circle $O$, tell whether each line segment is a radius, a chord, a diameter, or no special line segment.

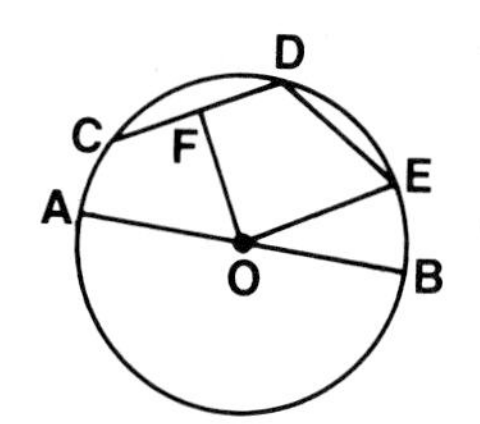


| | *Line Segment* | *Part of Circle* | *Explanation* |
|---|---|---|---|
| **a.** | $AB$ | diameter | connects two points on the circle and goes through the center |
| **b.** | $DE$ | chord | connects two points on the circle |

| | *Line Segment* | *Part of Circle* | *Explanation* |
|---|---|---|---|
| **c.** | *CD* | chord | connects two points on the circle |
| **d.** | *OE* | radius | connects center to a point on the circle |
| **e.** | *OA* | radius | connects center to a point on the circle |
| **f.** | *OF* | no special line segment | *F* is not a point on the circle. |

**EXAMPLE 2** Find the measure of:

**a.** the radius of a circle whose diameter measures 10 cm.

$$\text{radius} = \frac{1}{2} \times \text{diameter}$$
$$= \frac{1}{2} \times 10$$
$$= 5 \text{ cm} \quad \textit{Ans.}$$

**b.** the diameter of a circle whose radius measures 4 in.

$$\text{diameter} = 2 \times \text{radius}$$
$$= 2 \times 4$$
$$= 8 \text{ in.} \quad \textit{Ans.}$$

## CLASS EXERCISES

**1.** Tell whether each of the line segments is a radius, a chord, a diameter, or no special line segment.

**a.** *AB* **b.** *CD* **c.** *EF* **d.** *OF* **e.** *OG* **f.** *CH*

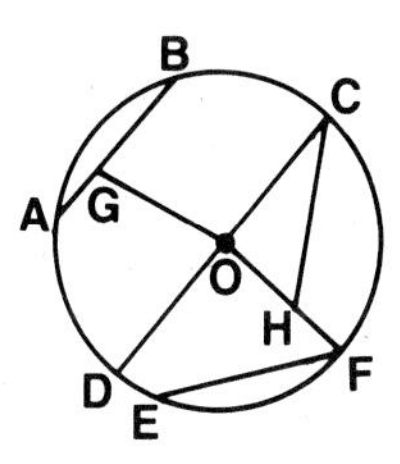


**2. a.** Explain why chord *CD* is not a diameter.

**b.** Explain why line segment *OA* is not a radius.

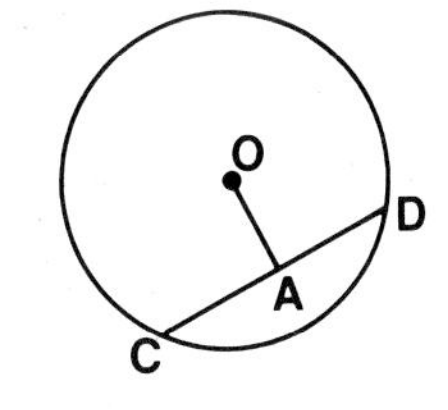


**3.** Find the measure of the radius of a circle whose diameter measures:

**a.** 6 cm **b.** 18 in. **c.** 12 m **d.** $\frac{3}{4}$ in. **e.** 5 yd. **f.** 3.7 in.

**4.** Find the measure of the diameter of a circle whose radius measures:

**a.** 3 in. **b.** 9 cm **c.** 11 m **d.** $\frac{5}{8}$ in. **e.** 2.8 yd. **f.** $3\frac{1}{2}$ cm

**5.** What are the measures of the radius and the diameter of each circle?

a. 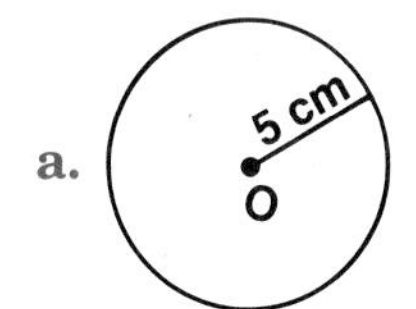


b. 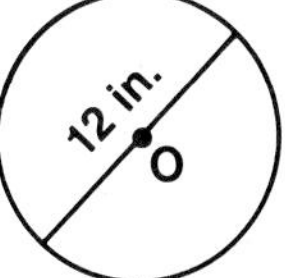


c. 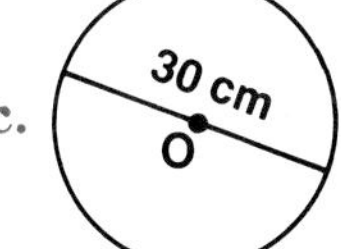


d. 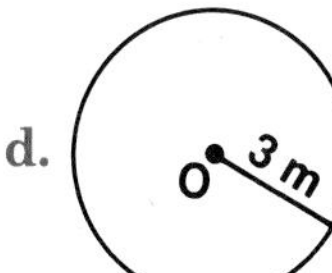

## THE CIRCUMFERENCE OF A CIRCLE

### THE MAIN IDEA

1. The ***circumference*** of a circle is the distance around the circle.
2. The ratio of the circumference of a circle to the diameter of that circle is the same for all circles. The Greek letter $\pi$ (pi) is used to represent this number:

$$\frac{C \leftarrow \text{circumference}}{d \leftarrow \text{diameter}} = \pi \quad \text{or} \quad C = \pi d$$

3. The number $\pi$ is approximately equal to 3.14 or $\frac{22}{7}$.
4. To find the circumference of a circle:
   a. Use the formula $C = \pi d$ when you are given the measure of the diameter.
   b. Use the formula $C = 2\pi r$ when you are given the measure of the radius.

**EXAMPLE 3** Using $\pi = 3.14$, find the circumference of a circle if:

**a.** the diameter measures 40 in.

$$\begin{aligned} C &= \pi d \\ &= 3.14 \times 40 \\ &= 125.6 \text{ in.} \quad \textit{Ans.} \end{aligned}$$

**b.** the radius measures 10 cm

$$\begin{aligned} C &= 2\pi r \\ &= 2 \times 3.14 \times 10 \\ &= 62.8 \text{ cm} \quad \textit{Ans.} \end{aligned}$$

**EXAMPLE 4** Using $\pi = \frac{22}{7}$, find the circumference of each circle.

a.

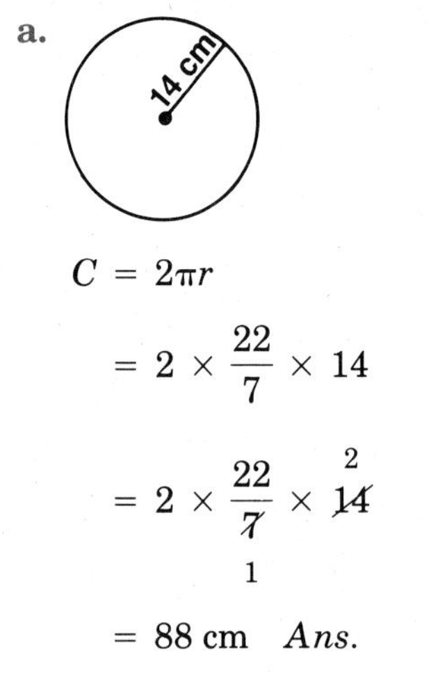


$$\begin{aligned} C &= 2\pi r \\ &= 2 \times \frac{22}{7} \times 14 \\ &= 2 \times \frac{22}{\cancel{7}_1} \times \cancel{14}^{2} \\ &= 88 \text{ cm} \quad \textit{Ans.} \end{aligned}$$

b.

70 in.

$$\begin{aligned} C &= \pi d \\ &= \frac{22}{7} \times 70 \\ &= \frac{22}{\cancel{7}_1} \times \cancel{70}^{10} \\ &= 220 \text{ in.} \quad \textit{Ans.} \end{aligned}$$

**EXAMPLE 5** Using $\pi = 3.14$, find the measure of the diameter of a circle whose circumference is 15.70 cm.

Using $C = \pi d$, you can find the measure of the diameter by dividing the circumference by $\pi$.

$$d = \frac{C}{\pi}$$

$$d = \frac{15.70}{3.14}$$

$$\begin{array}{r} 5 \\ 3.14\overline{)15.70.} \\ \underline{15\ 70} \\ 0 \end{array}$$

$d = 5$ cm *Ans.*

## CLASS EXERCISES

**1.** Using $\pi = 3.14$, find the circumference of a circle if the length of the radius is:

**a.** 20 cm **b.** 15 m **c.** 8.6 in. **d.** 30.5 ft.

**2.** Using $\pi = \frac{22}{7}$, find the circumference of a circle if the length of the radius is:

**a.** 14 in. **b.** 35 m **c.** $3\frac{1}{2}$ ft. **d.** $1\frac{3}{11}$ yd.

**3.** Using $\pi = 3.14$, find the circumference of a circle if the length of the diameter is:

**a.** 10 cm **b.** 22 in. **c.** 9.8 m **d.** 24.5 yd.

**4.** Using $\pi = \frac{22}{7}$, find the circumference of a circle if the length of the diameter is:

**a.** 21 in. **b.** 49 cm **c.** $1\frac{3}{4}$ ft. **d.** $2\frac{1}{3}$ yd.

**5.** Using $\pi = 3.14$, find the circumference of each circle.

a. 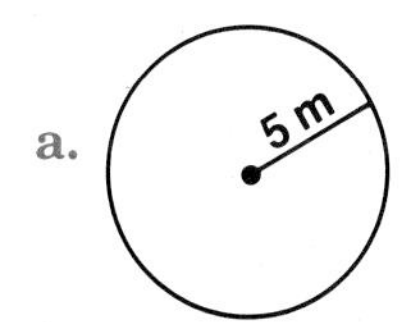


b. 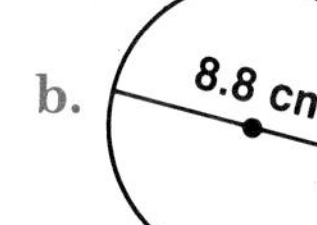


c. 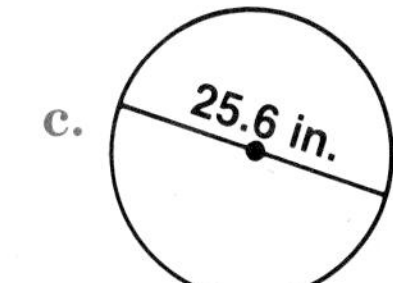


d. 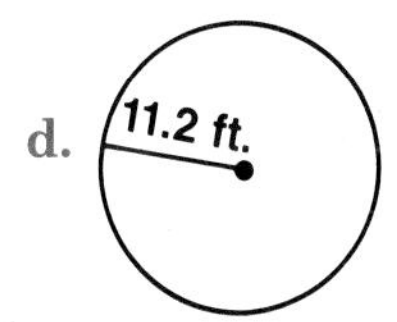


**6.** Using $\pi = 3.14$, find the measure of the diameter of a circle whose circumference is 31.4 m.

**7.** Using $\pi = \frac{22}{7}$, find the measure of the diameter of a circle whose circumference is 66 cm.

**8.** A circular swimming pool has a radius of length 4.2 m. Find the circumference of the pool. (Use $\pi = \frac{22}{7}$.)

## THE AREA OF A CIRCLE

### THE MAIN IDEA

1. As with any flat, closed surface, the area of a circle is the measure of the space contained inside.

2. $A_{\text{circle}} = \pi r^2$

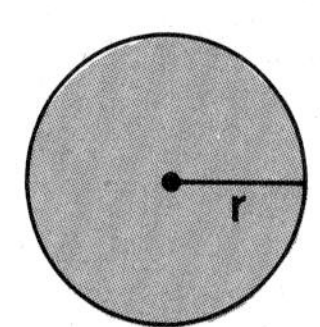


**EXAMPLE 6** Using $\pi = 3.14$, find the area of a circle whose radius measures 5 cm.

In the formula for the area of a circle, substitute 3.14 for $\pi$ and 5 for $r$.

$$A_{\text{circle}} = \pi r^2$$
$$= 3.14 \times 5^2$$

Evaluate, following the order of operations.

$$= 3.14 \times 5 \times 5$$
$$= 3.14 \times 25$$
$$= 78.5 \text{ cm}^2 \quad \textit{Ans.}$$

**EXAMPLE 7** Using $\pi = 3.14$, find the area of a circle whose diameter measures 8 in.

Since you are given the diameter, divide by 2 to find the measure of the radius.

In the formula for the area of a circle, substitute 3.14 for $\pi$ and 4 for $r$.

$$A_{\text{circle}} = \pi r^2$$
$$= 3.14 \times 4^2$$

Evaluate, following the order of operations.

$$= 3.14 \times 4 \times 4$$
$$= 3.14 \times 16$$
$$= 50.24 \text{ m}^2 \quad \textit{Ans.}$$

**EXAMPLE 8** Using $\pi = \frac{22}{7}$, find the area of a circle whose radius measures 7 ft.

In the formula for the area of a circle, substitute $\frac{22}{7}$ for $\pi$ and 7 for $r$.

$$A_{\text{circle}} = \pi r^2$$
$$= \frac{22}{7} \times 7^2$$
$$= \frac{22}{\cancel{7}_1} \times \overset{1}{\cancel{7}} \times 7$$
$$= 154 \text{ ft.}^2 \quad \textit{Ans.}$$

## CLASS EXERCISES

1. Using $\pi = 3.14$, find the area of a circle whose radius is:

   **a.** 10 in. **b.** 4 cm **c.** 8 yd. **d.** 2.3 m

2. Using $\pi = \frac{22}{7}$, find the area of a circle whose radius is:

   **a.** 14 cm **b.** 28 in. **c.** 2.1 m **d.** $3\frac{1}{2}$ ft.

3. Using $\pi = 3.14$, find the area of a circle whose diameter is:

   **a.** 2 cm **b.** 8 in. **c.** 20 m **d.** 9.6 ft.

4. Using $\pi = \frac{22}{7}$, find the area of a circle whose diameter is:

   **a.** 14 in. **b.** 42 cm **c.** 2.8 m **d.** 7 yd.

5. Using $\pi = 3.14$, find the area of each circle.

   a. 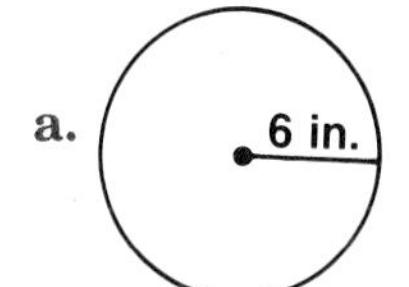


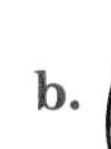

   b. 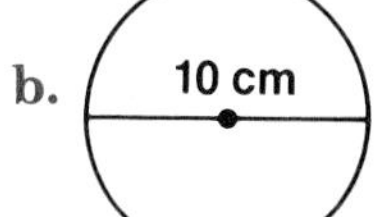


   c. 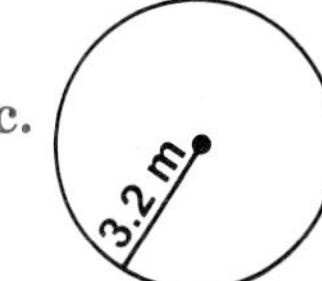


   d. 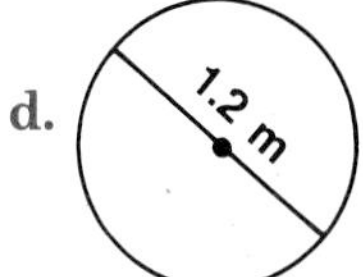


## HOMEWORK EXERCISES

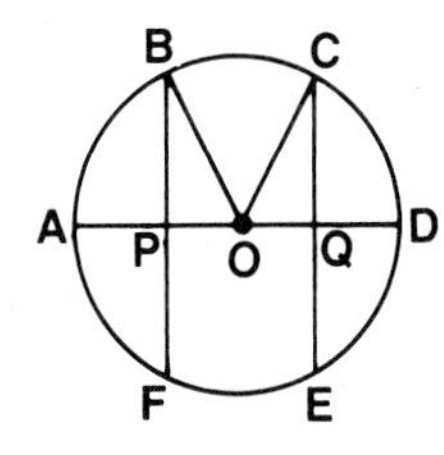


1. Tell whether each of the line segments is a radius, a chord, a diameter, or no special line segment.

   **a.** $AD$ **b.** $BF$ **c.** $OB$ **d.** $OC$ **e.** $CE$ **f.** $CQ$

2. **a.** Explain why chord $AC$ is not a diameter.

   **b.** Explain why line segment $OP$ is not a radius.

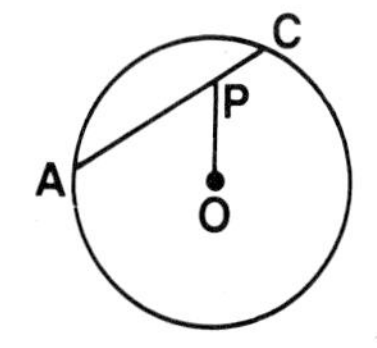


3. Find the measure of the radius of a circle whose diameter measures:

   **a.** 4 in. **b.** 12 m **c.** 28 cm **d.** 8.2 in. **e.** $\frac{5}{16}$ in. **f.** 9.6 yd.

   **g.** 14.7 ft. **h.** $4\frac{1}{3}$ in.

4. Find the measure of the diameter of a circle whose radius measures:

   **a.** 5 cm **b.** 11 in. **c.** 14.7 m **d.** $\frac{3}{4}$ ft. **e.** 2.95 m **f.** $4\frac{5}{8}$ in.

   **g.** 7.5 yd. **h.** $9\frac{1}{4}$ ft.

**5.** What are the measures of the radius and the diameter of each circle?

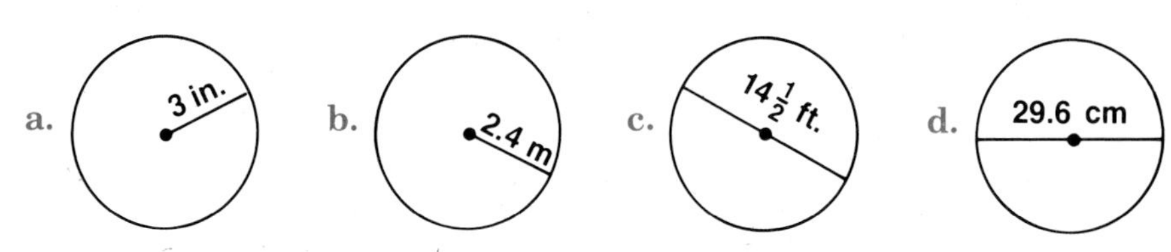


**6.** Using $\pi = 3.14$, find the circumference of a circle if the length of the radius is:

**a.** 30 cm **b.** 55 in. **c.** 9.4 m **d.** 28.3 ft.

**7.** Using $\pi = \frac{22}{7}$, find the circumference of a circle if the length of the radius is:

**a.** 28 cm **b.** 42 in. **c.** $2\frac{1}{3}$ ft. **d.** $4\frac{3}{8}$ yd.

**8.** Using $\pi = 3.14$, find the circumference of a circle if the length of the diameter is:

**a.** 50 in. **b.** 100 m **c.** 10.2 cm **d.** 9.8 yd.

**9.** Using $\pi = \frac{22}{7}$, find the circumference of a circle if the length of the diameter is:

**a.** 7 cm **b.** 56 in. **c.** $5\frac{1}{4}$ yd. **d.** $8\frac{3}{4}$ ft.

**10.** Using $\pi = 3.14$, find the circumference of each circle.

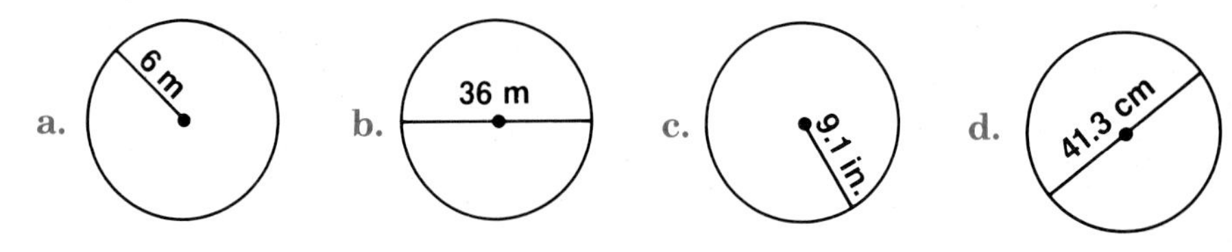


**11.** Using $\pi = 3.14$, find the measure of the diameter of a circle whose circumference is 628 cm.

**12.** Using $\pi = \frac{22}{7}$, find the measure of the diameter of a circle whose circumference is 110 m.

**13.** A circular driveway has a diameter of length 50.6 m. Find the circumference of the driveway. (Use $\pi = 3.14$.)

**14.** Using $\pi = 3.14$, find the area of a circle whose radius is:

**a.** 3 cm **b.** 9 in. **c.** 10 m **d.** 9 ft. **e.** 8.2 cm

**15.** Using $\pi = \frac{22}{7}$, find the area of a circle whose radius is:

**a.** 7 in. **b.** 14 m **c.** 2.8 cm **d.** 7.7 m **e.** $2\frac{1}{3}$ ft.

**16.** Using $\pi = 3.14$, find the area of a circle whose diameter is:

**a.** 4 in. **b.** 12 ft. **c.** 40 in. **d.** 1.8 cm **e.** 5.8 m

**17.** Using $\pi = \frac{22}{7}$, find the area of a circle whose diameter is:

**a.** 28 cm **b.** 70 in. **c.** 4.2 m **d.** 9.8 m **e.** .14 cm

**18.** Using $\pi = 3.14$, find the area of each circle.

a. 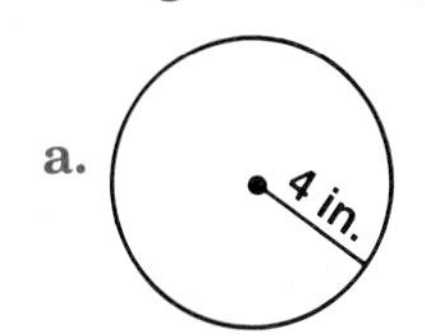


b. 14 ft.

c. 20.8 m

d. 9.4 ft.

**19.** A circular swimming pool has a diameter that measures 28 meters. Find the area of the pool. (Use $\pi = \frac{22}{7}$.)

## SPIRAL REVIEW EXERCISES

**1. a.** Find the perimeter of the rectangle.

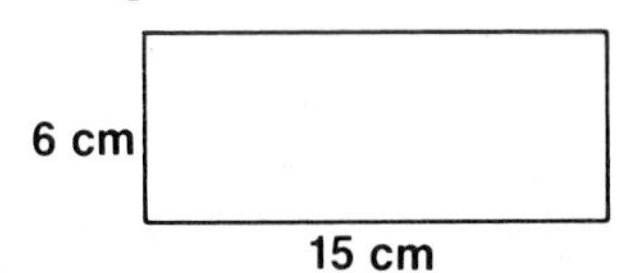


**b.** Find the area of the rectangle.

**2.** What is the length of a side of a square whose perimeter is 50.8 meters?

**3.** A square has an area of 144 square inches. What is the measure of each side?
(a) 18 inches (c) 24 inches
(b) 3 feet (d) 1 foot

**4.** Find the area of each polygon.

a. 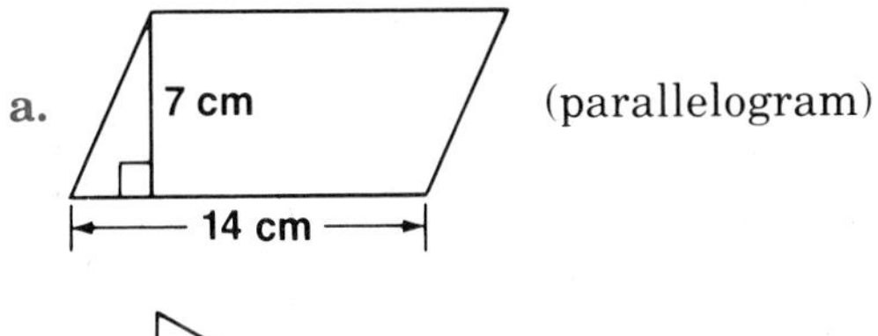


b. 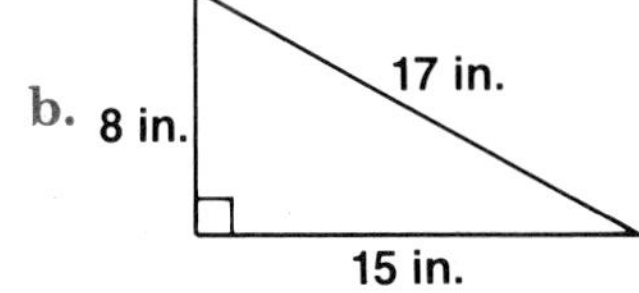


**5.** On the graph, which point has the coordinates $(-2, -4)$?
(a) $Q$ 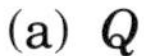 (b) $R$ 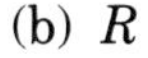 (c) $S$ (d) $T$ 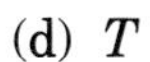

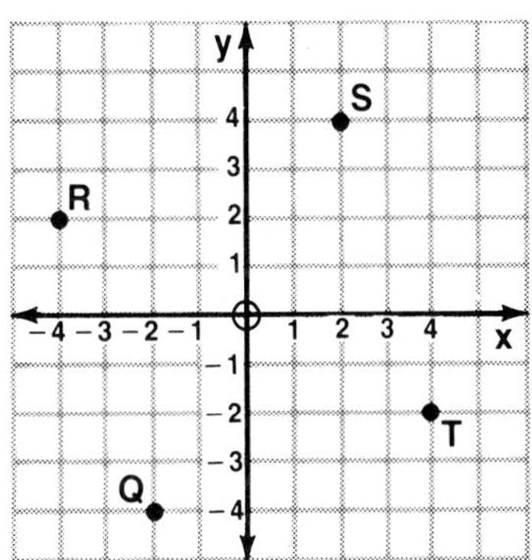


**6.** Which open sentence is represented by the graph?

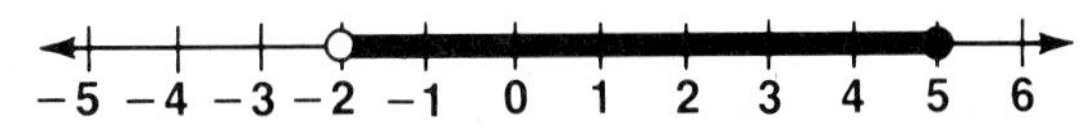


(a) $x > -2$
(b) $x \leq 5$
(c) $-2 \leq x \leq 5$
(d) $-2 < x \leq 5$

### CHALLENGE QUESTION

In the diagram, the radius of the circle measures 6 cm. Find the perimeter of the square.

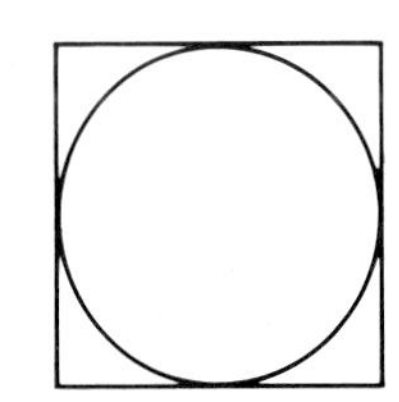

# UNIT 20–6 Volume

## THE VOLUME OF A RECTANGULAR SOLID AND OF A CUBE

### THE MAIN IDEA

1. *Volume* means the measurement of the amount of space contained within a solid figure.
2. Volume is measured in *cubic units*. A cubic centimeter (written 1 $cm^3$) is an example of a unit of volume. It is the space contained in a cube that measures 1 centimeter on each side.

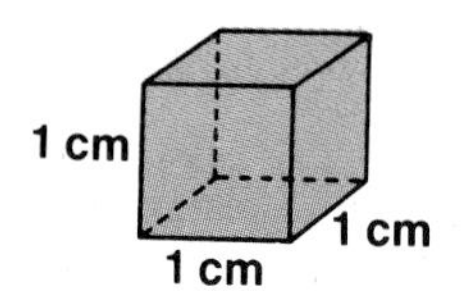


3. Other units of volume include the cube of any unit used to measure a line. For example: cubic meter ($m^3$) and cubic inch (cu. in. or in.$^3$).
4. To find the volume of a *rectangular solid,* multiply the measure of its length by the measure of its width and then multiply by the measure of its height.

$$V_{\text{rectangular solid}} = \ell wh$$

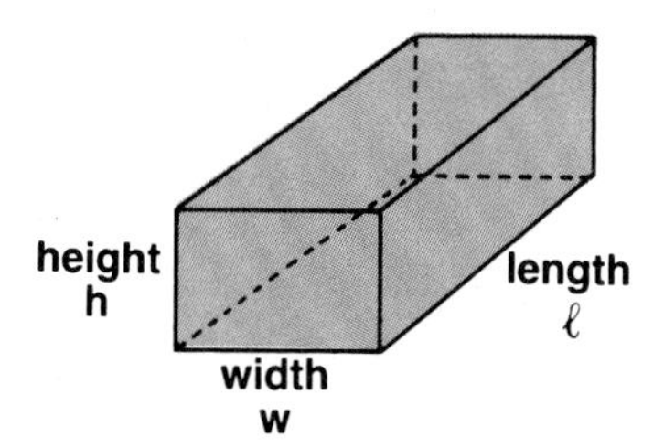


5. A *cube* is a rectangular solid that has all 12 edges equal in measure. To find the volume of a cube, use the measure of an edge as a factor three times.

$$V_{\text{cube}} = e^3$$

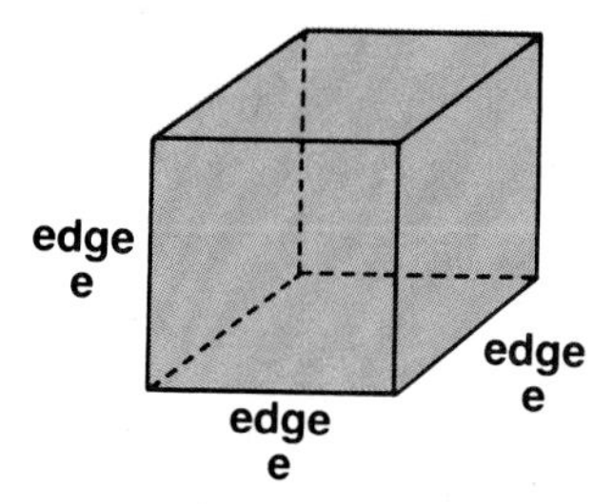


**EXAMPLE 1** Find the volume of the rectangular solid.

In the formula for the volume of a rectangular solid, substitute 8 for $\ell$, 5 for $w$, and 3 for $h$.

$$\begin{aligned} V_{\text{rectangular solid}} &= \ell wh \\ &= 8 \times 5 \times 3 \\ &= 120 \text{ cm}^3 \quad \textit{Ans.} \end{aligned}$$

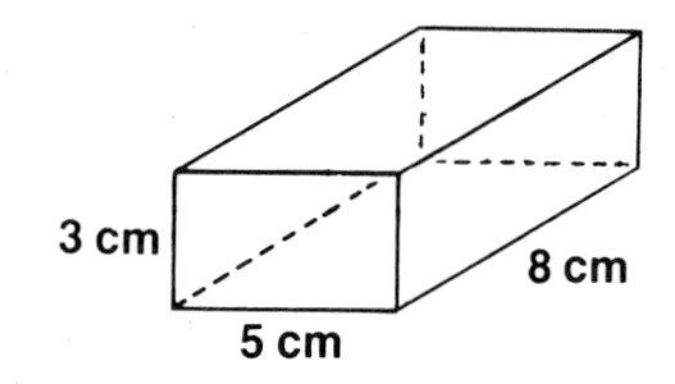

**EXAMPLE 2** Find the volume of the cube.

In the formula for the volume of a cube, substitute 2 for $e$.

$$V_{\text{cube}} = e^3$$
$$= 2^3$$
$$= 8 \text{ m}^3 \quad \textit{Ans.}$$

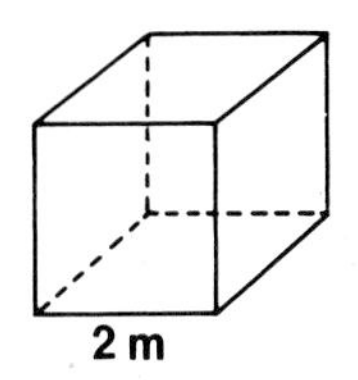


## CLASS EXERCISES

**1.** Find the volume of each rectangular solid.

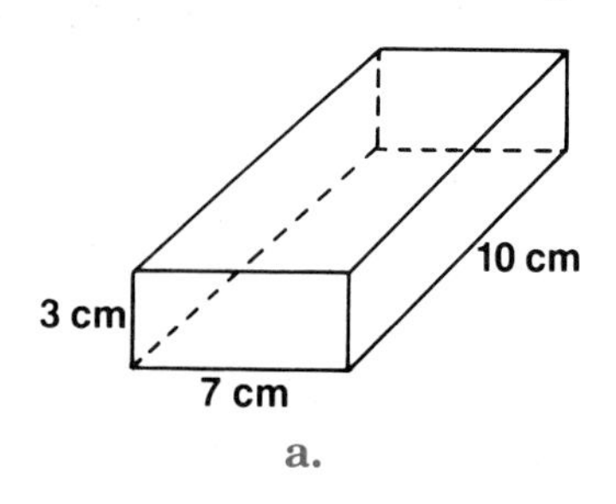


a.

12 in.
8 in.
4 in.

b.

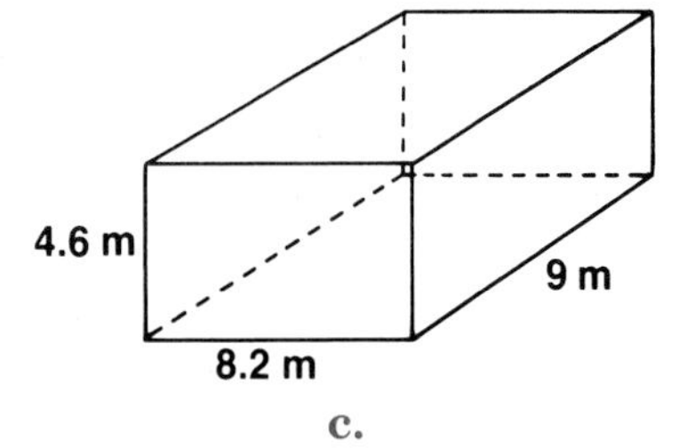


c.

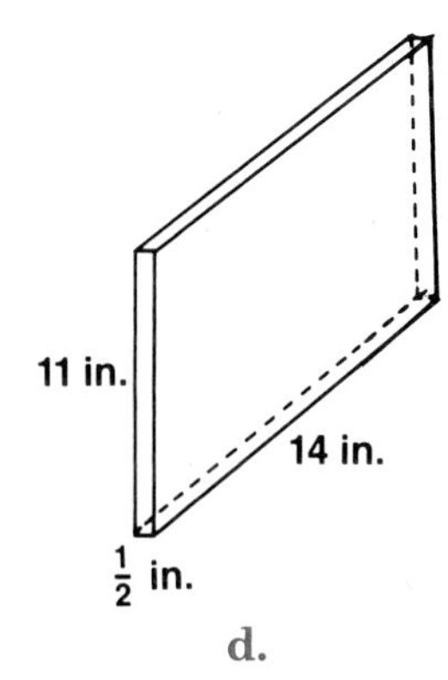


d.

6 ft.
$1\frac{1}{2}$ ft.
4 ft.

e.

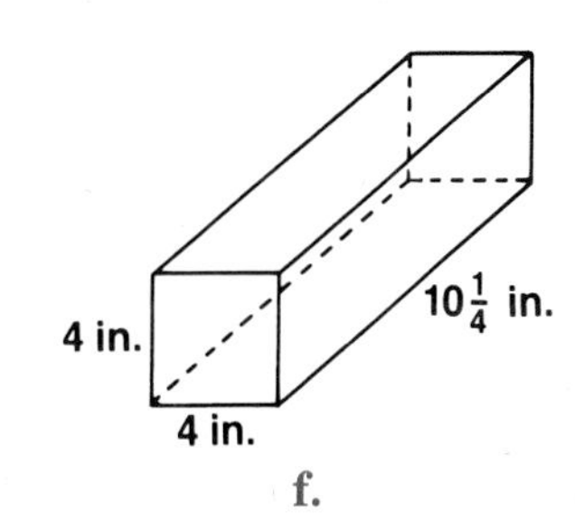


f.

**2.** Find the volume of the rectangular solids having dimensions:

**a.** $\ell = 2$ cm, $w = 1$ cm, $h = 6$ cm   **b.** $\ell = 8$ in., $w = 5$ in., $h = 10$ in.

**c.** $\ell = 14$ m, $w = 10$ m, $h = 1.5$ m   **d.** $\ell = 8$ ft., $w = 6$ ft., $h = 3\frac{1}{2}$ ft.

**e.** $\ell = 10$ cm, $w = 4.5$ cm, $h = 2.8$ cm   **f.** $\ell = 6$ yd., $w = 2$ yd., $h = 9\frac{1}{4}$ yd.

**3.** What is the volume of a rectangular box that is 12 inches long, 6 inches wide, and 3 inches high?

**4.** How many cubic feet are there in a rectangular shipping case 4 feet long, 2 feet wide, and $5\frac{1}{2}$ feet high?

**5.** Find the volume of a cube each of whose edges measures:

**a.** 2 in.   **b.** 4 cm   **c.** 10 m   **d.** 8 ft.   **e.** $\frac{3}{4}$ in.   **f.** 1.5 cm

## THE VOLUME OF A CYLINDER AND OF A SPHERE

### THE MAIN IDEA

1. A ***right circular cylinder*** is a solid figure that has two parallel circular bases.

   $V_{cylinder} = \pi r^2 h$

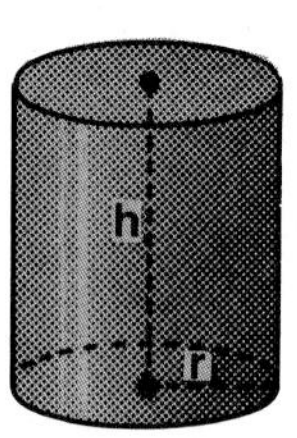


2. A ***sphere*** is a solid figure in the shape of a globe.

   $V_{sphere} = \frac{4}{3}\pi r^3$

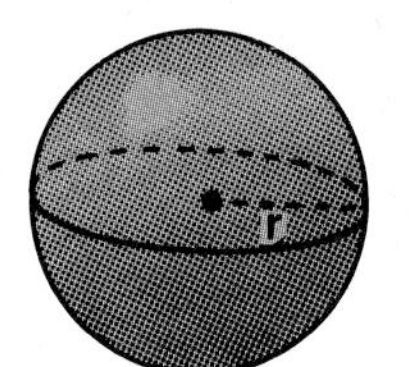


**EXAMPLE 3** Using $\pi = 3.14$, find the volume of the cylinder.

In the formula for the volume of a cylinder, substitute 3.14 for $\pi$, 5 for $r$, and 10 for $h$.

$$\begin{aligned} V_{cylinder} &= \pi r^2 h \\ &= 3.14 \times 5^2 \times 10 \\ &= 3.14 \times 25 \times 10 \\ &= 785 \text{ cm}^3 \quad \textit{Ans.} \end{aligned}$$

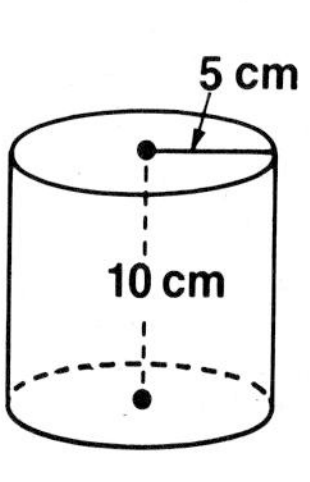


**EXAMPLE 4** Using $\pi = \frac{22}{7}$, find the volume of the cylinder.

Since you are given the measure of the diameter, divide by 2 to find the measure of the radius.

In the formula for the volume of a cylinder, substitute $\frac{22}{7}$ for $\pi$, 7 for $r$, and 20 for $h$.

$$\begin{aligned} V_{cylinder} &= \pi r^2 h \\ &= \frac{22}{7} \times 7^2 \times 20 \\ &= \frac{22}{\cancel{7}_1} \times \cancel{7}^1 \times 7 \times 20 \\ &= 3{,}080 \text{ in.}^3 \quad \textit{Ans.} \end{aligned}$$

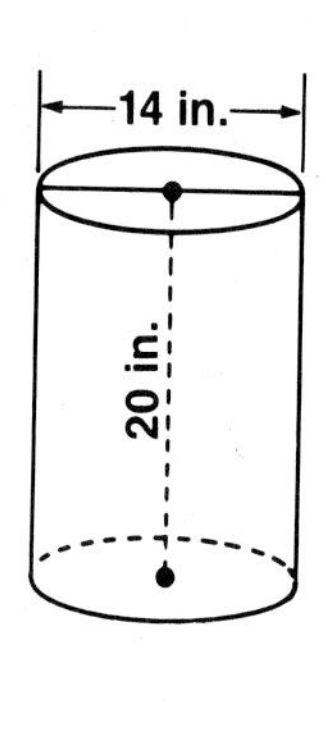


**EXAMPLE 5** Using $\pi = 3.14$, find the volume of the sphere.

In the formula for the volume of a sphere, substitute 3.14 for $\pi$ and 6 for $r$.

$$\begin{aligned} V_{sphere} &= \frac{4}{3}\pi r^3 \\ &= \frac{4}{3} \times 3.14 \times 6^3 \\ &= \frac{4}{\cancel{3}_1} \times 3.14 \times \cancel{216}^{72} \\ &= 904.32 \text{ cm}^3 \quad \textit{Ans.} \end{aligned}$$

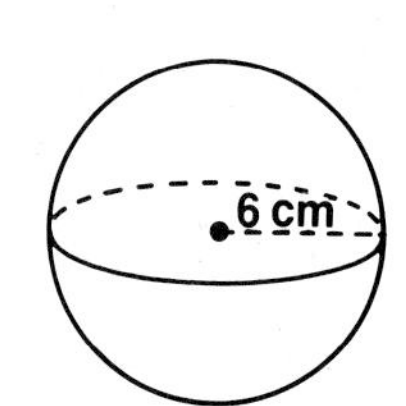

**EXAMPLE 6** Using $\pi = \frac{22}{7}$, find the volume of a sphere that has a diameter of 42 m.

Divide the diameter by 2 to find the measure of the radius.

In the formula for the volume of a sphere, substitute $\frac{22}{7}$ for $\pi$ and 21 for $r$.

$$V_{sphere} = \frac{4}{3}\pi r^3$$

$$= \frac{4}{3} \times \frac{22}{7} \times 21^3$$

$$= \frac{4}{\cancel{3}_1} \times \frac{22}{\cancel{7}_1} \times \overset{7}{\cancel{21}} \times \overset{3}{\cancel{21}} \times 21$$

$$= 38{,}808 \text{ m}^3 \quad \textit{Ans.}$$

## CLASS EXERCISES

1. Using $\pi = 3.14$, find the volume of each cylinder.

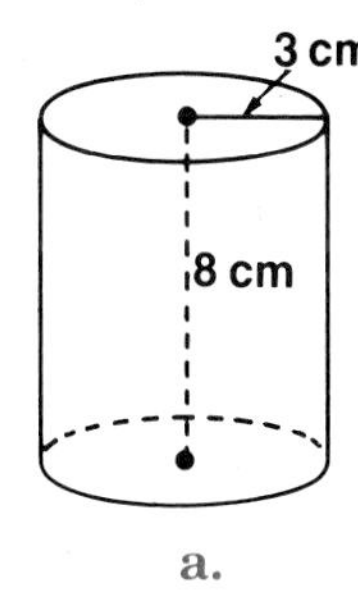


a.

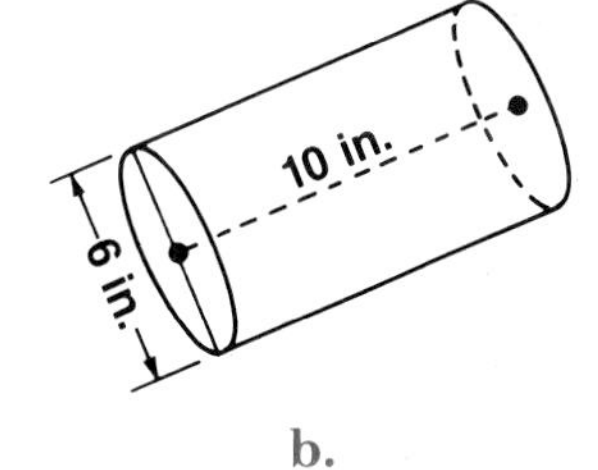


b.

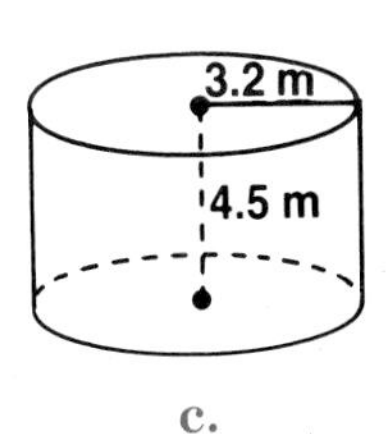


c.

2. Using $\pi = \frac{22}{7}$, find the volume of each cylinder, given the measures:

   **a.** $r = 7$ cm, $h = 8$ cm  **b.** $r = 14$ in., $h = 20$ in.

   **c.** $r = 28$ cm, $h = 1\frac{1}{2}$ cm  **d.** $d = 14$ ft., $h = 10$ ft.

3. Using $\pi = 3.14$, find the volume of a cylindrical can whose radius measures 2 inches and whose height measures 5 inches.

4. Using $\pi = 3.14$, find the volume of each sphere.

a. 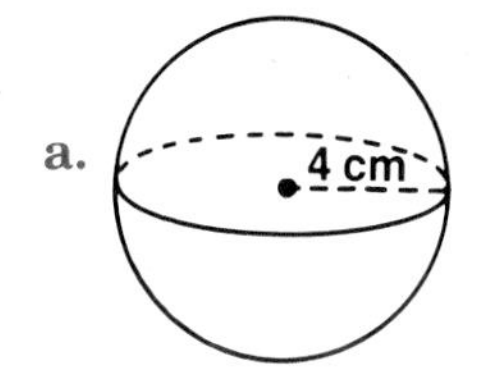


b. 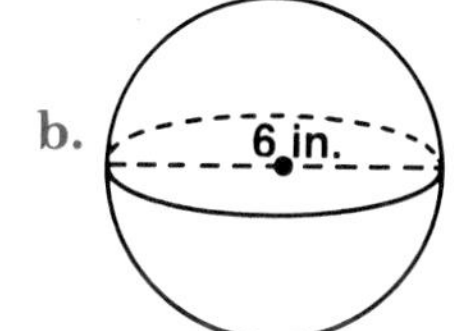


c. 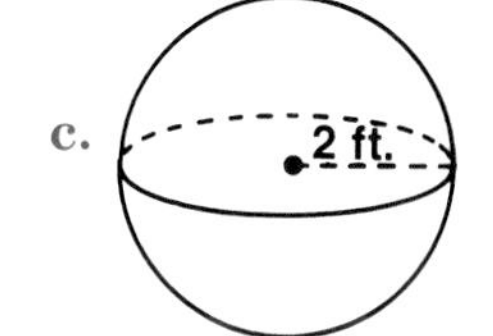


d. 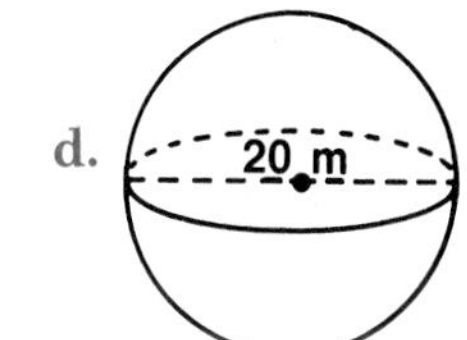


5. Using $\pi = \frac{22}{7}$, find the volume of a sphere that has a diameter of 84 centimeters.

## THE VOLUME OF A PYRAMID AND OF A CONE

### THE MAIN IDEA

1. A ***rectangular pyramid*** is a solid figure whose base is a rectangle and whose sides are triangles.

   The volume of such a pyramid is one-third the volume of a rectangular solid with the same dimensions.

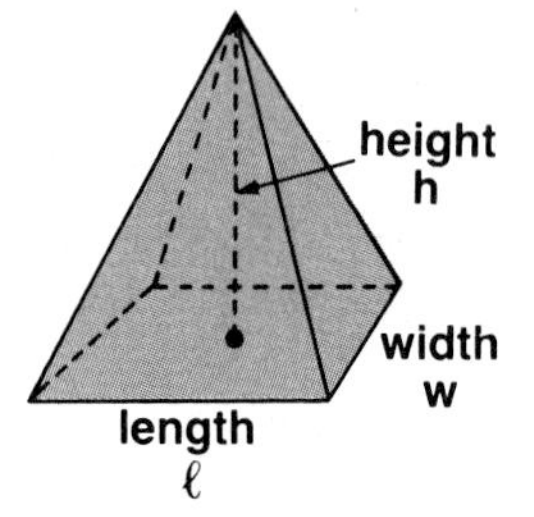


$$V_{\text{pyramid}} = \frac{1}{3} \times V_{\text{rectangular solid}}$$

$$V_{\text{pyramid}} = \frac{1}{3}\ell wh$$

2. A ***right circular cone*** has a base that is a circle.

   The volume of such a cone is one-third the volume of a right circular cylinder with the same dimensions.

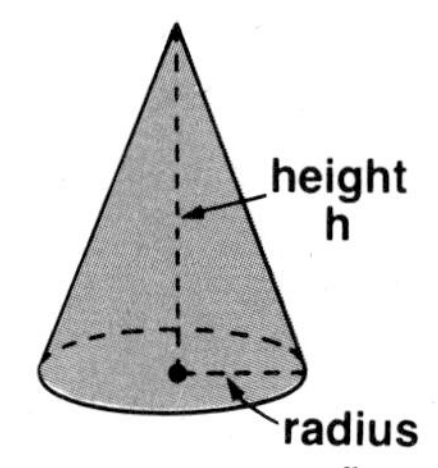


$$V_{\text{cone}} = \frac{1}{3} \times V_{\text{cylinder}}$$

$$V_{\text{cone}} = \frac{1}{3}\pi r^2 h$$

**EXAMPLE 7** Find the volume of the rectangular pyramid.

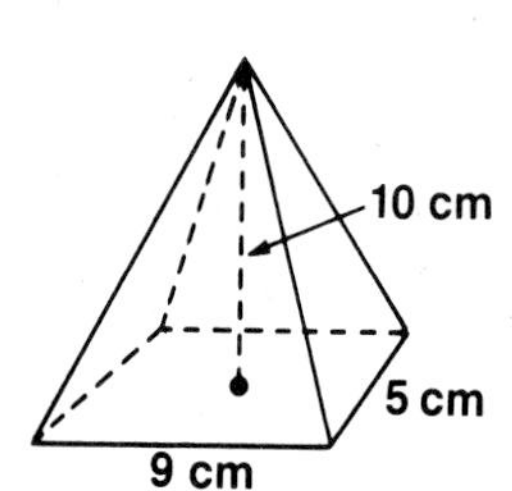


In the formula for the volume of a pyramid, substitute 9 for $\ell$, 5 for $w$, and 10 for $h$.

$$V_{\text{pyramid}} = \frac{1}{3}\ell wh$$

$$= \frac{1}{3} \times 9 \times 5 \times 10$$

$$= \frac{1}{\cancel{3}_{1}} \times \cancel{9}^{3} \times 5 \times 10$$

$$= 150 \text{ cm}^3 \quad \textit{Ans.}$$

**EXAMPLE 8** Using $\pi = 3.14$, find the volume of the cone.

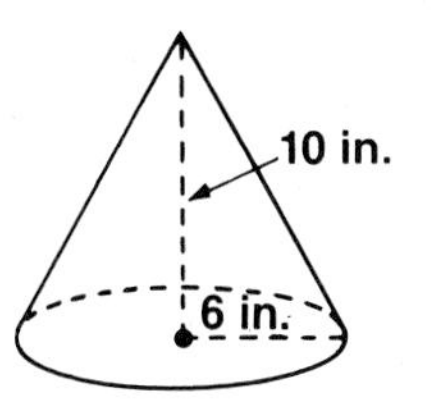


In the formula for the volume of a cone, substitute 3.14 for $\pi$, 6 for $r$, and 10 for $h$.

$$V_{cone} = \frac{1}{3}\pi r^2 h$$

$$= \frac{1}{3} \times 3.14 \times 6^2 \times 10$$

$$= \frac{1}{3} \times 3.14 \times 36 \times 10$$

$$= \frac{1}{\cancel{3}_1} \times 3.14 \times \overset{12}{\cancel{36}} \times 10$$

$$= 376.8 \text{ in.}^3 \quad \textit{Ans.}$$

## CLASS EXERCISES

**1.** Find the volume of each rectangular pyramid.

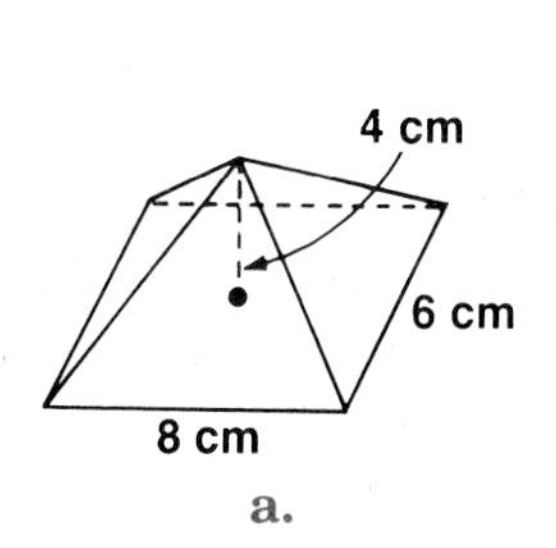


a.

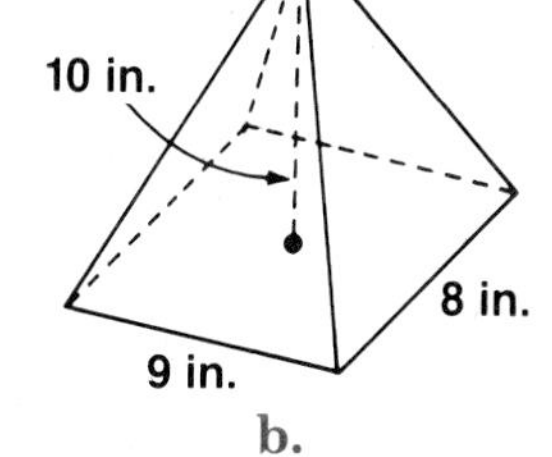


b.

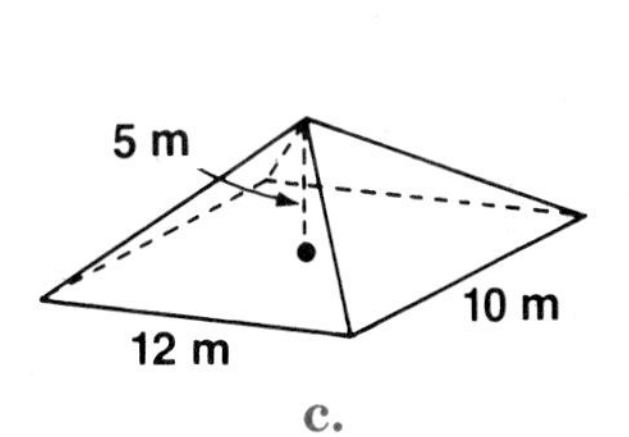


c.

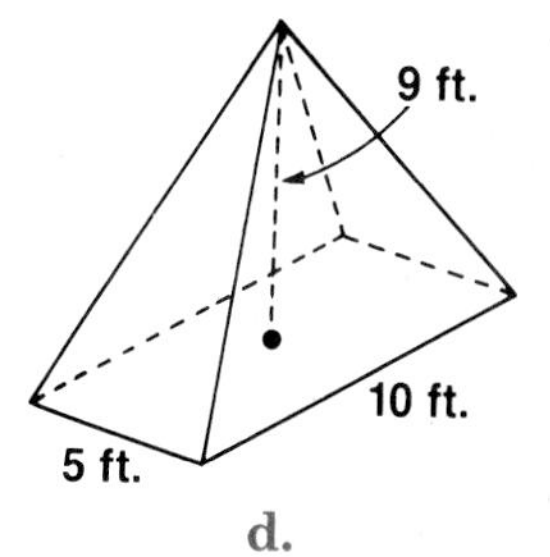


d.

**2.** Find the volume of each rectangular pyramid, given the dimensions:

**a.** $\ell = 20$ in., $w = 10$ in., $h = 6$ in. **b.** $\ell = 8$ cm, $w = 4$ cm, $h = 15$ cm

**c.** $\ell = 9$ m, $w = 5\frac{1}{2}$ m, $h = 2$ m **d.** $\ell = 1.2$ ft., $w = 8$ ft., $h = 10$ ft.

**3.** Using $\pi = 3.14$, find the volume of each cone.

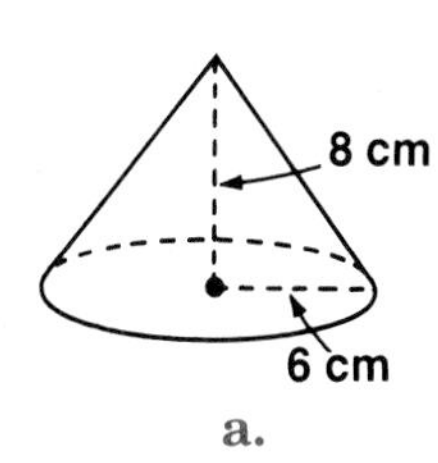


a.

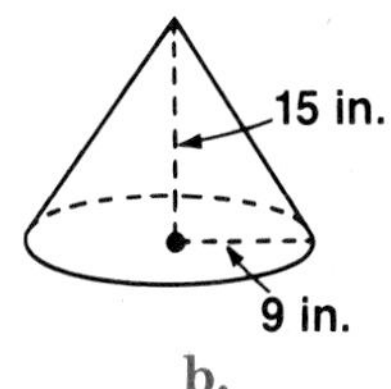


b.

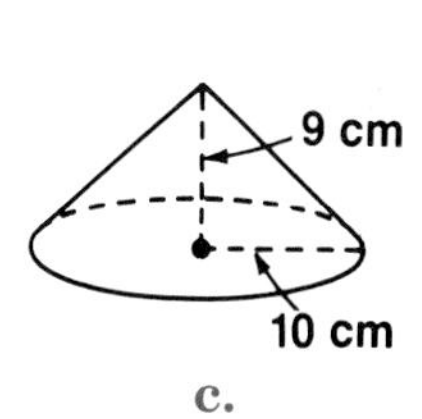


c.

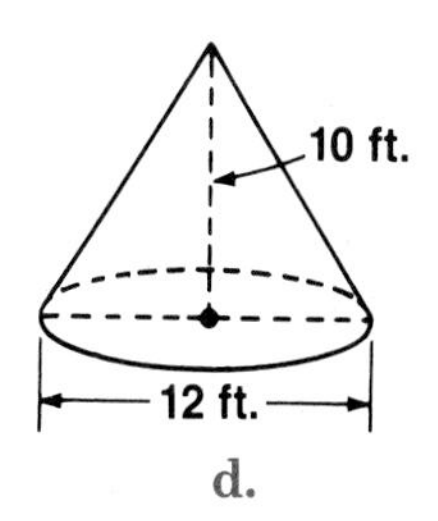


d.

**4.** Using $\pi = 3.14$, find the volume of each cone, given the dimensions:

**a.** $r = 5$ cm, $h = 3$ cm **b.** $r = .6$ in., $h = 20$ in. **c.** $r = 1\frac{1}{2}$ m, $h = 6$ m

**d.** $r = 4$ ft., $h = 12$ ft.

## HOMEWORK EXERCISES

1. Find the volume of each rectangular solid.

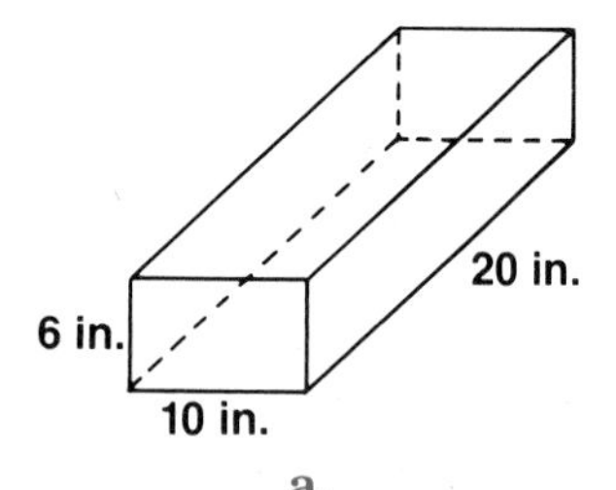

a.

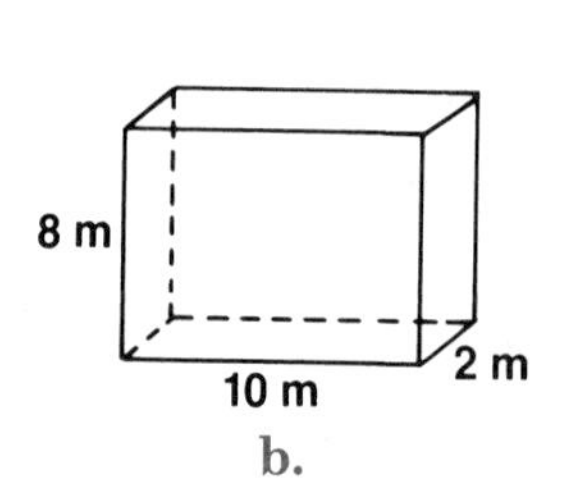

b.

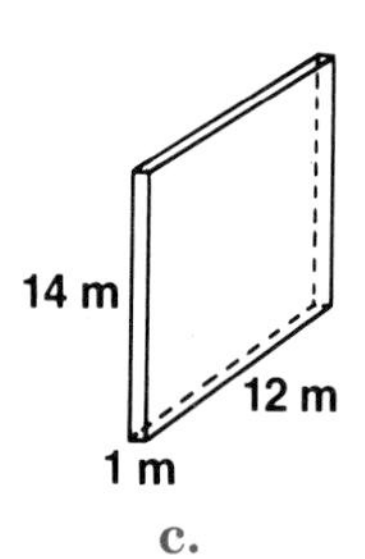

c.

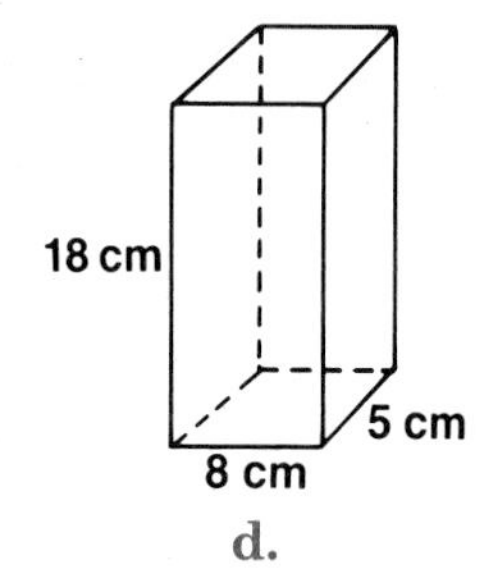

d.

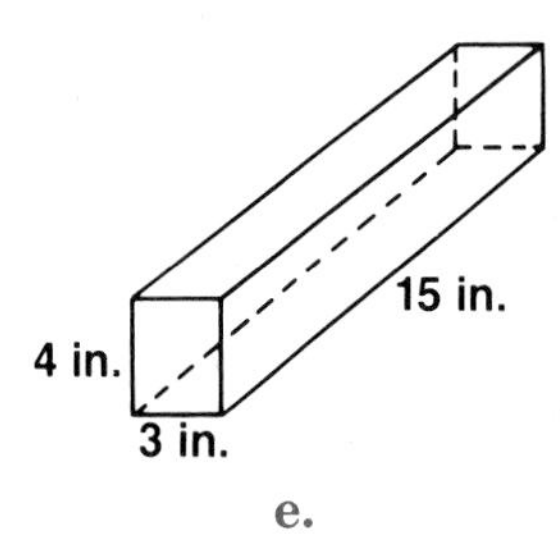

e.

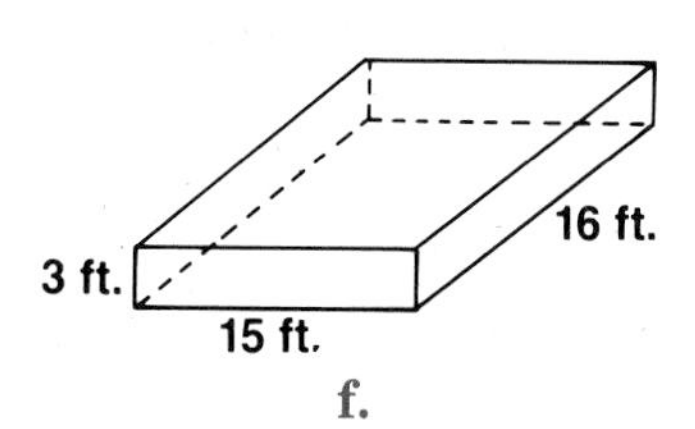

f.

2. Find the volume of each rectangular solid, given the dimensions:

**a.** $\ell = 3$ in., $w = 2$ in., $h = 8$ in. **b.** $\ell = 10$ cm, $w = 6$ cm, $h = 12$ cm

**c.** $\ell = 9$ m, $w = 8$ m, $h = 7$ m **d.** $\ell = 10$ ft., $w = 3\frac{1}{2}$ ft., $h = 8$ ft.

**e.** $\ell = 14$ cm, $w = 8.2$ cm, $h = 4.1$ cm **f.** $\ell = 12$ m, $w = 10.6$ m, $h = .8$ m

3. What is the volume of a rectangular fish tank that is 20 inches long, 12 inches high, and 10 inches deep?

4. How many cubic feet are there in a rectangular room that is 9 feet wide, 12 feet long, and 8 feet high?

5. What is the volume of a box that is 20 cm long, 10 cm wide, and 6.5 cm high?

6. Find the volume of a cube each of whose edges measures:

**a.** 1 m **b.** 3 ft. **c.** 7 yd. **d.** 8 cm **e.** .6 m **f.** $\frac{3}{4}$ in.

**g.** 1.2 cm **h.** 5 in.

7. Using $\pi = 3.14$, find the volume of each cylinder.

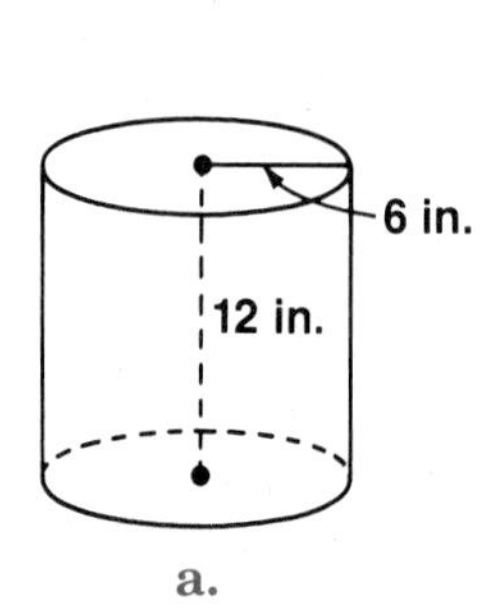

a.

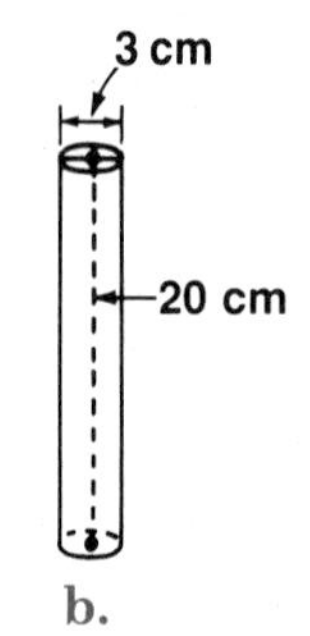

b.

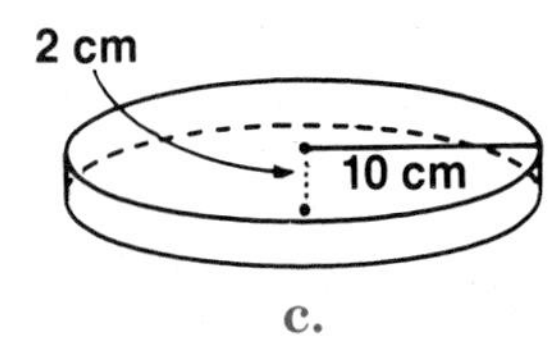

c.

**8.** Using $\pi = \frac{22}{7}$, find the volume of each cylinder, given the measures:

**a.** $r = 14$ cm, $h = 10$ cm  **b.** $r = 28$ cm, $h = 6$ cm  **c.** $d = 14$ in., $h = 5.2$ in.

**d.** $d = 42$ ft., $h = 3\frac{1}{2}$ ft.

**9.** Using $\pi = 3.14$, find the volume of a cylindrical can whose radius measures 3 inches and whose height measures 8 inches.

**10.** Using $\pi = 3.14$, find the volume of each sphere.

a.
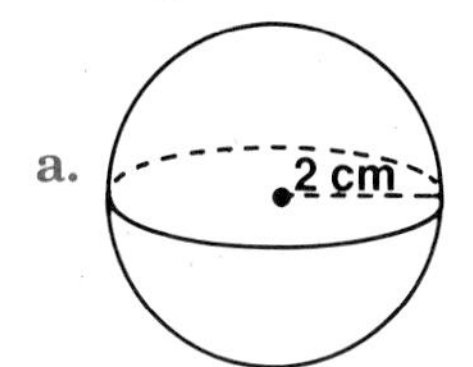


b. 8 in.

c.
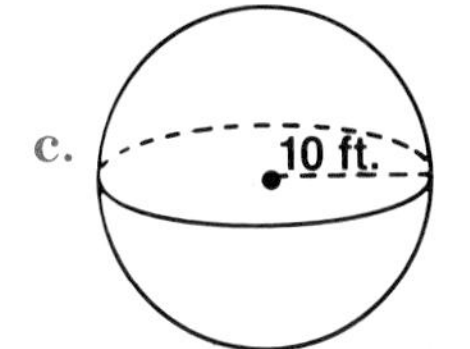


d. 18 in.

**11.** Using $\pi = \frac{22}{7}$, find the volume of a globe that has a radius of 21 cm.

**12.** Find the volume of each rectangular pyramid.

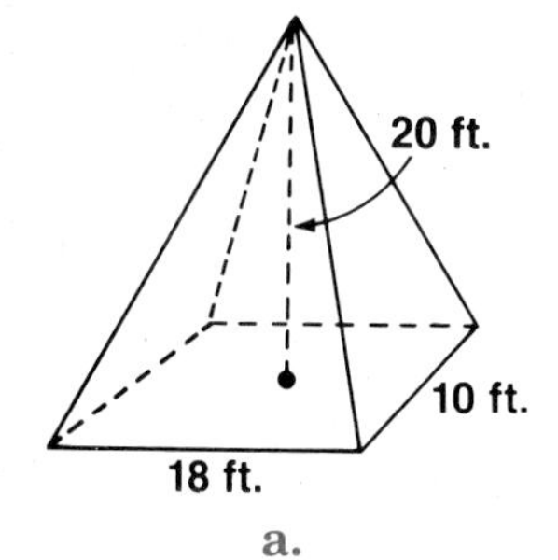

a.

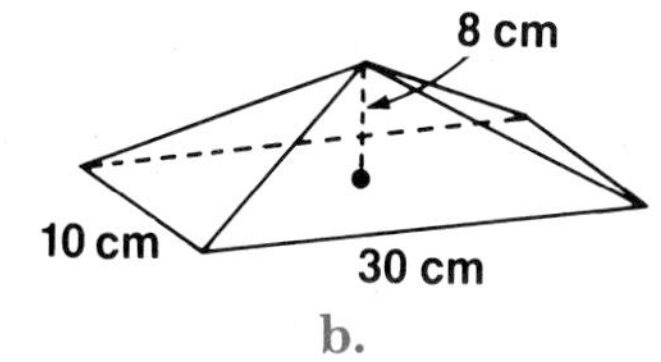

b.

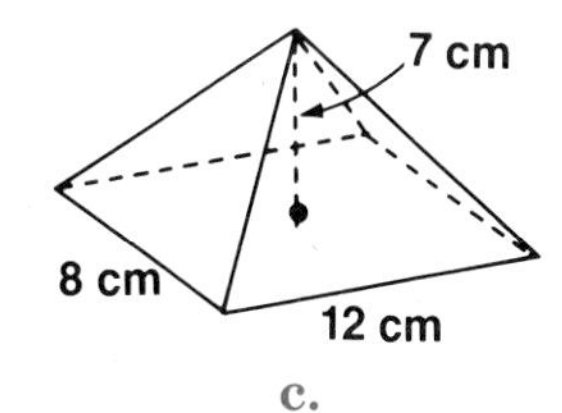

c.

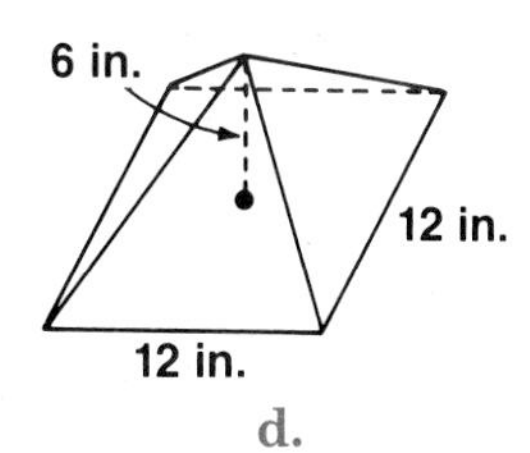

d.

**13.** Find the volume of each rectangular pyramid, given the dimensions:

**a.** $\ell = 6$ cm, $w = 4$ cm, $h = 2$ cm  **b.** $\ell = 9$ in., $w = 6$ in., $h = 20$ in.

**c.** $\ell = 15$ ft., $w = 8$ ft., $h = 10$ ft.  **d.** $\ell = 24$ m, $w = 10$ m, $h = 30$ m

**e.** $\ell = 18$ cm, $w = 6$ cm, $h = 4$ cm  **f.** $\ell = 20$ yd., $w = 10$ yd., $h = 30$ yd.

**14.** Using $\pi = 3.14$, find the volume of each cone.

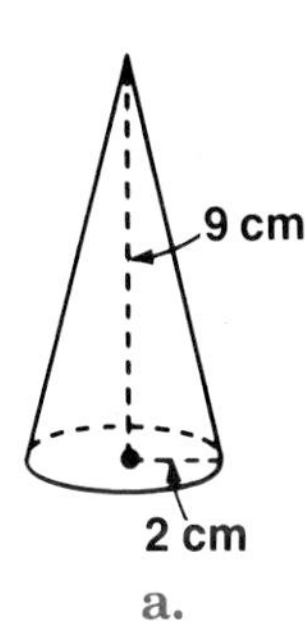

a.

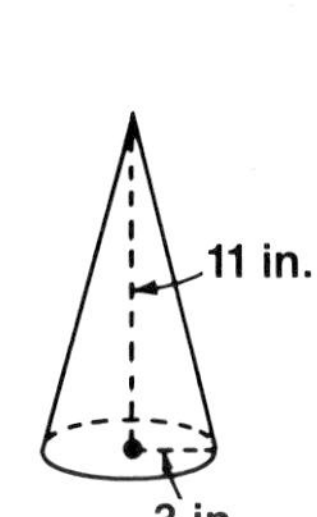

b.

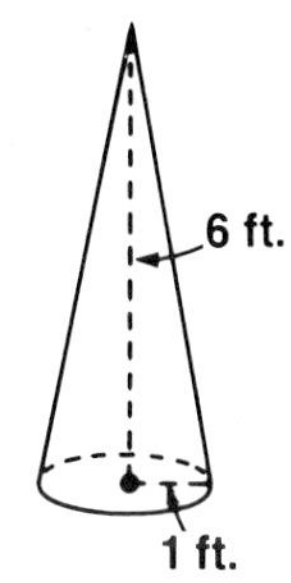

c.

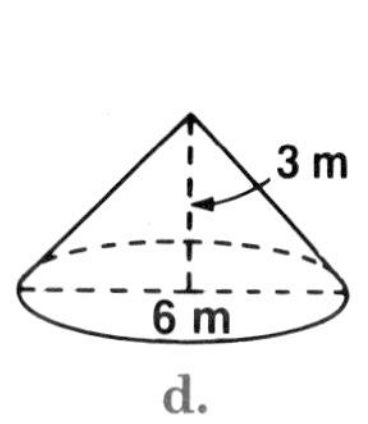

d.

**15.** Using $\pi = 3.14$, find the volume of each cone, given the dimensions:

**a.** $r = 2$ cm, $h = 6$ cm  **b.** $r = 6$ cm, $h = 2$ cm  **c.** $r = 8$ in., $h = 3$ in.

## SPIRAL REVIEW EXERCISES

1. a. Find the area of the trapezoid.

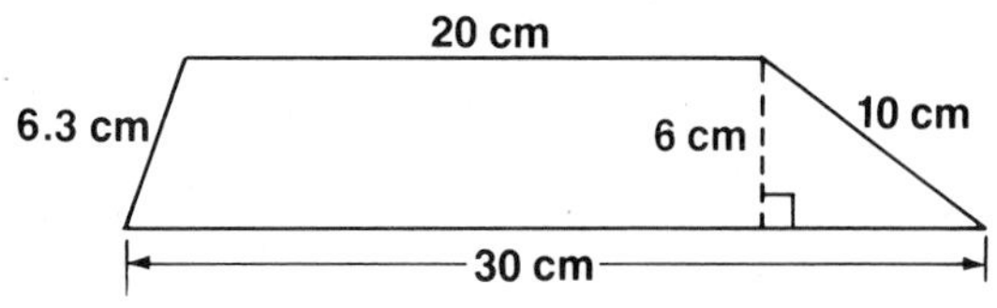


b. Find the perimeter of the trapezoid.

2. A $120 jacket is on sale at 20% off. What is the sale price of the jacket?

3. .57 is equivalent to

(a) 5.7% (b) 57% (c) $\frac{57}{1000}$ (d) $\frac{1}{2}$

4. $\frac{2}{3} - \frac{1}{4}$ is

(a) $\frac{5}{12}$ (b) 1 (c) $\frac{1}{6}$ (d) $\frac{3}{8}$

5. 9 is what percent of 36?

6. If $a = 5$ and $b = 2$, the value of $3a - 5b$ is
(a) 10 (b) 5 (c) 20 (d) 22

7. Which pair of lines is drawn parallel?

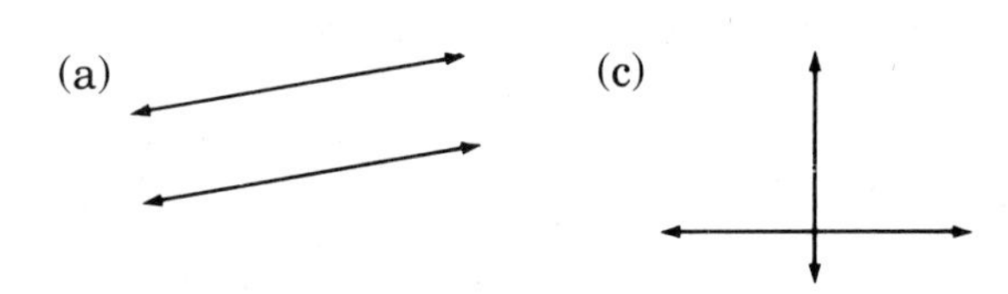


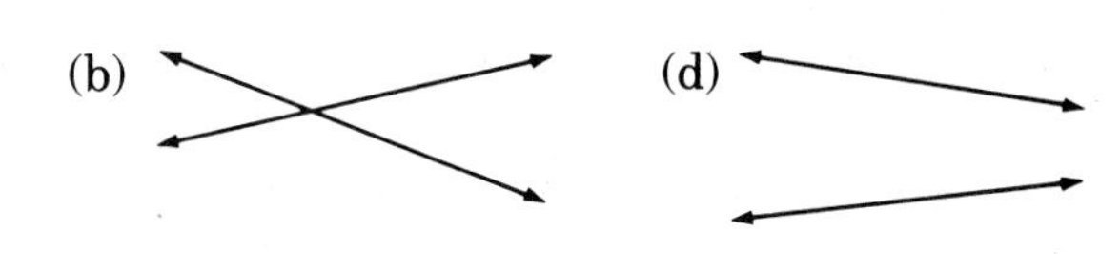


### CHALLENGE QUESTION

A ball placed in a cubical box just touches each side of the box. If the volume of the box is 343 $cm^3$, find the volume of the ball. (Use $\pi = \frac{22}{7}$.)

# GLOSSARY

These informal definitions are intended to be brief descriptions of the terms listed.

**acute angle** an angle that measures more than 0° and less than 90°
**acute triangle** a triangle that contains three acute angles
**addend** a number to be added. In 2 + 3, 2 and 3 are the addends.
**angle** a figure formed by two rays with a common endpoint
**annual** yearly
**area** the number of square units contained within a closed, two-dimensional figure
**average** the value found by adding a set of values and dividing by the number of values; the mean

**balance** an amount of money kept in a financial account
**bar graph** a graph in which numerical facts are shown as bars of different lengths
**base** the number that is raised to a power in an expression with an exponent. In $3^5$, 3 is the base.
**base** the side of a polygon from which the height is measured
**budget** a plan for organizing expenses

**carrying charge** the increase in cost of an item purchased on an installment plan
**Celsius** a scale for measuring temperature
**center of a circle** the point inside a circle that is the same distance from any point on the circle
**check** a written order authorizing a bank to pay out money from an account
**checking account** a financial account for drawing money against deposits
**chord** a line segment that has both endpoints on a circle
**circle** a flat, closed curve that has all its points the same distance from an inside point called the center
**circle graph** a graph in which numerical facts are represented by sectors of a circle
**circumference** the distance around a circle
**commission** a percentage of the money received for a sale, paid to the salesperson who made the sale
**common factor** a number that divides evenly into two or more given numbers. 3 is a common factor of 9 and 12.
**composite number** a whole number that has factors other than 1 and itself. Since $6 = 3 \times 2$, the number 6 is a composite.
**coordinates** the numbers describing the location of a point on a graph
**cross-multiply** to multiply the diagonally opposite numerators and denominators (means and extremes) in a proportion
**cross products** the results of cross-multiplying
**cube** a three-dimensional figure whose sides are all squares
**cube** to raise a number to the third power (use a number three times as a factor)
**cubic unit** a unit used to measure volume
**Customary measures** a system of measures, commonly used in the U.S., that includes units such as the foot, the pound, and the gallon

**data** numerical facts
**decimal** a way of writing a fraction whose denominator is a power of 10. The fraction $\frac{3}{100}$ can be written as the decimal .03.
**degree** a unit of measure for angles or for temperatures
**denominator** the nonzero numeral below the division line in a fraction, showing into how many parts the whole is divided. In the fraction $\frac{3}{5}$, the denominator is 5.
**deposit** an amount of money added to the balance in a bank account
**diagonal** a line segment connecting two nonconsecutive vertices of a polygon

**diameter** a chord that passes through the center of a circle
**dice** cubes with six sides that represent the numbers 1 through 6 (singular: die)
**difference** the result of a subtraction. In $7 - 2 = 5$, the difference is 5.
**dimension** a measure of length, width, or height
**directed number** a signed number (positive, negative, or zero)
**discount** the amount by which the price of an item is lowered
**dividend** a number that is divided. In $16 \div 2$, 16 is the dividend.
**divisor** a number by which another number is divided. In $16 \div 2$, the divisor is 2.
**down payment** an original payment at the time of purchase on an installment plan

**equation** a mathematical sentence stating that two quantities are equal. $5 - 2 = 3$ and $x + 1 = 7$ are equations.
**equilateral** a word used to describe a polygon whose sides are all equal in measure
**equivalent** equal in value
**estimate** to find an approximate value
**evaluate** to find the value of a mathematical expression by carrying out the indicated operations
**exact divisor** a number that divides into a quantity evenly, leaving a remainder of 0; a factor. 4 is an exact divisor of 12.
**exponent** a number that shows how many times a base is used as a factor. In $3^5$, the exponent is 5.
**extremes** the first and fourth numbers in a proportion. In the proportion $\frac{2}{3} = \frac{8}{12}$, 2 and 12 are the extremes.

**factor** an exact divisor. 4 is a factor of 12.
**Fahrenheit** a scale for measuring temperature
**formula** a mathematical sentence that shows how variables are related to each other. $A = bh$ is a formula.
**fraction** a mathematical expression, such as $\frac{5}{6}$, that shows the quotient of two numbers
**frequency** the number of times a value appears when a tally is made
**frequency table** a list of the frequencies of a group of values

**graph** a visual way to organize and present data
**greatest common factor** the largest exact divisor of two or more numbers. 6 is the greatest common factor of 12 and 18.

**height** the perpendicular distance between a point and a line segment
**hexagon** a six-sided polygon
**horizontal** in a direction of side to side; across
**hypotenuse** the side opposite the right angle in a right triangle

**improper fraction** a fraction in which the numerator is greater than or equal to the denominator. $\frac{7}{5}$ and $\frac{5}{5}$ are improper fractions.
**inequality** a mathematical sentence stating that two quantities are not equal. $2 < 3$, $3 > 2$, and $2 \neq 3$ are inequalities.
**installment plan** the purchase of an item by making regular equal payments over a period of time
**integers** the set of numbers consisting of the whole numbers and their opposites. $\{\ldots, -2, -1, 0, 1, 2, \ldots\}$
**interest** a charge for money that is borrowed or a return on an investment
**inverse operations** two operations, such as addition and subtraction, that "undo" each other
**investment** the use of money to earn money
**isosceles triangle** a triangle that contains two sides equal in measure

**least common multiple** the smallest whole number that is a multiple of two or more whole numbers. 24 is the least common multiple of 8 and 12.
**like fractions** fractions that have the same denominator. $\frac{2}{5}$ and $\frac{3}{5}$ are like fractions.
**linear unit** a unit used to measure distance
**line graph** a graph in which numerical facts are represented by points with connecting line segments
**line segment** a part of a line with two endpoints
**list price** the original price of an item

**lowest common denominator** the least common multiple of the denominators of two or more fractions. 24 is the lowest common denominator of the fractions $\frac{1}{8}$ and $\frac{5}{12}$.
**lowest terms** the form of a fraction found by dividing the numerator and denominator by their greatest common factor. $\frac{12}{20}$, in lowest terms, is $\frac{3}{5}$.

**mean** the value found by adding a set of values and dividing by the number of values; the average
**means** the second and third numbers in a proportion. In the proportion $\frac{2}{3} = \frac{8}{12}$, 3 and 8 are the means.
**median** the middle value in a set of values that are arranged in order of size
**metric measures** a system of measures that uses powers-of-ten multiples of the meter, the gram, and the liter
**minuend** a number from which another number is subtracted. In $7 - 2$, the minuend is 7.
**mixed number** a numeral consisting of a whole number and a proper fraction. $7\frac{1}{3}$ is a mixed number.
**mode** the value that has the greatest frequency in a set of data
**multiplicand** a number to be multiplied. In $1{,}436 \times 5$, the number 1,436 is the multiplicand.

**negative number** a number less than 0
**number line** a graph that represents numbers by marked points on a line
**numeral** the written symbol for a number
**numerator** the numeral above the division line in a fraction. In the fraction $\frac{3}{5}$, 3 is the numerator.
**numerical expression** a written combination of numerals and symbols for operations

**obtuse angle** an angle that measures more than 90° and less than 180°
**obtuse triangle** a triangle that contains an obtuse angle
**octagon** an eight-sided polygon
**open sentence** a sentence that contains a variable. $x + 2 = 10$ and $x > 4$ are open sentences.
**opposites** two signed numbers that are the same distance from 0 on a number line. $+5$ and $-5$ are opposites.
**ordered pair** the coordinates that represent the horizontal and vertical locations of a point on a graph. $(1, -4)$ is an ordered pair.
**order of operations** rules that decide how a numerical expression is evaluated
**origin** the point where the $x$-axis and $y$-axis of a graph intersect
**overtime** the money earned for working extra hours, generally at the rate of $1\frac{1}{2}$ or 2 times the regular rate of pay

**parallel lines** two lines that extend in the same direction and never meet
**parallelogram** a quadrilateral that has both pairs of opposite sides parallel
**pentagon** a five-sided polygon
**per** for each
**percent** hundredths
**perfect square** a rational number whose square root is also a rational number. 9 is a perfect square since $9 = 3^2$ and $\sqrt{9} = 3$.
**perimeter** the distance around a polygon; the sum of the measures of its sides
**perpendicular lines** two lines that meet at right angles
**pictograph** a graph in which numerical facts are represented by pictures
**pie graph** a graph in which numerical facts are represented by sectors of a circle (also called a circle graph)
**place value** the number by which a digit is multiplied to find its value in a numeral. The place value of 3 in 35 is 10.
**polygon** a flat, closed figure whose sides are line segments
**positive number** a number greater than 0
**power** the product obtained when a number is multiplied by itself a given number of times. 8, or $2^3$, is the third power of 2.
**prime number** a whole number that has 1 and itself as its only factors. 2, 3, 5, and 7 are prime numbers.
**principal** the amount of money that is invested or borrowed
**probability** a number describing the likelihood that an event will occur

**product** the result of a multiplication. In $5 \times 2 = 10$, the product is 10.
**proper fraction** a fraction in which the numerator is less than the denominator. $\frac{3}{5}$ is a proper fraction.
**proportion** a statement that two ratios are equal. $\frac{2}{3} = \frac{8}{12}$ is a proportion.
**protractor** an instrument that is used to measure angles
**Pythagorean relationship** the relation between the lengths of the sides of a right triangle. In a right triangle, the square of the hypotenuse equals the sum of the squares of the other two sides.

**quadrilateral** a four-sided polygon
**quotient** the result of a division. In $16 \div 2 = 8$, the quotient is 8.

**radius** a line segment that has one endpoint at the center of a circle and the other endpoint on the circle (plural: radii)
**random** chosen in no particular order
**rate of discount** the percent by which the price of an item is lowered
**rate of interest** the percent of principal paid on a loan or an investment
**ratio** a comparison between two numbers. 2 to 3, also written as 2:3 or $\frac{2}{3}$, is a ratio.
**rational number** a number that can be written as a fraction whose numerator and denominator are integers and whose denominator is not 0. Some examples are 5 or $\frac{5}{1}$, $-\frac{2}{3}$, 397, and 0.
**ray** a part of a line consisting of a fixed point and all points to one side of it (a ray has a beginning, but no end)
**reciprocal** the result of inverting a fraction, or exchanging the numerator and denominator of the fraction. $\frac{7}{4}$ is the reciprocal of $\frac{4}{7}$.
**rectangle** a parallelogram that has four right angles
**rectangular solid** a three-dimensional figure whose sides are all rectangles; a box
**remainder** the quantity left after division. 2 is the remainder when 17 is divided by 3.
**repeating decimal** a decimal that results when a division never ends. $\frac{1}{3} = .3333\ldots$ and $\frac{5}{11} = .454545\ldots$ are examples.
**rhombus** a parallelogram that has all four sides equal in measure
**right angle** an angle that measures exactly 90°
**right triangle** a triangle that contains one right angle
**rounded number** an approximation for a number, to a given place value. To the nearest hundred, 324 rounds to 300.

**salary** the amount of money earned by an employee
**sale price** the price of an item after it has been discounted
**sales tax** a state or local tax on purchases
**scale drawing** a drawing in which the actual dimensions of an object are enlarged or reduced proportionally
**scalene triangle** a triangle that contains no two sides equal in measure
**scientific notation** a form of writing very large or very small numbers as a product in which the first factor is a number between 1 and 10 and the second factor is a power of 10. $9.4 \times 10^5$ is the scientific notation for 940,000.
**sector** a section of a circle between two radii
**sentence** a statement that expresses a complete thought
**signed numbers** numbers that are positive, negative, or zero
**similar figures** figures that have the same shape
**simplify** to express in lowest terms
**solution** the answer to a problem
**solving an equation** to find the values of the variable for which an equation is true
**sphere** a three-dimensional figure that has all of its points the same distance from its center; a ball
**square** a rectangle that has all four sides equal in measure, or a rhombus that has four right angles
**square** to multiply a number by itself; to raise to the second power (use a number two times as a factor). 5 squared, or $5^2$, is 25.
**square root** one of two equal factors of a number. 5 is a square root of 25.
**square unit** a unit used to measure area

**statistics** the study of numerical facts, or data

**stepped rates** rates that decrease when larger quantities are purchased. A long-distance telephone cost is a stepped rate. There is an initial charge followed by a decrease in charge for additional time.

**straight angle** an angle that measures exactly 180°

**subtrahend** a number to be subtracted. In $7 - 2$, the subtrahend is 2.

**sum** the result of an addition. In $2 + 3 = 5$, 5 is the sum.

**tally** to count, using tally marks to keep score

**tax** money collected by governments to pay for public services

**terminating decimal** a decimal that results when a division has a zero remainder. For example, $\frac{1}{4} = .25$.

**trapezoid** a quadrilateral that has only one pair of opposite sides parallel

**triangle** a three-sided polygon

**unit** a quantity chosen as a standard by which other quantities are to be expressed

**unlike fractions** fractions that do not have the same denominator. $\frac{1}{2}$ and $\frac{1}{3}$ are unlike fractions.

**variable** a symbol, usually a letter, used to represent an unknown number

**vertex** the point at which the two rays of an angle meet (plural: vertices)

**vertical** in a direction of up and down; upright

**volume** the number of cubic units contained within a closed, three-dimensional figure

**wages** an amount of money earned, usually by the hour

**whole numbers** numbers that are greater than or equal to zero and that contain no fractions. {0, 1, 2, 3, . . .}

***x*-axis** the horizontal number line on a coordinate graph

***y*-axis** the vertical number line on a coordinate graph

# INDEX